Algebraic Structures and Graph Theory

Algebraic Structures and Graph Theory

Editors

Irina Cristea
Hashem Bordbar

MDPI • Basel • Beijing • Wuhan • Barcelona • Belgrade • Manchester • Tokyo • Cluj • Tianjin

Editors
Irina Cristea
University of Nova Gorica
Nova Gorica, Slovenia

Hashem Bordbar
University of Nova Gorica
Nova Gorica, Slovenia

Editorial Office
MDPI
St. Alban-Anlage 66
4052 Basel, Switzerland

This is a reprint of articles from the Special Issue published online in the open access journal *Mathematics* (ISSN 2227-7390) (available at: https://www.mdpi.com/journal/mathematics/special_issues/Algebraic_Structures_Graph_Theory).

For citation purposes, cite each article independently as indicated on the article page online and as indicated below:

LastName, A.A.; LastName, B.B.; LastName, C.C. Article Title. *Journal Name* **Year**, *Volume Number*, Page Range.

ISBN 978-3-0365-8440-9 (Hbk)
ISBN 978-3-0365-8441-6 (PDF)

© 2023 by the authors. Articles in this book are Open Access and distributed under the Creative Commons Attribution (CC BY) license, which allows users to download, copy and build upon published articles, as long as the author and publisher are properly credited, which ensures maximum dissemination and a wider impact of our publications.

The book as a whole is distributed by MDPI under the terms and conditions of the Creative Commons license CC BY-NC-ND.

Contents

About the Editors vii

Irina Cristea and Hashem Bordbar
Preface to the Special Issue "Algebraic Structures and Graph Theory"
Reprinted from: *Mathematics* 2023, 11, 3259, doi:10.3390/math11153259 1

Natalia Agudelo Muñetón, Agustín Moreno Cañadas, Pedro Fernando Fernández Espinosa and Isaías David Marín Gaviria
{0,1}-Brauer Configuration Algebras and Their Applications in Graph Energy Theory
Reprinted from: *Mathematics* 2021, 9, 3042, doi:10.3390/math9233042 5

Mohammad Ashraf, Jaber H. Asalool, Abdulaziz M. Alanazi and Ahmed Alamer
An Ideal-Based Dot Total Graph of a Commutative Ring
Reprinted from: *Mathematics* 2021, 9, 3072, doi:10.3390/math9233072 23

Štěpán Křehlík, Michal Novák and Jana Vyroubalová
From Automata to Multiautomata via Theory of Hypercompositional Structures
Reprinted from: *Mathematics* 2022, 10, 1, doi:10.3390/math10010001 37

Saifur Rahman, Maitrayee Chowdhury, Firos A. and Irina Cristea
Knots and Knot-Hyperpaths in Hypergraphs
Reprinted from: *Mathematics* 2022, 10, 424, doi:10.3390/math10030424 53

Hashem Bordbar
The Structure of the Block Code Generated by a BL-Algebra
Reprinted from: *Mathematics* 2022, 10, 692, doi:10.3390/math10050692 67

Linming Qi, Lianying Miao, Weiliang Zhao and Lu Liu
A Lower Bound for the Distance Laplacian Spectral Radius of Bipartite Graphs with Given Diameter
Reprinted from: *Mathematics* 2022, 10, 1301, doi:10.3390/math10081301 79

Yongde Feng, Yanting Xie, Fengxia Liu and Shoujun Xu
The Extendability of Cayley Graphs Generated by Transposition Trees
Reprinted from: *Mathematics* 2022, 10, 1575, doi:10.3390/ math10091575 89

Michal Staš
Parity Properties of Configurations
Reprinted from: *Mathematics* 2022, 10, 1998, doi:10.3390/math10121998 97

Gerasimos G. Massouros and Christos G. Massouros
State Machines and Hypergroups
Reprinted from: *Mathematics* 2022, 10, 2427, doi:10.3390/math10142427 111

Yuzheng Ma, Yubin Gao and Yanling Shao
Upper and Lower Bounds for the Spectral Radius of Generalized Reciprocal Distance Matrix of a Graph
Reprinted from: *Mathematics* 2022, 10, 2683, doi:10.3390/math10152683 137

Gabriela Diaz-Porto, Ismael Gutierrez and Armando Torres-Grandisson
The t-Graphs over Finitely Generated Groups and the Minkowski Metric
Reprinted from: *Mathematics* 2022, 10, 3030, doi:10.3390/math10173030 149

Obaidullah Wardak, Ayushi Dhama and Deepa Sinha
On Some Properties of Addition Signed Cayley Graph $\Sigma_n^\wedge$
Reprinted from: *Mathematics* **2022**, *10*, 3492, doi:10.3390/math10193492 . . . . **161**

Dmitry Solovyev
Congruence for Lattice Path Models with Filter Restrictions and Long Steps
Reprinted from: *Mathematics* **2022**, *10*, 4209, doi:10.3390/math10224209 . . . . **173**

Narjes Firouzkouhi, Reza Ameri, Abbas Amini and Hashem Bordbar
Semihypergroup-Based Graph for Modeling International Spread of COVID-n in Social Systems
Reprinted from: *Mathematics* **2022**, *10*, 4405, doi:10.3390/math10234405 . . . . **199**

Bo-Ye Zhang
Hopf Differential Graded Galois Extensions
Reprinted from: *Mathematics* **2023**, *11*, 128, doi:10.3390/math11010128 . . . . **213**

Dawid Edmund Kedzierski, Alessandro Linzi and Hanna Stojałowska
Characteristic, C-Characteristicand Positive Cones in Hyperfields
Reprinted from: *Mathematics* **2023**, *11*, 779, doi:10.3390/math11030779 . . . . **223**

M. Palanikumar, Chiranjibe Jana, Omaima Al-Shanqiti and Madhumangal Pal
A Novel Method for Generating the M-Tri-Basis of an Ordered Γ-Semigroup
Reprinted from: *Mathematics* **2023**, *11*, 893, doi:10.3390/math11040893 . . . . **245**

Kitsanachai Sripon, Ekkachai Laysirikul and Worachead Sommanee
Left (Right) Regular Elements of Some Transformation Semigroups
Reprinted from: *Mathematics* **2023**, *11*, 2230, doi:10.3390/math11102230 . . . . **259**

About the Editors

Irina Cristea

Irina Cristea is an associate professor in Mathematics and head of the Centre for Information Technologies and Applied Mathematics at the University of Nova Gorica, Slovenia. She is conducting research in Hypercompositional Algebra, having published more than 70 articles indexed by WoS and Scopus, one book in Springer and two edited books for MDPI. In addition, Dr. Cristea has served as a reviewer for more than 60 international journals and currently is chief editor of the Italian Journal of Pure and Applied Mathematics, associated editor for Heliyon Mathematics, Elsevier, and member of the editorial board of Mathematics, MDPI, and seven other international journals. She is a guest editor of several Special Issues related to algebraic structures and graph theory.

Hashem Bordbar

Hashem Bordbar graduated with a Ph.D. in Algebra, working on Ordered Algebra and Hypercompositional Algebra, under the supervision of Prof. Young Bae Jun as part of his doctoral studies at Gyeongsang National University, Jinju, South Korea. He continued his research in applying algebraic structures in computer science as a postdoc at Shahid Beheshti University, Tehran, Iran, and then he moved on Hypercompositional Algebra under the supervision of Prof. Irina Cristea at the University of Nova Gorica, Slovenia. Here he is an Assistant Professor in the Center for Information Technologies and Applied Mathematics, investigating algebraic coding theory and his previous field of research in Algebra.

Editorial

Preface to the Special Issue "Algebraic Structures and Graph Theory"

Irina Cristea * and Hashem Bordbar

Centre for Information Technologies and Applied Mathematics, University of Nova Gorica, Vipavska Cesta 13, 5000 Nova Gorica, Slovenia; hashem.bordbar@ung.si
* Correspondence: irina.cristea@ung.si

1. Introduction

Connections between algebraic structure theory and graph theory have been established in order to solve open problems in one theory with the help of the tools existing in the other, emphasizing the remarkable properties of one theory with techniques involving the second. This has provided new methods for solving several open problems, and has proposed new ones. One remarkable example in this direction is the contribution of Artur Cayley, who defined the concept of a group in 1854 (the composition table of the operation on the group takes his name, i.e., the Cayley table) and described in 1878 the structure of a group with a special graph, called a Cayley graph. There are many ways to define an algebraic structure (as a group, ring, hypergroup, hyperfield, lattice, etc.) starting from a graph and also vice versa, with the algebraic structures leading to the various types of graph. Many such constructions are discussed and illustrated with several non-trivial examples in [1–4].

This Special Issue aims to collect recent theoretical and applied studies on the interrelations between algebraic structures and graphs. This topic has attracted the interest of many researchers from different branches of algebra, and among the 63 submissions, 18 articles have been selected and published in this book. In the next section, we will briefly summarize their findings; for more detail, we recommend the readers to consult the original articles and the related bibliographies.

2. Contributions

The articles published in this Special Issue present new and up-to-date theoretical and applied research topics related to the following: (1) algebraic structures, (2) graphs and hypergraphs, and (3) connections between graphs and algebraic structures. We will start with the first group, containing seven articles dealing with semigroups, differential graded algebras, BL-algebras, hypergroups and hyperfields.

Regular elements in a semigroup play a fundamental role, being the key focus of the study of regular semigroups. As an example of such a structure, we recall the total transformation semigroup $T(X)$ on a nonempty set X. In 1975, Symons [5] introduced a subsemigroup of $T(X)$, defined as $T(X,Y) = \{\alpha \in T(X) \mid X\alpha \subseteq Y\}$, for a nonempty subset Y of X, determining all its automorphisms. The regularity of this subsemigroup and its implications for the computation of the number of the left/right regular elements of $T(X,Y)$ are discussed in [6]. A second paper [7] related to the theory of semigroups presents a novel method for generating the M-tri basis of an ordered Γ-semigroup. Here, the authors showed how the elements and subsets of an ordered Γ-semigroup yield to M-tri-ideals and the M-tri basis. Another two articles are related to the theory of algebras. The first [8] regards the differential graded algebras. Based on the notion of the Hopf Galois extension, the author emphasizes the relationships between the derived categories $\mathcal{D}(R\#H)$ of the smash product $R\#H$ and $\mathcal{D}(R^H)$, where H is a finite dimensional semisimple Hopf algebra and R a left H-module algebra. The second paper [9] studies the structure of a block code generated by a BL-algebra. In particular, H. Bordbar defines a new order for the generated code associated with a BL-algebra, and shows that the structure of the BL-algebra with its initial order and the one of the corresponding generated code with the new defined order coincide.

The last three manuscripts within this first group are in the framework of hypercompositional algebra. This is a relatively new field of abstract algebra, studying algebraic structures endowed with at least one hyperoperation, i.e., a multivalued operation associating with any pair of elements a subset of the support set. Thus, the hypercompositional structures are natural generalizations of the classical algebraic structures. For example, the hypergroups are a generalization of groups, while hyperfields are a generalization of the concept of fields. In [10] the authors work with the most well-known class of hyperfields, Krasner hyperfields, studying the notions of positive cone, characteristic and C-characteristic. Using these notions, they provide a criterion for deciding whether certain hyperfields cannot be obtained via Krasner's quotient construction. Furthermore, they prove that for any positive integer n greater than 1, there exists an infinite quotient hyperfield of characteristic n. A similar result holds for the C-characteristic. The manuscripts [11,12] cover some applications of hypercompositional algebra to automata theory. Massouros et al. [11] study the binary state machines with magma of two elements as their environment. Another aspect of automata theory is discussed in [12], where the authors propose several conditions for simplifying the verification of the GMAC condition for systems of quasi-multiautomata. Furthermore, using the concatenation, they construct quasi-multiautomata corresponding to the deterministic automata of the theory of formal languages.

We continue our presentation of the second group of contributions published in this Special Issue. They discuss innovative aspects in graph and hypergraph theory. Addition signed Cayley graphs are investigated in [13] with respect to the balancing, clusterability and sign compatibility properties. Here, the author presents the necessary conditions such that an addition signed Cayley graph is balanced, sign-compatible, clusterable, a line signed graph or C-consistent. The Cayley graphs are also the focus in [14]. In particular, it is shown that the Cayley graphs generated by transposition trees on the set $\{1, 2, \ldots, n\}$ are $n-2$-extendable and their extendability number is $n-2$ for any integer $n \geq 3$. Based on the notion of the trace norm of a $\{0,1\}$-Brauer configuration, the authors of [15] compute the graph energy of some families of graphs defined by Brauer configuration algebras. In [16], the author counts the crossing number of the joint product $G^* \, D_n$ of the disconnected graph G^* consisting of two components isomorphic to K_2 and K_3 and the discrete graph D_n with n isolated vertices. Moreover, a lower bound of the distance of the Laplacian spectral radius of the n-vertex bipartite graphs, with a diameter equal to 4, is determined in [17]. The paper concludes with the conjecture that the graph $G(1, \ldots, 1, n-d, 1, \ldots, 1)$ is unique, minimizing the distance of the Laplacian spectral radius among the n-vertex bipartite graphs, with all having a diameter greater than or equal to 4. A similar argument is posed by the authors of [18], who find upper and lower bounds on the spectral radius of the generalized reciprocal distance matrix of a connected graph with n vertices. The main goal of Solovyev's paper [19] is to establish a counting formula for a 2-dimensional lattice path model with filter restrictions in the presence of long steps. The last manuscript [20] within this group is in the area of hypergraphs, originally introduced by Berge as extensions of graphs, where the edges are substituted by hyperedges, being nonempty subsets of the set of vertices. Using the innovative concepts of knot and knot-hyperpath, the authors study the behaviour of the hyperpaths under hyper-continuous mappings and pseudo-open mappings, and find the sufficient conditions under which a hypergraph becomes a hypertree. The paper ends with an algorithm extracting a host graph from a hypertree.

We conclude this section with a description of the third group of manuscripts, presenting different relationships between algebraic structures and graph theory. The first paper [21] introduces a construction of a new graph associated with a semihypergroup, using the fundamental relation γ^*. Several properties such as completeness, regularity, being Eulerian or Hamiltonian, and Cartesian products are studied. Much the same direction is followed in the second paper [22], where a t-graph associated with a finitely generated group, using the Minkowski metric, is defined. The groups involved here are the two-generator finite groups, and the authors characterize the chromatic number of a t-graph depending exclusively on the parity of t. Finally, the study presented in the eighteenth paper [23] of this collection is focused on the construction and properties of an ideal-based dot total graph associated with a commutative ring with nonzero unity.

3. Conclusions

Based on the number of views and citations received, we are confident that the selected manuscripts of this edited book have aroused considerable interest among researchers in this field, and will open new lines of investigation not only in the domain of algebraic structure theory or graph theory, but also in other research topics. Therefore, this Special Issue will continue with a second edition, edited by Irina Cristea and Alessandro Linzi.

Funding: This work was partially funded by the Slovenian Research Agency (research core funding No. P1-0285).

Acknowledgments: The Guest Editors would like to thank all authors for their contributions to this Special Issue, and all involved reviewers for their valuable suggestions of improvements for the reviewed manuscripts. We also acknowledge the entire MDPI staff for their great cooperation with this Special Issue and in particular, we express our gratitude to the Contact Editor, Ursula Tian, for her high level of professionalism in managing the publication of this book.

Conflicts of Interest: The authors declare no conflict of interest.

References

1. Ribenboim, P. Algebraic structures on graphs. *Alg. Univ.* **1983**, *16*, 105–123. [CrossRef]
2. Budden, F. Cayley graphs for some well-known groups. *Math. Gazette* **1985**, *69*, 271–278. [CrossRef]
3. Bertram, E.A.; Herzog, M.; Mann, A. On a graph related to conjugacy classes of groups. *Bull. London Math. Soc.* **1990**, *22*, 569–575. [CrossRef]
4. Anderson, D.F.; Badawi, A. The total graph of a commutative ring. *J. Algebra* **2008**, *320*, 2706–2719. [CrossRef]
5. Symons, J.S.V. Some result concerning a transformation semigroup. *J. Aust. Math. Soc.* **1975**, *19*, 135–141. [CrossRef]
6. Sripon, K.; Laysirikul, E.; Sommanee, W. Left (Right) Regular Elements of Some Transformation Semigroups. *Mathematics* **2023**, *11*, 3042. [CrossRef]
7. Palanikumar, M.; Jana, C.; Al-Shanqiti, O.; Pal, M. A Novel Method for Generating the M-Tri-Basis of an Ordered Γ-Semigroup. *Mathematics* **2023**, *11*, 893. [CrossRef]
8. Zhang, B. Hopf Differential Graded Galois Extensions. *Mathematics* **2023**, *11*, 128. [CrossRef]
9. Bordbar, H. The Structure of the Block Code Generated by a BL-Algebra. *Mathematics* **2022**, *10*, 692. [CrossRef]
10. Kedzierski, D.E.; Linzi, A.; Stojalowska, H. Characteristic, C-Characteristic and Positive Cones in Hyperfields. *Mathematics* **2023**, *11*, 779. [CrossRef]
11. Massouros, G.G.; Massouros, C.G. State Machines and Hypergroups. *Mathematics* **2022**, *10*, 2427. [CrossRef]
12. Krehlik, S.; Novak, N.; Vyroubalova, J. From Automata to Multiautomata via Theory of Hypercompositional Structures. *Mathematics* **2022**, *10*, 1. [CrossRef]
13. Obaidullah, W.; Ayushi, D.; Deepa, S. On Some Properties of Addition Signed Cayley Graph. *Mathematics* **2022**, *10*, 3492.
14. Yongde, F.; Yanting, X.; Fengxia, L.; Shoujun, X. The Extendability of Cayley Graphs Generated by Transposition Trees. *Mathematics* **2022**, *10*, 1575.
15. Agudelo Muneton, N.; Canadas, A.M.; Espinosa, P.F.F.; Gaviria, I.D.M. {0,1}-Brauer Configuration Algebras and Their Applications in Graph Energy Theory. *Mathematics* **2021**, *9*, 3042. [CrossRef]
16. Stas, M. Parity Properties of Configurations. *Mathematics* **2022**, *10*, 1998. [CrossRef]
17. Qi, L.; Miao, L.; Zhao, W.; Liu, L. A Lower Bound for the Distance Laplacian Spectral Radius of Bipartite Graphs with Given Diameter. *Mathematics* **2022**, *10*, 1301. [CrossRef]
18. Ma, Y.; Gao, Y.; Shao, Y. Upper and Lower Bounds for the Spectral Radius of Generalized Reciprocal Distance Matrix of a Graph. *Mathematics* **2022**, *10*, 2683. [CrossRef]
19. Solovyev, D. Congruence for Lattice Path Models with Filter Restrictions and Long Steps. *Mathematics* **2022**, *10*, 4209. [CrossRef]
20. Rahman, S.; Chowdhury, F.A.M.; Cristea, I. Knots and Knot-Hyperpaths in Hypergraphs. *Mathematics* **2022**, *10*, 424. [CrossRef]

21. Firouzkouhi, N.; Ameri, R.; Amini, A.; Bordbar, H. Semihypergroup-Based Graph for Modeling International Spread of COVID-n in Social Systems. *Mathematics* **2022**, *10*, 4405. [CrossRef]
22. Diaz-Porto, G.; Gutierrez, I.; Torres-Grandisson, A. The t-Graphs over Finitely Generated Groups and the Minkowski Metric. *Mathematics* **2022**, *10*, 3030. [CrossRef]
23. Ashraf, M.; Asalool, J.H.; Alanazi, A.M.; Alamer, A. An Ideal-Based Dot Total Graph of a Commutative Ring. *Mathematics* **2021**, *9*, 3072. [CrossRef]

Disclaimer/Publisher's Note: The statements, opinions and data contained in all publications are solely those of the individual author(s) and contributor(s) and not of MDPI and/or the editor(s). MDPI and/or the editor(s) disclaim responsibility for any injury to people or property resulting from any ideas, methods, instructions or products referred to in the content.

Article

{0,1}-Brauer Configuration Algebras and Their Applications in Graph Energy Theory

Natalia Agudelo Muñetón [†], Agustín Moreno Cañadas [*,†], Pedro Fernando Fernández Espinosa [†] and Isaías David Marín Gaviria [†]

Departamento de Matemáticas, Universidad Nacional de Colombia, Edificio Yu Takeuchi 404, Kra 30 No 45-03, Bogotá 11001000, Colombia; nagudel83@gmail.com (N.A.M.); pffernandeze@unal.edu.co (P.F.F.E.); imaringa@unal.edu.co (I.D.M.G.)
* Correspondence: amorenoca@unal.edu.co
† Authors contributed equally to this work.

Abstract: The energy $\mathcal{E}(G)$ of a graph G is the sum of the absolute values of its adjacency matrix. In contrast, the trace norm of a digraph Q, which is the sum of the singular values of the corresponding adjacency matrix, is the oriented version of the energy of a graph. It is worth pointing out that one of the main problems in this theory consists of determining appropriated bounds of these types of energies for significant classes of graphs, digraphs and matrices, provided that, in general, finding out their exact values is a problem of great difficulty. In this paper, the trace norm of a $\{0,1\}$-Brauer configuration is introduced. It is estimated and computed by associating suitable families of graphs and posets to Brauer configuration algebras.

Keywords: brauer configuration algebra; graph energy; path algebra; poset; spectral radius; trace norm; wild representation type

1. Introduction

Brauer configuration algebras (BCAs) were introduced recently by Green and Schroll [1]. These algebras are multiserial symmetric algebras whose theory of representation is based on combinatorial data.

Since its introduction, BCAs have been a tool in the research of different fields of mathematics. Its role in algebra, combinatorics, and cryptography is remarkable. For instance, Malić and Schroll [2] associated a Brauer configuration algebra to some dessins d'enfants used to study Riemann surfaces, Cañadas et al. investigated the structure of the keys related to the Advanced Encryption Standard (AES) by using some so-called polygon-mutations in BCAs. On the other hand, BCAs were a helpful tool for Espinosa et al. to describe the number of perfect matchings in some snake graphs. We point out that Schiffler et al. used perfect matchings of snake graphs to provide a formula for the cluster variables associated with appropriated cluster algebras of surface type. In their doctoral dissertation, Espinosa used the notion of the message of a Brauer configuration to obtain the results [3,4]. According to him, each polygon in a Brauer configuration has associated a word. The concatenation of such words constitutes a message after applying a suitable specialization.

Perhaps, the message associated with a Brauer configuration is one of the most helpful tools to obtain applications of BCAs. In this work, we use Brauer configuration messages, some results of the theory of posets (partially ordered sets) and integer partitions to obtain the trace norm of some $\{0,1\}$-Brauer configurations, which are Brauer configurations whose sets of vertices consist only of 0's and 1's.

It is worth pointing out that the research on trace norm has its roots in chemistry within the Hückel molecular orbital theory (HMO) [5]. Afterwards, Gutman [6] founded an independent line of investigation in spectral graph theory based on graph energy, which is the sum $\mathcal{E}(G) = \sum_{\lambda \in spect(M_G)} |\lambda|$, where $spect(M_G)$ is the set of eigenvalues of

the adjacency matrix M_G of a graph G. The trace norm associated with the adjacency matrix of a digraph or quiver Q denoted $||Q||_*$ is a generalization of the graph energy. It is also called the Schatten 1-norm, Ky Fan n-norm or nuclear norm. If $\sigma_1, \sigma_2, \ldots, \sigma_n$ are the singular values of the $m \times n$- adjacency matrix M_Q, with $\sigma_1 \geq \sigma_2 \geq \cdots \geq \sigma_n$ then $||Q||_* = \sum_{i=1}^{\min\{m,n\}} \sigma_i$. Relationships between energy graph and trace norm were investigated first by Nikiforov [7].

One of the main problems in graph energy theory is giving the extremal values of the energy of significant classes of graphs. For instance, Gutman [6] proved that if T_n is a tree with n vertices then the following identity holds:

$$\mathcal{E}(S_n) \leq \mathcal{E}(T_n) \leq \mathcal{E}(\mathbb{A}_n) \tag{1}$$

where, S_n ($\mathbb{A}_n$) denotes the star (the Dynkin diagram of type $\mathbb{A}$) with n vertices.

Graph energy associated with digraphs was investigated first by Kharaghani–Tayfeh–Rezaie [8], afterwards by Agudelo–Nikiforov [9], who found bounds of extremal values of the trace norm for $(0,1)$-matrices. It is worth noticing that if the adjacency matrix of a graph G is normal, then the graph energy equals the trace norm. In particular, if the adjacency matrix M_G of a graph G is symmetric, then $\mathcal{E}(G) = ||M_G||_*$.

Contributions

In this paper, we introduce the notion of trace norm of a $\{0,1\}$-Brauer configuration. Bounds and explicit values of these trace norms are given for significant classes of graphs induced by this kind of configuration. In particular, the dimension of the associated algebras and their centers are obtained. These results give a relationship between Brauer configuration algebras and graph energy theories with an open problem in the field of integer partitions proposed by Andrews in 1986. Such a problem asks for sets of integer numbers S, T for which $P(S, n) = P(T, n + a)$, where $P(X, n)$ denote the number of integer partitions of n into parts within the set X with a being a fixed positive integer [10].

As a consequence of their investigations regarding Andrews's problem, Cañadas et al. [11,12] introduced and enumerated a particular class of integer compositions (i.e., partitions for which the order of the parts matter) of type $\mathcal{D}_n$, for which the Andrews's problem holds if $a = 1$. For each n, compositions of type $\mathcal{D}_n$ constitute a partially ordered set whose number of two-point antichains is given by the integer sequence encoded in the OEIS (On-Line Encyclopedia of Integer Sequences) A344791 [13]. The following identity (2) gives the nth term $(A344791)_n$ of this sequence:

$$(A344791)_n = \sum_{i=1}^{n}\sum_{j=0}^{\lfloor \frac{i}{2}\rfloor} h_{ij}(t_i - 2t_j). \tag{2}$$

where t_k denotes the kth triangular number, and

$$h_{ij} = \begin{cases} n+1-i, & \text{if } i = 2j \text{ and } 1 \leq j \leq \lfloor \frac{n}{2} \rfloor, \\ 0, & \text{if } i = n \text{ and } j = 0, \\ 1, & \text{otherwise.} \end{cases}$$

This paper uses this sequence to estimate eigenvalues sums of matrices associated with polygons of some $\{0,1\}$-Brauer configurations.

It is worth noting that the relationships introduced in this paper between the theory of Brauer configuration algebras and the graph energy theory do not appear in the current literature devoted to these topics.

This paper is distributed as follows; in Section 2, we recall definitions and notation used throughout the document. In particular, we introduce the notion of trace norm of a $\{0,1\}$-Brauer configuration. In Section 3, we give our main results, we compute and estimate the trace norm and graph energy of some families of graphs defined by Brauer configuration algebras. Concluding remarks are given in Section 4. Examples of trace norm values associated with some Brauer configurations are given in Appendix A.

The following diagram (3) shows how the notions of Brauer configuration and trace norm are related to some of the main results presented in this paper.

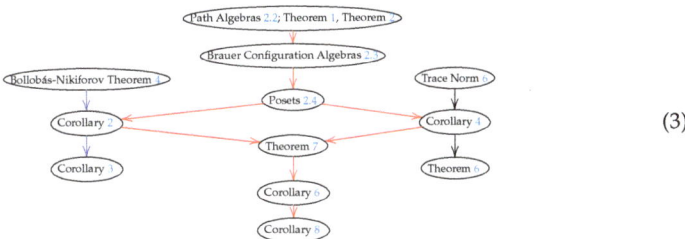

(3)

2. Background and Related Work

In this section, we introduce some definitions and notations to be used throughout the paper. In particular, it is given a brief overview regarding the development of the research of graph energy theory, path algebras, and Brauer configuration algebras.

Henceforth, the symbol A^* will denote the adjoint of a matrix A, and $\|A\|_F$ the Frobenius norm of a matrix A. Furthermore, $\mathbb{F}$ is a field, $\mathbb{N}^+$ is the set of positive integers, and t_n denotes the nth triangular number.

2.1. Graph Energy

The notion of graph energy as the sum of the absolute values of an adjacency matrix was introduced in 1978 by Gutman based on a series of lectures held by them in Stift Rein, Austria [6]. As we explained in the introduction, he was motivated by earlier results regarding the Hückel orbital total π-electron energy. According to Gutman and Furtula [14], the results were proposed at that time in good hope that the mathematical community would recognize its significance. However, there was no interest in the subject despite Gutman's efforts, perhaps due to the restrictions imposed on the studied graphs.

The interest in graph energy was renewed at the earliest 2000 when a plethora of results started appearing. Since then, more than one hundred variations of the initial notion have been introduced with applications in different sciences fields. In the same work, Gutman and Furtula claim that an average of two papers per week (more than one hundred in 2017) are written regarding the subject.

Some of the graph energy variations are:

1. The *Nikiforov energy* of a matrix M, which is the sum of the singular values of a matrix.
2. The *Laplacian energy* of a graph G of order n and size m defined as the sum of the absolute values of the eigenvalues of the matrix $L(G) - \frac{2m}{n} I_n$, where I_n is the identity matrix of order n, and $L(G)$ is the Laplacian matrix associated with G whose entries $L(G)_{ij}$ are given by the following identities:

$$(L(G))_{ij} = \begin{cases} \deg(v_i) & \text{if } i = j, \\ -1 & \text{if } i \neq j \text{ and } v_i \text{ is adjacent to } v_j, \\ 0 & \text{otherwise.} \end{cases}$$

where $\deg(v)$ denotes the degree of a vertex v in G.

3. The *Randić energy*, which is the sum of the absolute values of the Randić matrix $R(G) = (R(G)_{ij})$ of a graph G, with

$$(R(G))_{ij} = \begin{cases} 0 & \text{if } i = j, \\ \frac{1}{\sqrt{\deg(v_i)\deg(v_j)}} & \text{if } v_i \text{ is adjacent to } v_j, \\ 0 & \text{otherwise.} \end{cases}$$

Although the notion of graph energy was introduced only for theoretical purposes, currently, its applications embrace a broad class of sciences. The following Table 1 shows some examples of different works devoted to the applications of graph energy and its variations. The authors refer the reader to [14] for more examples of these types of applications.

Table 1. Works devoted to the applications of the graph energy theory. In the case of pattern recognition, the applications deal with military purposes.

Subject	Work
Chemistry	[15]
Biology	[16]
Crystallography	[17]
Epidemics	[18]
Pattern Recognition	[19]
Computer Vision	[20]
Satellite Communication	[21]
Spacecrafts Construction	[22]
Neural Networks	[23]

2.2. Path Algebras

This section recalls some facts regarding quivers, their associated path algebras, and corresponding module categories. It is worth noting that the quiver or pass graph technique is used in representation theory, and it is an important tool to solve many ring problems, as Belov-Kanel et al. report in [24].

A *quiver* $Q = (Q_0, Q_1, s, t)$ is a quadruple consisting of two sets Q_0 whose elements are called *vertices* and Q_1 whose elements are called arrows, s and t are maps $s, t : Q_1 \to Q_0$ such that if α is an arrow, then $s(\alpha)$ is called the *source* of α, whereas $t(\alpha)$ is called the *target* of α [25]. The adjacency matrix M_Q and the spectral radius $\rho(Q) = \rho(M_Q) = \max|\lambda|$ (where λ runs over all the eigenvalues of $M_{\overline{Q}}$) of a quiver Q are given by those defined by its underlying graph $\overline{Q}$.

Recall that the adjacency matrix M_G associated with a graph G is defined by the following identities:

$$(M_G)_{ij} = \begin{cases} \text{number of edges between } i \text{ and } j, & \text{if } i \neq j, \\ \text{two times the number of loops at } i, & \text{if } i = j. \end{cases}$$

A *path* of length $l \geq 1$ with source a and target b is a sequence $(a \mid \alpha_1, \alpha_2, \ldots, \alpha_l \mid b)$ where $t(\alpha_i) = s(\alpha_{i+1})$ for any $1 \leq i < l$. Vertices are paths of length 0 [25–27].

If Q is a quiver and $\mathbb{F}$ is an algebraically closed field, then the *path algebra* $\mathbb{F}Q$ of Q is the $\mathbb{F}$-algebra whose underlying $\mathbb{F}$-vector space has as basis the set of all paths of length $l \geq 0$ in Q, the natural graph concatenation is the product of two paths [25,26].

An $\mathbb{F}$-algebra Λ is said to be *basic* if it has a complete set $\{e_1, e_2, \ldots, e_l\}$ of primitive orthogonal idempotents such that $e_i A \not\cong e_j A$ for all $i \neq j$.

A *relation* for a quiver Q is a linear combination of paths of length ≥ 2 with the same starting points and same endpoints, not all coefficients being zero [25,26].

Let Q be a finite and connected quiver. The two-sided ideal of the path algebra $\mathbb{F}Q$ generated by the arrows of Q is called the *arrow ideal* of $\mathbb{F}Q$ and is denoted by R_Q, R_Q^l is the ideal of $\mathbb{F}Q$ generated as an $\mathbb{F}$-vector space, by the set of all paths of length $\geq l$. A two-sided ideal I of the path algebra $\mathbb{F}Q$ is said to be *admissible* if there exists $m \geq 2$ such that $R_Q^m \subseteq I \subseteq R_Q^2$.

If I is an admissible ideal of $\mathbb{F}Q$, the pair (Q, I) is said to be a *bound quiver*. The quotient algebra $\mathbb{F}Q/I$ is said to be a *bound quiver algebra*.

Gabriel [28] proved that any basic algebra is isomorphic to a bound quiver algebra. He also showed the finiteness criterion for these algebras. Taking into account that one of the main problems in the theory of representation of algebras consists of giving a complete description of the indecomposable modules and irreducible morphisms of the category of finitely generated modules mod Λ of a given algebra Λ.

According to the number of indecomposable modules an algebra Λ can be of finite, tame or wild representation type. We recall that if $\mathcal{C}$ is a category of finitely generated modules over an $\mathbb{F}$-algebra Λ (in this case, $\mathbb{F}$ is an algebraically closed field). Then a one-parameter family in $\mathcal{C}$ is a set of modules of the form:

$$\overline{\mathcal{M}} = \{\mathcal{M}/(x-a)\mathcal{M} \mid a \in \mathbb{F}\} \tag{4}$$

where $\mathcal{M}$ is a $\Lambda - \mathbb{F}[x]$-bimodule, which is finitely generated and free over $\mathbb{F}[x]$ [29].

Category $\mathcal{C}$ is said to be of *tame representation type* or *tame type*, if $\mathcal{C} = \bigcup_n \mathcal{C}_n$, and for every n, the indecomposable modules form a *one-parameter* family with maybe finitely many exceptions equivalently in each dimension d, all but a finite number of indecomposable Λ-modules of dimension d belong to a finite number of one-parameter families. On the other hand, $\mathcal{C}$ is of *wild representation type* or *wild type* if it contains n-parameter families of indecomposable modules for arbitrarily large n [29].

It is worth noting that Drozd in 1977 and Crawley-Boevey in 1988 proved the following result known as the trichotomy theorem.

Theorem 1 ([30,31], Corollary C). *Let Λ be a finite-dimensional algebra over an algebraically closed field. Then Λ-mod has either tame type or wild type, and not both.*

The following result proved by Smith establishes a relationship between the theory of representation of algebras and the spectra graph theory.

Theorem 2 ([32]). *Let G be a finite simple graph with the spectral radius (index) $\rho(G)$. Then $\rho(G) = 2$ if and only if each connected component of G is one of the extended Dynkin diagram $\widetilde{\mathbb{A}}_n, \widetilde{\mathbb{D}}_n, \widetilde{\mathbb{E}}_6, \widetilde{\mathbb{E}}_7, \widetilde{\mathbb{E}}_8$. Moreover, $\rho(G) < 2$ if and only if each connected component of G is one of Dynkin diagrams $\mathbb{A}_n, \mathbb{D}_n, \mathbb{E}_6, \mathbb{E}_7, \mathbb{E}_8$.*

Remark 1. *Note that if Q is a connected quiver without oriented cycles, then Theorem 2 allows concluding that Q is of finite type (tame type) if and only if $\rho(Q) < 2$ ($\rho(Q) = 2$). Otherwise, Q is of wild type. A quiver Q has one of these three properties means that the corresponding path algebra $\mathbb{F}Q$ also does.*

2.3. {0,1}-Brauer Configuration Algebras

In this section, we discuss some results regarding $\{0,1\}$-Brauer configuration algebras, we refer the reader to [1] for a more detailed study of general Brauer configuration algebras.

$\{0,1\}$-Brauer configuration algebras are bound quiver algebras induced by a Brauer configuration $\Gamma = (\Gamma_0, \Gamma_1, \mu, \mathcal{O})$ with the following characteristics:

- $\Gamma_0 = \{0, 1\}$ is said to be the set of *vertices*.
- $\Gamma_1 = \{U_1, U_2, \ldots, U_{n-1}, U_n\,; n \geqslant 1\}$ is a collection of multisets U_i consisting of vertices called *polygons*.
- The *word* w_i defined by the polygon U_i has the form;

$$w_i = w_{i,1} w_{i,2} \ldots w_{i,\delta_i}.$$

where $w_{i,j} \in \{0,1\}$, $\alpha_i = occ(0, U_i)$ is the number of times that the vertex 0 occurs in the polygon U_i, $\delta_i - \alpha_i = occ(1, U_i)$ is the number of times that the vertex 1 appears in the same polygon with $\delta_i = |U_i| \geq 2$.

- μ is a map $\mu : \Gamma_0 \to \mathbb{N}^+$, such that $\mu(0) = \mu(1) = 1$. μ is said to be the *multiplicity function* associated with Γ.
- Successor sequences S_0 and S_1 associated with the vertices are defined by an orientation $\mathcal{O}$, which is an ordering on the polygons of the form:

$$S_0 : \underbrace{U_1 < \cdots < U_1}_{\alpha_1-times} < \underbrace{U_2 < \cdots < U_2}_{\alpha_2-times} < \cdots < \underbrace{U_{n-1} < \cdots < U_{n-1}}_{\alpha_{n-1}-times} < \underbrace{U_n < \cdots < U_n}_{\alpha_n-times}$$

$$S_1 : \underbrace{U_1 < \cdots < U_1}_{(\delta_1-\alpha_1)-times} < \underbrace{U_2 < \cdots < U_2}_{(\delta_2-\alpha_2)-times} < \cdots < \underbrace{U_{n-1} < \cdots < U_{n-1}}_{(\delta_{n-1}-\alpha_{n-1})-times} < \underbrace{U_n < \cdots < U_n}_{(\delta_n-\alpha_n)-times}$$

Successor sequences is a way of recording how vertices appear in the polygons.

The construction of the quiver Q_Γ (or simply Q, if no confusion arises) goes as follows:

- Add a circular relation $U_n < U_1$, to each successor sequence S_0 and S_1. $C^i = S_i \cup \{U_n < U_1\}$, $i \in \{0,1\}$ is said to be a *special cycle* associated with i.
- Define Γ_1 as the set of vertices Q_0 of Q.
- Each cover $U_i < U_j$ in a special cycle C^i defines an arrow $U_i \to U_j \in Q_1$.

Note that there are different special cycles associated with a vertex $i \in \{0,1\}$ in a polygon U_i.

Figure 1 shows the Brauer quiver Q_Γ induced by a $\{0,1\}$-Brauer configuration Γ.

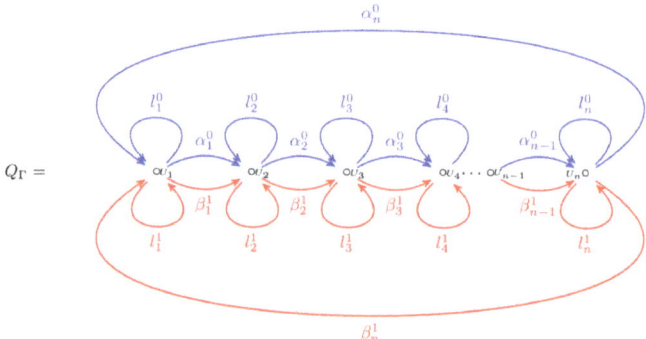

Figure 1. Brauer quiver induced by a $\{0,1\}$-Brauer configuration. Symbols l_j^i, $i \in \{0,1\}$, $j \in \{1,2,\ldots,n\}$ mean that the corresponding vertex U_j has associated $l_j^i = occ(i, U_j) - 1$ different loops.

The *valency* $val(i)$ of a vertex $i \in \{0,1\}$ is given by the identity:

$$val(i) = \sum_{j=1}^{n} occ(i, U_j). \tag{5}$$

$val(i)$ is the number of arrows in the i-cycles. A vertex $i \in \{0,1\}$ is said to be *truncated* if $val(i) = 1$, otherwise i is *non-truncated*. Vertices 0 and 1 are non-truncated in a $\{0,1\}$-Brauer configuration algebra.

The Brauer configuration algebra Λ_Γ (or Λ) defined by the quiver Q is the path algebra $\mathbb{F}Q$ bounded by the admissible ideal I_Γ (or I) generated by the following set of relations:

1. If a polygon $U_k \in \Gamma_1$ contains vertices i, j and C^i, C^j are special cycles then $C^i - C^j \in I$.
2. If a is the first arrow of a special cycle C^i then $C^i a \in I$.

3. $\alpha_i^0 \beta_j^1$, $\beta_h^1 \alpha_s^0$, for all possible values of i, j, h, and s.
4. $\alpha_i^0 \alpha_{i+1}^0$, $\beta_i^1 \beta_{i+1}^1$, for all possible values of i.
5. $\alpha_i^0 l_j^1$, $\beta_i^1 l_j^0$, for all possible values of i and j.
6. $l_i^0 l_i^1$, $l_j^0 \beta_i^1$, $l_j^1 \alpha_i^0$ for all possible values of i and j.
7. $(l_i^0)^2$, $(l_j^1)^2$, for all possible values of i and j.

If there exists a word-transformation T such that $w_i = T(w_{i-1})(R_i)$, for instance, if $w_i = w_{i-1} R_i$ with R_i a suitable $\{0,1\}$-word, then the *cumulative message* $M(\Gamma)$ of Γ is defined in such a way that $M(\Gamma) = w_1 w_2 \ldots w_n$ and the *reduced message* $M_R(\Gamma)$ is defined by the concatenation word:

$$M_R(\Gamma) = w_1 R_2 R_3 \ldots R_n$$

If $M_R(\Gamma)$ can be written as a $m \times n$ matrix, then $\rho(M_R(\Gamma))$ denotes the *spectral radius of the Brauer configuration* Γ and *the trace norm of the Brauer configuration* Γ is defined as:

$$||M_R(\Gamma)||_* = \sum_{k=1}^{\min\{m,n\}} \sigma_k(M_R(\Gamma)). \tag{6}$$

where $\sigma_1(M_R(\Gamma)) \geqslant \sigma_2(M_R(\Gamma)) \geqslant \cdots \geqslant \sigma_n(M_R(\Gamma)) \geqslant 0$ are the singular values of $M_R(\Gamma)$, i.e., the square roots of the eigenvalues of $M_R(\Gamma) M_R(\Gamma)^*$.

The following Proposition 1 and Theorem 3 prove that the dimension and the center of a Brauer configuration algebra can easily be computed from its Brauer configuration [1,33].

Proposition 1 ([1], Proposition 3.13). *Let Λ be a Brauer configuration algebra associated with the Brauer configuration Γ and let $\mathcal{C} = \{C_1, \ldots, C_t\}$ be a full set of equivalence class representatives of special cycles. Assume that for $i = 1, \ldots, t$, C_i is a special α_i-cycle where α_i is a non-truncated vertex in Γ. Then*

$$\dim_{\mathbb{F}} \Lambda = 2|Q_0| + \sum_{C_i \in \mathcal{C}} |C_i|(n_i |C_i| - 1),$$

where $|Q_0|$ denotes the number of vertices of Q, $|C_i|$ denotes the number of arrows in the α_i-cycle C_i and $n_i = \mu(\alpha_i)$.

Theorem 3 ([33], Theorem 4.9). *Let Γ be a reduced and connected Brauer configuration and let Q be its induced quiver and let Λ be the induced Brauer configuration algebra such that $\mathrm{rad}^2 \Lambda \neq 0$ then the dimension of the center of Λ denoted $\dim_{\mathbb{F}} Z(\Lambda)$ is given by the formula:*

$$\dim_{\mathbb{F}} Z(\Lambda) = 1 + \sum_{\alpha \in \Gamma_0} \mu(\alpha) + |\Gamma_1| - |\Gamma_0| + \#(Loops\ Q) - |\mathcal{C}_\Gamma|.$$

where $|\mathcal{C}_\Gamma| = \{\alpha \in \Gamma_0 \mid val(\alpha) = 1, \text{ and } \mu(\alpha) > 1\}$.

In this case, rad M denotes the radical of a module M, rad M is the intersection of all the maximal submodules of M.

The following are properties of $\{0,1\}$-Brauer configuration algebras based on Proposition 1 and Theorem 3.

Corollary 1. *Let Λ be a Brauer configuration algebra induced by a $\{0,1\}$-Brauer configuration $\Gamma = (\Gamma_0, \Gamma_1, \mu, \mathcal{O})$ with $\mathrm{rad}^2 \Lambda \neq 0$. Then the following statements hold:*

1. *Λ is reduced and connected.*
2. *$\dim_{\mathbb{F}} \Lambda = 2n + 2t_{val(0)-1} + 2t_{val(1)-1}$, where t_j denotes the jth triangular number.*
3. *$\dim_{\mathbb{F}} Z(\Lambda) = 1 + n + \sum_{i=0}^{1} \sum_{j=1}^{n} l_j^i$.*

2.4. Posets

A partially ordered set (or *poset*) is an ordered pair $(\mathcal{P}, \leq)$ where $\mathcal{P}$ is a not empty set, and $\leq$ is a partial order over the elements of $\mathcal{P}$, i.e., $\leq$ is reflexive, antisymmetric, and transitive. Henceforth, if no confusion arises we will write $\mathcal{P}$ instead of $(\mathcal{P}, \leq)$ to denote a partially ordered set.

For each $x, y \in \mathcal{P}$, if $x \leq y$ or $y \leq x$, we say that x and y are *comparable points*, whereas if $x \not\leq y$ and $y \not\leq x$, we say that x and y are *incomparable points* (the subset $\{x, y\}$ is a *two-point antichain*), this situation is denoted by $x \parallel y$. An ordered set $\mathcal{C}$ is called a *chain* (or a totally ordered set or a linearly ordered set) if and only if for all $x, y \in \mathcal{C}$ we have $x \leq y$ or $y \leq x$ (i.e., x and y are comparable points).

A relation $x \leq y$ in a poset $\mathcal{P}$ is said to be a covering, if for any $z \in \mathcal{P}$ such that $x \leq z \leq y$ it holds that $x = z$ or $y = z$ [34].

3. Applications

In this section, we give applications of $\{0,1\}$-Brauer configuration algebras in graph energy. We start by defining some suitable $\{0,1\}$- Brauer configuration algebras, dimensions of these algebras and corresponding centers are given as well. We also compute and estimate eigenvalues and trace norm of their reduced messages $M_R(\Gamma)$.

1. For $n \geq 2$ fixed, let us consider the $\{0,1\}$-Brauer configuration $\Delta^n = (\Delta_0^n, \Delta_1^n, \mu, \mathcal{O})$, such that:

$$\Delta_0^n = \{0,1\}.$$
$$\Delta_1^n = \{D_1, D_2, \ldots, D_n\}, \quad \text{for } 1 \leq i \leq n, \quad |D_i| = (t_{i+2} - 1)^2. \tag{7}$$
$$\mu(0) = \mu(1) = 1.$$

The orientation $\mathcal{O}$ is defined in such a way that in successor sequences associated with vertices 0 and 1, it holds that $D_i < D_{i+1}$, for $1 \leq i \leq n$.

Polygons D_i can be seen as $(t_{i+2} - 1) \times (t_{i+2} - 1)$-matrices over $\mathbb{Z}_2$ or as $(t_{i+2} - 1) \times 1$-matrices over the vector space $P_{t_{i+2}-2}$ of polynomials of degree $\leq t_{i+2} - 2$. Its construction goes as follows:

(a) For any $i, 1 \leq i \leq n$, D_i is a symmetric matrix,

(b) $$D_1 = \begin{bmatrix} 1 & 1 & 0 & 1 & 1 \\ 1 & 1 & 1 & 0 & 1 \\ 0 & 1 & 1 & 1 & 1 \\ 1 & 0 & 1 & 1 & 1 \\ 1 & 1 & 1 & 1 & 1 \end{bmatrix} = \begin{bmatrix} t^4 + t^3 + t + 1 \\ t^4 + t^3 + t^2 + 1 \\ t^3 + t^2 + t + 1 \\ t^4 + t^2 + t + 1 \\ t^4 + t^3 + t^2 + t + 1 \end{bmatrix},$$

(c) $$D_i = \begin{bmatrix} D_{i-1} & \begin{matrix} B_1^{i+1} \\ B_2^{i+1} \\ \vdots \end{matrix} \\ * & \begin{matrix} B_i^{i+1} \\ B_{i+1}^{i+1} \end{matrix} \end{bmatrix}$$

(d) Blocks B_j^{i+k}, with $k > 1$ are defined as follows:

i. Over $\mathbb{Z}_2$, $B_j^j \in M_{(j+1)\times(j+1)}$, $B_j^{j+s} \in M_{(j+1)\times(j+s+1)}$, $0 \leq s \leq j+1$,

ii. Over P_{j+s+2}, $B_j^{i+k} = \begin{bmatrix} p_1^{i+k}(t) \\ p_2^{i+k}(t) \\ \vdots \\ p_{j+1}^{i+k}(t) \end{bmatrix}$,

$$p_h^{i+k}(t) = \begin{cases} \sum_{l=0}^{j-h+1} x^l, & \text{if } 1 \leq h \leq k, \\ \sum_{l=0}^{j-h+1} x^l + \sum_{j=0}^{h-k-1} x^{j+k-h+1}, & \text{if } h > k \text{ and } 2 \leq k \leq i+2, \\ p_h^m(t), & \text{if } m > i+2. \end{cases}$$

Corollary 2. *If $\mathfrak{D}^n = \mathbb{F}Q_{\Delta^n}^n / I_{\Delta^n}^n$ is the Brauer configuration algebra induced by the {0,1}-Brauer configuration Δ^n then the following statements hold:*

$$\dim_\mathbb{F} \mathfrak{D}^n = (e_n - d_n)^2 + (e_n - 1)^2 + (d_n - 1), \tag{8}$$
$$\dim_\mathbb{F} Z(\mathfrak{D}^n) = (t_{n+2})^2 + n + 3.$$

where

$$a_n = \frac{1 - (-1)^n - 8n - 4n^2 + 8n^3 + 2n^4}{32} = (A344791)_n,$$

$$b_{n+2} = \sum_{i=1}^{n+2} t_i^2 - 10,$$

$$c_{n+2} = -\frac{(n+2)(n+3)(n+4)}{3} + 8, \tag{9}$$

$$d_n = b_{n+2} + c_{n+2} + n, \quad n \geq 1,$$

$$e_n = 2\sum_{i=1}^{n} a_{i+1}.$$

Proof. For $n > 1$ fixed, consider the following set:

$$\mathcal{P}_n = \{x_{1,1}, x_{1,2}, x_{2,1}, x_{2,2}, x_{2,3}, \ldots, x_{i,1}, \ldots, x_{i,i+1}, \ldots, x_{n,1}, \ldots, x_{n,n+1}\} \tag{10}$$

$\mathcal{P}_n$ is endowed with a partial order $\unlhd$, which defines the following coverings:

$$\begin{aligned} x_{j,k} &\unlhd x_{j,k+1}, & 1 \leq j \leq n, & \quad 1 \leq k \leq j, \\ x_{j,k} &\unlhd x_{j+1,k+1}, & 1 \leq j < n, & \quad 1 \leq k \leq j+1, \\ x_{r,k} &\unlhd x_{r-1,k+1}, & 1 < r \leq n, & \quad 1 \leq k \leq r. \end{aligned} \tag{11}$$

$(\mathcal{P}_n, \unlhd)$ defines a matrix M_n whose entries $m_{i,j}$ are given by the following identities:

$$m_{i,j} = \begin{cases} 1, & \text{if } x_{i,r} \unlhd x_{j,s} \text{ or } x_{j,s} \unlhd x_{i,r} \\ 0, & \text{otherwise.} \end{cases}$$

Clearly M_n is a $(t_{n+1} - 1) \times (t_{n+1} - 1)$ symmetric matrix with $M_n = D_{n-1} \in \Delta_1^n$, that is, M_n is the matrix associated with the polygon $D_{n-1} \in \Delta_1^n$. Thus, $\frac{1}{2}\text{occ}(0, D_n)$ equals the number of two-point antichains in $(\mathcal{P}_n, \unlhd)$. Therefore, $\text{occ}(0, D_n)$ is twice the nth term of the sequence A344791 (see (2), (9)), and $\text{occ}(1, D_n) = (t_{n+1} - 1)^2 - \text{occ}(0, D_n)$. Since $\dim_\mathbb{F} \mathfrak{D}^n = 2n + val(0)(val(0) - 1) + val(1)(val(1) - 1)$. The result holds. Since $\text{rad}^2 \mathfrak{D}^n \neq 0$, then $\dim_\mathbb{F} Z(\mathfrak{D}^n) = 1 + n + \#(Loops\ Q_{\Delta^n})$ with $\#(Loops\ Q_{\Delta^n}) = (t_{n+2})^2 + 2$. We are done. □

Now we are interested in estimating the eigenvalues of M_n. Since the polygons $D_n \in \Delta_1^n$ can be seen as $(t_{n+1} - 1)$ square symmetric matrices described in the previous proof as $D_{n-1} = M_n$. We will assume that for each n, the real eigenvalues of a matrix M_n are indexed in the following decreasing order:

$$\mu_{max}(M_n) = \mu_1(M_n) \geq \mu_2(M_n) \geq \cdots \geq \mu_{t_{n+1}-1}(M_n) = \mu_{min}(M_n).$$

The next result, which derives two inequalities for the eigenvalues of Hermitian matrices, was proved by Bollobás and Nikiforov [35].

Theorem 4 ([35], Theorem 2). *Suppose that $2 \leq k \leq n$ and let $A = (a_{ij})$ be a Hermitian matrix of size n. For every partition $\{1, 2, \ldots, n\} = N_1 \cup \cdots \cup N_k$ we have*

$$\mu_1(A) + \cdots + \mu_k(A) \geq \sum_{r=1}^{k} \frac{1}{|N_r|} \sum_{i,j \in N_r} a_{ij}$$

and

$$\mu_{k+1}(A) + \cdots + \mu_n(A) \leq \sum_{r=1}^{k} \frac{1}{|N_r|} \sum_{i,j \in N_r} a_{ij} - \frac{1}{n} \sum_{i,j \in \{1,2,\ldots,n\}} a_{ij}.$$

The following result on the eigenvalues of M_n can be obtained by applying Theorem 4 to the matrix M_n associated with the polygon $D_{n-1} \in \Delta_1^n$.

Corollary 3. *For $n > 1$ and $k = n$. Let $M_n = (m_{ij})$ be the matrix associated with the polygon $D_{n-1} \in \Delta_1^n$. For partition $\{1, 2, \ldots, t_{n+1} - 1\} = N_1 \cup \cdots \cup N_n$ where $N_i = \left\{\frac{i(i+1)}{2}, \ldots, \frac{i(i+3)}{2}\right\}$. We have*

$$\sum_{i=1}^{n} \mu_i(M_n) \geq t_{n+1} - 1 \qquad (12)$$

and

$$\sum_{i=n+1}^{t_{n+1}-1} \mu_i(M_n) \leq \frac{2(A344791)_n}{t_{n+1} - 1}. \quad \text{(see (2))}. \qquad (13)$$

Proof. Since $N_i = \left\{\frac{i(i+1)}{2}, \ldots, \frac{i(i+3)}{2}\right\}$, for each $i = \{1, 2, \ldots, n\}$ then $|N_i| = i + 1$, besides each set N_i can be seen as a subset of the set $\mathcal{P}_n$ defined in (10) as follows:

$$N_i = \{x_{i,1}, \ldots, x_{i,i+1}\}.$$

On the other hand, to compute $\sum_{i,j \in N_i} a_{ij}$, we will use the coverings defined in (11) and the fact that $\mathcal{P}_n$ is a partial order, so we obtain:

$$\sum_{i,j \in N_i} m_{ij} = 2 \sum_{j=1}^{i} (x_{i,j} \trianglelefteq x_{i,j+1}) + \sum_{j=1}^{i+1} (x_{i,j} \trianglelefteq x_{i,j}) + 2 \sum_{j=1}^{i-1} (x_{i,j} \trianglelefteq x_{i,j+2})$$

$$= 2i + (i+1) + 2t_{i-1}$$

$$= (i+1)^2$$

Therefore:

$$\sum_{i=1}^{n} \frac{1}{|N_i|} \sum_{i,j \in N_i} m_{ij} = i + 1 \text{ and}$$

$$\frac{1}{t_{n+1}-1} \sum_{i,j \in \{1,2,\ldots,t_{n+1}-1\}} m_{i,j} = \frac{1}{t_{n+1}-1} \|M_n\|_F^2 = \frac{1}{t_{n+1}-1}((t_{n+1}-1)^2 - 2(A344791)_n)$$

Hence, applying Theorem 4 we obtain (12) and (13). □

2. For $n \geq 1$ fixed, let $\Gamma^n = \{\Gamma_0^n, \Gamma_1^n, \mu, \mathcal{O}\}$ be a $\{0,1\}$-Brauer configuration such that:

$$\Gamma_0^n = \{0,1\}.$$
$$\Gamma_1^n = \{U_1, U_2, \ldots, U_n\}, \quad \text{for } 1 \leq i \leq n, \quad |U_i| = 2^{2n}. \quad (14)$$
$$\mu(0) = \mu(1) = 1.$$

The orientation $\mathcal{O}$ is defined in such a way that in successor sequences associated with vertices 0 and 1, it holds that $U_i < U_{i+1}$.

Polygons U_i can be seen as $2^n \times 2^n$-matrices over $\mathbb{Z}_2$ using the Kronecker product, denoted by $\otimes$, as follows:

$$U_1 = \begin{bmatrix} 1 & 0 \\ 1 & 1 \end{bmatrix}$$
$$U_2 = U_1 \otimes U_1 \qquad (15)$$
$$\vdots$$
$$U_i = U_1 \otimes U_{i-1}.$$

Corollary 4. *For $n \geq 1$, if $\mathfrak{G}^n = \mathbb{F}Q_{\Gamma^n}^n / I_{\Gamma^n}^n$ is the Brauer configuration algebra induced by the $\{0,1\}$-Brauer configuration Γ^n then the following statements hold:*

$$\dim_{\mathbb{F}} \mathfrak{G}^n = 2n + 2r_n(r_n - 1) + 2s_n(s_n - 1)$$
$$\dim_{\mathbb{F}} Z(\mathfrak{G}^n) = \begin{cases} 6, & \text{if } n = 1 \\ 1 - n + r_n + s_n, & \text{if } n \geq 2. \end{cases} \quad (16)$$

where r_n and s_n are the nth term of the OEIS sequences A016208 and A029858, respectively.

Proof. Given $n \in \mathbb{N}$, let $\mathcal{P}_n = \{A : A \subseteq \{1, 2, \ldots, n\}\}$. For $x, y \in \mathcal{P}_n$, define $x < y$ if $x \subseteq y$. In this case the poset $(\mathcal{P}_n, \subseteq)$ consists of all subsets of $\{1, 2, \ldots, n\}$ ordered by inclusion.

We associate to each finite poset $\mathcal{P}_n$ of size n the following $2^n \times 2^n$-matrix:

$$[M_{\mathcal{P}_n}]_{ij} = \begin{cases} 1, & \text{if } i, j \text{ are comparable} \\ 0, & \text{if } i, j \text{ are incomparable}. \end{cases}$$

Under appropriate labeling of poset points $\mathcal{P}_n$, the matrix $M_{\mathcal{P}_n}$ can be viewed using the Kronecker product as follows:

$$M_{\mathcal{P}_1} = \begin{bmatrix} 1 & 0 \\ 1 & 1 \end{bmatrix}$$
$$M_{\mathcal{P}_2} = \left[\begin{array}{c|c} M_{\mathcal{P}_1} & 0 \\ \hline M_{\mathcal{P}_1} & M_{\mathcal{P}_1} \end{array}\right] = M_{\mathcal{P}_1} \otimes M_{\mathcal{P}_1}$$
$$M_{\mathcal{P}_3} = \left[\begin{array}{c|c} M_{\mathcal{P}_2} & 0 \\ \hline M_{\mathcal{P}_2} & M_{\mathcal{P}_2} \end{array}\right] = M_{\mathcal{P}_1} \otimes M_{\mathcal{P}_2}$$
$$\vdots$$
$$M_{\mathcal{P}_n} = \left[\begin{array}{c|c} M_{\mathcal{P}_{n-1}} & 0 \\ \hline M_{\mathcal{P}_{n-1}} & M_{\mathcal{P}_{n-1}} \end{array}\right] = M_{\mathcal{P}_1} \otimes M_{\mathcal{P}_{n-1}}$$

matrices $M_{\mathcal{P}_n}$ can be seen as pavements, cells with 1's are colored black and those with 0's are colored white. Figure 2 shows examples of these types of matrices.

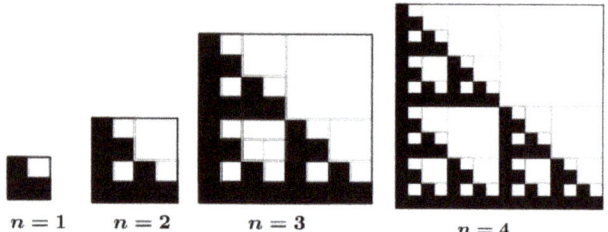

Figure 2. Matrices $M_{\mathcal{P}_n}$ for $n = 1, 2, 3$ and 4.

$M_{\mathcal{P}_n}$ is the matrix associated with the polygon $U_n \in \Gamma_1^n$, thus $\text{occ}(0, U_n)$ can be computed in the following fashion:

$$\begin{aligned} \text{occ}(0, U_1) &= 1 \\ \text{occ}(0, U_n) &= 3(\text{occ}(0, U_{n-1})) + 2^{2n-2} \end{aligned} \quad (17)$$

Therefore $\text{occ}(0, U_n) = \sum_{k=1}^n 3^{n-k} 2^{2(k-1)}$ and $\text{occ}(1, U_n) = 3^n$ thus the result holds. □

Now we are interested in computing the trace norm of the {0,1}-Brauer configuration Γ^n. For this, we recall the following theorem about the singular values of the Kronecker product:

Theorem 5 ([36], Theorem 4.2.15). *Let $A \in M_{m,n}$ and $B \in M_{p,q}$ have singular value decompositions $A = V_1 \Sigma_1 W_1^*$ and $B = V_2 \Sigma_2 W_2^*$ and let $\text{rank} A = r_1$ and $\text{rank} B = r_2$. Then $A \otimes B = (V_1 \otimes V_2)(\Sigma_1 \otimes \Sigma_2)(W_1 \otimes W_2)^*$. The nonzero singular values of $A \otimes B$ are the $r_1 r_2$ positive numbers $\{\sigma_i(A)\sigma_j(B) : 1 \leqslant i \leqslant r_1, 1 \leqslant j \leqslant r_2\}$ (including multiplicities).*

The following Lemma 1 is helpful to prove Theorem 6.

Lemma 1. *Let $A \in M_n(\mathbb{C})$ be a given matrix. If $B = \begin{bmatrix} A & 0 \\ A & A \end{bmatrix} \in M_{2n}(\mathbb{C})$ then the singular values of B are $\phi \sigma_i(A)$ and $\phi^{-1} \sigma_i(A)$ for $i = 1, \ldots, n$, where $\phi = \dfrac{1 + \sqrt{5}}{2}$ is the golden ratio.*

Proof. Note that $B = \begin{bmatrix} A & 0 \\ A & A \end{bmatrix} = \begin{bmatrix} 1 & 0 \\ 1 & 1 \end{bmatrix} \otimes A$. The singular values for $\begin{bmatrix} 1 & 0 \\ 1 & 1 \end{bmatrix}$ are ϕ and ϕ^{-1}, then by Theorem 5 the result holds. □

Theorem 6. *For each $n \geqslant 1$, if $M_R(\Gamma^n) = M_{\mathcal{P}_n}$ is the matrix associated with the polygon $U_n \in \Gamma_1^n$ then*

$$\|M_{\mathcal{P}_n}\|_* = 5^{n/2} \quad (18)$$

Proof. By induction on n. For $n = 1$, $\|M_{\mathcal{P}_1}\|_* = \phi + \phi^{-1} = \sqrt{5}$. Let us suppose that $\|M_{\mathcal{P}_n}\|_* = (2\phi - 1)^n = 5^{n/2}$ and let us see that the result is fulfilled for $n+1$, i.e.,

$$\|M_{\mathcal{P}_n}\|_* = (2\phi - 1)^{n+1} = 5^{\frac{n+1}{2}}$$

Since $M_{\mathcal{P}_{n+1}} = M_{\mathcal{P}_1} \otimes M_{\mathcal{P}_n}$, then for the Lemma 1 the singular values of $M_{\mathcal{P}_{n+1}}$ are

$$\phi \sigma_i(M_{\mathcal{P}_n}) \text{ and } \phi^{-1} \sigma_i(M_{\mathcal{P}_n})$$

for $i = 1, \ldots, 2^n$. Thus,

$$\|M_{\mathcal{P}_{n+1}}\|_* = \sum_{i=1}^{2^{n+1}} \sigma_i(M_{\mathcal{P}_{n+1}})$$
$$= \sum_{i=1}^{2^n} \phi\sigma_i(M_{\mathcal{P}_n}) + \sum_{i=1}^{2^n} \phi^{-1}\sigma_i(M_{\mathcal{P}_n})$$
$$= \phi\|M_{\mathcal{P}_n}\|_* + \phi^{-1}\|M_{\mathcal{P}_n}\|_*$$
$$= \|M_{\mathcal{P}_n}\|_*\left(\phi + \phi^{-1}\right) = \|M_{\mathcal{P}_n}\|_*(2\phi - 1)$$
$$= (2\phi - 1)^{n+1} = 5^{\frac{n+1}{2}}$$

□

Corollary 5.
$$\sum_{n=2}^{\infty} \frac{1}{\|M_{\mathcal{P}_n}\|_*} = \frac{1}{2(3 - \phi)}$$

Proof. By Theorem 6, we have:
$$\sum_{n=2}^{\infty} \frac{1}{\|M_{\mathcal{P}_n}\|_*} = \sum_{n=2}^{\infty} \frac{1}{(2\phi - 1)^n}$$

which is a convergent geometric series with $r = \frac{1}{(2\phi-1)} < 1$ and $a = \frac{1}{(2\phi-1)^2}$, therefore:

$$\sum_{n=2}^{\infty} \frac{1}{\|M_{\mathcal{P}_n}\|_*} = \frac{\frac{1}{(2\phi-1)^2}}{1 - \frac{1}{2\phi-1}} = \frac{1}{2(3-\phi)}$$

□

3. For $n \geq 1$ fixed, let $\Phi^n = \{\Phi_0^n, \Phi_1^n, \mu, \mathcal{O}\}$ be a {0,1}-Brauer configuration such that:

$$\Phi_0^n = \{0, 1\}.$$
$$\Phi_1^n = \{U_1, U_2, \ldots, U_n\}, \quad \text{for } 1 \leq i \leq n, \quad |U_i| = (i+5)^2. \tag{19}$$
$$\mu(0) = \mu(1) = 1.$$

For $i \geq 1$, the word w_i associated with the polygon U_i has the form $w_i = w_{i,1}w_{i,2}\ldots w_{i,\delta_i}$, $w_{i,j} \in \{0,1\}$, $occ(0, U_i) = (i+5)(i+3)$, $occ(1, U_i) = 2(i+5)$.
The orientation $\mathcal{O}$ is defined in such a way that for successor sequences associated with vertices 0 and 1, it holds that $U_i < U_{i+1}$.

Polygons U_i can be seen as $(i+5) \times (i+5)$-matrices over $\mathbb{Z}_2$. Each row R_j is defined by coefficients of a polynomial $P_j^i(t)$ with the form $P_j^i(t) = u_{j,1}^i + u_{j,2}^i t + \cdots + u_{j,i+4}^i t^{i+4}$, $u_{j,k}^i \in \{0,1\}$.

$$U_1 = \begin{bmatrix} 0 & 1 & 1 & 0 & 0 & 0 \\ 1 & 0 & 0 & 1 & 0 & 0 \\ 1 & 0 & 0 & 1 & 0 & 0 \\ 0 & 1 & 1 & 0 & 1 & 0 \\ 0 & 0 & 0 & 1 & 0 & 1 \\ 0 & 0 & 0 & 0 & 1 & 0 \end{bmatrix}$$

$$u_{j,k}^i = u_{j,k}^{i-1}, \quad 1 \leq j,k \leq i+4,$$
$$u_{j,i+5}^i = 0, \quad 1 \leq j \leq i+3,$$
$$u_{i+4,i+5}^i = 1,$$
$$u_{i+5,i+4}^i = 1,$$
$$u_{i+5,i+5}^i = 0.$$
(20)

Theorem 7. *For $n \geq 1$, if $\mathfrak{F}^n = \mathbb{F}Q_{\Phi^n}^n / I_{\Phi^n}^n$ is the Brauer configuration algebra induced by the $\{0,1\}$-Brauer configuration Φ^n, $\alpha_n = 2(t_{n+5} - 6)$, and $\beta_n = \varepsilon_{n+5} - \varepsilon_5$, with $\varepsilon_i = \frac{i(i+1)(2i+6)}{6}$ for $i \geq 1$ then the following statements hold:*

1. $\dim_{\mathbb{F}} \mathfrak{F}^n = 2n + 2t_{\alpha_n - 1} + 2t_{\beta_n - 1}$,
2. $\dim_{\mathbb{F}} Z(\mathfrak{F}^n) = 1 + n + \varepsilon_{n+4} - 2n$,
3. $\lim_{n \to \infty} \rho(M_R(\Phi^n)) = \sqrt{2 + 2\sqrt{2}}$.

Proof. The Formulas (1) and (2). for the dimension of the algebra $\mathfrak{F}^n$ and its center $Z(\mathfrak{F}^n)$ are consequences of the definition of a Brauer configuration Φ^n and Corollary 1.

Let us prove identity 3. Firstly, we note that the characteristic polynomials $P_n(\lambda)$ associated with matrices U_n can be obtained recursively. They obey the following general rules according to the size of the corresponding matrices.

$$P_3(\lambda) = \lambda^3 - 2\lambda,$$
$$P_4(\lambda) = \lambda^4 - 4\lambda^2,$$
$$P_n(\lambda) = \sum_{j=1}^{n} a_j^n \lambda^j, \quad \text{if } n \geq 5,$$
$$a_n^n = 1, \quad a_{n-1}^n = 0, \quad a_1^n = (-1)^{n+1} 2,$$
$$a_s^n = a_{s-1}^{n-1} - a_s^{n-2}, \quad \text{for the remaining vertices.}$$

$P_3(\lambda)$, $P_4(\lambda)$ and $P_5(\lambda)$ are characteristic polynomials of the following matrices T_3, T_4, and T_5, respectively:

$$T_3 = \begin{bmatrix} 0 & 1 & 1 \\ 1 & 0 & 0 \\ 1 & 0 & 0 \end{bmatrix}, \quad T_4 = \begin{bmatrix} 0 & 1 & 1 & 0 \\ 1 & 0 & 0 & 1 \\ 1 & 0 & 0 & 1 \\ 0 & 0 & 1 & 0 \end{bmatrix}, \quad T_5 = \begin{bmatrix} 0 & 1 & 1 & 0 & 0 \\ 1 & 0 & 0 & 1 & 0 \\ 1 & 0 & 0 & 1 & 0 \\ 0 & 1 & 1 & 0 & 1 \\ 0 & 0 & 0 & 1 & 0 \end{bmatrix}.$$

For any $k \geq 6$, $P_k(\lambda)$ is the characteristic polynomial of $U_{k-5} \in \Phi^n$.

We note that for $k \geq 5$, $|\sqrt{2 + 2\sqrt{2}} - \rho(M_R(\Phi^{(2^k - 1)}))| \leq \frac{1}{10^{\delta_k}}$, where

$$\delta_k = \begin{cases} \lceil s_k \sqrt{2} Ln(2^k - 1) \rceil, & \text{if } k \text{ is odd,} \\ \lfloor s_k \sqrt{2} Ln(2^k - 1) \rfloor, & \text{if } k \text{ is even.} \end{cases}$$

$$s_k = \begin{cases} k-4, & \text{if } 5 \leq k \leq 7, \\ 62^{k-8}, & \text{if } k \geq 8. \end{cases}$$

Then $\lim_{k \to \infty} |\sqrt{2 + 2\sqrt{2}} - \rho(M_R(\Phi^{(2^k-1)}))| = 0$. Thus, $\rho(M_R(\Phi^{(2^k-1)}))$ is a Cauchy subsequence of the sequence $\rho(M_R(\Phi^n))$, $n \geq 5$ converging to $\sqrt{2 + 2\sqrt{2}}$. □

Corollary 6. *For any $n \geq 5$, an n-vertex quiver Q_n with underlying graph $\overline{Q_n}$ of the form:*

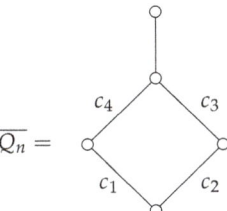

is of wild type.

Proof. Since $\rho(\overline{Q_5}) = \frac{\sqrt{\sqrt{17}+5}}{2}$, then the result holds as a consequence of Theorem 2, Remark 1, and Theorem 7. □

The following results [37] regarding some relationship between graph operations and energy graph allow finding upper and lower bounds for $\|M_R(\Phi^n)\|_*$.

Theorem 8 (Theorema 4.18 [37]). *Let G, H, and $G \circ H$ be graphs as specified above. Then*

$$\|G \circ H\|_* \leq \|G\|_* + \|H\|_*$$

Equality is attained if and only if either u is an isolated vertex of G or v is an isolated vertex of H or both.

Corollary 7 (Corollary 4.6 [37]). *If $\{e\}$ is a cut edge of a simple graph G, then $\|G - \{e\}\|_* < \|G\|_*$.*

As a consequence of these results, we obtain the following Corollary 8.

Corollary 8. *For $n \geq 6$.*

$$2\sqrt{n-1} < \|M_R(\Phi^{n-5})\|_* < 2 + \begin{cases} 2\csc(\frac{\pi}{2(n-2)}), & \text{if } n - 3 \equiv 0 \pmod{2}, \\ 2\cot(\frac{\pi}{2(n-2)}), & \text{if } n - 3 \equiv 1 \pmod{2}. \end{cases} \quad (21)$$

Proof. The inequality at right hand holds as a consequence of Theorem 8 taking into account that $\overline{Q_n}$ is the coalescence [37] between the cycle $\mathbb{C}_4$ and $\mathbb{A}_{n-3}$, and that:

$$\|\mathbb{C}_4\|_* = 4 \text{ and } \|\mathbb{A}_{n-3}\|_* = \begin{cases} 2\csc(\frac{\pi}{2(n-2)}) - 2, & \text{if } n - 3 \equiv 0 \pmod{2}, \\ 2\cot(\frac{\pi}{2(n-2)}) - 2, & \text{if } n - 3 \equiv 1 \pmod{2}. \end{cases}$$

To prove the left hand inequality, we remove edges c_1 and c_2 in $\overline{Q_n}$, obtaining in this fashion a connected tree. Since among all trees of order n, S_n attains the minimal energy. The result holds as a consequence of Corollary 7. □

4. Concluding Remarks

$\{0,1\}$-Brauer configuration algebras give rise to the so-called trace norm of a Brauer configuration. Such Brauer configurations are a source of a great variety of graphs and posets via its reduced message. The structure of the adjacency matrices associated with these graphs allows estimating the corresponding trace norm or graph energy values. In line with the main problem in the graph energy theory, we give explicit formulas for the trace norm of some $(0,1)$-matrices associated with these families of graphs and posets. On the other hand, bounds for the energy of some families of graphs can be obtained via graph coalescence. It is worth pointing out that some of these graphs underlying quivers of wild type.

An interesting task for the future will be to find the trace norms of a wide variety of Brauer configuration algebras.

Author Contributions: Investigation, N.A.M., A.M.C., P.F.F.E. and I.D.M.G.; writing—review and editing, N.A.M., A.M.C., P.F.F.E. and I.D.M.G. All authors have read and agreed to the published version of the manuscript.

Funding: MinCiencias-Colombia and Seminar Alexander Zavadskij on Representation of Algebras and their Applications, Universidad Nacional de Colombia.

Institutional Review Board Statement: Not applicable.

Informed Consent Statement: Not applicable.

Data Availability Statement: Not applicable.

Acknowledgments: N. Agudelo and A.M. Cañadas thanks to MinCiencias and Universidad Nacional de Colombia, sede Bogotá (Convocatoria 848- Programa de estancias Postdoctorales 2019) for their support.

Conflicts of Interest: The authors declare no conflict of interest.

Abbreviations

The following abbreviations are used in this manuscript:

$\dim_{\mathbb{F}} \Lambda_\Gamma$	(Dimension of a Brauer configuration algebra)
$\dim_{\mathbb{F}} Z(\Lambda_\Gamma)$	(Dimension of the center of a Brauer configuration algebra)
Γ_0	(Vertices in a Brauer configuration Γ)
$M(\Gamma)$	(Message of a Brauer configuration Γ)
$M_R(\Gamma)$	(Reduced message of a Brauer configuration Γ)
$occ(\alpha, V)$	(Number of occurrences of a vertex α in a polygon V)
$V_i^{(\alpha)}$	(Ordered sequence of polygons)
$val(\alpha)$	(Valency of a vertex α)
$w(U)$	(Word associated with a polygon of a Brauer configuration)
$\|M\|_F$	(Frobenius norm of matrix M)
$\|M\|_*$	(Trace norm of matrix M)
$\otimes$	(Kronecker product)
ϕ	(Golden ratio)
$\mu_i(M)$	(Eigenvalues of matrix M)
$\rho(G)$	(Spectral radius of a graph G)
$\sigma_i(M)$	(Singular values of matrix M)
t_j	(The jth triangular number)
$M_{\mathcal{P}_n}$	(Matrix associated with the polygon U_n)

Appendix A

Table A1. This table shows the graphical representation of reduced messages of the Brauer configurations Γ^3 (14), Δ^2 (7) and Φ^1 (19). The dimension of the corresponding Brauer configuration algebras and their centers together with trace norm values.

$M_R(\Gamma)$	n	$\dim_\mathbb{F} \Lambda$	$\dim_\mathbb{F} Z(\Lambda)$	$\|M_R(\Lambda)\|_*$
$M_R(\Gamma^3)$	3	96,630	230	$\sqrt{5^3} \approx 11.1803$
$M_R(\Delta^2)$	2	7358	105	$\sum_{i=1}^{2} \mu_i(M_3) \geq 5$ $\sum_{i=3}^{9} \mu_i(M_3) \leq \frac{4}{5}$
$M_R(\Phi^1)$	1	2942	80	$4.4721 \leq \|M_R(\Phi^1)\|_* \leq 6.8284$

References

1. Green, E.L.; Schroll, S. Brauer configuration algebras: A generalization of Brauer graph algebras. *Bull. Sci. Math.* **2017**, *121*, 539–572. [CrossRef]
2. Malić, G.; Schroll, S. Dessins d'enfants and Brauer configuration algebras. In *Galois Covers, Grothendieck-Teichmüller Theory and Dessins d'Enfants, Proceedings of the LMS Midlands Regional Meeting & International Workshop, Leicester, UK, 4–7 June 2018*; Springer: Cham, Switzerland, 2020; pp. 205–225.
3. Cañadas, A.M.; Gaviria, I.D.M.; Vega, J.D.C. Relationships between the Chicken McNugget Problem, Mutations of Brauer Configuration Algebras and the Advanced Encryption Standard. *Mathematics* **2021**, *9*, 1937. [CrossRef]
4. Espinosa, P.F.F. Categorification of Some Integer Sequences and Its Applications. Ph.D. Thesis, Universidad Nacional de Colombia, Bogotá, Colombia, 2021.
5. Coulson, C.A.; O'Leary, B.; Mallion, R.B. *Hückel Theory for Organic Chemists*; Academic Press: London, UK, 1978.
6. Gutman, I. The energy of a graph. *Ber. Math. Statist. Sekt. Forschungszentrum Graz.* **1978**, *103*, 1–22.
7. Nikiforov, V. The energy of graphs and matrices. *J. Math. Anal. Appl.* **2007**, *326*, 1472–1475. [CrossRef]
8. Kharaghany, H.; Tayfeh-Rezaie, B. On the energy of (0, 1)-matrices. *Linear Algebra Appl.* **2008**, *429*, 2046–2051. [CrossRef]
9. Nikiforov, V.; Agudelo, N. On the minimum trace norm/energy of (0, 1)-matrices. *Linear Algebra Appl.* **2017**, *526*, 42–59. [CrossRef]
10. Andrews, G.E. Unsolved problems; further problems on partitions. *Am. Math. Mon.* **1987**, *94*, 437–439. [CrossRef]
11. Cañadas. A.M.; Gaviria, I.D.M.; Giraldo, H. Representation of equipped posets to generate Delannoy numbers. *Far East J. Math. Sci.* **2017**, *8*, 1677–1695.
12. Gaviria, I.D.M. The Auslander-Reiten Quiver of Equipped Posets of Finite Growth Representation Type, Some Functorial Descriptions and Its Applications. Ph.D. Thesis, Universidad Nacional de Colombia, Bogotá, Colombia, 2020.
13. Sloane, N.J.A. OEIS. Available online: https://oeis.org/search?q=A344791 (accessed on 30 June 2021).
14. Gutman, I.; Furtula, B. Graph energies and their applications. *Bulletin* **2019**, *44*, 29–45.
15. Dhanalakshmi, A.; Rao, K.S.; Sivakumar, K. Characterization of α-cyclodextrin using adjacency and distance matrix. *Indian J. Sci.* **2015**, *12*, 78–83.

16. Giuliani, A.; Filippi, S.; Bertolaso, M. Why network approach can promote a new way of thinking in biology. *Front. Genet.* **2014**, *5*, 83. [CrossRef]
17. Yuge, K. Graph representation for configuration properties of crystalline solids. *J. Phys. Soc. Jpn.* **2017**, *86*, 024802. [CrossRef]
18. Van Mieghem P.; Van de Bovenkamp R. Accuracy criterion for the mean-field approximation in susceptible-infected-susceptible epidemics on networks, *Phys. Rev. E* **2015**, *91*, 032812.
19. Angadi, S.A.; Hatture, S.M. Face recognition through symbolic modelling of face graphs and texture. *Int. J. Pattern Rec. Artif. Intell.* **2019**, *33*, 1956008. [CrossRef]
20. Bai, Y.; Dong, L.; Hunag, X.; Yang, W.; Liao, M. Hierarchical segmentation of polarimetric SAR image via non-parametric graph entropy. In Proceedings of the 2014 IEEE Geoscience and Remote Sensing Symposium, Quebec City, QC, Canada, 13–18 July 2014.
21. Akram, M.; Naz, S. Energy of pythagorean fuzzy graphs with applications. *Mathematics* **2018**, *6*, 136. [CrossRef]
22. Pugliese, A.; Nilchiani, R. Complexity analysis of fractionated spacecraft architectures. *Am. Inst. Aeronaut. Astronaut. Space Forum* **2017**, *33*, 2721275.
23. Bolaños, M.E.; Aviyente, S. Quantifying the functional importance of neural assemblies in the brain using Laplacian Hückel graph energy. In Proceedings of the 2011 IEEE International Conference on Acoustics, Speech and Signal Processing, Prague, Czech Republic, 22–27 May 2011; pp. 753–756.
24. Belov-Kanel, A.; Rowen, L.H.; Vishne, U. Full quivers of representations of algebras. *Trans. Am. Math. Soc.* **2021** *364*, 5525–5569. [CrossRef]
25. Assem, I.; Skowronski, A.; Simson, D. *Elements of the Representation Theory of Associative Algebras*; Cambridge University Press: Cambridge, UK, 2006.
26. Auslander, M.; Reiten, I.; Smalo, S.O. *Representation Theory of Artin Algebras*; Cambridge University Press: Cambridge, UK, 1995.
27. Simson, D. *Linear Representations of Partially Ordered Sets and Vector Space Categories*; Gordon and Breach: London, UK, 1992.
28. Gabriel, P. Unzerlegbare Darstellungen I. *Manuscripta Math.* **1972**, *6*, 71–103. [CrossRef]
29. Crabbe, A.; Leuschke, G. Wild hypersurfaces. *J. Pure Appl. Algebra* **2011**, *215*, 2884–2891. [CrossRef]
30. Drozd, J. On tame and wild matrix problems. In *Matrix Problems, Kiev.*; Istitute of Mathematics of SA of Ukr. SSR: Kyiv, Ukraine, 1977; pp. 104–114. (In Russian)
31. Crawley-Boevey, W. On tame algebras and bocses. *Proc. Lond. Math. Soc.* **1988**, *56*, 451–483. [CrossRef]
32. Smith, J.H. Some properties of the spectrum of a graph. In *Combinatorial Structures and Their Applications*; Guy, R., Hanani, H., Sauer, N., Schönheim, J., Eds.; Gordon and Breach: New York, NY, USA, 1970; pp. 403–406.
33. Sierra, A. The dimension of the center of a Brauer configuration algebra. *J. Algebra* **2018**, *510*, 289–318. [CrossRef]
34. Davey, B.A.; Priestley, H.A. *Introduction to Lattices and Order*, 2nd ed.; Cambridge University Press: Cambridge, UK, 2002.
35. Bollobás, B.; Nikiforov, V. Graphs and Hermitian matrices: eigenvalue interlacing. *Discret. Math.* **2004**, *289*, 119–127. [CrossRef]
36. Horn, R.; Johnson, C. *Topics in Matrix Analysis*; Cambridge University Press: Cambridge, UK, 1991.
37. Li, X.; Shi, Y.; Gutman, I. *Graph Energy*; Springer: New York, NY, USA, 2012.

Article

An Ideal-Based Dot Total Graph of a Commutative Ring

Mohammad Ashraf [1], Jaber H. Asalool [1], Abdulaziz M. Alanazi [2,*] and Ahmed Alamer [2]

[1] Department of Mathematics, Aligarh Muslim University, Aligarh 202002, India; mashraf80@hotmail.com (M.A.); asalooljaber@gmail.com (J.H.A.)
[2] Department of Mathematics, University of Tabuk, Tabuk 71491, Saudi Arabia; aalamer@ut.edu.sa
* Correspondence: am.alenezi@ut.edu.sa

Abstract: In this paper, we introduce and investigate an *ideal-based dot total graph* of commutative ring $\mathcal{R}$ with nonzero unity. We show that this graph is connected and has a small diameter of at most two. Furthermore, its vertex set is divided into three disjoint subsets of $\mathcal{R}$. After that, connectivity, clique number, and girth have also been studied. Finally, we determine the cases when it is Eulerian, Hamiltonian, and contains a Eulerian trail.

Keywords: commutative rings; zero-divisor graph; dot total graph; ideal-based; zero-divisors

MSC: 97K30; 68R10; 94C15; 13A15

Citation: Ashraf, M.; Asalool, J.H.; Alanazi, A.M.; Alamer, A. An Ideal-Based Dot Total Graph of a Commutative Ring. *Mathematics* **2021**, *9*, 3072. https://doi.org/10.3390/math9233072

Academic Editors: Irina Cristea and Hashem Bordbar

Received: 6 November 2021
Accepted: 25 November 2021
Published: 29 November 2021

Publisher's Note: MDPI stays neutral with regard to jurisdictional claims in published maps and institutional affiliations.

Copyright: © 2021 by the authors. Licensee MDPI, Basel, Switzerland. This article is an open access article distributed under the terms and conditions of the Creative Commons Attribution (CC BY) license (https://creativecommons.org/licenses/by/4.0/).

1. Introduction

Let $Z(\mathcal{R})$ and $Reg(\mathcal{R})$ be a set of zero-divisors and a set of regular elements of commutative ring $\mathcal{R}$ with $1 \neq 0$, respectively. In [1], Mohammad Ashraf et al. defined the dot total graph of $\mathcal{R}$, denoted by $T_{Z(\mathcal{R})}(\Gamma(\mathcal{R}))$, as an (undirected) graph, which consists of all elements of $\mathcal{R}$ as vertex set $V(T_{Z(\mathcal{R})}(\Gamma(\mathcal{R})))$ and includes all edges such that for distinct $x, y \in \mathcal{R}$, $e = xy \in E(T_{Z(\mathcal{R})}(\Gamma(\mathcal{R})))$ if and only if $xy \in Z(\mathcal{R})$. In this paper, we replace $Z(\mathcal{R})$ by an ideal I, and we introduce and investigate an *ideal-based dot total graph of* $\mathcal{R}$ denoted $T_I(\Gamma(\mathcal{R}))$. In addition, Redmond [2] defined $\Gamma_I(\mathcal{R})$ as an undirected graph. It has vertices $\{x \in \mathcal{R} \setminus I \mid xy \in I \text{ for some } y \in \mathcal{R} \setminus I\}$. In this case, x and y are vertices that are both distinct and adjacent if and only if $xy \in I$, i.e., $\Gamma_I(\mathcal{R})$ is subgraph of $T_I(\Gamma(\mathcal{R}))$. It will also appear in this paper. Further, if $I = (0)$ in $\Gamma_I(\mathcal{R})$, then $\Gamma_I(\mathcal{R}) = \Gamma(\mathcal{R})$; this graph is studied by Anderson et al. [3], and they were interested in studying the interplay of ring-theoretic properties of $\mathcal{R}$ with graph-theoretic properties of $\Gamma(\mathcal{R})$. Moreover, they associated a (simple) graph $\Gamma(\mathcal{R})$ to $\mathcal{R}$, which consists of a vertex set $V(\Gamma(\mathcal{R})) = Z(\mathcal{R})^* = Z(\mathcal{R}) \setminus \{0\}$ and edge set $E(\Gamma(\mathcal{R}))$ such that for all distinct $x, y \in Z(\mathcal{R})^*$, $e = xy \in E(\Gamma(\mathcal{R}))$ if and only if $xy = 0$. Furthermore, if $I = (0)$ in $T_I(\Gamma(\mathcal{R}))$, then $T_I(\Gamma(\mathcal{R})) = \Gamma_0(\mathcal{R})$; this graph is studied by Beck [4], in which he considered $\mathcal{R}$ as a simple graph for which its vertex set is the set of all elements of $\mathcal{R}$ and edge set such that for all distinct $x, y \in \mathcal{R}$, $e = xy \in E(\Gamma_0(\mathcal{R}))$ if and only if $xy = 0$. In addition, some fundamentals of Laplacian eigenvalues and energy of graphs can be identified in [5–7].

Assuming G to be a graph, G can be said to be connected when a path connects every pair of its distinctive vertices. Denoting distinct vertices of graph G to be x and y, $d(x, y)$ will indicate the shortest distance between the two vertices. However, where no such path exists, it will be represented by $d(x, y) = \infty$. Similarly, the diameter of G is $diam(G) = sup\{d(x, y) \mid x \text{ and } y \text{ are distinct vertices of } G\}$. The girth of G, denoted by $gr(G)$, is defined as the length of shortest cycle in G ($gr(G) = \infty$ if G contains no cycle). Note that if G contains a cycle, then $gr(G) \leq 2 \, diam(G) + 1$. The degree of vertex v, written $deg_G(v)$ or $deg(v)$, is the number of edges incident to v (or the degree of the vertex v is the number of vertices adjacent to v). In a connected graph G, a vertex v is said to be a cut-vertex of G if and only if $G \setminus \{v\}$ is disconnected. Let $V(G)$ be a vertex set of G. Then, the subset $U \subseteq V(G)$ is called a vertex-cut if $G \setminus U$ is disconnected. The connectivity of a graph G denoted by $k(G)$ and is defined as the cardinality of a minimum vertex-cut of G, which is also the same concepts we have in the edges. In a connected graph G, an edge e is said to be a bridge of G if and only if $G \setminus \{e\}$ is disconnected. Let $E(G)$ be an edge set of G. If $G \setminus X$ is disconnected, it will have a subset $X \subseteq E(G)$ as its edge-cut. Let $\lambda(G)$ denote the edge-connectivity of a connected graph G which is the size of the smallest set of edges for which removal disconnects G. Moreover, a clique is a complete subgraph of a graph G. The clique number denoted by $\omega(G)$ is the greatest integer $n \geqslant 1$ such that $K_n \subseteq G$, and $\omega(G) = \infty$ if $K_n \subseteq G$ for all $n \geqslant 1$. A nontrivial connected graph G is Eulerian if every vertex of G has an even degree. Moreover, G contains a Eulerian trail if exactly two vertices of G have an odd degree. In addition, let G be a graph of order $n \geq 3$. If $deg(u) + deg(v) \geq n$ for each pair u and v of vertices of G that are not adjacent, then G is Hamiltonian.

The present paper is organized as follows:

In Section 2, we define an ideal-based dot total graph of $\mathcal{R}$ and study the most basic results of $T_I(\Gamma(\mathcal{R}))$. We provide many examples and show that $T_I(\Gamma(\mathcal{R}))$ is always connected with $diam(T_I(\Gamma(\mathcal{R}))) \leqslant 2$ and $gr(T_I(\Gamma(\mathcal{R}))) \leqslant 5$, and we determine when $T_I(\Gamma(\mathcal{R}))$ is a complete graph and a regular graph. Moreover, we find the degree of each vertex of $T_I(\Gamma(\mathcal{R}))$. Furthermore, in Section 3, we study the connectivity of $T_I(\Gamma(\mathcal{R}))$ when $T_I(\Gamma(\mathcal{R}))$ has a no cut-vertex, and $T_I(\Gamma(\mathcal{R}))$ has a bridge. We shall also find the $k(T_I(\Gamma(\mathcal{R})))$. On the other hand, in Section 4, we study the clique number and girth of $T_I(\Gamma(\mathcal{R}))$, and we determine the clique number when $T_I(\Gamma(\mathcal{R}))$ has a cycle. Furthermore, we find the girth of $T_I(\Gamma(\mathcal{R}))$, i.e., $gr(T_I(\Gamma(\mathcal{R})))$. Finally, in Section 5, we study the traversability of $T_I(\Gamma(\mathcal{R}))$ when $T_I(\Gamma(\mathcal{R}))$ is Eulerian or contains a Eulerian trail, and $T_I(\Gamma(\mathcal{R}))$ is Hamiltonian.

2. Definition and Basic Structure of $T_I(\Gamma(\mathcal{R}))$

In this section, we define an ideal-based dot total graph, denoted by $T_I(\Gamma(\mathcal{R}))$, and show that this graph is always connected and has a small diameter of at most two. By dividing the element of $\mathcal{R}$ into three disjoint subsets, we study the basic results on the structure of this graph and the relationship between $T_I(\Gamma(\mathcal{R}))$ and $T_{Z(\mathcal{R})}(\Gamma(\mathcal{R}))$, $T_{Z(\mathcal{R}/I)}(\Gamma(\mathcal{R}/I))$, $\Gamma_I(\mathcal{R})$, or $\Gamma(\mathcal{R}/I)$ with some examples clarification. Moreover, we find the degree of each vertex of $T_I(\Gamma(\mathcal{R}))$ that depends on the three sets I, X, and Y. Furthermore, we determine the case when $T_I(\Gamma(\mathcal{R}))$ is a complete graph or a regular graph.

Definition 1. *Let $\mathcal{R}$ be a commutative ring with $1 \neq 0$ and ideal I. Then, a simple graph that is not directed is defined as $T_I(\Gamma(\mathcal{R}))$, possessing vertices of $\mathcal{R}$. In this case, the graph has vertices x and y that are both distinct and adjacent if and only if $xy \in I$.*

Proposition 1. (a) *If $I = (0)$, then $T_I(\Gamma(\mathcal{R})) = \Gamma_0(\mathcal{R})$.*
(b) *If $Z(\mathcal{R})$ is an ideal and $I = Z(\mathcal{R})$, then $T_I(\Gamma(\mathcal{R})) = T_{Z(\mathcal{R})}(\Gamma(\mathcal{R}))$.*
(c) *If $I = \mathcal{R}$, then $T_I(\Gamma(\mathcal{R})) = K_n$, where $n = |\mathcal{R}|$.*
(d) *Let $\mathcal{R}$ have a proper nonzero ideal I. Consequently, this means that $T_{Z(\mathcal{R}/I)}(\Gamma(\mathcal{R}/I)) = K_{1,n}$. The value of n is given by $n = |\mathcal{R}/I| - 1$ if and only if the prime ideal of $\mathcal{R}$ is I.*

Proof. The proofs of (a) and (b) follow by the definition of the zero-divisor graph of $\mathcal{R}$, which appeared in Beck [4], and definition of the dot total graph of $\mathcal{R}$, which appeared recently in Ashraf et al. [1], respectively.

(c) Let x and y be distinct vertices of $\mathcal{R}$. Then, $xy \in \mathcal{R} = I$. Thus, x is adjacent to y for all $x, y \in \mathcal{R}$. Hence, $T_I(\Gamma(\mathcal{R})) = K_n$, where $n = |I| = |\mathcal{R}|$. This completes the proof.

(d) Let I be a prime ideal of $\mathcal{R}$. Then, $\mathcal{R}/I$ is an integral domain. Thus, $T_{Z(\mathcal{R}/I)}(\Gamma(\mathcal{R}/I))$ is a star graph. Hence, $T_{Z(\mathcal{R}/I)}(\Gamma(\mathcal{R}/I)) = K_{1,n}$, where $n = |\mathcal{R}/I| - 1$.

Conversely, let $T_{Z(\mathcal{R}/I)}(\Gamma(\mathcal{R}/I)) = K_{1,n}$, where $n = |\mathcal{R}/I| - 1$. Then, $T_{Z(\mathcal{R}/I)}(\Gamma(\mathcal{R}/I))$ is a star graph, and we have the following two cases:

Case(i) If $Z(\mathcal{R}/I) = 1$, then $\mathcal{R}/I$ is an integral domain.

Case(ii) If $Z(\mathcal{R}/I) > 1$, then there exists at least two vertices in $Z(\mathcal{R}/I)$. Therefore, $T_{Z(\mathcal{R}/I)}(\Gamma(\mathcal{R}/I))$ is not a star graph, which is a contradiction.

Thus, we obtain $\mathcal{R}/I$ as an integral domain. Hence, I is the prime ideal of $\mathcal{R}$. This completes the proof. □

In view of the following examples, we shall find $T_I(\Gamma(\mathcal{R}))$ and $T_{Z(\mathcal{R}/I)}(\Gamma(\mathcal{R}/I))$ with several ideals I of the same ring $\mathcal{R}$.

Example 1. Let $\mathcal{R} = \mathbb{Z}_8$. Then, $(0), (2), (4)$, and $\mathbb{Z}_8$ are ideals of $\mathcal{R}$:

(i)   Let $I = (0)$, then $\mathcal{R}/I = \mathbb{Z}_8$ (see Figure 1).

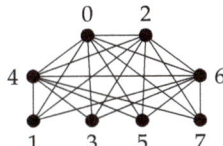
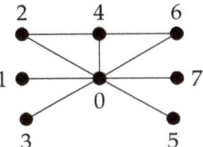

Figure 1. (**left**) $T_{Z(\mathcal{R}/I)}(\Gamma(\mathcal{R}/I))$ and (**right**) $T_I(\Gamma(\mathcal{R}))$, when $\mathcal{R} = \mathbb{Z}_8$ and $I = (0)$.

(ii)   Let $I = (2)$, then $\mathcal{R}/I = \mathbb{Z}_2$ (see Figure 2).

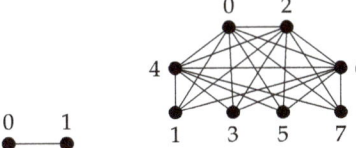

Figure 2. (**left**) $T_{Z(\mathcal{R}/I)}(\Gamma(\mathcal{R}/I))$ and (**right**) $T_I(\Gamma(\mathcal{R}))$, when $\mathcal{R} = \mathbb{Z}_8$ and $I = (2)$.

(iii)   Let $I = (4)$, then $\mathcal{R}/I = \mathbb{Z}_4$ (see Figure 3).

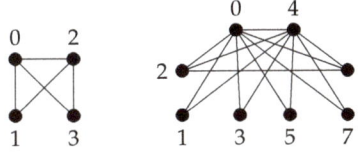

Figure 3. (**left**) $T_{Z(\mathcal{R}/I)}(\Gamma(\mathcal{R}/I))$ and (**right**) $T_I(\Gamma(\mathcal{R}))$, when $\mathcal{R} = \mathbb{Z}_8$ and $I = (4)$.

(iv)   Let $I = \mathbb{Z}_8$, then $\mathcal{R}/I = (0)$ (see Figure 4).

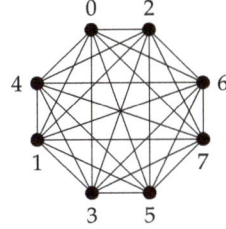

Figure 4. (**left**) $T_{Z(\mathcal{R}/I)}(\Gamma(\mathcal{R}/I))$ and (**right**) $T_I(\Gamma(\mathcal{R}))$, when $\mathcal{R} = \mathbb{Z}_8$ and $I = \mathbb{Z}_8$.

Example 2. Let $\mathcal{R} = \mathbb{Z}_6$. Then, $(0), (2), (3)$, and $\mathbb{Z}_6$ are ideals of $\mathcal{R}$:

(i)   Let $I = (0)$, then $\mathcal{R}/I = \mathbb{Z}_6$ (see Figure 5).

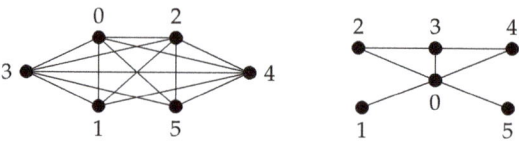

Figure 5. (**left**) $T_{Z(\mathcal{R}/I)}(\Gamma(\mathcal{R}/I))$ and (**right**) $T_I(\Gamma(\mathcal{R}))$, when $\mathcal{R} = \mathbb{Z}_6$ and $I = (0)$.

(ii)   Let $I = (2)$, then $\mathcal{R}/I = \mathbb{Z}_2$ (see Figure 6).

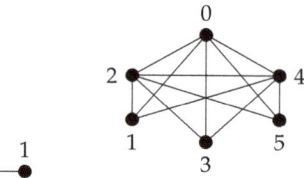

Figure 6. (**left**) $T_{Z(\mathcal{R}/I)}(\Gamma(\mathcal{R}/I))$ and (**right**) $T_I(\Gamma(\mathcal{R}))$, when $\mathcal{R} = \mathbb{Z}_6$ and $I = (2)$.

(iii)   Let $I = (3)$, then $\mathcal{R}/I = \mathbb{Z}_3$ (see Figure 7).

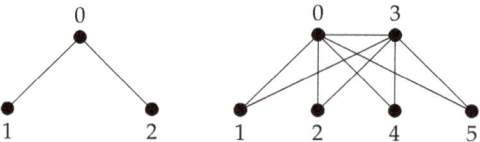

Figure 7. (**left**) $T_{Z(\mathcal{R}/I)}(\Gamma(\mathcal{R}/I))$ and (**right**) $T_I(\Gamma(\mathcal{R}))$, when $\mathcal{R} = \mathbb{Z}_6$ and $I = (3)$.

(iv)   Let $I = \mathbb{Z}_6$, then $\mathcal{R}/I = (0)$ (see Figure 8).

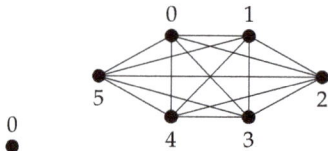

Figure 8. (**left**) $T_{Z(\mathcal{R}/I)}(\Gamma(\mathcal{R}/I))$ and (**right**) $T_I(\Gamma(\mathcal{R}))$, when $\mathcal{R} = \mathbb{Z}_6$ and $I = \mathbb{Z}_6$.

Theorem 1. $T_I(\Gamma(\mathcal{R}))$ is connected and $diam(T_I(\Gamma(\mathcal{R}))) \leq 2$. Therefore, where $T_I(\Gamma(\mathcal{R}))$ has a cycle, it implies $gr(T_I(\Gamma(\mathcal{R}))) \leq 5$.

Proof. Assume that x and y are distinct vertices of $T_I(\Gamma(\mathcal{R}))$. In such a scenario, various cases will be true as shown:

Case(i): If $x \in I$ and $y \in I$, then $xy \in I$. Thus, $x - y$ is a path of length one in $T_I(\Gamma(\mathcal{R}))$.

Case(ii): If $x \in I$ and $y \notin I$, then $xy \in I$. Thus, $x - y$ is a path of length one in $T_I(\Gamma(\mathcal{R}))$.

Case(iii): If $x \notin I$ and $y \in I$, then this will result in a path of length one as in the previous case.

Case(iv): If $x \notin I$ and $y \notin I$, then we will consider the following subcases:

Subcase(a): If $xy \in I$, then $x - y$ is a path of length one in $T_I(\Gamma(\mathcal{R}))$.

Subcase(b): If $xy \notin I$, then there is some $z \in I$ such that $xz \in I$ and $zy \in I$. Thus, $x - z - y$ is a path of length two in $T_I(\Gamma(\mathcal{R}))$. Hence, $T_I(\Gamma(\mathcal{R}))$ is connected and $diam(T_I(\Gamma(\mathcal{R}))) \leq 2$. Since for any undirected graph, H contains a cycle, $gr(H) \leq 2\ diam(H) + 1$ (for reference see [8]). Thus, $gr(T_I(\Gamma(\mathcal{R}))) \leq 5$. □

Suppose that $\mathcal{R}$ is a commutative ring, and I is an ideal of $\mathcal{R}$. We construct a graph $T_I(\Gamma(\mathcal{R}))$ with the following method:

First, the set of vertices of $T_I(\Gamma(\mathcal{R}))$ can be classified into three disjoint subsets of $\mathcal{R}$:
(i) $I = (a)$ is the subset of $\mathcal{R}$ such that I is the ideal generated by the element a.
(ii) $X = \{x \in \mathcal{R} \setminus I : xb \notin I, \text{for all } b \in \mathcal{R} \setminus I\}$.
(iii) $Y = \{y \in \mathcal{R} \setminus I : yb \in I, \text{for some } b \in \mathcal{R} \setminus I\}$.

Second, we will connect the edges between the vertices defined in the three previous sets as follows:

We define a complete graph (K_n, where $n = |I|$) by using the first set $I = (a)$ as its vertex set. Thus, we have an edge between each vertex of I (i.e., $ab \in I$). Then, join each vertex of the second set X to all vertices of the complete graph K_n, and similarly, join each vertex of the third set Y to all vertices of the complete graph K_n. Thus, we have an edge between each vertex of the sets X and Y with all vertex of I (i.e., $xa \in I$ and $ya \in I$). Finally, in this part of the edges the relationship between $T_I(\Gamma(\mathcal{R}))$ and $\Gamma_I(\mathcal{R})$ is identical. Thus, for distinct $y_1, y_2 \in Y$, y_1 is adjacent to y_2 in $T_I(\Gamma(\mathcal{R}))$ if and only if y_1 is adjacent to y_2 in $\Gamma_I(\mathcal{R})$ (i.e., $y_1 y_2 \in I$).

Henceforth, we shall rely on the three sets I, X, and Y defined above in this paper.

Theorem 2. *Let y_1 and y_2 be any two distinct vertices of the third set Y and y_1 is adjacent to y_2 in $T_I(\Gamma(\mathcal{R}))$. If $y_1 + I \neq y_2 + I$, then $y_1 + I$ is adjacent to $y_2 + I$ in $\Gamma(\mathcal{R}/I)$ and if $y_1 + I = y_2 + I$, then $y_1^2, y_2^2 \in I$.*

Corollary 1. *Let $T_I(\Gamma(\mathcal{R}))$ have two vertices (u and v) that are both distinct and adjacent. This implies that the elements, $u + I$ and $v + I$, are adjacent in $T_I(\Gamma(\mathcal{R}))$. Assuming that $u^2 \in I$, this implies that $T_I(\Gamma(\mathcal{R}))$ has all the distinct elements of $u + I$ adjacent to it.*

Corollary 2. *Let I be an ideal of $\mathcal{R}$. Then, $\Gamma(\mathcal{R}/I)$ and $\Gamma_I(\mathcal{R})$ are subgraphs of $T_I(\Gamma(\mathcal{R}))$.*

Corollary 3. *Assume that a nonzero ideal of $\mathcal{R}$ is I. Then, if and only if the prime ideal of $\mathcal{R}$ is I, $T_{Z(\mathcal{R}/I)}(\Gamma(\mathcal{R}/I))$ is subgraph of $T_I(\Gamma(\mathcal{R}))$.*

Corollary 4. *$T_I(\Gamma(\mathcal{R}))$ contains $|I|$ disjoint subgraphs isomorphic to $\Gamma(\mathcal{R}/I)$.*

Proof. Since $\Gamma_I(\mathcal{R})$ is subgraph of $T_I(\Gamma(\mathcal{R}))$ (see Corollary 2) and $\Gamma_I(\mathcal{R})$ contains $|I|$ disjoint subgraphs isomorphic to $\Gamma(\mathcal{R}/I)$, $T_I(\Gamma(\mathcal{R}))$ contains $|I|$ disjoint subgraphs isomorphic to $\Gamma(\mathcal{R}/I)$. □

Remark 1. *Let I, X, and Y be three disjoint sets defined above, and $u, v \in \mathcal{R}$. Then, we have the following results:*
(i) *If $u + I$ is adjacent to $v + I$ in $\Gamma(\mathcal{R}/I)$, then u is adjacent to v in $T_I(\Gamma(\mathcal{R}))$.*
(ii) *If $u + I$ is adjacent to $v + I$ in $T_{Z(\mathcal{R}/I)}(\Gamma(\mathcal{R}/I))$, then u may or may not be adjacent to v in $T_I(\Gamma(\mathcal{R}))$ (see Example 1(iii); Figure 3).*
(iii) *If I is a prime ideal, then the set Y defined above will vanish.*
(iv) *If $I = \mathcal{R}$, then the sets X and Y defined above will vanish, i.e., X and Y are empty. Thus, $T_I(\Gamma(\mathcal{R})) = K_n$, where $n = |\mathcal{R}|$.*
(v) *If $u \in I$, then u is adjacent to each vertex $v \in \mathcal{R}$.*
(vi) *If $u \in X$, then u is adjacent to $v \in I$ only.*
(vii) *If $u \in Y$, then u is adjacent to $v \in I$ and some $v \in Y$.*
(viii) *Any two distinct vertices of X are not adjacent in $T_I(\Gamma(\mathcal{R}))$, i.e., if $u, v \in X$, and $u \neq v$, then $uv \notin E(T_I(\Gamma(\mathcal{R})))$.*
(ix) *If Y is a subgraph of $T_I(\Gamma(\mathcal{R}))$, then each pair of distinct vertices u and v of Y is connected by a path with a length of at most three.*
(x) *There are no adjacencies between elements of X and elements of Y.*

Theorem 3. *Let I be a prime ideal of $\mathcal{R}$. Then, $T_I(\Gamma(\mathcal{R}))$ contains a subgraph isomorphic to K_n, where $n = |I| + 1$.*

Proof. Since I is a prime ideal, Y will vanish, and there is at least one element v in X. Moreover, we have a complete subgraph K_n, where $n = |I|$ and the vertex v is adjacent to each vertex of K_n. Thus, we have a complete subgraph of order $n = |I| + 1$. □

Theorem 4. *Let I be a non-prime ideal of $\mathcal{R}$, and there exists $u \in Y$ such that $u^2 \in I$. Then, $T_I(\Gamma(\mathcal{R}))$ contains a subgraph isomorphic to K_m, where m is at least $2|I|$.*

Proof. Since there exists $u \in Y$ such that $u^2 \in I$, all the distinct elements of $u + I$ are adjacent in $T_I(\Gamma(\mathcal{R}))$ (see Corollary 1). Thus, we have a complete subgraph of order $n = |I|$. Moreover, by element of the set I, we have a complete subgraph K_n, where $n = |I|$ and all the distinct elements of $u + I$ are adjacent to each vertex of K_n. Thus, we have a complete subgraph of order $m = n + n = |I| + |I| = 2|I|$. Therefore, if Y consists of the elements $u + I$ only, then $m = 2|I|$. If there exists $v \in Y$ other than the elements $u + I$, then v is adjacent to all the elements $u + I$ or there exists an element $w \in Y$ such that v is adjacent to w, and w is adjacent to all the elements $u + I$. Hence, in both cases we have a complete subgraph of order $|I| + 1$, and all elements of this subgraph are adjacent to each vertex of K_n. Thus, we have a complete subgraph of order $m = |I| + 1 + |I| = 2|I| + 1$. □

Theorem 5. *Let I be an ideal of $\mathcal{R}$ that is not prime, and $u^2 \notin I$ for all $u \in Y$. Then, $T_I(\Gamma(\mathcal{R}))$ contains a subgraph isomorphic to K_m, where m is at least $|I| + 1$.*

Proof. Assume that $u^2 \notin I$ for all $u \in Y$. By the same arguments as used in Theorems 3 and 4, we obtain the result. □

Corollary 5. *$T_I(\Gamma(\mathcal{R}))$ is a complete graph if and only if either $I = \mathcal{R}$ or $\mathcal{R} \cong \mathbb{Z}_2$.*

Theorem 6. *Let I be a prime ideal of $\mathcal{R}$. Then, the degree of each vertex of $T_I(\Gamma(\mathcal{R}))$ is either $|\mathcal{R}| - 1$ or $|I|$.*

Proof. Since I is a prime ideal of $\mathcal{R}$, the set Y will vanish. Thus, $\mathcal{R}$ consists of two disjoint subsets I and X. Then, we have the following two cases:
    Case (i): If $u \in I$, then u is adjacent to each vertex in $T_I(\Gamma(\mathcal{R}))$ except u; that is, u is adjacent to $(|\mathcal{R}| - 1)$ vertices, and hence the degree of u is $|\mathcal{R}| - 1$.
    Case (ii): If $u \in X$, then u is adjacent to the vertices, which belongs to I; that is, u is adjacent to $|I|$ vertices and, hence, $deg(u) = |I|$. □

Corollary 6. *Let I be an ideal of $\mathcal{R}$ and $u \in Y$. Then, the number of elements of $T_I(\Gamma(\mathcal{R}))$ adjacent to u is either $|I|$ or at least $|I| + 1$.*

Proof. Let $u \in Y$. Then, we have two types of adjacencies. First, u is adjacent to each element of I, i.e., u is adjacent to $|I|$ vertices. Second, if $|Y| \geq 2$, then u is adjacent to some elements of Y, i.e., u is adjacent to at least one vertex of Y. Thus, u is adjacent to at least $|I| + 1$ vertices. If $|Y| = 1$, then u is adjacent to $|I|$ elements only. Hence, $deg(u) \geq |I| + 1$ or $deg(u) = |I|$. □

Corollary 7. *Let I be an ideal of $\mathcal{R}$ and $u \in T_I(\Gamma(\mathcal{R}))$. Then, the degree of u depends on the three sets I, X, and Y defined earlier as follows.*

$$deg(u) = \begin{cases} |\mathcal{R}| - 1 & if\ u \in I \\ |I| & if\ u \in X\ or\ u \in Y\ and\ |Y| = 1 \\ at\ least\ |I| + 1 & if\ u \in Y\ and\ |Y| \geq 2. \end{cases}$$

Corollary 8. $T_I(\Gamma(\mathcal{R}))$ is regular graph if and only if either $I = \mathcal{R}$ or $\mathcal{R} \cong \mathbb{Z}_2$ (i.e., $T_I(\Gamma(\mathcal{R}))$ is a complete graph).

Remark 2. Minimum degree of $T_I(\Gamma(\mathcal{R}))$ is $\delta(T_I(\Gamma(\mathcal{R}))) = |I|$, and maximum degree of $T_I(\Gamma(\mathcal{R}))$ is $\Delta(T_I(\Gamma(\mathcal{R}))) = |\mathcal{R}| - 1$.

3. Connectivity of $T_I(\Gamma(\mathcal{R}))$

In this section, we study the connectivity of $T_I(\Gamma(\mathcal{R}))$.

Theorem 7. Let I be a nonzero ideal of $\mathcal{R}$. Then, $T_I(\Gamma(\mathcal{R}))$ has no cut-vertex.

Proof. Assume that the vertex u of $T_I(\Gamma(\mathcal{R}))$ is a cut-vertex. Then, there exist $x, y \in T_I(\Gamma(\mathcal{R}))$ such that u lies on every path from x to y. Thus, we have the following cases.

Case (i): If x is adjacent to y, then there is a path from x to y in $T_I(\Gamma(\mathcal{R}))$ such that u does not lie on it. Hence, we obtain a contradiction.

Case (ii): If x is not adjacent to y, then $x, y \notin I$. Since I is nonzero ideal, I has at least two elements, and we have the following subcases.

Subcase(a): If $x, y \in X$, then x is not adjacent to y, and there exist $w_1, w_2 \in I$ such that x is adjacent to w_1, w_2, and similarly, y is adjacent to w_1, w_2. Therefore, if u is equal to w_1 or w_2, then there is at least one path from x to y, and u does not lie on it, which is a contradiction. Moreover, we obtain the same contradiction when u is not equal to w_1 and w_2.

Subcase(b): If $x \in X$ and $y \in Y$, then x is not adjacent to y, and by the same arguments as used in the above subcase, we obtain the contradiction.

Subcase(c): If $x, y \in Y$, then x may or may not adjacent to y. Thus in both the cases and by the same arguments as used in the subcase(a), we obtain a contradiction. □

Corollary 9. Let I be an ideal of $\mathcal{R}$. Then, $T_I(\Gamma(\mathcal{R}))$ has a cut-vertex if and only if I is a zero ideal.

Remark 3. If $T_I(\Gamma(\mathcal{R}))$ has a cut-vertex, then 0 is the cut-vertex of $T_I(\Gamma(\mathcal{R}))$.

Theorem 8. $k(T_I(\Gamma(\mathcal{R}))) = |I|$.

Proof. By Remark 2, $\delta(T_I(\Gamma(\mathcal{R}))) = |I|$. Moreover, for any graph G, $k(G) \leq \lambda(G) \leq \delta(G)$. Therefore, $k(T_I(\Gamma(\mathcal{R}))) \leq |I|$. Now, if $u \in I$, then u is adjacent to each vertex $v \in T_I(\Gamma(\mathcal{R}))$. Thus, the minimum vertex-cut is the set of all those vertices in I. Therefore, $k(T_I(\Gamma(\mathcal{R}))) \geq |I|$, and hence $k(T_I(\Gamma(\mathcal{R}))) = |I|$. □

Remark 4. For any commutative ring $\mathcal{R}$ with $1 \neq 0$, the elements of I with some elements of Y form a vertex-cut of $T_I(\Gamma(\mathcal{R}))$. However, only the elements of I is the minimum vertex-cut of $T_I(\Gamma(\mathcal{R}))$.

Theorem 9. $T_I(\Gamma(\mathcal{R}))$ has a bridge if and only if either $T_I(\Gamma(\mathcal{R}))$ is a graph with two vertices (i.e., $T_I(\Gamma(\mathcal{R})) \cong T_I(\Gamma(\mathbb{Z}_2)))$, or I is the zero ideal of $\mathcal{R}$.

Proof. Suppose that $T_I(\Gamma(\mathcal{R}))$ has a bridge. Now, we have the following cases.

Case(i): If $|\mathcal{R}| = 2$, then it is clear that $T_I(\Gamma(\mathcal{R})) \cong K_2$, which has a bridge. Hence, $T_I(\Gamma(\mathcal{R})) \cong T_I(\Gamma(\mathbb{Z}_2))$.

Case(ii): If $|\mathcal{R}| \geq 3$, then either $V(T_I(\Gamma(\mathcal{R}))) \subseteq I$, $V(T_I(\Gamma(\mathcal{R}))) \subseteq X$, or $V(T_I(\Gamma(\mathcal{R}))) \subseteq Y$. Let $e = uv$ be the bridge of $T_I(\Gamma(\mathcal{R}))$. Since there is no edge neither between any two elements of X nor between any element of X with element of Y, we have the following subcases.

Subcase(a): If $u, v \in I$, and $|\mathcal{R}| \geq 3$, then there exists $w \in \mathcal{R}$ such that u and v are adjacent to w. We note that $u - v - w - u$ is a cycle, and there is no bridge between them; we obtain a contradiction.

Subcase(b): If $u, v \in Y$, and $|\mathcal{R}| \geq 3$, then there exists $w \in I$ such that u and v are adjacent to w. We note that $u - v - w - u$ is a cycle, and there is no bridge between them; we obtain a contradiction.

Subcase(c): If $u \in I, v \in X$, and $|\mathcal{R}| \geq 3$, then there are two possibilities.

If $|I| = 1$ (i.e., I is a zero ideal of $\mathcal{R}$), then v is adjacent to $u = 0$ only, and uv is a bridge of $T_I(\Gamma(\mathcal{R}))$. Moreover, each vertex of X with u forms a bridge of $T_I(\Gamma(\mathcal{R}))$ (i.e., we have $|X|$ bridges).

If $|I| \geq 2$ (i.e., I is not zero ideal of $\mathcal{R}$), then there exists $u \neq w \in I$ such that u and v are adjacent to w. We note that $u - v - w - u$ is a cycle, and there is no bridge between them; we obtain a contradiction.

Subcase(d): If $u \in I, v \in Y$, and $|\mathcal{R}| \geq 3$, then there are three possibilities.

If $|I| = 1$ (i.e., I is a zero ideal of $\mathcal{R}$) and $|Y| = 1$, then v is adjacent to $u = 0$ only, and uv is a bridge of $T_I(\Gamma(\mathcal{R}))$.

If $|I| = 1$ (i.e., I is a zero ideal of $\mathcal{R}$) and $|Y| \geq 2$, then there exists at least one element $v \neq w \in Y$ such that v and w are connected vertices by a path P. Since each elements of Y are adjacent to elements of I, $v - P - w - 0 - v$ is a cycle and there is no bridge between them. This is a contradiction.

If $|I| \geq 2$ (i.e., I is not zero ideal of $\mathcal{R}$), then there exists $u \neq w \in I$ such that v is adjacent to w. Since each element of I is adjacent, $u - v - w - u$ is a cycle, and there is no bridge between them; we obtain a contradiction.

Conversely, suppose that $T_I(\Gamma(\mathcal{R}))$ is a graph with two vertices. Then, it is clear that $T_I(\Gamma(\mathcal{R}))$ has a bridge. Let us suppose that I is the zero ideal of $\mathcal{R}$ (i.e., $I = \{0\}$) and $|\mathcal{R}| \geq 3$. Then, we have at least one element u in X. Hence, $0u$ is a bridge in $T_I(\Gamma(\mathcal{R}))$. □

Remark 5. *If the ring $\mathcal{R} \cong \mathbb{Z}_2$ or I is the zero ideal of $\mathcal{R}$, then $T_I(\Gamma(\mathcal{R}))$ has a bridge and vice versa, i.e., if $T_I(\Gamma(\mathcal{R}))$ has a bridge, then the ring $\mathcal{R} \cong \mathbb{Z}_2$ or I is the zero ideal of $\mathcal{R}$.*

4. Clique Number and Girth of $T_I(\Gamma(\mathcal{R}))$

In this section, we study the clique number and girth of $T_I(\Gamma(\mathcal{R}))$.

Theorem 10. *Let I be a prime ideal of $\mathcal{R}$ and $|I| = m$. Then, $T_I(\Gamma(\mathcal{R}))$ has cliques of the form $K_1, K_2, K_3, ..., K_{m+1}$. Moreover, $\omega(T_I(\Gamma(\mathcal{R}))) = m + 1$.*

Proof. Suppose that I is a prime ideal of $\mathcal{R}$ and $|I| = m$. Then, by using Theorem 3, $T_I(\Gamma(\mathcal{R}))$ contains a complete subgraph of order $|I| + 1$, and this order is the greatest integer $n = |I| + 1 \geq 2$ such that $K_n \subseteq T_I(\Gamma(\mathcal{R}))$. Hence, $\omega(T_I(\Gamma(\mathcal{R}))) = |I| + 1 = m + 1$. □

Theorem 11. *Let I be an ideal of $\mathcal{R}$, which is not prime and there exists $u \in Y$ such that $u^2 \in I$. Then, $\omega(T_I(\Gamma(\mathcal{R}))) \geq 2|I|$.*

Proof. Suppose that I is not a prime ideal of $\mathcal{R}$. Then, by using Theorem 4, $T_I(\Gamma(\mathcal{R}))$ contains a complete subgraph of order at least $2|I|$. Hence, $\omega(T_I(\Gamma(\mathcal{R}))) \geq 2|I|$. □

Theorem 12. *Let I be an ideal of $\mathcal{R}$, which is not prime, and $u^2 \notin I$ for all $u \in Y$. Then, $\omega(T_I(\Gamma(\mathcal{R}))) \geq |I| + 1$.*

Proof. Suppose that I is not a prime ideal of $\mathcal{R}$. Then, by using Theorem 5, $T_I(\Gamma(\mathcal{R}))$ contains a complete subgraph of order at least $|I| + 1$. Hence, $\omega(T_I(\Gamma(\mathcal{R}))) \geq |I| + 1$. □

Corollary 10. *Let I be a non-prime ideal of $\mathcal{R}$ and $|I| = n$. Then, $\omega(T_I(\Gamma(\mathcal{R}))) = \omega(\Gamma_I(\mathcal{R})) + n$.*

Remark 6. *Let I be an ideal of a ring $\mathcal{R}$. Then, $\omega(\Gamma(\mathcal{R}/I)) \leq \omega(\Gamma_I(\mathcal{R})) \leq \omega(T_I(\Gamma(\mathcal{R})))$. Moreover, we know that if $\Gamma_I(\mathcal{R})$ has no connected columns (i.e., if $u^2 \notin I$ for all $u \in Y$), then $\omega(\Gamma(\mathcal{R}/I)) = \omega(\Gamma_I(\mathcal{R}))$ (for reference see Theorem 4.5 [2]).*

Theorem 13. *Let I be a nonzero ideal of a ring $\mathcal{R}$. If $|\mathcal{R}| \geq 3$, then $T_I(\Gamma(\mathcal{R}))$ has a cycle. Moreover, $gr(T_I(\Gamma(\mathcal{R}))) = 3$.*

Proof. Since I is a nonzero ideal, I has at least two elements (say u, v). Moreover, each element of $\mathcal{R}$ is adjacent to the elements of I and $|\mathcal{R}| \geq 3$, i.e., there exists $w \in \mathcal{R}$ such that u and v are adjacent to w. Thus, $u - w - v - u$ is a cycle of length three, which is the smallest cycle in $T_I(\Gamma(\mathcal{R}))$. Hence, $gr(T_I(\Gamma(\mathcal{R}))) = 3$. □

Corollary 11. *Let I be an ideal of a ring $\mathcal{R}$. If $|\mathcal{R}| \leq 2$, then $gr(T_I(\Gamma(\mathcal{R}))) = \infty$.*

Corollary 12. *Let I be a zero ideal of a ring $\mathcal{R}$. Then,*

$$gr(T_I(\Gamma(\mathcal{R}))) = \begin{cases} 3 & \text{if } |Y| \geq 2 \\ \infty & \text{if } |Y| \leq 1. \end{cases}$$

Remark 7. *Let I be a non-prime ideal of a ring $\mathcal{R}$. Then,*

$$gr(T_I(\Gamma(\mathcal{R}))) \leq gr(\Gamma_I(\mathcal{R})) \leq gr(\Gamma(\mathcal{R}/I)).$$

5. $T_I(\Gamma(\mathcal{R}))$ Is Eulerian and Hamiltonian

In this section, we determine when $T_I(\Gamma(\mathcal{R}))$ is Eulerian, Hamiltonian, and $T_I(\Gamma(\mathcal{R}))$ contains a Eulerian trail.

Theorem 14. *Let I be an ideal of a ring $\mathcal{R}$ such that $|\mathcal{R}| = n \geq 3$. If $I = \mathcal{R}$ and $|\mathcal{R}|$ are odd, then $T_I(\Gamma(\mathcal{R}))$ is a Eulerian.*

Proof. Suppose that $I = \mathcal{R}$ and $|\mathcal{R}|$ is odd. Then, by Corollary 5, $T_I(\Gamma(\mathcal{R}))$ is a complete graph K_n of odd vertices. Thus, the degree of each vertex of $T_I(\Gamma(\mathcal{R}))$ is even. Hence, $T_I(\Gamma(\mathcal{R}))$ is Eulerian. □

Theorem 15. *Let I be a zero ideal of a ring $\mathcal{R}$ such that $|X| = 1$ and $|Y|$ are even. If each vertex of $\Gamma(\mathcal{R}) = \Gamma_0(\mathcal{R})$ has odd degree, then $T_I(\Gamma(\mathcal{R}))$ contains a Eulerian trail.*

Proof. Suppose that each vertex of $\Gamma(\mathcal{R}) = \Gamma_0(\mathcal{R})$ has odd degree. Then, each vertex of Y in $T_I(\Gamma(\mathcal{R}))$ has even degree. Since $|X| = 1$ and $|Y|$ are even, the vertex 0 of I has odd degree. Moreover, the degree of the vertex of X has degree one. Thus, each vertex of $T_I(\Gamma(\mathcal{R}))$ has an even degree except for two vertices that have odd degrees. Hence, $T_I(\Gamma(\mathcal{R}))$ contains a Eulerian trail. □

Remark 8. *In view of Theorem 15, the element of X is the unity of the ring $\mathcal{R}$. Moreover, the Eulerian trail of $T_I(\Gamma(\mathcal{R}))$ begins at unity and ends at zero of $\mathcal{R}$ or begins at zero and ends at unity of $\mathcal{R}$ (for example, $\mathcal{R} = \mathbb{Z}_2 \times \mathbb{Z}_2$ and $I = (0)$).*

Theorem 16. *Assume that a prime ideal of Ring $\mathcal{R}$ is I. Thus, if and only if $|I|$ is even and $|X|$ is odd, then $T_I(\Gamma(\mathcal{R}))$ is Eulerian. $|\mathcal{R}|$ also is odd.*

Proof. Suppose that $T_I(\Gamma(\mathcal{R}))$ is Eulerian. Then, every vertex of $T_I(\Gamma(\mathcal{R}))$ has an even degree. Since I is a prime ideal of $\mathcal{R}$, the degree of each vertex of $T_I(\Gamma(\mathcal{R}))$ either ($|\mathcal{R}| - 1$) or $|I|$ (Theorem 6). Therefore, we have the following cases.
  Case(i): If $u \in X$, then $deg(u) = |I|$, which is even. Thus, $|I|$ is even.
  Case(ii): If $u \in I$, then $deg(u) = |\mathcal{R}| - 1$, which is even, and we obtain $|\mathcal{R}|$ as odd.
Now, we have $|\mathcal{R}|$ as odd and $|I|$ as even. Therefore, $|X|$ is odd.
  Conversely, suppose that $|I|$ is even and $|X|$ is odd. Then, $|\mathcal{R}|$ is odd. Thus, $|\mathcal{R}| - 1$ is even, and $|I|$ is also even. Since I is a prime ideal of $\mathcal{R}$, the degree of each vertex of

$T_I(\Gamma(\mathcal{R}))$ is either $|\mathcal{R}| - 1$ or $|I|$. Thus, the degree of each vertex of $T_I(\Gamma(\mathcal{R}))$ is even. Hence, $T_I(\Gamma(\mathcal{R}))$ is Eulerian. □

Theorem 17. *Let I be a prime ideal of a ring $\mathcal{R}$. Then, $T_I(\Gamma(\mathcal{R}))$ contains a Eulerian trail if and only if either $|I| = 2$ or $|X|$ is even or $|X| = 2$ and $|I|$ are odd.*

Proof. Suppose that $T_I(\Gamma(\mathcal{R}))$ contains a Eulerian trail. Then, exactly two vertices of $T_I(\Gamma(\mathcal{R}))$ have odd degree. Since I is a prime ideal of $\mathcal{R}$, the vertex set of $T_I(\Gamma(\mathcal{R}))$ consists of I and X only, and Y will vanish. Let u and v be the two vertices of odd degree and let $w_1, w_2, ..., w_n$ be the vertices of even degree. Then, we have the following cases.

Case(i): If $u, v \in I$ and $w_i \in X$ for all $1 \leq i \leq n$, then $deg(u) = deg(v)$ is odd and $deg(w_i)$ for all $1 \leq i \leq n$ is even. Therefore, $|\mathcal{R}| - 1$ is odd, and $|I| = 2$ is even; thus, $|\mathcal{R}|$ is even and $|I| = 2$. Hence, $|I| = 2$ and $|X|$ is even. Moreover, $|\mathcal{R}|$ is even.

Case(ii): If $u, v \in I$ and there exists at least one $w_j \in I$, then $deg(u) = deg(v) = deg(w_j)$ is odd. Hence, there are more than two odd vertices in $T_I(\Gamma(\mathcal{R}))$, and we obtain a contradiction.

Case(iii): If $u, v \in X$ and $w_i \in I$ for all $1 \leq i \leq n$, then $deg(u) = deg(v)$ is odd and $deg(w_i)$ for all $1 \leq i \leq n$ is even. Note that $|I|$ is odd, and $|\mathcal{R}| - 1$ is even. We obtain $|I|$ as odd and $|\mathcal{R}|$ as odd. Since $u, v \in X$ only, we have $|X| = 2$. Hence, $|X| = 2$ and $|I|$ are odd. Moreover, $|\mathcal{R}|$ is odd.

Case(iv): If $u, v \in X$ and there exists at least one $w_j \in X$, then $deg(u) = deg(v) = deg(w_j)$ is odd. Thus, there are more than two odd vertices in $T_I(\Gamma(\mathcal{R}))$, we obtain a contradiction.

Case(v): If $u \in I$ and $v \in X$, then $deg(u) = deg(v) = deg(w_i)$ for all $1 \leq i \leq n$ is odd. Thus, all the vertices of $T_I(\Gamma(\mathcal{R}))$ have odd degree, and we obtain a contradiction. Therefore in all the cases, we obtain either $|I| = 2$ and $|X|$ as even or $|X| = 2$ and $|I|$ as odd.

Conversely, suppose that either $|I| = 2$ and $|X|$ are even or $|X| = 2$ and $|I|$ are odd. We first assume that $|I| = 2$ and $|X|$ are even. Thus, $|\mathcal{R}|$ is even, and let u be any vertex of $T_I(\Gamma(\mathcal{R}))$. Therefore, we have the following cases.

Case(i): If $u \in I$, then $deg(u) = |\mathcal{R}| - 1$, which is odd. Since $|I| = 2$ and $|X|$ are even, and there are only two vertices in I posessing an odd degree, and each vertices in X has an even degree. Hence, $T_I(\Gamma(\mathcal{R}))$ contains a Eulerian trail.

Case(ii): If $u \in X$, then $deg(u) = |I| = 2$, which is even by the same argument; there are only two vertices $w_1, w_2 \in I$ such that w_1 and w_2 are adjacent to each vertices in X and w_1 adjacent to w_2 and $deg(w_1) = deg(w_2) = |X| + 1$, which is odd. Therefore, there are only two vertices in I that have an odd degree, and each other vertices in X has even degree. Hence, $T_I(\Gamma(\mathcal{R}))$ contains a Eulerian trail.

Now, we assume that $|X| = 2$ and $|I|$ are odd. Thus, $|\mathcal{R}|$ is odd, and let u be any vertex of $T_I(\Gamma(\mathcal{R}))$. Then, we have the following cases.

Case(i): If $u \in I$, then $deg(u) = |\mathcal{R}| - 1$, which is even. Since $|X| = 2$ and $|I|$ are odd, there are only two vertices in X that have an odd degree, and each other vertices in I has an even degree. Hence, $T_I(\Gamma(\mathcal{R}))$ contains a Eulerian trail.

Case(ii): If $u \in X$, then $deg(u) = |I|$, which is odd; thus, $|\mathcal{R}|$ is odd. By the same argument, there are only two vertices in X possessing an odd degree and each other vertices in I has degree $|\mathcal{R}| - 1$, which is even. Hence, $T_I(\Gamma(\mathcal{R}))$ contains a Eulerian trail.

From all the above cases, we conclude that $T_I(\Gamma(\mathcal{R}))$ contains a Eulerian trail. Hence, if either $|I| = 2$ and $|X|$ are even or $|X| = 2$ and $|I|$ are odd, then $T_I(\Gamma(\mathcal{R}))$ contains a Eulerian trail. □

Remark 9. *In view of Theorem 17, if $|I| = 2$, then Eulerian trail of $T_I(\Gamma(\mathcal{R}))$ begins at one of these two elements of I and ends at other(for example $\mathcal{R} = \mathbb{Z}_6$ and $I = (3)$ see Example 2(iii) Figure 7). Moreover, if $|X| = 2$, then the Eulerian trail of $T_I(\Gamma(\mathcal{R}))$ begins at one of these two elements of X and ends at the other (for example $\mathcal{R} = \mathbb{Z}_3$ and $I = (0)$).*

Theorem 18. *Let I be an ideal of a ring $\mathcal{R}$ that is not prime such that $|I|$ is even, $|X|$ is odd, and $|Y|$ is even. Then, we have the following case:*
  (a) $T_I(\Gamma(\mathcal{R}))$ *is Eulerian if and only if* $\Gamma_I(\mathcal{R})$ *is Eulerian.*
  (b) $T_I(\Gamma(\mathcal{R}))$ *contains a Eulerian trail if and only if* $\Gamma_I(\mathcal{R})$ *contains a Eulerian trail.*

Theorem 19. *Let I be an ideal of a ring $\mathcal{R}$ that is not prime such that $|I|$ is even, $|X|$ is even, and $|Y|$ is odd. Then, we have the following case.*
  (a) $T_I(\Gamma(\mathcal{R}))$ *is Eulerian if and only if* $\Gamma_I(\mathcal{R})$ *is Eulerian.*
  (b) $T_I(\Gamma(\mathcal{R}))$ *contains a Eulerian trail if and only if* $\Gamma_I(\mathcal{R})$ *contains a Eulerian trail.*

Remark 10. *In view of Theorems 16, 18, and 19, since $|I|$ is even and $|\mathcal{R}|$ is odd, there is no graph on n vertices that can be realized as $T_I(\Gamma(\mathcal{R}))$ for some ring $\mathcal{R}$ and an ideal I of $\mathcal{R}$.*

Theorem 20. *Let I be a non-prime ideal of a ring $\mathcal{R}$ such that $|I| = 2$, $|X|$ is even, and $|Y|$ is even. Then, $T_I(\Gamma(\mathcal{R}))$ contains a Eulerian trail if and only if $\Gamma_I(\mathcal{R})$ is Eulerian.*

Proof. Suppose that $T_I(\Gamma(\mathcal{R}))$ contains a Eulerian trail. Then, each vertex of $T_I(\Gamma(\mathcal{R}))$ has an even degree except for two vertices that have odd degrees. Since I is not a prime ideal of $\mathcal{R}$, the vertex set of $T_I(\Gamma(\mathcal{R}))$ consists of I, X, and Y. Let u and v be the two vertices of odd degree and let $w_1, w_2, ..., w_n$ be the vertices of even degree. Since $|X|$ and $|Y|$ are even and $|I| = 2$, $u, v \in I$. If we assume that at least one of $u, v \in X$, then $|I|$ is odd, which is a contradiction, and if we assume that at least one of $u, v \in Y$, then we have more than two odd vertices in $T_I(\Gamma(\mathcal{R}))$. Therefore, $T_I(\Gamma(\mathcal{R}))$ has no Eulerian trail, which is a contradiction. Thus, the two elements of I are the only ones that are odd and all other elements of X and Y are even. Now, we know that $\Gamma_I(\mathcal{R})$ is a subgraph of $T_I(\Gamma(\mathcal{R}))$, i.e., the set Y is the set of vertex of $\Gamma_I(\mathcal{R})$, but to obtain the exact number of edges that incident on each vertex of $\Gamma_I(\mathcal{R})$, we remove the edges join all vertex of I with each vertex of Y. Thus, the degree of each vertex of $\Gamma_I(\mathcal{R})$ is same as the degree of each vertex of Y in $T_I(\Gamma(\mathcal{R}))$ subtract $|I|$. Therefore, the degree of each vertex of $\Gamma_I(\mathcal{R})$ is even. Hence, $\Gamma_I(\mathcal{R})$ is Eulerian.

Conversely, suppose that $\Gamma_I(\mathcal{R})$ is Eulerian. Then, each vertex of $\Gamma_I(\mathcal{R})$ has an even degree. Thus, each vertex of Y in $\Gamma_I(\mathcal{R})$ has an even degree. Now, we will construct $T_I(\Gamma(\mathcal{R}))$. First, we have two vertices of the set I adjacent to all vertex of $\mathcal{R}$, i.e., adjacent to $(|I| - 1 + |X| + |Y|)$ vertices. Since $|I| = 2$, $|X|$ and $|Y|$ are even, and the degree of each vertex of I is odd, i.e., we have two vertices of I being odd. Now, if we have an even vertex of set X, which is adjacent to all vertices of I only. Thus, the degree of each vertex of X is even. Finally, we have an even vertex of set Y adjacent to all vertex of I and at least one vertex of Y. Since $|I| = 2$ is even and each vertex of Y in $\Gamma_I(\mathcal{R})$ has an even degree, the degree of each vertex of Y in $T_I(\Gamma(\mathcal{R}))$ is even. Thus, each vertex of $T_I(\Gamma(\mathcal{R}))$ has an even degree except for two vertices, which have odd degrees. Hence, $T_I(\Gamma(\mathcal{R}))$ contains a Eulerian trail. □

Theorem 21. *Let I be an ideal of a ring $\mathcal{R}$ that is not prime such that $|I| = 2$, $|X|$ is odd, and $|Y|$ is odd. Then, $T_I(\Gamma(\mathcal{R}))$ contains a Eulerian trail if and only if $\Gamma_I(\mathcal{R})$ is Eulerian.*

Proof. By the same arguments used in the above Theorem 18, the proof is clear. □

Example 3. *Let $\mathcal{R} = \mathbb{Z}_{12}$ and $I = (6) = \{0, 6\}$. Then, $Y = \{2, 3, 4, 8, 9, 10\}$ and $X = \{1, 5, 7, 11\}$. We observe that $\Gamma_I(\mathcal{R})$ is Eulerian and $T_I(\Gamma(\mathcal{R}))$ contains a Eulerian trail (see Figure 9).*

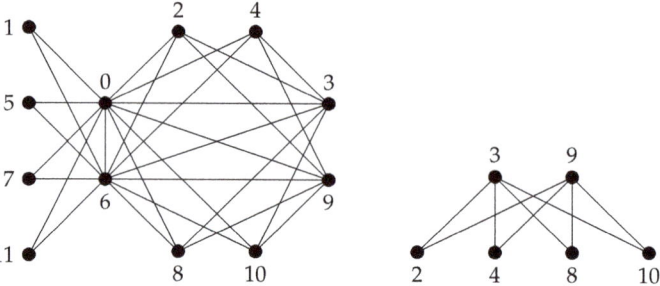

Figure 9. (**left**)$T_I(\Gamma(\mathcal{R}))$ and (**right**) $\Gamma_I(\mathcal{R})$, when $\mathcal{R} = \mathbb{Z}_{12}$ and $I = (6)$.

Theorem 22. *Let I be an ideal of a ring $\mathcal{R}$ such that $|\mathcal{R}| = n \geq 3$. If $I = \mathcal{R}$, then $T_I(\Gamma(\mathcal{R}))$ is Hamiltonian.*

Proof. Suppose that $I = \mathcal{R}$. Then, by Corollary 5, $T_I(\Gamma(\mathcal{R}))$ is a complete graph. Hence, $T_I(\Gamma(\mathcal{R}))$ is Hamiltonian. □

Theorem 23. *Let I be a zero ideal of a ring $\mathcal{R}$. Then, $T_I(\Gamma(\mathcal{R}))$ cannot be Hamiltonian.*

Proof. Suppose that $I = (0)$. Then, by Corollary 9, $T_I(\Gamma(\mathcal{R}))$ has a cut-vertex. Hence $T_I(\Gamma(\mathcal{R}))$ cannot be Hamiltonian. □

Theorem 24. *Let I be a nonzero ideal of a ring $\mathcal{R}$ such that $|\mathcal{R}| = n \geq 3$. If $|I| \geq \frac{n}{2}$, then $T_I(\Gamma(\mathcal{R}))$ is Hamiltonian.*

Proof. Suppose that u and v are any two vertices of $T_I(\Gamma(\mathcal{R}))$. Then, we have the following cases.
   Case(*i*): If u and v are adjacent for all $u, v \in T_I(\Gamma(\mathcal{R}))$, then $T_I(\Gamma(\mathcal{R}))$ is complete. Therefore, $T_I(\Gamma(\mathcal{R}))$ is Hamiltonian.
   Case(*ii*): If u and v are nonadjacent for some $u, v \in T_I(\Gamma(\mathcal{R}))$, then by Remarks 1 (*viii*), (*ix*), and (*x*), we have the following subcases:
   Subcase(*a*): If $u, v \in X$, then by Corollary 7, $deg(u) = deg(v) = |I|$. Thus, $deg(u) + deg(v) = |I| + |I| \geq \frac{n}{2} + \frac{n}{2} = n$. Hence, $T_I(\Gamma(\mathcal{R}))$ is Hamiltonian.
   Subcase(*b*): If $u, v \in Y$, then by Corollary 7, $deg(u) = deg(v) \geq |I|$. Thus, $deg(u) + deg(v) \geq |I| + |I| \geq \frac{n}{2} + \frac{n}{2} = n$. Hence, $T_I(\Gamma(\mathcal{R}))$ is Hamiltonian.
   Subcase(*c*): If $u \in X$ and $v \in Y$, then by Corollary 7, $deg(u) = |I|$ and $deg(v) \geq |I|$. Thus, $deg(u) + deg(v) \geq |I| + |I| \geq \frac{n}{2} + \frac{n}{2} = n$. Hence, $T_I(\Gamma(\mathcal{R}))$ is Hamiltonian. □

Corollary 13. *Let I be a nonzero ideal of a ring $\mathcal{R}$ such that $|\mathcal{R}| = n \geq 3$. If $|I| \geq \frac{n}{2}$ for each pair u, v of X, then $T_I(\Gamma(\mathcal{R})) + uv$ is Hamiltonian if and only if $T_I(\Gamma(\mathcal{R}))$ is Hamiltonian.*

Corollary 14. *Let I be a nonzero ideal of a ring $\mathcal{R}$. Then, I is a maximal ideal of a ring $\mathcal{R}$ if and only if $T_I(\Gamma(\mathcal{R}))$ is Hamiltonian.*

6. Conclusions

We considered a generalization of dot total graph of $\mathcal{R}$ as well as an ideal-based zero-divisor graph. We showed that $T_I(\Gamma(\mathcal{R}))$ is connected and has a small diameter of at most two. Furthermore, we studied the connectivity, clique number and the girth of $T_I(\Gamma(\mathcal{R}))$. In addition, the cases when $T_I(\Gamma(\mathcal{R}))$ is Eulerian, Hamiltonian, and $T_I(\Gamma(\mathcal{R}))$ contains a Eulerian trail. For future work, the application of this graph to the study on Laplacian eigenvalues of an ideal-based dot total graph, which is closely related to the work in the paper [6], can be investigated. Additionally, the energy of an ideal-based dot total graph, which is related to the recent work in [5,7], requires more consideration. The purpose of studying this type of graph is beneficial from its application point of view in practical life, such as networks, especially communication networks, which will be studied in an independent manuscript.

Author Contributions: Investigation, M.A., J.H.A., A.M.A. and A.A.; Writing—original draft, M.A., J.H.A., A.M.A. and A.A.; Writing—review and editing, M.A., J.H.A., A.M.A. and A.A. All authors have read and agreed to the published version of the manuscript.

Funding: This research received no external funding.

Institutional Review Board Statement: Not applicable.

Informed Consent Statement: Not applicable.

Data Availability Statement: Not applicable.

Acknowledgments: The authors are grateful to the referees for their valuable comments and suggestions.

Conflicts of Interest: The authors declare no conflict of interest.

References

1. Mohammad, A.; Asalool, J.H.; Mohit, K. The total graph of commutative ring with respect to multiplication. *Italian J. Pure Appl. Math.* **2022**, in press.
2. Redmond, S.P. An ideal-based zero-divisor graph of a commutative ring. *Comm. Algebra* **2003**, *31*, 4425–4443. [CrossRef]
3. Anderson, D.F.; Livingston, P.S. The zero-divisor graph of a commutative ring. *J. Algebra* **1999**, *217*, 434–447. [CrossRef]
4. Beck, I. Coloring of commutative rings. *J. Algebra* **1988**, *116*, 208–226. [CrossRef]
5. Nath, R.K.; Fasfous, W.N.T.; Das, K.C.; Shang, Y. Common Neighborhood Energy of Commuting Graphs of Finite Groups. *Symmetry* **2021**, *13*, 1651. [CrossRef]
6. Rather, B.A.; Pirzada, S.; Naikoo, T.A.; Shang, Y. On Laplacian eigenvalues of the zero-divisor graph associated to the ring of integers modulo n. *Mathematics* **2021**, *9*, 482. [CrossRef]
7. Shang, Y. A note on the commutativity of prime near-rings. *Algebra Colloq.* **2015**, *22*, 361–366. [CrossRef]
8. Diestel, R. *Graph Theory*; Springer: New York, NY, USA, 1997.

Article

From Automata to Multiautomata via Theory of Hypercompositional Structures

Štěpán Křehlík [1], Michal Novák [2,*] and Jana Vyroubalová [2]

[1] CDV—Transport Research Centre, Líšeňská 33a, 636 00 Brno, Czech Republic; stepan.krehlik@cdv.cz
[2] Faculty of Electrical Engineering and Communication, Brno University of Technology, 616 00 Brno, Czech Republic; 143690@vut.cz
* Correspondence: novakm@vut.cz or novakm@vutbr.cz; Tel.: +420-541-146-077

Abstract: In this paper, we study two important problems related to quasi-multiautomata: the complicated nature of verification of the GMAC condition for systems of quasi-multiautomata, and the fact that the nature of quasi-multiautomata has deviated from the original nature of automata as seen by the theory of formal languages. For the former problem, we include several new conditions that simplify the procedure. For the latter problem, we close this gap by presenting a construction of quasi-multiautomata, which corresponds to deterministic automata of the theory of formal languages and is based on the operation of concatenation.

Keywords: automata theory; hypergroups; quasi-automata; quasi-multiautomata; semiautomata

MSC: 20N20; 68Q70

1. Introduction

The theory of formal languages is closely linked to the theory of automata. An automaton is a finite representation of a formal language that can consist of an infinite number of words. Automata are often classified by means of a class of formal languages that they can accept; see, e.g., [1], a paper dealing with the automata theory from the point of view of our present paper, or papers such as [2,3]. The algebraic theory of automata studies various types of such structures, which are linked to actions of groups on sets. In the case of algebraic automata, the way of functioning is rather straightforward and simple: the automaton has a set of states and a set of inputs and, after we apply a certain input on a certain state, the automaton switches to a new state as specified by the transition function. However, in the theory of formal languages, automata are regarded as devices reading strings of words instead of single input symbols only. In other words, inputs (or characters) are catenated and one-by-one put into the automaton, which causes changes of states. In the algebraic definition, this can be seen in the definition of automaton, where the input alphabet is a free monoid over the input set, i.e., for each nonempty set A, we denote A^* the set of all finite sequences $a_1 a_2 \ldots a_n$, $a_i \in A$, i.e., finite words made of symbols from A. Moreover, A^* regards the usual binary operation of concatenation: $u = a_1 \ldots a_n, v = b_1 \ldots b_k, uv = a_1 \ldots a_n b_1 \ldots b_k$. With this, A^* is a free monoid over A with a neutral element e, the empty word.

In the course of time, the algebraic theory of automata began to regard automata without output, the operation of concatenation has been replaced by an arbitrary group operation, and monoid or a group have been used instead of the free monoid. Definition 2 complies with traditional books such as [4–7] or recent papers such as [8]. One can see that the conditions are constructed in such a way that both monoid and free monoid are applicable.

Since the late 1930s, the group theory has been generalized in the sense that the synthesis of elements of the carrier set need no longer to be an element of that carrier

set. When subsets are permitted (such as a line segment being a result of an "operation" on its endpoints), we arrive at a concept of hypercompositional structures. For a basic introduction to the topic stressing its context from a historical perspective—see, e.g., an easy-to-follow overview paper [9].

The generalizations of algebraic automata in the sense of the theory of hypercompositional structures first focused on constructions of commutative hypergroups on their state sets. Properties of automata have been described by means of properties of such hypergroups over their state sets; see, e.g., [5,10–12]. The next step is to construct hypercompositional structures on the input sets and generalize the MAC condition to GMAC. Since there are no unique neutral elements in hypercompositional structures, the UC condition is omitted; see, e.g., [13,14]. The concept of quasi-multiautomaton originated in conference proceedings [15] while the GMAC condition was used for the first time (in the context of dynamical systems) in [12]—e.g., in [16,17]. In this respect, notice also suffix automata, which accept all suffixes of a given string and belong to the basic stringologic principles. When generalizing their transitions to include specific buffer operations, we obtain new subtree pushdown automata, which accept all subtrees of a given tree in the prefix notation; see, e.g., [18].

In 1959, M. O. Rabin and D. Scott introduced the concept of nondeterministic finite automata in [19] and proved their equivalence to deterministic finite automata. A nondeterministic automaton, such as a deterministic one, consumes a string of input symbols. It enters a new state for each input symbol until all input symbols are consumed. Unlike in deterministic finite automata, the way symbols are consumed is nondeterministic, i.e., for a state and an input symbol, the next state may be the original or one, two, or more possible states. Thus, in the formal definition, another state is an element of a potential set of states, which is a set of states that must be considered simultaneously. In this respect, the connection with the theory of hypercompositional structures is rather obvious. However, introducing hypergroups on input sets does not lead to nondeterministic quasi-multiautomata because the transition function $\delta : H \times S \longrightarrow S$ used in Definition 5 of a quasi-multiautomaton maps to the state set S instead of the set of its subsets $\mathcal{P}(S)$.

2. Basic Definitions

In order to clarify terminology used throughout the paper, in this section, we collect all basic definitions.

Definition 1. *[1] A deterministic automaton is a 5-tuple (A, S, s_0, δ, F), where A is input alphabet, S is a finite nonempty set of states, $s_0 \in S$ is the initial (or start) state, $\delta : A \times S \longrightarrow S$ is the state transition function, and $F \subseteq S$ is the set of final states. Sometimes it is convenient to use, instead of δ, the extended transition function $\delta^* : A^* \times S \longrightarrow S$, where A^* is the set of words over the alphabet A, which is defined recursively as follows:*

1. $(\forall s \in S)(\forall a \in A) \quad \delta^*(a, s) = \delta(a, s)$,
2. $(\forall s \in S) \quad \delta^*(a, \lambda) = s$ *where λ is the empty string*,
3. $(\forall s \in S)(\forall x \in A^*)(\forall a \in A) \quad \delta^*(ax, s) = \delta^*(\delta(a, s), x)$.

One can see that the conditions of the following definition are constructed in such a way that both monoid and free monoid are applicable.

Definition 2. *By automaton, we mean a structure $\mathbb{A} = (I, S, \delta)$ such that $I \neq \emptyset$ is a free monoid, $S \neq \emptyset$, and $\delta : I \times S \rightarrow S$ satisfies the following condition:*

1. *There exists an element $e \in I$ such that $\delta(e, s) = s$ for any state $s \in S$.*
2. $\delta(y, \delta(x, s)) = \delta(xy, s)$ *for any pair $x, y \in I$ and any state $s \in S$.*

The set I is called the input set or input alphabet, the set S is called the state set, and the mapping δ is called next-state or transition function.

Remark 1. *Condition 1 is often called the unit condition (UC) while condition 2 is often called the Mixed Associativity Condition (MAC).*

Notice that, in our paper, we write "xy" instead of "$x \cdot y$". However, in order to stress the difference between concatenation and arbitrary operation, we write "$x \cdot y$" in Definition 3. For a deeper insight including historical perspective and terminology issues (e.g., "quasi-automaton" vs. "semiautomaton"), see [16].

Definition 3. *By quasi-automaton, we mean a structure $\mathbb{A} = (I, S, \delta)$ such that $I \neq \emptyset$ is a monoid, $S \neq \emptyset$, and $\delta : I \times S \to S$ satisfies the following condition:*
1. *There exists an element $e \in I$ such that $\delta(e, s) = s$ for any state $s \in S$.*
2. *$\delta(y, \delta(x, s)) = \delta(x \cdot y, s)$ for any pair $x, y \in I$ and any state $s \in S$.*

The set I is called the input set or input alphabet, the set S is called the state set, and the mapping δ is called next-state or transition function.

The following definition is a standard introductory definition of the theory of hypercompositional structures (or algebraic hyperstructures as they are also known).

Definition 4. *A hypergroupoid is a pair $(H, *)$, where H is a nonempty set and the mapping $* : H \times H \longrightarrow \mathcal{P}^*(H)$ is a binary hyperoperation (or hypercomposition) on H (here, $\mathcal{P}^*(H)$ denotes the system of all nonempty subsets of H). If $a * (b * c) = (a * b) * c$ holds for all $a, b, c \in H$, then $(H, *)$ is called a semi-hypergroup. Moreover, if the reproduction axiom—i.e., relation $a * H = H = H * a$ for all $a \in H$—is satisfied, then the semi-hypergroup $(H, *)$ is called hypergroup.*

The following definition transfers the concept of quasi-automaton into the theory of hypercompositional structures.

Definition 5. *[15] A quasi–multiautomaton is a triad $\mathbb{MA} = (H, S, \delta)$, where $(H, *)$ is a semi-hypergroup, S is a nonempty set and $\delta : H \times S \to S$ is a transition function satisfying the following condition:*

$$\delta(b, \delta(a, s)) \in \delta(a * b, s) \text{ for all } a, b \in H, s \in S. \tag{1}$$

The semi-hypergroup $(H, *)$ is called the input semi-hypergroup of the quasi–multiautomaton $\mathbb{A}$ (H alone is called the input set or input alphabet), the set S is called the state set of the quasi–multiautomaton $\mathbb{A}$, and δ is called next-state or transition function. Elements of the set S are called states; elements of the set H are called input symbols or letters. Condition (1) is called Generalized Mixed Associativity Condition (abbr. as GMAC).

Finally, we recall the notion of nondeterministic automaton.

Definition 6. *If, in Definition 1, we have $\delta : A \times S \longrightarrow \mathcal{P}(S)$ instead of $\delta : A \times S \longrightarrow S$, then the 5-tuple (A, S, s_0, δ, F) is called nondeterministic automaton.*

3. GMAC Condition in the Definition of Quasi-Multiautomata

This section aims at facilitating verifications of the GMAC condition. When doing so, we use ideas included in [1], where the notion of *order* of a state is defined in the context of deterministic automata of Definition 1. Notice that in [1], the term *word length*—meaning the number of concatenated input symbols—is used. For example, if we consider the word $aaabba$ on the set a, b, then its length is 6.

Definition 7. *[1] The order of a state $s \in S$ of the deterministic automaton in Definition 2, denoted by $\operatorname{ord} s$, is the minimum of the lengths of words that lead from the start state s_0 to s.*

Example 1. *Consider an automaton as defined in Definition 1. Now, let s_0 be the initial state and s_5 the final state. The input alphabet is a free monoid over the set $\{a, b\}$. It is clear from Figure 1 that ord $s_0 = 0$, ord $s_1 =$ ord $s_2 = 1$, ord $s_3 =$ ord $s_4 = 2$, and ord $s_5 = 3$.*

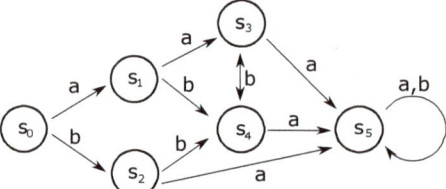

Figure 1. Finite quasi-automaton.

Obviously, the order of a state is related to the operation of concatenation of words. Thus, in Example 1, we have that *ord* $s_3 = 2$ because the shortest word taking us from s_0 to s_3 is *aa*. However, if we generalize the concatenation operation to an arbitrary associative operation, i.e., move to Definition 3, we have that $a \cdot a$ is an element of I (say a). Thus, the "length of the word" becomes either 1 or 0 depending on whether the element is or is not isolated. The situation becomes even more complicated for quasi-multiautomata of Definition 5.

However, in both cases, we observe a discrepancy when transferring the intuitive notion of order tailored to the classical case of deterministic automata to quasi-automata or quasi-multiautomata of Definitions 3 and 5. The reason for such a discrepancy lies in the visualization of the algebraic concept by graphs and the fact that we no longer distinguish between start and end states. Therefore, further on, we will focus on the "descriptions of graphs" by "counting arrows" rather than attempts to stress the algebraic part of the notion. As a result, we cannot use the notion of *order* (based on *word length*) anymore (because, technically speaking, there are no words anymore). Notice that the forthcoming definitions once again enable us to "count arrows" of the graphs.

Definition 8. *By the transition number of states $s, t \in S$ (in this order), denoted by $tn(s, t)$, we mean the smallest number of transitions that take us from the state s to the state t.*

It is easy to see that in Figure 1, $tn(s_0, s_5) = 2$. Indeed, applying input b takes us from s_0 to s_2; then, applying input a takes us from s_2 to s_5, which means that the smallest number of transitions that take us from s_0 to s_5 is 2.

Notice that, from Definition 8, it does not follow that $tn(s, t) = tn(t, s)$. This is evident in Figure 2, where $tn(s_1, s_3) = 2$ while $tn(s_3, s_1) = 1$. Next, if there is no input that would take us from one state to another, we say that the respective transition number is 0. An example is depicted in Figure 1, where $tn(s_3, s_1) = 0$.

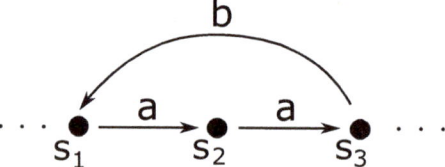

Figure 2. Noncommutativity of the order of two states.

Before introducing Theorem 1, recall the definition of a reversible automaton from [5].

Definition 9. *An automaton $\mathbb{A} = (A, S, \delta)$ is called reversible, if for every state $s \in S$ and every input $a \in A$ (or word $a \in A^*$) there exists an input $b \in A$ (or a word $b \in A^*$) such that $\delta(b, \delta(a, s)) = s$ (or $\delta(ab, s) = s$).*

Now, we generalize the notion to the case of quasi-multiautomata.

Definition 10. *A quasi-multiautomaton* $\mathbb{MA} = (H, S, \delta)$ *is called reversible, if for every state* $s \in S$ *and every input* $a \in H$ *there exists an input* $b \in H$ *such that* $\delta(b, \delta(a, s)) = s$.

Remark 2. *Notice that in every reversible quasi-multiautomaton there is* $tn(s, t) = tn(t, s)$ *for all* $s, t \in S$.

To fully clarify the above notion and its application to quasi-multiautomata, we present the following example, in which two multiautomata are given. The first one is not reversible while the second one, with a modified input set, is.

Example 2. *Consider the interval of real numbers* $I_1 = [1, \infty)$ *and the hyperoperation* $\circ_1$: $I_1 \times I_1 \to \mathcal{P}^*(I_1)$ *defined by*

$$a \circ b = \{c \in I, c \geq a \cdot b\}, \text{for all } a, b \in I_1.$$

It is obvious that the associative law holds. Therefore, the structure $(I_1, \circ_1)$ *is a semi-hypergroup. Then, the triad* $((I_1, \circ_1), \mathbb{R}^+ \setminus \{0\}, \delta_1)$, *where the transition function* $\delta_1 : I_1 \times \mathbb{R}^+ \setminus \{0\} \to \mathbb{R}^+ \setminus \{0\}$ *is defined by*

$$\delta(a, r) = a \cdot r, \text{for all } a \in I_1, r \in \mathbb{R}^+ \setminus \{0\}.$$

which is a quasi-multiautomaton. Now, for input $a = 5.2$ *and state* $r = 10$, *we have* $\delta_1(5.2, 20) = 104$. *Thus, the quasi-multiautomaton is not reversible, because there is no input* b *such that* $\delta_1(b, 104) = 5.2$.

Now, consider the interval $I_2 = (0, \infty)$ *instead and the hyperoperation* "$\circ_2$" *defined in the same way as* "$\circ_1$". *For the transition function* δ_2, *defined in the same way as* δ_1, *the triad* $((I_2, \circ_2), \mathbb{R}^+ \setminus \{0\}, \delta_2)$ *is a reversible quasi-multiautomaton because, for each input symbol* a, *there exists an input symbol* $\frac{1}{a}$ *such that* $\delta_2\left(\frac{1}{a}, \delta_2(a, r)\right) = r$.

At this point, using the notion of a reversible quasi-multiautomaton and the transition number of two states, we can provide the following theorem regarding the validity of the GMAC condition 1.

Theorem 1. *In every reversible quasi-multiautomaton* (H, S, δ), *there is* $tn(s, t) = tn(t, s) = 1$ *for every two states* $s, t \in S$.

Proof. Recall that the GMAC condition (1) is $\delta(b, \delta(a, s)) \in \delta(a * b, s)$ for all inputs and states. In a reversible quasi-multiautomaton, there is $tn(s, t) = tn(t, s)$ for all $s, t \in S$. Suppose that there exists at least one pair of states $(t, s) \in S$, such that $tn(t, s) = 2$—i.e., from the state t we can reach the state s after application of at least two inputs. Denote such inputs as $a, b \in H$. Therefore, on the left-hand side of the condition GMAC, we have $\delta(a, \delta(b, t)) = r$. However, on the right-hand side, for all $c \in a * b$, we never obtain the state r, because this would mean that $tn(t, s) = 1$, which would be a contradiction to the assumption that $tn(t, s) = 2$. Naturally, the same is true for transition numbers greater than 2. □

In the following example, we show that the implication in Theorem 1 cannot be reversed, i.e., it is not true that if for every two states there is $tn(s, t) = tn(t, s) = 1$, we obtain a reversible quasi-multiautomaton (or a quasi-multiautomaton).

Example 3. *Consider the structure* (H, S, δ). *In this structure, we show that there is a path between each two states, yet the GMAC condition does not hold, i.e., from the fact that* $tn(s, t) = 1$ *for all* $s, t \in S$, *it cannot be deduced that* (H, S, δ) *is a quasi-multiautomaton. Consider the set*

$H = \{h_1, h_2\}$, the hyperoperation "∘" defined by the following Figure 3, and the transition function δ defined by the transition diagram in Figure 4.

∘	h_1	h_2
h_1	h_1	H
h_2	H	H

Figure 3. Definition of hyperoperation "∘".

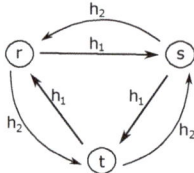

Figure 4. Transition number of every pair of states is 1, yet, GMAC does not hold.

It is clear from the transition diagram that for each pair of states $r, s \in S$ there is $tn(r,s) = 1$. However, the GMAC condition does not hold. Indeed, on the left-hand side, we have $\delta(h_2, \delta(h_1, r)) = \delta(h_1, t) = r$ while on the right-hand side, we have $\delta(h_1 \circ h_2, r) = \delta(h_1, r) \cup \delta(h_2, r) = \{s, t\}$ and obviously the left-hand side is not included in the right-hand side.

Notice that the above example also shows that the validity of the GMAC condition depends on the transition function as well as on the definition of the hyperoperation.

Theorem 2. *If in a quasi-multiautomaton $((H, *), S, \delta)$ there is $tn(r,s) = 1$ and $tn(s,t) = 1$ for some $r, s, t \in S$, then there is $tn(r,t) = 1$.*

Proof. Suppose that there is $tn(r,s) = 1$ and also $tn(s,t) = 1$ for some $r, s, t \in S$. Then, there exists such an input $i \in H$ for which there is $\delta(i,r) = s$, and also an input $j \in H$ for which there is $\delta(j,s) = t$. Since the GMAC condition holds, i.e., the state $t = \delta(j, \delta(i,r))$ is included in the right-hand side $\delta(i * j, r) = \bigcup_{c \in i*j} \delta(c, r)$, there is an input $k \in i * j$, where $\bigcup_{c \in i*j} \delta(c, r) \ni \delta(k, r) = t$. Thus, there must be $tn(r,t) = 1$. □

In the following Example 4, we present a trivial quasi-multiautomaton, where the order of each pair of states is 1. The nontrivial quasi-multiautomaton is presented in Example 5.

Example 4. *Consider a quasi-multiautomaton, which was first presented in Example 5 of [16]. By coincidence, the order of each pair of states is 1, i.e., Theorem 2 holds for an arbitrary triad of elements.*

One can see that in the automaton with a free monoid, in the MAC condition of Definition 2, we have links in the sequence of states that coincide with the "links" of strings, i.e., concatenation. In other words, in order to reach a given state, the automaton passes through the same states regardless of whether we regard the left- or the right-hand side of the MAC condition of Definition 2. However, this is not the case for the quasi-multiautomaton, where the GMAC condition suggests that there must exist a shorter, or more efficient input that enables us to reach the same state as when applying two different catenated inputs. Indeed, in Figure 5, we can get from b_1 to b_2 and from b_2 to b_3 or directly from b_1 to b_3. Notice that without the input e_4 applied to b_1, the GMAC condition would not hold.

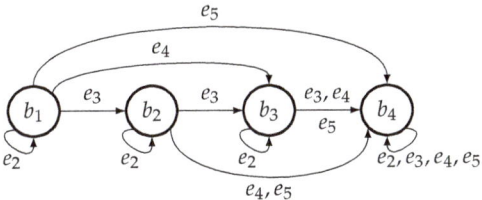

Figure 5. A quasi-multiautomaton where the transition number of each pair of states is 1.

Example 5. *In this example, we summarize our above considerations and also explain why Theorems 1 and 2 cannot be given as one even though they are semantically similar. In order to maintain clarity, the quasi-multiautomaton in Figure 6 does not have evaluated transitions. However, it is evident that the inputs could be easily supplemented as in Example 4. For this quasi-multiautomaton, it is obvious that Theorem 2 applies. Furthermore, it is obvious that the multi-automaton in Figure 6 is not reversible because there are no arrows in the opposite direction in the transition diagram—i.e., from the state s_1, we go to the state s_2, but from the state s_2 it is not possible go to s_1. If we considered bidirectional arrows in Figure 6, the quasi-multiautomaton would be reversible. In such a case, we would be able to get from state s_2 to state s_4 via state s_1, which corresponds to the left side of the GMAC condition. This must be met, so we must go directly from the state s_2 to the state s_4.*

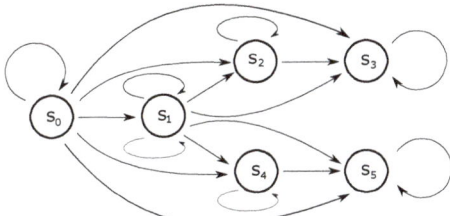

Figure 6. A quasi-multiautomaton that is not reversible.

Another indicator that would help to decide whether the GMAC condition is met or not are identities (neutral elements) of the input hyperstructure. Recall that by an *identity* of a semi-hypergroup $(H, *)$, we mean such an element $e \in H$ that there is $x \in x \circ e \cap e \circ x$ for all $x \in H$. Although, in the definition of the quasi-multiautomaton, the UC condition of Definition 2 is not required, it does not mean that no elements fulfilling it exist. If they do, they are identities of the input semi-hypergroup.

Theorem 3. *If in the quasi-multiautomaton $((H, *), S, \delta)$ there exists an element $e \in H$ with the property $\delta(e, s) = s$ for all $s \in S$, then there is $a \in a * e \cap e * a$ for all $a \in H$—i.e., e is an identity of $(H, *)$.*

Proof. Suppose that there is $\delta(e, s) = s$ for some $e \in H$ and all $s \in S$ and that the GMAC condition $\delta(a, \delta(b, s)) \in \delta(a * b, s)$ is satisfied for all $a, b \in H$ and for all $s \in S$.

For an element $e \in H$, and arbitrary $a \in H$ and $s \in S$, we have

$$\delta(a, \delta(e, s)) \in \delta(a * e, s) \quad \wedge \quad \delta(e, \delta(a, s)) \in \delta(e * a, s)$$
$$\delta(a, s)) \in \delta(a * e, s) \quad \wedge \quad \delta(a, s)) \in \delta(e * a, s)$$
$$\delta(a, s)) \in \bigcup_{c \in a * e} \delta(c, s) \quad \wedge \quad \delta(a, s)) \in \bigcup_{d \in e * a} \delta(d, s).$$

It is obvious that the state $\delta(a, s)$ belongs to the set of states on the right-hand side if and only if $a \in a * e \wedge a \in e * a$, i.e., $a \in a * e \cap a \in e * a$, for all $a \in H$. □

When the Cartesian composition of two quasi-multiautomata was constructed in [17], the necessary condition of Theorem 1 was not satisfied. As a result, the GMAC condition was not satisfied. The authors solved the problem by modifying the condition by adding an extension to its right-hand side. Notice that the Cartesian composition of two automata were introduced by Dörfler in [20], the composition was subsequently generalized to the case of quasi-multiautomata in [17].

We conclude this section with the definition of products of automata introduced by Dörfler.

Definition 11. *[21] Let $\mathbb{A}_1 = (I, S, \delta)$, $\mathbb{A}_2 = (I, R, \tau)$, and $\mathbb{B} = (J, T, \sigma)$ be quasi-automata. By the homogeneous product $\mathbb{A}_1 \times \mathbb{A}_2$, we mean the quasi-automaton $(I, S \times R, \delta \times \tau)$, where $\delta \times \tau : I \times (S \times R) \to S \times R$ is a mapping satisfying, for all $s \in S, r \in R, a \in H$, $(\delta \times \tau)(a, (s, r)) = (\delta_1(a, s), \tau(a, r))$, while the heterogeneous product $\mathbb{A}_1 \otimes \mathbb{B}$ is the quasi-automaton $(I \times J, S \times T, \delta \otimes \sigma)$, where $\delta \otimes \tau : (I \times J) \times (S \times T) \to S \times T$ is a mapping satisfying, for all $a \in I, b \in J, s \in S, t \in T$, $\delta \otimes \sigma((a, b), (s, t)) = (\delta(a, s), \sigma(b, t))$. For I, J disjoint, by $\mathbb{A} \cdot \mathbb{B}$, we denote the Cartesian composition of $\mathbb{A}$ and $\mathbb{B}$, i.e., the quasi-automaton $(I \cup J, S \times T, \delta \cdot \sigma)$, where $\delta \cdot \sigma : (I \cup J) \times (S \times T) \to S \times T$ is defined, for all $x \in I \cup J$, $s \in S$, and $t \in T$, by*

$$(\delta \cdot \sigma)(x, (s, t)) = \begin{cases} (\delta(x, s), t) & \text{if } x \in I, \\ (s, \sigma(x, t)) & \text{if } x \in J. \end{cases}$$

Generalizing the homogeneous or heterogeneous products of quasi-automata to the case of quasi-multiautomata is straightforward because, in these two cases, the condition used in Theorem 2 holds. However, in the case of the Cartesian composition, the situation is different. Since in the definition of the Cartesian composition the state set is created as the Cartesian product of the state set of the respective quasi-multiautomata, it is obvious from Figure 7 that the necessary condition $tn(r, t) = 1$ is not satisfied in the resulting quasi-multiautomaton, as there is no direct path from state (s_0, t_0) to state (s_1, t_1) because the respective input elements can affect one component only. For a deeper insight into this issue, we refer the reader to Example 1 in [17], the proof of Theorem 2 in [22], or Example 4 in [16], where the GMAC condition is not satisfied anywhere and we consider modified GMAC conditions, called E-GMAC.

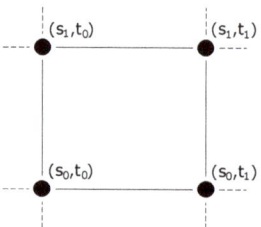

Figure 7. Cartesian product of state sets.

4. Nondeterministic Quasi-Multiautomata

First of all, we provide an example of a nondeterministic automaton.

Example 6. *Consider a nondeterministic finite automaton depicted in Figure 8. For input a applied to the state s_0 there is a transition to s_1, or the automaton can remain in the state s_0. As a result, the machine must "decide" how to behave. The same situation applies to the input b and the state s_1. Nondeterminism means ambiguity—from a given state, more transitions can lead to the same symbol. A nondeterministic automaton always chooses (sometimes we also say "guess") the transition that will lead to accepting the word—if that is possible. If we insert the word aba into the automaton in Figure 8, then it has the following options for dealing with it:*

$$s_0 \longrightarrow s_0 \longrightarrow s_0 \longrightarrow s_0$$
$$s_0 \longrightarrow s_1 \longrightarrow s_3 \longrightarrow s_3$$
$$s_0 \longrightarrow s_1 \longrightarrow s_2 \longrightarrow s_3$$

In the first case, the automaton will remain in the state of s_0; it certainly can because the rules of transition allow this. In the second case, the automaton moves to other states and eventually reaches s_3, which is the final one. The third case has the same final state. In the second and third cases, the automaton accepts the word. Notice that a nondeterministic automaton always automatically selects the branch in which it accepts a word, if such a branch exists.

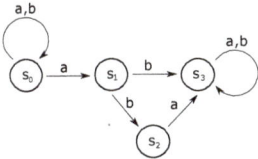

Figure 8. Nondeterministic automaton.

The following definition transfers the notion of nondeterministic automaton from the context of automata to the context of quasi-multiautomata.

Definition 12. *By nondeterministic quasi-multiautomaton (denoted by $_N$MA), we mean a triad $_N\mathrm{MA} = (\mathcal{C}(H), S, \delta)$, where $(H, *)$ is a semi-hypergroup, S is a nonempty set, and $\delta : \mathcal{C}(H) \times S \to \mathcal{P}(S)$, where $\mathcal{C}(H) \subseteq \mathcal{P}^*(H)$ is a transition function satisfying the following condition:*

$$\delta(B, \delta(A, s)) \subseteq \delta(A * B, s) \text{ for all } A, B \in \mathcal{C}(H), s \in S. \quad (2)$$

Notation 1. *We will call the condition (2) big-GMAC.*

In Figure 9, we can see the basic concepts of deterministic quasi-automata, i.e., automaton with the free monoid, automaton with a monoid, and quasi-multiautomaton.

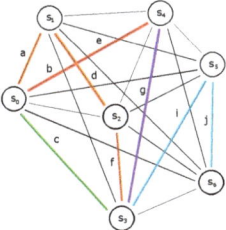

Figure 9. Deterministic automaton.

In the case of an *automaton with a free monoid*, if we apply the string adf to s_0 then $\delta(adf, s_0) = s_3$, and the string ij to s_3, we reach the state s_6. This is the same as applying $adfij$ directly to s_0.

In the case of an *automaton with a monoid*, if we apply the input (character) b to the state s_0, we reach the state s_4, where we apply g, which gets us to the state s_3. In order for the MAC condition to hold, there must exist an input (character) by which the automaton goes directly from s_0 to s_3. In our case, such an input (character) is c (on condition that $b \cdot g$ is defined as c).

The case of a *quasi-multiautomaton* is similar with the difference that there must be $c \in b * g$. Therefore, in Theorem 1, this is only a necessary condition, as it does not take into account the fact that $c \in b * g$.

While in nondeterministic finite automata of the formal language theory nondeterminism occurs in the transition function, i.e., after we apply one input we can reach multiple states (see Figure 8), nondeterministic quasi-multiautomata of Definition 12 provide nondeterminism for the input set and leave the transition function single-valued.

In Figure 10, we consider the *minimal extensive hyperoperation*, where, for all a, b, there is $a \circ b = \{a, b\}$. Using Definition 6, we regard the element of the potent set, which we—by means of the transition function—apply on a state. Thus, in Figure 10, we apply on s_0 the input in the form of the hypercomposition (hyperproduct) $e * f$, which takes us to two states: s_3 and s_6. Thus, we obtain a similar result to that after application of input a to s_0 in Figure 8.

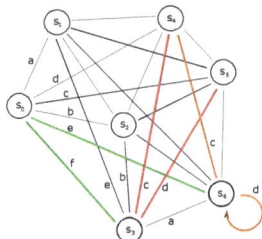

Figure 10. Nondeterministic quasi-multiautomaton.

Now, consider input $e * f = \{e, f\}$ applied on state s_0 in Figure 10. Application of condition (2) brings $_N$MA into two states: s_3 and s_6. If we apply $c * d = \{c, d\}$ on each of these two states, $_N$MA turns into states s_4, s_5, s_6, which are the results of the left-hand side of the GMAC condition. On the right-hand side of this condition, we first evaluate $\{e, f\} * \{c, d\} = \{e, f, c, d\}$, which takes us to the set of states $\{s_3, s_4, s_5, s_6\}$. Obviously, the set of states on the left-hand side is the subset of the set of states on the right-hand side of GMAC.

5. Quasi-Multiautomata with the Input Semi-Hypergroup Based on Concatenation

In this section, inspired by [23], we present the construction of a quasi-multiautomata, in which the input semi-hypergroup is based on the original concatenation operation, as is the case of the classical concept of automata. For this type of construction, the necessary condition of Theorem 2, $tn(r, s) = 1$ for all $r, s \in S$, is not required.

First, we recall the necessary concepts from the theory of formal languages. *String length* $|x|$ is the total number of symbols in the string x. A *substring* of a string is a sequence of symbols that is contained in the original string—i.e., if x and y are strings, then x is a substring of y if there exist strings z, z' such that $zxz' = y$. *Prefix* of the string a, denoted by $pref(a)$, is such a substring of the string a that there exists a substring z of a (which can be empty, however) such that $pref(a)z = a$. *Suffix* of the string b, denoted $suf(b)$, is such a substring of the string b that there exists a substring z of a (which can be empty, however) where $zsuf(a) = a$. The set of all prefixes of the string x will be denoted $S_{pref}(x)$; the empty word will be denoted by ε (see also notation used for binary trees in [24]).

Now, denote H^* as the set of all strings over the set of symbol H and define a hyperoperation $\star : H^* \times H^* \longrightarrow \mathcal{P}^*(H^*)$ by

$$x \star y = \left\{ ab \in H^* \mid a \in S_{pref}(x), b \in S_{pref}(y) \right\} \quad (3)$$

In other words, $x \star y$ is in fact a set of all mutual concatenations of prefixes of x and y.

Example 7. *Consider set $M = \{0, 1, 2\}$ and the set M^* of all strings over M. Further, consider strings $a, b \in M^*$, where $a = 1010$ and $b = 22$. For these, we have*

$$S_{pref}(a) = \{\varepsilon, 1, 10, 101, 1010\} \quad \text{and} \quad S_{pref}(b) = \{\varepsilon, 2, 22\}.$$

Thus, we obtain

$$a \star b = \{\varepsilon, 2, 22, 1, 12, 122, 10, 102, 1022, 101, 1012, 10122, 1010, 10102, 101022\}.$$

Theorem 4. *Let H^* by an arbitrary nonempty set of strings over H and let "$\star$" be a defined by (3). Then, $(H^*, \star)$ is a hypergoup.*

Proof. First, we show that the associative law applies. For all strings $a, b, c \in H^*$, we have

$$(a \star b) \star c = \bigcup_{\substack{x \in S_{pref}(a) \\ y \in S_{pref}(b)}} xy \star c = \bigcup_{\substack{x \in S_{pref}(a) \\ y \in S_{pref}(b) \\ z \in S_{pref}(c)}} xyz = a \star \bigcup_{\substack{y \in S_{pref}(b) \\ z \in S_{pref}(c)}} yz = a \star (b \star c).$$

The reproductive axiom holds automatically because "$\star$" is extensive, i.e., $a, b \in a \star b$ for all $a, b \in H^*$. Indeed, each set of prefixes contains an empty word and the original word, if we perform the concatenation operation of the empty string and the original string from the second set of prefixes, we obtain the original string. Thus, the structure $(H^*, \star)$ is a hypergoup. □

In the following two examples, Examples 8 and 10, we use the above hypergroup as the input sets for two quasi-multiautomata. We will consider two types of transition function. In the first case, it has the role of a "pointer", i.e., it points to the follower of s_0, which is the result of the transition $s_1 = \delta(a, s_0)$. In this case, the transition function is usually specified by a table or a transition diagram as in Figure 4 and there is no formula or rule to calculate the transition. In the second case, the transition function has the form of an "operation", i.e., we obtain the new state by means of calculation (as in Example 2).

Example 8. *Consider the hypergroup $(M^*, \star)$ from Example 7 and the set of states $T = \{a, b, c, d, \ldots, m\}$. The transition function δ_T is defined by means of the transition diagram in Figure 8. It is easy to verify that the structure $\mathbb{MA} = ((M^*, \star), T, \delta_T)$ is a multiautomaton satisfying the GMAC condition. In Figure 11, we use different colors to highlight the following: processing the input word 1010, i.e., $\delta_T(1010, a)$, (blue); the left-hand side of GMAC $\delta_T(22, \delta_T(1010, a)$ (blue and red); right-hand side of GMAC $\delta_T(1010 \star 20, a)$ (yellow).*

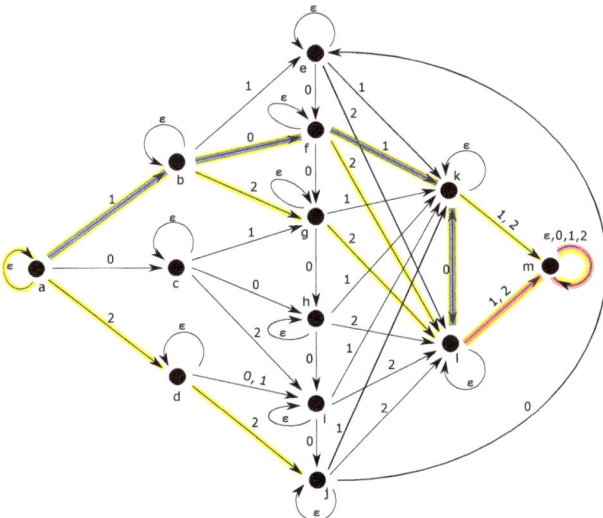

Figure 11. Quasi-multiautomaton based on a concatenation hypercomposition.

Theorem 5. *Let $(H^*, \star)$ be a hypergroup from Theorem 4 and S be a set of states. Then, it is possible to define a transition function δ such that $((H^*, \star), S, \delta)$ is a quasi-multiautomaton.*

Proof. Proof of the condition GMAC is obvious from the presented scheme in Figure 8 and from the definition of hyperoperation, where for two strings a and b, there is $ab \in a \star b$. Indeed, suppose that $a = a_1 \ldots a_n$, $b = b_1 \ldots b_n$. Then, the left-hand side of GMAC is

$$\delta(a, \delta(b, s)) = \delta(a_1 \ldots a_n, \delta(b_1 \ldots b_n, s)) = \delta(a_n, \delta(a_{n-1}, \ldots \delta(a_1, \delta(b_n, \delta(b_{n-1, \ldots \delta(b_1, s)}))))))$$

while on the right-hand side, we have

$$\delta(b \ast a, s) = \delta(b_1, s) \cup \delta(b_1 a_1, s) \cup \delta(b_1 a_2, s) \cup \ldots \cup \delta(b_1 b_2 a_1, s) \cup \delta(b_1 b_2 a_1 a_2, s) \cup \ldots \cup \delta(b_1 \ldots b_n a_1 \ldots a_n, s),$$

where the last term of the union is $\delta(a_n, \delta(a_{n-1}, \delta(a_1, \delta(b_n \ldots, \delta(b_1, s))))))$, which is the left-hand side of the GMAC condition. □

Of course, the transition function cannot be arbitrary.

Example 9. *Consider a quasi-multiautomaton with the same input hyprergroup $(M^*, \star)$. However, instead of the state set T, consider the set of all natural numbers $\mathbb{N}$. Next, define the transition function $\delta_O : M^* \times \mathbb{N} \longrightarrow \mathbb{N}$ by*

$$\delta_O(a, r) = a \cdot r$$

for all $a \in M^*$ and $r \in \mathbb{N}$. We can afford to define the transition function δ in such a way because we treat numeric strings (1010 and 2020 below) as numbers. The GMAC condition is not satisfied in this case. Indeed,

$$\delta_O(22, \delta_O(1010, 2)) = \delta_O(22, 2020) = 44440,$$

yet for the right-hand side of GMAC—i.e., $\delta_O(1010 \star 22, 2)$—we require the string 22220 to belong to $1010 \star 22$. Yet, we could see in Example 7 that $22220 \notin 1010 \star 22$.

Next, we will use the construction of a multiautomaton of Theorem 5 and construct a nondeterministic quasi-multiautomaton of Definition 12. There, the element of the power set will be used as the input word, which we will obtain as a result of two elements (strings) $a, b \in H^*$. In this context, on the right-side of the GMAC condition, the hypercomposition of two sets will be required. It is therefore desirable to first prove the following lemma.

Lemma 1. *In the hypergroup $(H^*, \star)$, where "$\star$" is defined by (3), there is $S_{pref}(a \star b) = a \star b$ for all $a, b \in H^*$.*

Proof. Obviously, there is $S_{pref}(x \star y) = \bigcup\limits_{z \in x \star y} S_{pref}(z)$. Moreover, it is obvious that $x \in a \star b$ implies that $x \in S_{pref}(a \star b)$. Proving the other inclusion is also simple. Indeed, the fact that $x \in S_{pref}(a \star b) = \bigcup\limits_{c \in a \star b} S_{pref}(c)$ implies that there exist words $y, z \in H^*$ such that $x = yz$, where $y \in S_{pref}(a)$, $z \in S_{pref}(b)$. Yet, this means that $x \in a \star b$. □

Example 10. *Consider the quasi-multiautomaton $\mathbb{MA} = ((M^*, \star), T, \delta_T)$ from Example 8 and sets $A = a \star b$ and $C = c \star d$, where $a, b, c, d \in M^*$. For $a = 1010, b = 22, c = 1, d = \varepsilon$, we have $A = \{\varepsilon, 2, 22, 1, 12, 122, 10, 102, 1022, 101, 1012, 10122, 1010, 10102, 101022\}$ (see Example 7) and $B = \{\varepsilon, 1\}$. Proving that $\mathbb{MA}$ is a nondeterministic quasi-multiautomaton is rather difficult because one needs to show validity of big-GMAC (2) for all states and inputs. However, we outline the idea of the proof for our specific choice of states and inputs.*

We need to show that there is $\delta(B, \delta(A, a)) \subseteq \delta(A \star B, a)$. From the transition diagram, we calculate the left-hand side of big-GMAC (2):

$$\delta(B, \delta(A, a)) = \delta(B, \{a, b, d, f, g, j, k, l, m\}) = \{a, b, d, e, f, g, i, j, k, l, m\} \quad (4)$$

Before calculating the right-hand side, we first establish $A \star B$. This is quite easy (given the specific choice of the set B and Lemma 1):

$A \star B = \{\varepsilon, 1, 2, 10, 12, 22, 101, 102, 122, 1010, 1012, 1022, 10102, 10122, 101022\} \cup$
$\{\varepsilon, 11, 21, 101, 121, 221, 1011, 1021, 1221, 10101, 10121, 10221, 101021, 101221, 1010221\} =$
$\{\varepsilon, 1, 2, 10, 11, 12, 21, 22, 101, 102, 121, 122, 221, 1010, 1011, 1012, 1021, 1022, 1221, 10101,$
$10102, 10121, 10221, 101021, 101221, 101022, 1010221\}$

Now, again using the transition diagram, we compute

$$\delta(A \star B, a)) = \{a, b, d, e, f, g, i, j, k, l, m\}, \quad (5)$$

and we can see that $\delta(B, \delta(A, a)) \subseteq \delta(A \star B, a)$; in this case, even $\delta(B, \delta(A, a)) = \delta(A \star B, a)$.

Lemma 2. *A set $H \subseteq H^*$ is reflexive in a hypergroup $(H^*, \star)$.*

Proof. Reflexivity of a subset H of H^*, where $(H^*, *)$ is a hypergroupoid, is defined by validity of implication $x \star y \cap A \neq \emptyset \Rightarrow y \star x \cap A \neq \emptyset$ for all $x, y \in H^*$.
Suppose that $x = a_1 \ldots a_n$ and $y = b_1 \ldots b_m$, where $a_i, b_j \in H$ for all $i \in \{1, \ldots, n\}$ and $j \in \{1, \ldots, m\}$. Obviously, $a_1 \in S_{pref}(x)$ and $b_1 \in S_{pref}(y)$. Next, thanks to the fact that $\varepsilon \in H^*$, there is $S_{pref}(x) \subseteq x \star y$ and $S_{pref}(y) \subseteq x \star y$. Even though "$\star$" is not commutative, there is $S_{pref}(x) \subseteq y \star x$ and $S_{pref}(y) \subseteq y \star x$. Thus, we have the two-element sets $\{a_1, b_1\} = x \star y \cap H$ and $\{a_1, b_1\} = y \star x \cap H$. □

In the end of this section, we are going to discuss nondeterminism, which is caused by the input structure as stated in Definition 12. In order to do so, we are going to use the hypergroup constructed using Theorem 4. We want to show that for such a structure, there exists a nondeterminism that is "controlled" due to the nature of the hyperoperation. (Recall that in the theory of formal languages, nondeterminism is caused by the transition function.) In order to do so, we are going to use the hypergroup constructed using Theorem 4. The following example shall thus be read within the context of Definition 12 and Theorem 4.

Example 11. *Regard a string $a = a_1 \ldots a_n \in H^*$. The transition function produces*

$$\delta(a, r) = \delta(a_1 \ldots a_n, r) = \delta(a_n, \delta(a_{n-1} \ldots \delta(a_1, r))). \quad (6)$$

If we now regard a nondeterministic quasi-multiautomaton, where the nondeterminism is provided in the input by hyperoperation (3), the nondeterminism is "controlled" because from each state in the sequence followed by the automaton, there are at most two paths. Indeed, for $a = a_1 \ldots a_n$ and $b = b_1 \ldots b_n$, their hypercomposition is a set of strings, which are concatenations of prefixes of a and b. Thus e.g., at the second position (which of course exists), the symbol a_1 is followed by a_2 or b_1. As Figure 12 suggests, this idea holds for all positions.

Figure 12 shows that the first position of an arbitrary string from $a * b$ (which can be regarded as input) will be occupied by a_1 or b_1, the second position by a_2 or b_1, etc. Thus, given the input $a * b$, the quasi-multiautomaton will pass at most $2n$ paths (where $a_1 b_1$ is included in $a_1 b_1 b_2$, i.e., these two are counted as one path).

Showing that the big-GMAC condition holds for all strings $a \in H^*$ such as in Figure 10, where the transition function is given by a diagram (or by a table yet not a rule) is complicated. There exists Light's associativity test invented by F. W. Light for testing whether a binary operation defined on a finite set is associative. Miyakawa, Rosenberg, and Tatsumi [25] generalized this test for semi-hypergroups. We are not aware of any such test for finite quasi-multiautomata with a transition diagram or table. Finding such tests might be our next objective and the subject of further research.

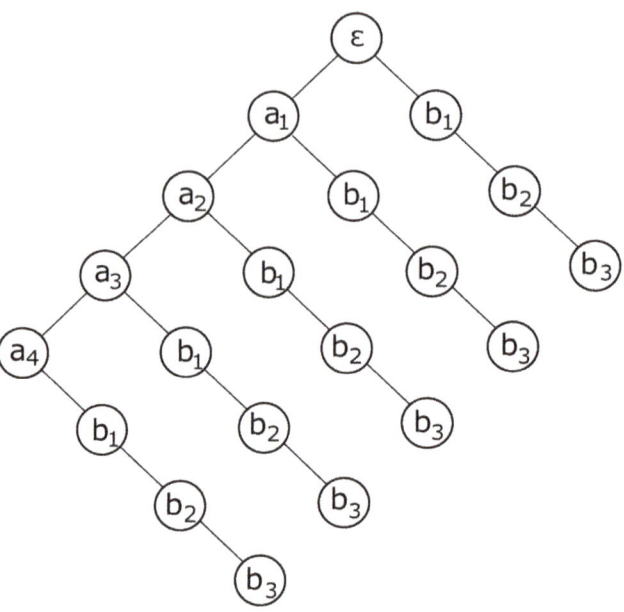

Figure 12. Scheme of possible concatenations of $a = a_1a_2a_3a_4$ and $b = b_1b_2b_3$.

6. Discussion

Currently, the combinations of algebraic multiautomata into higher entities, using various rules suggested by Dörfler [20,26], are studied—see, e.g., [27]. Such combinations seem to be suitable tools for modeling various real-life systems—see, e.g., [16,21,28]—or are even tools to control such systems [22]. However, two main problems appeared in this respect:

1. When constructing algebraic quasi-multiautomata of Definition 5, one needs to show that the GMAC condition (1) is satisfied. This in fact means that one has to prove that if two arbitrary inputs are applied sequentially to a certain state s_1, we obtain a state that is contained in the subset of states given by the application potential determined by the hyperoperation of the inputs and the state s_1. In this respect, the proof of validity of the GMAC condition is not always straightforward. In [17], the proof is computationally demanding. If we consider various combinations of such quasi-multiautomata, *the proofs become even more complicated*. For these reasons, in Section 3, we search for conditions that could facilitate such proofs. Moreover, we show why, in some previous cases, the GMAC condition of the composition failed even though each separate quasi-multiautomata fulfilled it. This approach constitutes a sufficiently solid base for further research and the identification of conditions equivalent to GMAC. Note that the way transition number of states (see Definition 8), or even the whole concept of quasi-multiautomaton, is understood is similar to considerations of [29], where, in oriented graphs, one of the vertices is considered as the initial state while selected vertices are final states and the path from state q_0 to state q_f (denoted $q_0 \to^{a^m} q_f$) is simply a sequence of edges in the transition graph without any specific structure.

2. *The abovementioned generalizations of the original concept of automaton have deviated from the original idea of concatenation of input symbols.* In Definition 5, H is a semi-hypergroup, which is called input alphabet. However, under this consideration, this "alphabet" cannot create words. Indeed, if a, b are elements of H, i.e., letters, then $a * b$ is a subset of H; so, it is a set of letters, not words. This might be seen as a weak point of the theory, which diminishes its applicability. Therefore, in Section 4 and especially in Section 5,

we construct quasi-multiautomata based on standard techniques of the theory of formal languages. In Section 4, we modify Definition 5 in such a way that quasi-multiautomata will work nondeterministically. However, being aware of the fact that nondeterministic automata are of no real added value, our concept is designed to include a limited degree of nondeterminism only. Moreover, since this paper deals with the generalization of the automaton, the quasi-multiautomaton can be further generalized by considering H to be an arbitrary hypergroupoid. This enhances possibilities to create input hypercompositional structures reflecting needs of automata of the theory of formal languages. In other words, weakening requirements of the input structure provides us a wider range of choices to construct quasi-multiautomata based on concatenation. For example, consider that $(H, *)$ is a hypergroupoid if $x * y = \{z, w\}$, where z is formed by deleting the odd-positioned letters from the word xy and w is formed by deleting the even-positioned letters from the word xy. However, one needs to discuss the impact of losing associativity on GMAC, which is based on it. In Section 5, we show a construction of quasi-multiautomata, which corresponds to automata of the theory of formal languages and is based on the idea of concatenation of strings with associativity preserved. For quasi-automata, this is possible thanks to the free monoid. For quasi-multiautomata, i.e., structures making use of hypercompositional structures, we concatenate words for input. We present a specific example. However, thanks to the multivalued nature of the hypercomposition, a whole range of similar schemes might be thought of.

7. Conclusions

In our paper, we defined conditions that the GMAC condition must satisfy. We also constructed algebraic a quasi-multiautomaton related to automata of the theory of formal languages. On top of that, we constructed a quasi-multiautomaton with a limited degree of nondeterminism. In contrast to nondeterministic automata of the theory of formal languages, where nondeterminism is caused by the transition function, the nondeterminism in our construction follows from the input hyperstructure. It is properties of the input hyperstructure that have the potential to yield interesting results in such nondeterministic quasi-multiautomata, which suggest a potential line of further research.

Author Contributions: Investigation, Š.K.; writing—original draft preparation, Š.K., M.N., J.V.; writing—review and editing, M.N. All authors have read and agreed to the published version of the manuscript.

Funding: The first author was supported by the Ministry of Transport within the program of long-term conceptual development of research organizations. The third author was supported by the FEKT-S-20-6225 grant of Brno University of Technology.

Institutional Review Board Statement: Not applicable.

Informed Consent Statement: Not applicable.

Data Availability Statement: Not applicable.

Acknowledgments: The authors would like to express their thanks to Christos Massouros for providing context and background for their thoughts.

Conflicts of Interest: The authors declare no conflict of interest. The funders had no role in the design of the study; in the collection, analyses, or interpretation of data; in the writing of the manuscript, or in the decision to publish the results.

References

1. Massouros, G. Hypercompositional structures in the theory of languages and automata. *An. Şt. Univ. Al Çuza Iaşi Sect. Inform.* **1994**, *III*, 65–73.
2. Taş, M.K.; Kaya, K.; Yenigün, H. Synchronizing billion-scale automata. *Inf. Sci.* **2021**, *574*, 162–175. [CrossRef]
3. Jing, M.; Yang, Y.; Lu, N.; Shi, W.; Yu, C. Postfix automata. *Theor. Comput. Sci.* **2015**, *562*, 590–605. [CrossRef]
4. Bavel, Z. The source as tool in automata. *Inform. Control* **1971**, *18*, 14–155. [CrossRef]

5. Chvalina, J. *Functional Graphs, Quasi-Ordered Sets and Commutative Hypergroups*; Masaryk University: Brno, Czech Republic, 1995.
6. Gécseg, F.; Peák, I. *Algebraic Theory of Automata*; Akadémia Kiadó: Budapest, Hungary, 1972.
7. Ginzburg, A. *Algebraic Theory of Automata*; Academic Press: New York, NY, USA, 1968.
8. Atani, S.E.; Bazari, M.S.S. Lattice structures of automata. *Commun. Fac. Sci. Univ. Ank.—Ser. A1 Math. Stat.* **2020**, *69*, 1133–1145. [CrossRef]
9. Massouros, C.; Massouros, G. An Overview of the Foundations of the Hypergroup Theory. *Mathematics* **2021**, *9*, 1014. [CrossRef]
10. Chvalina, J.; Chvalinová, L. State hypergroups of automata. *Acta Math. Inform. Univ. Ostrav.* **1996**, *4*, 105–120.
11. Ashrafi, A.R.; Madanshekaf, A. Generalized action of a hypergroup on a set. *Ital. J. Pure Appl. Math.* **1998**, *15*, 127–135.
12. Hošková, Š.; Chvalina, J. Discrete transformation hypergroups and transformation hypergroups with phase tolerance space. *Discret. Math.* **2008**, *308*, 4133–4143. [CrossRef]
13. Chvalina, J.; Hošková-Mayerová Š.; Dehghan Nezhad, A. General actions of hyperstructures and some applications. *An. Şt. Univ. Ovidius Constanţa* **2013**, *21*, 59–82. [CrossRef]
14. Chvalina, J. Infinite multiautomata with phase hypergroups of various operators. In Proceedings of the 10th International Congress on Algebraic Hyperstructures and Applications, Brno, Czech Republic, 3–9 September 2008; Hošková, Š., Ed.; University of Defense: Brno, Czech Republic, 2009; pp. 57–69.
15. Chvalina, J.; Moučka, J.; Vémolová, R. Funktoriální přechod od kvaziautomatů k multiautomatům (En Functorial passage from quasiautomata to multiautomata). In *XXIV International Colloqium on the Acquisition Process Management Proceedings*; Univerzita Obrany: Brno, Czech Republic, 2006.
16. Novák, M.; Křehlík, Š.; Staněk, D. n-ary Cartesian composition of automata. *Soft Comput.* **2020**, *24*, 1837–1849. [CrossRef]
17. Chvalina, J.; Křehlík, Š.; Novák, M. Cartesian composition and the problem of generalising the MAC condition to quasi-multiautomata. *An. Şt. Univ. Ovidius Constanţa* **2016**, *24*, 79–100.
18. Janoušek, J. String Suffix Automata and Subtree Pushdown Automata. In Proceedings of the Prague Stringology Conference 2009, 4th Prague Stringology Conference (PSC), Prague, Czech Republic, 31 August–2 September 2009; Department of Computer Science & Engineering, Czech Technical University: Prague, Czech Republic, 2009; pp. 160–172.
19. Rabin, M.; Scott, D. Finite Automata and Their Decision Problems. *IBM J. Res. Dev.* **1959**, *3*, 115–125. [CrossRef]
20. Dörfler, W. The cartesian composition of automata. *Math. Syst. Theory* **1978**, *11*, 239–257. [CrossRef]
21. Křehlík, Š.; Novák, M. Modified Product of Automata as a Better Tool for Description of Real-Life Systems. In Proceedings of the International Conference on Numerical Analzsis and Applied Mathematics ICNAAM 2019, Rhodes, Greece, 23–28 September 2019; AIP Conference Proceedings; AIP Publishing: Melville, NJ, USA 2020.
22. Křehlík, Š. n-Ary Cartesian Composition of Multiautomata with Internal Link for Autonomous Control of Lane Shifting. *Mathematics* **2020**, *8*, 835. [CrossRef]
23. Mousavi, S.; Jafarpour, M. On Free and Weak Free (Semi)Hypergroups. *Algebra Colloq.* **2011**, *18*, 873–880. [CrossRef]
24. Brough, T.; Cain, A.J. Automaton semigroups constructions. *Semigr. Forum* **2015**, *90*, 763–774. [CrossRef]
25. Miyakawa, M.; Rosenberg, I.G.; Tatsumi, H. Associativity Test in Hypergroupoids. In Proceedings of the 36th International Symposium on Multiple-Valued Logic, Singapore, 17–20 May 2006.
26. Dörfler, W. The direct product of automata and quasi-automata. In *Mathematical Foundations of Computer Science, Proceedings of the 5th Symposium, Gdansk, Poland, 6–10 September 1976*; Mazurkiewicz, A., Ed.; Springer: Berlin/Heidelberg, Germany, 1976.
27. Dutta, M.; Kalita, S.; Saikia, H.K. Cartesian product of automata. *Adv. Math. Sci. J.* **2020**, *9*, 7915–7924. [CrossRef]
28. Novák, M.; Křehlík, Š.; Ovaliadis, K. Elements of hyperstructure theory in UWSN design and data aggregation. *Symmetry* **2019**, *11*, 734. [CrossRef]
29. Pighizzini, G. Investigations on Automata and Languages Over a Unary Alphabet. *Int. J. Found. Comput. Sci.* **2015**, *26*, 827–850. [CrossRef]

Article

Knots and Knot-Hyperpaths in Hypergraphs

Saifur Rahman [1], Maitrayee Chowdhury [1], Firos A. [2] and Irina Cristea [3,*]

[1] Department of Mathematics, Rajiv Gandhi University, Rono Hills, Itanagar 791112, India; saifur.rahman@rgu.ac.in (S.R.); maitrayee.chowdhury@rgu.ac.in (M.C.)
[2] Department of Computer Science and Engineering, Rajiv Gandhi University, Rono Hills, Itanagar 791112, India; firos.a@rgu.ac.in
[3] Centre for Information Technologies and Applied Mathematics, University of Nova Gorica, 5000 Nova Gorica, Slovenia
* Correspondence: irina.cristea@ung.si or irinacri@yahoo.co.uk; Tel.: +386-0533-15-395

Abstract: This paper deals with some theoretical aspects of hypergraphs related to hyperpaths and hypertrees. In ordinary graph theory, the intersecting or adjacent edges contain exactly one vertex; however, in the case of hypergraph theory, the adjacent or intersecting hyperedges may contain more than one vertex. This fact leads to the intuitive notion of knots, i.e., a collection of explicit vertices. The key idea of this manuscript lies in the introduction of the concept of the knot, which is a subset of the intersection of some intersecting hyperedges. We define knot-hyperpaths and equivalent knot-hyperpaths and study their relationships with the algebraic space continuity and the pseudo-open character of maps. Moreover, we establish a sufficient condition under which a hypergraph is a hypertree, without using the concept of the host graph.

Keywords: hypergraph; hyperpath; hypertree; knot; hypercontinuity; equivalent hyperpaths

Citation: Rahman, S.; Chowdhury, M.; A., F.; Cristea, I. Knots and Knot-Hyperpaths in Hypergraphs. *Mathematics* **2022**, *10*, 424. https://doi.org/10.3390/math10030424

Academic Editor: Mikhail Goubko

Received: 21 December 2021
Accepted: 27 January 2022
Published: 28 January 2022

Publisher's Note: MDPI stays neutral with regard to jurisdictional claims in published maps and institutional affiliations.

Copyright: © 2022 by the authors. Licensee MDPI, Basel, Switzerland. This article is an open access article distributed under the terms and conditions of the Creative Commons Attribution (CC BY) license (https://creativecommons.org/licenses/by/4.0/).

1. Introduction

Being a generalization of graphs and yet having its own unique complexity and utility, hypergraph theory has emerged as a completely new dynamic research area. The fundamental concepts of path, tree, trail, cycle and their different well-known properties have already found plenty of applications in real-world problems in networking systems [1,2] of different types or in the field of bioinformatics [3–5]. The concept of the *hyperpath*, called also the path (both terms being used in a synonymous way), in a hypergraph represents the foundation of many research works. In the majority of these studies, the hypergraphs are considered to be directed, though there are papers related to paths in the case of undirected hypergraphs as well. Nguyen and Pallottino [6], in their work based on directed hypergraphs, have given some efficient algorithms in connection to some shortest path properties. In the same direction, we recall the work of Nielsen, Andersen and Pretolani [7], where the authors present the procedures for finding the K-shortest hyperpaths in a directed hypergraph. It is worth underlining that the area of research related to hyperpaths, shortest hyperpaths [6] and their links with vehicle navigation [1], network systems based on transit schedules [2], cellular networks [3], etc., is flourishing.

In this paper, we deal with two different problems related to hypergraphs. One concerns the behavior of hyperpaths under hyper-continuous mappings and pseudo-open mappings, while the other one is related to hyperpaths and hypertrees. Our study was motivated by the definition of the so-called *algebraic space* [8], introduced as a pair (X, S_X), where X is a non-empty arbitrary set and $S_X \subseteq \mathcal{P}(X)$ a non-empty family of subsets of X. An algebraic space can be seen as an extended version of a topological space but without having any closure property with respect to union or intersection, and it recalls the definition of the hypergraph to a great extent. As a result, the concept of pseudo-map or pseudo-continuity could be then defined between two hypergraphs. The key element of this parallel study is the new concept of the *knot*, which is a subset of hyperedge intersection vertices. Since, in a hypergraph, the hyperedges appear as some subsets of the vertex set, it is trivial to note that the intersections of all possible adjacent hyperedges may contain more than one vertex. This fact leads to the intuitive notion of the knot that is the collection of explicit vertices. This notion further changes the dimension of perceiving the different concepts of hypergraphs such as walk, trail, path, tree, etc., where each of the adjacent hyperedge intersections gives rise to knots.

In graph theory, another important concept is that of the tree, which has been extensively used in networking, especially in theoretical computer science [9]. A graph G is a tree if there exists a unique path between any two vertices. Recall that the concept of the hypertree was introduced in hypergraph theory in terms of its host graph, as the hypergraph that admits a host graph that is a tree [10]. We emphasize that this fundamental characterization of trees is not generalized in hypergraph theory, in the sense that there is no characterization of hypertrees merely in terms of hyperpaths. This motivated us to present, in the second part of the paper, a characteristic of hypertrees in terms of hyperpaths, without using the concept of the host graph.

The structure of this work can be summarized as follows. First, in Section 2, we introduce the new concepts of point-hyperwalk, point-hypertrail and point-hyperpath, showing their differences in one illustrative example. Next, the key concepts of the knot and knot-hyperpath are defined. In Section 3, the notions of the hyper-continuous map, strictly hyper-continuous map and pseudo-open map between two hypergraphs are introduced and the behavior of point-hyperpaths and knot-hyperpaths under these notions is observed. In particular, we prove that the image of a point-hyperpath under an injective pseudo-open mapping is a point-hyperpath, while the image of a knot-hyperpath under a pseudo-open map is again a knot-hyperpath. Regarding the inverse image, we show that the inverse image of a knot-hyperpath under a surjective hyper-continuous map is a weak knot-hyperpath, or a knot-hyperpath if the map is surjective and strictly hyper-continuous. Section 4 is dedicated to the study of hypertrees. Based on the concept of equivalent entire knot-hyperpaths, we establish a sufficient condition under which a hypergraph becomes a hypertree. Moreover, we present an algorithm that extracts a host graph from a hypertree. A concluding section ends our study.

2. Preliminaries

Many definitions of hypergraphs exist; here, we will adopt the original one, given by Berge [11]. A *hypergraph* is a couple $H = (V, E)$ defined by a finite set of *vertices* (called also *nodes*) $V = \{v_1, \ldots, v_n\}$, with $n \in \mathbb{N}$, and the set $E = \{E_i\}_{i \in \mathbb{N}}$ of non-empty subsets of V, called *hyperedges*. Two hyperedges $E_j, E_k \in E$, with $j \neq k$, such that $E_j = E_k$ are called *repeated hyperedges* [12]. In this paper, all hypergraphs are considered to be with no repeated hyperedges.

Definition 1 ([13])**.** *Let $H = (V, E)$ be a hypergraph. By a hyperpath between two distinct vertices v_1 and v_k in V, we mean a sequence $v_1 E_1 v_2 E_2 \ldots v_{k-1} E_{k-1} v_k$ of vertices and hyperedges having the following properties:*

(i) *k is a positive integer;*
(ii) *$v_1, v_2, \ldots, v_k$ are distinct vertices;*
(iii) *$E_1, E_2, \ldots, E_{k-1}$ are hyperedges (not necessarily distinct);*
(iv) *$v_j, v_{j+1} \in E_j$ for $j = 1, 2, \ldots, k-1$.*

We call this sequence a $v_1 - v_k$–hyperpath.

Definition 2 ([13])**.** *A hypercycle in a hypergraph $H = (V, E)$ on a vertex v_1 is a sequence $v_1 E_1 v_2 E_2 \ldots v_{k-1} E_{k-1} v_k E_k v_1$, having the following properties:*

(i) *k is a positive integer ≥ 3;*
(ii) *$v_1 E_1 v_2 E_2 \ldots v_{k-1} E_{k-1} v_k$ is a $v_1 - v_k$–hyperpath;*
(iii) *at least one of the hyperedges $E_1, E_2, \ldots, E_{k-1}$ is distinct from E_k;*
(iv) *$v_j, v_{j+1} \in E_j$ for $j = 1, \ldots, k-1$.*

It is important to note that a path in a graph does not contain repeated edges, while this property is not retained in the definition of a hyperpath in a hypergraph as it appears in Definition 1. Since, in some cases, it is necessary to distinguish this special case; we define the following types of hyperpaths.

Definition 3. *A point-hyperwalk in a hypergraph $H = (V, E)$ is a hyperpath as defined in Definition 1, where the vertices may be repeated. A point-hyperwalk where no hyperedge is repeated (but vertices may be repeated) is called a point-hypertrail. A point-hyperpath is a point-hypertrail in which vertices are not repeated.*

In other words, a point-hyperpath is a point-hyperwalk where neither the edges nor the vertices are repeated.

The above definitions are illustrated in the following example.

Example 1. *Let $H = (V, E)$ be a hypergraph with the vertex set $V = \{v_i | i = 1, 2, \ldots, 50\}$ and hyperedges $E = \{E_1, E_2, \ldots, E_{10}\}$ such that*
$E_1 = \{v_1, v_2, v_3, v_4, v_5, v_{47}, v_9, v_{10}\}$,
$E_2 = \{v_{12}, v_{11}, v_{15}, v_9, v_{10}, v_8, v_6, v_{16}, v_7\}$,
$E_3 = \{v_{11}, v_{12}, v_{15}, v_{13}, v_{46}, v_{14}, v_{30}\}$,
$E_4 = \{v_{14}, v_{30}, v_{31}, v_{34}, v_{33}\}$,
$E_5 = \{v_{14}, v_{30}, v_{20}, v_{32}, v_{22}, v_{21}\}$,
$E_6 = \{v_{21}, v_{22}, v_{44}, v_{48}, v_{49}, v_{50}, v_{43}\}$,
$E_7 = \{v_{50}, v_{43}, v_{41}, v_{42}, v_{36}, v_{45}, v_{37}\}$,
$E_8 = \{v_{36}, v_{45}, v_{37}, v_{46}, v_{40}, v_{27}, v_{29}\}$,
$E_9 = \{v_{28}, v_{23}, v_{24}, , v_{27}, v_{29}, v_{25}, v_{26}\}$,
$E_{10} = \{v_{16}, v_7, v_{18}, v_{17}, v_{28}, v_{23}, v_{24}\}$.
We represent this hypergraph in Figure 1.

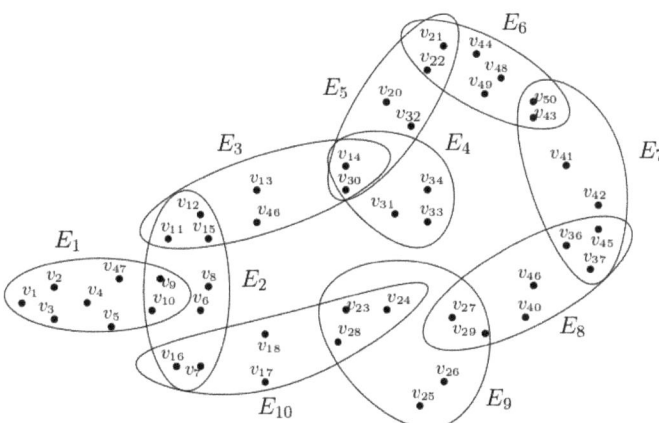

Figure 1. Hypergraph explaining point-hyperwalk, point-hypertrail and point-hyperpath notions.

We notice that
- $P \equiv v_1 E_1 v_9 E_2 v_{12} E_3 v_{14} E_5 v_{21} E_6 v_{50} E_7 v_{37} E_8 v_{20} E_9 v_{28} E_{10} v_7 E_2 v_{12}$ *is a point-hyperwalk, but not a point-hypertrail, because the hyperedge E_2 is repeated.*
- $P \equiv v_{12} E_3 v_{30} E_5 v_{22} E_6 v_{43} E_7 v_{37} E_8 v_{29} E_9 v_{28} E_{10} v_7 E_2 v_{12}$ *is a point-hypertrail. Here, the vertex v_{12} is repeated, but there is no repetition of the hyperedges.*
- $P \equiv v_{16} E_{10} v_{28} E_9 v_{27} E_8 v_{36} E_7 v_{50} E_6 v_{22}$ *is a point-hyperpath, since hyperedges and vertices are not repeated.*

Suppose that $H = (V, E)$ and $H' = (V', E')$ are two hypergraphs. Let $f : V \to V'$ be a mapping and let $P \equiv v_1 E_1 v_2 E_2 \ldots v_{k-1} E_{k-1} v_k$ denote an alternating sequence of vertices and edges in the hypergraph H. Then, we denote the *f-image* of this sequence as $f(P) \equiv f(v_1) f(E_1) f(v_2) f(E_2) \ldots f(v_{k-1}) f(E_{k-1}) f(v_k)$, where $f(E_i), i = 1, 2, \ldots, k$ is the *f*-image of $E_i, i = 1, 2, \ldots, k$, respectively.

Generalizing the notions in Definition 1, we are ready to introduce the concepts of the knot and knot-hyperpath, where the vertices are replaced by a cluster of vertices, each of them behaving in a significant manner.

Definition 4. *A knot K in a hypergraph $H = (V, E)$ is a non-empty subset of the intersections of some intersecting hyperedges. In other words, if $H = (V, E)$ is a hypergraph and K is a knot, then $K(\neq \emptyset) \subseteq \cap E_i$ for some intersecting hyperedges $E_i, i = 1, 2, \ldots, k$ and $k \geq 2$. In particular, if $K = \cap E_i$, then K is called an entire knot.*

Definition 5. *A knot-hyperpath in a hypergraph $H = (V, E)$ between two vertices v_1 and v_n is an alternating sequence of knots and hyperedges of the following type:*

$$\{v_1\} E_1 K_1 E_2 K_2 E_3 \ldots E_{n-1} K_{n-1} E_n \{v_n\}, \tag{1}$$

where $K_i \subseteq (E_i \cap E_{i+1}) \setminus (\cup_{t=1}^{i-1} K_t)$, with $i = 1, \ldots, n-1$, $v_1 \in E_1, v_n \in E_n$ and E_is are distinct hyperedges.

If $K_i = E_i \cap E_{i+1}$ for all $i = 1, 2, \ldots, n-1$, then the knot-hyperpath is called the entire knot-hyperpath.

Although the entire knot-hyperpath is a particular case of the knot-hyperpath, its significance can be seen in Section 4.

From the constructions of knots, it is clear that knots are mutually disjointed. Here, n is called the *length of the knot-hyperpath*.

Example 2. *By taking the hypergraph defined in Example 1, we can observe that*

$$\{v_4\} E_1 \{v_9, v_{10}\} E_2 \{v_{11}, v_{12}, v_{15}\} E_3 \{v_{14}, v_{30}\} E_5 \{v_{22}\} E_6 \{v_{50}\}$$

is a knot-hyperpath of length 5.

Definition 6. *Two knot-hyperpaths*

$$P_1 \equiv \{v_1\} E_1 K_1 E_2 K_2 E_3 \ldots E_{n-1} K_{n-1} E_n \{v_n\}$$

and

$$P_2 \equiv \{v_1\} E'_1 K'_1 E'_2 K'_2 E'_3 \ldots E'_{n-1} K'_{n-1} E'_n \{v_n\}$$

of the same length of a hypergraph $H = (V, E)$ are called equivalent or isomorphic if
(i) $E_i \cap E'_i \neq \emptyset$,
(ii) $K_i \cap K'_i \neq \emptyset$ for all $i = 1, 2, \ldots, n-1$.

The above definition further can be generalized to a finite number of knot-hyperpaths (entire knot-hyperpaths) $P_1, P_2, \ldots, P_k$, where $k \geq 2$ and the intersections in items (i) and (ii) are taken as follows:
(i) $\cap_{j=1}^{k} E_i^j \neq \emptyset$
(ii) $\cap_{j=1}^{k} K_i^j \neq \emptyset$ for all $i = 1, 2, \ldots, n-1$.

Example 3. *Consider the hypergraph H, with the vertex set*

$$V = \{v_1, v_2, v_3, v_4, v_5, v_6, v_7, v_8, v_9, v_{10}, v_{11}, v_{12}, v_{13}\}$$

and the hyperedges $E_1 = \{v_1, v_2, v_3, v_4\}$, $E_2 = \{v_3, v_4, v_5, v_7, v_9\}$, $E_3 = \{v_2, v_3, v_4, v_7, v_8, v_9\}$, $E_4 = \{v_8, v_9, v_{10}, v_{11}, v_{12}\}$, $E_5 = \{v_{11}, v_{12}, v_{13}\}$.
It can be easily verified that the following two knot-hyperpaths

$$P_1 \equiv \{v_1\} E_1 \{v_2, v_3, v_4\} E_3 \{v_8, v_9\} E_4 \{v_{11}, v_{12}\} E_5 \{v_{13}\}$$

and

$$P_2 \equiv \{v_1\} E_1 \{v_3, v_4\} E_2 \{v_9\} E_4 \{v_{11}, v_{12}\} E_5 \{v_{13}\}$$

are equivalent. We notice also that P_1 and P_2 are entire knot-hyperpaths, while

$$P_1' \equiv \{v_1\} E_1 \{v_2, v_3, v_4\} E_3 \{v_8, v_9\} E_4 \{v_{11}\} E_5 \{v_{13}\}$$

and

$$P_2' \equiv \{v_1\} E_1 \{v_3, v_4\} E_2 \{v_9\} E_4 \{v_{12}\} E_5 \{v_{13}\}$$

are not equivalent because the last two knots of the knot-hyperpaths P_1' and P_2' have empty intersections.

Definition 7 ([8]). *A mapping $f : V \to V'$ from the vertex set of a hypergraph $H = (V, E)$ to the vertex set of another hypergraph $K = (V', E')$ is said to be pseudo-open (in short, ps-open) if, for each hyperedge E_i in E, the corresponding image $f(E_i)$ is a hyperedge in E'.*

Example 4. Let $H = (V, E)$ and $K = (V', E')$ be two hypergraphs with the vertex sets $V = \{v_1, v_2, v_3, v_4, v_5, v_6\}$ and $V' = \{v_1', v_2', v_3', v_4', v_5'\}$ and the hyperedge sets $E = \{\{v_1, v_2\}, \{v_2, v_3, v_4\}, \{v_3, v_4, v_5\}\}$, $E' = \{\{v_1'\}, \{v_2', v_5'\}, \{v_1', v_2', v_5'\}\}$, respectively. Define the map $f : V \to V'$ such that $f(v_1) = v_1' = f(v_2)$, $f(v_3) = v_2' = f(v_5)$, $f(v_4) = v_5'$, $f(v_6) = v_3'$. Then, $f(\{v_1, v_2\}) = \{v_1'\}$, $f(\{v_2, v_3, v_4\}) = \{v_1', v_2', v_5'\}$, $f(\{v_3, v_4, v_5\}) = \{v_2', v_5'\}$. Thus, for each $E_i \in E$, we have $f(E_i) \in E'$. Hence, f is a ps-open mapping.

Definition 8. *A hypergraph $H = (V, E)$ is called connected if, for any two distinct vertices v_1 and v_2, there exists a hyperpath joining v_1 and v_2.*

Definition 9. *In a hypergraph $H = (V, E)$, a sequence*

$$\{v_1\} G_1 K_1 G_2 K_2 G_3 \ldots G_{n-1} K_{n-1} G_n \{v_n\}$$

is called a weak knot-hypergraph if each $G_i \supset E_i$, $(E_i \in E)$ with $K_i \subseteq (G_{i-1} \cap G_i) \setminus (\cup_{t=1}^{i-1} K_t)$ for all $i = 1, 2, \ldots, n-2$.

3. Hyperpaths and Hypercontinuity

In this section, we check whether the pseudo-open maps preserve the notion of the point-hyperpath and knot-hyperpath between two hypergraphs and under which conditions. Then, the notions of the hyper-continuous map and strictly hyper-continuous map between two hypergraphs are stated and various possible relationships between any two knot-hyperpaths under these notions are investigated.

Definition 10. *A mapping $f : V \to V'$ between the vertex sets of two hypergraphs $H = (V, E)$ and $K = (V', E')$ is called hyper-continuous if, for any $E_i' \in E'$, there is some $E_j \in E$ such that the corresponding inverse image satisfies $f^{-1}(E_i') \supseteq E_j$.*

Example 5. Suppose that $H = (V, E)$ and $K = (V', E')$ are two hypergraphs, where $V = \{v_1, v_2, v_3, v_4, v_5, v_6\}$ and $V' = \{v_1', v_2', v_3', v_4', v_5'\}$ and $E = \{\{v_1, v_2\}, \{v_3\}, \{v_3, v_4, v_5\}\}$, $E' = \{\{v_1'\}, \{v_2', v_3', v_4'\}, \{v_1', v_2', v_3'\}\}$. A map $f : V \to V'$ is defined such that $f(v_1) = v_1' = f(v_2)$, $f(v_3) = v_2' = f(v_5)$, $f(v_4) = v_5'$, $f(v_6) = v_3'$. Now, we have $\{v_1'\} \in E'$ and

$f^{-1}(\{v_1'\}) = \{v_1, v_2\} \supseteq \{v_1, v_2\} (\in E)$. Again, $\{v_2', v_3', v_4'\} \in E'$ and $f^{-1}(\{v_2', v_3', v_4'\}) = \{v_3, v_5, v_6\} \supseteq \{v_3\} (\in E)$. Moreover, $\{v_1', v_2', v_3'\} \in E'$ and $f^{-1}(\{v_1', v_2', v_3'\}) = \{v_1, v_2, v_3, v_5\} \supseteq \{v_1, v_2\}, \{v_3\} (\in E)$.

Thus, for each $E_i' \in E'$, $i = 1, 2, 3$, there is one $E_j \in E$, $j = 1, 2, 3$, such that $f^{-1}(E_i') \supseteq E_j (\in E)$. Thus, f is a hyper-continuous map from V to V'.

Definition 11. *A mapping $f : V \to V'$ between the vertex sets of two hypergraphs $H = (V, E)$ and $K = (V', E')$ is called strictly hyper-continuous if, for each $E_i' \in E'$, there is an $E_j \in E$, such that $f^{-1}(E_i') = E_j$.*

Example 6. Suppose that $H = (V, E)$ and $K = (V', E')$ are two hypergraphs, where $V = \{v_1, v_2, v_3, v_4, v_5, v_6\}$ and $V' = \{v_1', v_2', v_3', v_4', v_5'\}$ and $E = \{\{v_1, v_2\}, \{v_2, v_3, v_4\}, \{v_3, v_4, v_5\}\}$, $E' = \{\{v_1'\}, \{v_2', v_5'\}, \{v_1', v_4'\}\}$. A map $f : V \to V'$ is defined such that $f(v_1) = v_1' = f(v_2)$, $f(v_3) = v_2' = f(v_5)$, $f(v_4) = v_5'$, $f(v_6) = v_3'$. Now, we have $\{v_1'\} \in E'$ and $f^{-1}(\{v_1'\}) = \{v_1, v_2\} \in E$. Again, $\{v_2', v_5'\} \in E'$ and $f^{-1}(\{v_2', v_5'\}) = \{v_3, v_4, v_5\} \in E$. Moreover, we have $\{v_1', v_4'\} \in E'$ and $f^{-1}(\{v_1', v_4'\}) = \{v_1, v_2\} \in E$.

Thus, for each $E_i' \in E'$, $i = 1, 2, 3$, there exists an $E_j \in E$ such that $f^{-1}(E_i') = E_j$. Thus, f is strictly hyper-continuous.

Theorem 1. *Suppose that $H = (V, E)$ and $K = (V', E')$ are two hypergraphs and f is a mapping from V into V'. If f is a ps-open mapping, then the f-image of a point-hyperwalk in H is a point-hyperwalk in K.*

Proof. Let
$$P \equiv v_1 E_1 v_2 E_2 v_3 E_3 \ldots v_{n-1} E_n v_n$$
be a point-hyperwalk in H. Then, we obtain its f-image
$$f(P) \equiv f(v_1) f(E_1) f(v_2) f(E_2) f(v_3) f(E)_3 \ldots f(v_{n-1}) f(E_n) f(v_n).$$

Since P is a point-hyperwalk, it follows that $v_1 \in E_1, v_2 \in E_1 \cap E_2, \ldots, v_{n-1} \in E_{n-1} \cap E_n$ and $v_n \in E_n$. Thus, $f(v_1) \in f(E_1), f(v_2) \in f(E_1 \cap E_2), \ldots, f(v_{n-1}) \in f(E_{n-1} \cap E_n)$, $f(v_n) \in E_n$. Now, $E_1 \cap E_2 \subseteq E_1, E_2$ implies that $f(E_1 \cap E_2) \subseteq f(E_1), f(E_2)$, whence $f(E_1 \cap E_2) \subseteq f(E_1) \cap f(E_2)$. Therefore, $f(v_2) \in f(E_1 \cap E_2) \subseteq f(E_1) \cap f(E_2)$. Similarly, $f(v_3) \in f(E_2) \cap f(E_3), \ldots, f(v_n) \in f(E_n)$. Hence, $f(P)$ is a point-hyperwalk in K. □

Corollary 1. *In Theorem 1, if f is an injective mapping, then the f-image of a point-hyperpath in H is a point-hyperpath in K, too.*

Theorem 2. *Suppose that $H = (V, E)$ and $K = (V', E')$ are two hypergraphs and f is a ps-open mapping from V to V'. Then, the f-image of a knot-hyperpath in H is a knot-hyperpath in K, too.*

Proof. Let $P \equiv \{v_1\} E_1 K_1 E_2 K_2 E_3 K_3 \ldots K_{n-1} E_n \{v_n\}$ be a knot-hyperpath in H with $K_0 = \{v_1\} \subseteq E_1, K_n = \{v_n\} \subseteq E_n$ and $K_i \subseteq (E_{i+1} \cap E_i) \setminus (\cup_{t=1}^{i-1} K_t)$, $i = 1, 2, \ldots, n-1$. Then, we have the f-image
$$f(P) \equiv f(K_0) f(E_1) f(K_1) f(E_2) f(K_2) f(E_3) f(K_3) \ldots f(K_{n-1}) f(E_n) f(K_n).$$

In order to prove that $f(P)$ is a knot-hyperpath, we first show that $f(K_0) \subseteq f(E_1)$ and $f(K_n) \subseteq f(E_n)$. Since $K_0 \subseteq E_1$ and $K_n \subseteq E_n$, we have $f(K_0) \subseteq f(E_1)$ and $f(K_n) \subseteq f(E_n)$.

Since $K_2 \subseteq (E_2 \cap E_3) \setminus E_1$, we have $K_2 \subseteq (E_2 \cap E_3) \cap K_1^c$. It follows that $f(K_2) \subseteq f((E_2 \cap E_3) \cap K_1^c) \subseteq f(E_2 \cap E_3) \cap f(K_1^c) \subseteq f(E_3 \cap E_2) \cap (f(K_1))^c \subseteq f(E_3 \cap E_2) \setminus f(K_1)$. Hence, $f(K_2) \subseteq f(E_3 \cap E_2) \setminus f(K_1)$.

Similarly, $K_3 \subseteq (E_4 \cap E_3) \setminus (K_1 \cup K_2)$ implies that $f(K_3) \subseteq f(E_4 \cap E_3) \setminus f(K_1) \cup f(K_2)$ and so on. Thus, $K_i \subseteq (E_i \cap E_{i+1}) \setminus (\cup_{t=1}^{i-1} K_t)$ implies that

$$f(K_i) \subseteq f(E_i \cap E_{i+1}) \setminus \cup_{t=1}^{i-1} f(K_t) \tag{2}$$

for any $i = 1, 2, \ldots, n-1$. Hence, we conclude that $f(P)$ is a knot-hyperpath in K. □

Theorem 3. *Suppose that $H = (V, E)$ and $K = (V', E')$ are two hypergraphs. If f is a hyper-continuous map from V onto V', then the inverse image of a knot-hyperpath in K under f is a weak knot-hyperpath in H.*

Proof. Let $P' \equiv K_0' E_1' K_1' E_2' K_2' \ldots K_{n-1}' E_n' K_n'$ be a knot-hyperpath in K. As f is hyper-continuous, we have $f^{-1}(E_1') \supseteq E_1, f^{-1}(E_2') \supseteq E_2, \ldots, f^{-1}(E_n') \supseteq E_n$, for some hyperedges $E_1, E_2, \ldots, E_n \in E$. Moreover, the sets $f^{-1}(K_i), i = 0, 1, 2, \ldots, n$ are nonempty because f is an onto mapping. Now, the inverse image of the knot-hyperpath can be written as

$$f^{-1}(K_0') f^{-1}(E_1') f^{-1}(K_1') f^{-1}(E_2') \ldots f^{-1}(K_{n-1}') f^{-1}(E_n') f^{-1}(K_n'),$$

where $f^{-1}(E_i') \supseteq E_i$, for $i = 1, 2, 3, \ldots, n$. Since the inverse set function behaves well for union, intersection and complement, it follows that the conditions of a knot-hyperpath are easily satisfied. Hence, $f^{-1}(P')$ is a weak knot-hyperpath. □

Corollary 2. *The inverse image of an onto strictly hyper-continuous map of a knot-hyperpath is again a knot-hyperpath.*

Proof. Consider the knot-hyperpath

$$P' \equiv K_0' E_1' K_1' E_2' K_2' \ldots K_{n-1}' E_n' K_n'$$

as in the proof of Theorem 3. As f is strictly hyper-continuous, each $f^{-1}(E_i)$ belongs to E and, by using similar arguments, we can conclude that

$$f^{-1}(P') \equiv f^{-1}(K_0') f^{-1}(E_1') f^{-1}(K_1') \ldots f^{-1}(K_{n-1}') f^{-1}(E_n') f^{-1}(K_n')$$

is a knot-hyperpath. □

Theorem 4. *Let $f : V \to V'$ be a ps-open mapping from a hypergraph $H = (V, E)$ onto a hypergraph $K = (V', E')$. If H is connected, then K is connected, too.*

Proof. Let v_1' and v_2' be two any vertices in K. Since f is onto, there exists $v_1, v_2 \in V$ such that $f(v_1) = v_1'$ and $f(v_2) = v_2' \in V$. Moreover, since H is connected and $v_1, v_2 \in V$, there exists a knot-hyperpath P from v_1 to v_2. Because the image of a knot-hyperpath under a ps-open mapping is again a knot-hyperpath in K, starting at $f(v_1) = v_1'$ and ending at $f(v_2) = v_2'$, we immediately conclude that K is connected. □

4. Hyperpaths and Hypertrees

In this section, we will present a sufficient condition, only involving hyperpaths, under which a hypergraph is a hypertree. Till now, the definition of a hypertree has been based on the concept of the host graph.

Definition 12 ([14]). *Suppose that $H = (V, E)$ is a hypergraph and $G = (V, F)$ is a graph over the same vertex set V. We say that G is a host graph of H if each hyperedge $E_i \in E$ induces a connected subgraph in G.*

Lemma 1. *There exists at least one host graph G of the hypergraph H in which the induced subgraph obtained from any two equivalent knot-hyperpaths never forms a cycle.*

Proof. Let P_1 and P_2 be any two equivalent knot-hyperpaths of the hypergraph H, which may be denoted as follows:

$$P_1 \equiv K_0 = \{v_1\} E_1 K_1 E_2 K_2 E_3 K_3 \ldots K_{n-1} E_n K_n = \{v_n\}$$

and

$$P_2 \equiv K_0' = \{v_1\} E_1' K_1' E_2' K_2' E_3' K_3' \ldots K_{n-1}' E_n' K_n' = \{v_n\}$$

and graphically represented in Figure 2.

Since they are equivalent knot-hyperpaths, it follows that $K_i \cap K_i' \neq \emptyset$, $E_i \cap E_i' \neq \emptyset$, $K_i \cap K_{i+1} = \emptyset$, and $K_i' \cap K_{i+1}' = \emptyset$.

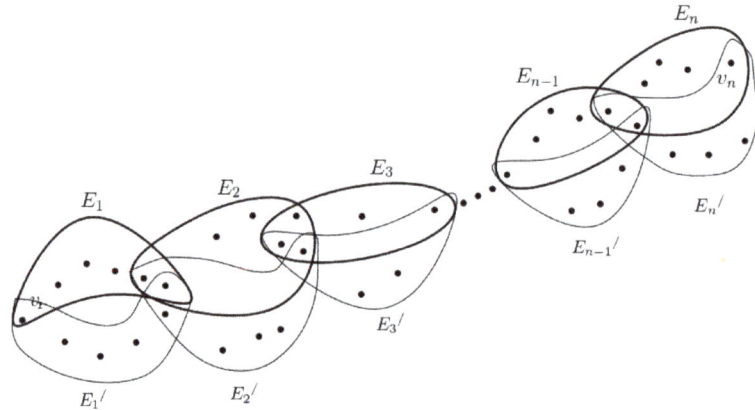

Figure 2. A schematic diagram of two equivalent knot-hyperpaths P_1 and P_2.

We note that $E_1 \cup E_1'$ can be expressed as the disjoint union of $E_1 \setminus E_1'$, $E_1' \setminus E_1$ and $E_1 \cap E_1'$. As we know that, in any host graph of a hypergraph, all the vertices in a hyperedge are connected, and since $E_1 \cap E_1'$ is contained in E_1 and E_1', it follows that all the vertices in $E_1 \cap E_1'$ can be connected to form a graph without cycles. Moreover, since $E_1 \cap E_1'$ and $E_1 \setminus E_1'$ are contained in E_1, a graph can be drawn by connecting all the vertices in $E_1 \setminus E_1'$ without forming a cycle, which can be further connected with the cycle-free graph drawn in $E_1 \cap E_1'$ in the previous step. By connecting vertices in such a manner, the resultant graph will never form a cycle. Similarly, a graph can be drawn by connecting the cycle-free graph drawn in $E_1 \cap E_1'$ with a cycle-free graph in $E_1' \setminus E_1$. All these constructions are depicted in Figure 3.

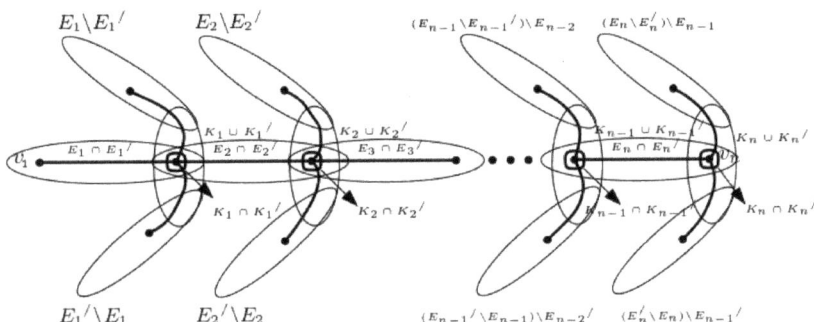

Figure 3. Model of cycle-free connected induced subgraph of a host graph of the hypergraph H.

The model is constructed in such a way that the vertex v_1 is connected to $K_1 \cap K_1'$ and $K_1 \cap K_1'$ is connected to both $E_1 \setminus E_1'$ and $E_1' \setminus E_1$ through $K_1 \cup K_1'$ without forming a cycle. Furthermore, it is to be noted that because $K_1 \cap K_1'$ is connected to $E_1 \setminus E_1'$, in the next step, $K_2 \cap K_2'$ will connect to those vertices of $E_2 \setminus E_2'$ that are not in E_1, in order to not create a cycle. Similarly, $K_2 \cap K_2'$ will connect to those vertices of $E_1' \setminus E_1$ that are not in E_1'. This further continues till the last vertex v_n, where v_n is connected to $K_{n-1} \cap K_{n-1}'$ and $K_{n-1} \cap K_{n-1}'$ is connected to $E_n \setminus E_n'$ and $E_n' \setminus E_n$ through $K_1 \cup K_1'$, without forming a cycle. In this manner, a host graph can be drawn from the hypergraph H, where the induced subgraph obtained from the vertices in the edges of the two paths is cycle-free. We conclude that there exists at least one host graph G of H in which the induced subgraph obtained from the two equivalent knot-hyperpaths will never form a cycle. □

Remark 1. *If the induced subgraph obtained from the vertex set of two knot-hyperpaths joining the same vertices of any host graph of a hypergraph always produces a cycle, then the knot-hyperpaths are not equivalent.*

Theorem 5. *Suppose that H is a connected hypergraph, which is a hypertree. Then, any entire knot-hyperpaths having the same length and connecting any two vertices are equivalent.*

Proof. Let P_1 and P_2 be any two entire knot-hyperpaths of the hypergraph H, which may be denoted as follows:

$$P_1 \equiv K_0 = \{v_1\} E_1 K_1 E_2 K_2 E_3 K_3 \ldots K_{n-1} E_n K_n = \{v_n\}$$

and

$$P_2 \equiv K_0' = \{v_1\} E_1' K_1' E_2' K_2' E_3' K_3' \ldots K_{n-1}' E_n' K_n' = \{v_n\}.$$

If P_1 and P_2 are equivalent knot-hyperpaths, then the result is proven.

On the contrary, if P_1 and P_2 are not equivalent, then there exists a pair of edges (E_{i_0}, E_{i_0}'), where E_{i_0} is from P_1 and E_{i_0}' is from P_2, such that $E_{i_0} \cap E_{i_0}' = \emptyset$. Since $K_{i_0-1}, K_{i_0} \subseteq E_{i_0}$ and $K_{i_0-1}', K_{i_0}' \subseteq E_{i_0}'$, we have $K_{i_0-1} \cap K_{i_0-1}' = \emptyset = K_{i_0} \cap K_{i_0}'$. Moreover, let E_{j_0}, E_{j_0}' be the edges such that $E_{j_0} \cap E_{j_0}' \neq \emptyset$, while $E_k \cap E_k' = \emptyset$, for any $k \in \{i_0, i_0 + 1, \ldots, j_0 - 1\}$. Then, the edges E_{i_0-1} to E_{j_0} and E_{i_0-1}' to E_{j_0}' will always form a cycle (see Figure 4) in any host graph of H, which is a contradiction. Therefore, P_1 and P_2 are equivalent. Thus, we can conclude that if H is a hypertree, then, between any two vertices, the entire knot-hyperpaths having the same length are unique up to isomorphism. □

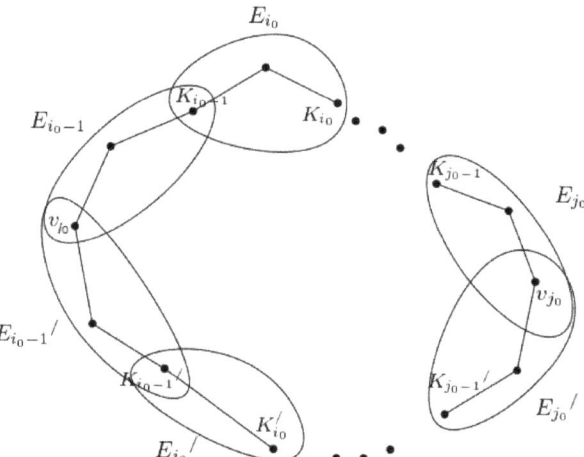

Figure 4. The cycle formed in a host graph of a hypergraph.

It is to be noted that two knot-hyperpaths joining two vertices in a hypertree may not always be equivalent. This can be observed in Example 7 by introducing an extra edge $\{v_6, v_7, v_{10}\}$ to the hypergraph, which subsequently produces two knot-hyperpaths joining v_0 and v_1, but with different lengths.

Theorem 6. *Suppose that H is a hypergraph such that, between any two vertices, there exists a unique entire knot-hyperpath up to isomorphism. Then, H is a hypertree.*

Proof. By hypothesis, between any two vertices v_1 and v_2 of H, there exists an entire knot-hyperpath, which is unique up to isomorphism. It follows that H is connected. To show that H is a hypertree, it is enough to show that H admits a host graph that is a tree. Let

$$P \equiv K_0 = \{v_1\} E_1 K_1 E_2 K_2 E_3 K_3 \ldots K_{n-1} E_n K_n = \{v_2\}$$

be an entire knot-hyperpath joining the vertices v_1 and v_2. Then, the vertices contained in the edges of this knot-hyperpath can be joined without forming a cycle, in such a way that the constructed graph G_1 is an induced subgraph with vertex set $V_1 = \cup E_i$ of some host graph G of the given hypergraph H. Now, if $\cup E_i = V$, then we can take $G = G_1$, which is a tree. Hence, in this case, H is a hypertree and the theorem is proven.

If $\cup E_i \neq V$, then let $v_3 \in V$ be such that $v_3 \notin \cup E_i$. Let

$$P' \equiv \{v_1\} E'_1 K'_1 E'_2 \ldots K'_{k-1} E'_k \{v_3\}$$

be an entire knot-hyperpath joining the vertices v_1 and v_3. We note that there may exist some hyperedges in P' that coincide with the hyperedges of P. Now, excluding these common hyperedges, the rest of the hyperedges of P' can be joined without forming a cycle. In this way, an induced subgraph G_2 can be formed with vertex set $\cup E'_j$ and the edges set as the union of those edges common with G_1 and the edges newly formed from hyperedges of P', which are not in P. It is clear from the construction that both subgraphs G_1 and G_2 are not cyclic and the union $G_1 \cup G_2$ is connected; otherwise, H would have two entire knot-hyperpaths joining the same vertices, but not equivalent (see proof of Theorem 5). Now, if $(\cup E_i) \cup (\cup E'_j) = V$, then $G = G_1 \cup G_2$ is the host graph of H that is a tree and hence H is again a hypertree.

If $(\cup E_i) \cup (\cup E'_j) \neq V$, then there exists a vertex $v_4 \in V$ that is not in $(\cup E_i) \cup (\cup E'_j)$. Then, we will have an entire knot-hyperpath P'' joining v_1 and v_4 as follows:

$$P'' \equiv \{v_1\} E''_1 K''_1 E''_2 \ldots K''_{l-1} E''_l \{v_4\}.$$

Now, excluding those hyperedges of P'' that are common with P and P', the rest of the hyperedges of P'' can be joined without forming a cycle. In this way, an induced subgraph G_3 can be formed with vertex set $\cup E''_l$ and the edges set as the union of those edges common with $G_1 \cup G_2$ and the edges newly formed from hyperedges of P'' that are not in P and P'. It is clear from the construction that all the subgraphs G_1, G_2 and G_3 are not cyclic and the union $G_1 \cup G_2 \cup G_3$ is connected. Now, if $(\cup E_i) \cup (\cup E'_j) \cup (\cup E''_l) = V$, then $G = G_1 \cup G_2 \cup G_3$ is the host graph of H that is a tree and hence H is a hypertree.

As the vertex set of the hypergraph is finite, the process has a finite number of steps. Thus, we can conclude that if H is a hypergraph such that, between any two vertices, there exists an entire knot-hyperpath unique up to isomorphism, then H is a hypertree. □

Remark 2. *We can notice that the hypergraph considered in Example 3 is a hypertree, but the two knot-hyperpaths P'_1 and P'_2 joining the vertices v_1 and v_{13} are not equivalent, even though they have the same length, while all the entire knot-hyperpaths (for example, P_1 and P_2) are equivalent. Hence, the property of knots of being entire, in the above two theorems, is an important hypothesis to be considered.*

To illustrate the algorithm stated in the proof of Theorem 6, we present the following example, where the considered hypergraph is a hypertree and a host graph is drawn using the technique used in the proof of Theorem 6. This hypertree is represented in Figure 5.

Example 7. *Consider the hypergraph $H = (V, E)$, where $V = \{v_0, v_1, v_2, \ldots, v_{16}\}$ and $E = \{E_1 = \{v_0, v_7, v_6\}, E_2 = \{v_6, v_{10}, v_{11}\}, E_3 = \{v_{11}, v_{14}, v_{15}, v_5, v_{16}\}, E_4 = \{v_3, v_1, v_{13}\}, E_5 = \{v_6, v_2\}, E_6 = \{v_5, v_{16}\}, E_7 = \{v_4, v_9, v_{12}\}, E_8 = \{v_5, v_8, v_{13}, v_9\}\}$. One can easily verify that H is a hypertree and, between any two vertices, there exists an entire knot-hyperpath, unique up to isomorphism. Now, we will use the technique used in the proof of Theorem 6, in order to obtain a host graph that is a tree.*

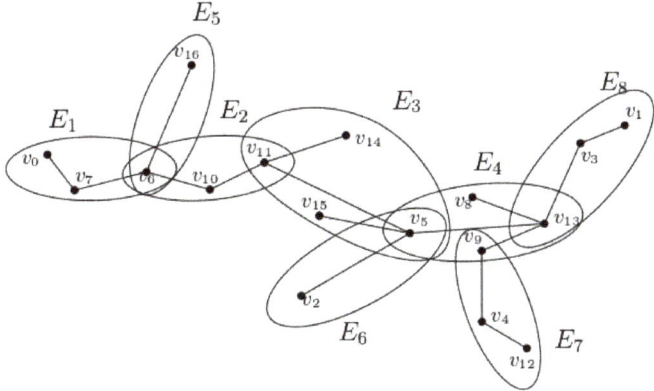

Figure 5. A hypergraph that is a hypertree.

Let us consider the vertices v_0 and v_1 and the knot-hyperpath

$$P \equiv \{v_0\} E_1 \{v_6\} E_2 \{v_{11}\} E_3 \{v_5\} E_4 \{v_{13}\} E_8 \{v_1\}$$

joining v_0 and v_1. Now, the vertices in all hyperedges are connected and form a graph G_1 in such a way that it is not cyclic and it is an induced subgraph with vertex set $V_1 = E_1 \cup E_2 \cup E_3 \cup E_4 \cup E_8$ of some host graph G of H.

Clearly, $V \neq V_1$, and so we consider the vertex $v_2 \in V$, which is not in V_1. Now, a hyperpath P' from v_0 to v_2 is constructed as follows:

$$P' \equiv \{v_0\} E_1 \{v_6\} E_2 \{v_{11}\} E_3 \{v_5\} E_6 \{v_2\}.$$

Clearly, except E_6, all other hyperedges of this knot-hyperpath appear in the previous knot-hyperpath, and so vertices of E_6 are joined in an acyclic way and represent a graph G_2 with vertex set $V_2 = E_1 \cup E_2 \cup E_3 \cup E_6$.

Here, we note that the union of the two graphs G_1 and G_2 is acyclic and connected. Moreover, $V_1 \cup V_2 \neq V$. Therefore, we consider an arbitrary vertex from v_4, v_{12}, v_{16} that is not in $V_1 \cup V_2$. Let us consider the vertex v_4 and the knot-hyperpath P'' constructed as follows:

$$P'' \equiv \{v_0\} E_1 \{v_6\} E_2 \{v_{11}\} E_3 \{v_5\} E_4 \{v_9\} E_7 \{v_4\}.$$

Clearly, except E_7, all other hyperedges of this knot-hyperpath appear in the previous knot-hyperpaths, and so vertices of E_7 are joined in an acyclic way that represents a graph G_3 with vertex set $V_3 = E_1 \cup E_2 \cup E_3 \cup E_4 \cup E_7$. Thus, $G_1 \cup G_2 \cup G_3$ is connected and acyclic. Since $V_1 \cup V_2 \cup V_3 \neq V$, we consider the vertex v_{16}, the only one that is not in this union and the knot-hyperpath

$$P''' \equiv \{v_0\} E_1 \{v_6\} E_5 \{v_{16}\}.$$

Clearly, except E_5, all other hyperedges of this knot-hyperpath appear in the previous knot-hyperpaths, and so vertices of E_5 are joined in an acyclic way that represents a graph G_4 with vertex set $V_4 = E_1 \cup E_5$. Now, $G_1 \cup G_2 \cup G_3 \cup G_4$ is connected and acyclic, and $V_1 \cup V_2 \cup V_3 \cup V_4 = V$. Therefore, $G = G_1 \cup G_2 \cup G_3 \cup G_4$ is the required host graph, which is a tree.

5. Conclusions

Based on the definition of a knot in a hypergraph H, which is a subset of the intersections of some intersecting hyperedges of H, we have introduced the notion of the knot-hyperpath, in order to better characterize the hyper-continuity and pseudo-continuity of functions between two hypergraphs. Moreover, in the second part of the paper, we have characterized the hypertrees without using the concept of a host graph. A sufficient condition is established to check whether or not a hypergraph is a hypertree. Furthermore, an algorithm is designed in order to extract from a hypertree a host graph that is a tree. This algorithm has the potential to determine whether a hypergraph is a hypertree or not. As we know, hypergraphs and hypertrees are extensively used in different branches of applied sciences, including networking and theoretical computer science, and therefore this investigation will give more future ideas towards the applicability of hypergraphs and hypertrees in these fields.

Author Contributions: Conceptualization, S.R. and M.C.; methodology, S.R., M.C. and I.C.; investigation, S.R., M.C., F.A. and I.C.; writing—original draft preparation, S.R., M.C. and F.A.; writing—review and editing, I.C.; funding acquisition, I.C. All authors have read and agreed to the published version of the manuscript.

Funding: The third author acknowledges the financial support of the Slovenian Research Agency (research core funding No. P1-0285).

Institutional Review Board Statement: Not applicable.

Informed Consent Statement: Not applicable.

Data Availability Statement: Not applicable.

Conflicts of Interest: The authors declare no conflict of interest.

References

1. Ma, J.; Fukuda, D. Faster hyperpath generating algorithms for vehicle navigation. *Transp. A Transp. Sci.* **2013**, *9*, 925–948. [CrossRef]
2. Noh, H.; Hickman, M.; Khani, A. Hyperpaths in Network Based on Transit Schedules. *Transp. Res. Rec.* **2012**, *2284*, 29–39. [CrossRef]
3. Klamt, S.; Haus, U.-U.; Theis, F. Hypergraphs and cellular networks. *PLoS Comput. Biol.* **2009**, *5*, e1000385. [CrossRef] [PubMed]
4. Ritz, A.; Avent, B.; Murali, M.T. Pathway Analysis with Signaling Hypergraphs. *IEEE/ACM Trans. Comput. Biol. Bioinform.* **2017**, *14*, 1042–1055. [CrossRef] [PubMed]
5. Ritz, A.; Tegge, N.A.; Kim, H.; Poirel, L.C.; Murali, T. Signalling hypergraphs. *Trends Biotechnol.* **2014**, *32*, 356–362. [CrossRef] [PubMed]
6. Nguyen, S.; Pallottino, S. Hyperpaths and shortest hyperpaths. In Proceedings of the Lectures Given at the 3rd Session of the Centro Internazionale Matematico Estivo (C.I.M.E), Como, Italy, 25 August–2 September 1986; Combinatorial Optimization, Springer: Berlin/Heidelberg, Germany, 1989.
7. Nielsen, L.; Andersen, K.; Pretolani, D. Finding the K-shortest hyperpaths. *Comput. Oper. Res.* **2005**, *32*, 1477–1497. [CrossRef]
8. Chowdhury, K.; Das, G. Some space-biased aspects of near-rings and near-ring groups. *Int. J. Modern Math.* **2007**, *2*, 103–124.
9. Miranda, G.; Luna, H.P.; Camargo, R.S.; Pinto, L.R. Tree network design avoiding congestion. *Appl. Math. Model.* **2011**, *35*, 4175–4188. [CrossRef]
10. Brandstädt, A.; Dragan, F.; Chepoi, V.; Voloshin, V. Dually chordal graphs. *Siam J. Discrete Math.* **1998**, *11*, 437–455. [CrossRef]
11. Berge, C. *Graphs and Hypergraphs*; North Holland Publishing Co.: Amsterdam, The Netherlands, 1973.
12. Ouvrard, X. Hypergraphs: An introduction and review. *arXiv* **2020**, arXiv:2002.05014.
13. Kannan, K.; Dharmarajan, R. Hyperpaths and Hypercycles. *Int. J. Pure Appl. Math.* **2015**, *98*, 309–312.
14. Bujtás, C.; Tuza, Z.; Voloshin, I.V. Color-bounded hypergraphs, V: Host graphs and subdivisions. *Discuss. Math. Graph Theory* **2011** *31*, 223–238. [CrossRef]

Article

The Structure of the Block Code Generated by a *BL*-Algebra

Hashem Bordbar

Centre for Information Technologies and Applied Mathematics, University of Nova Gorica,
5000 Nova Gorica, Slovenia; hashem.bordbar@ung.si; Tel.: +386-070-614-786

Abstract: Inspired by the concept of *BL*-algebra as an important part of the ordered algebra, in this paper we investigate the binary block code generated by an arbitrary *BL*-algebra and study related properties. For this goal, we initiate the study of the *BL*-function on a nonempty set P based on *BL*-algebra L, and by using that, l-functions and l-subsets are introduced for the arbitrary element l of a *BL*-algebra. In addition, by the mean of the l-functions and l-subsets, an equivalence relation on the *BL*-algebra L is introduced, and using that, the structure of the code generated by an arbitrary *BL*-algebra is considered. Some related properties (such as the length and the linearity) of the generated code and examples are provided. Moreover, as the main result, we define a new order on the generated code C based on the *BL*-algebra L, and show that the structures of the *BL*-algebra with its order and the correspondence generated code with the defined order are the same.

Keywords: *BL*-function; *BL*-code; binary linear block codes; coding theory; *BL*-algebra

Citation: Bordbar, H. The Structure of the Block Code Generated by a *BL*-Algebra. *Mathematics* **2022**, *10*, 692. https://doi.org/10.3390/math10050692

Academic Editor: Patrick Solé

Received: 27 January 2022
Accepted: 21 February 2022
Published: 23 February 2022

Publisher's Note: MDPI stays neutral with regard to jurisdictional claims in published maps and institutional affiliations.

Copyright: © 2022 by the authors. Licensee MDPI, Basel, Switzerland. This article is an open access article distributed under the terms and conditions of the Creative Commons Attribution (CC BY) license (https://creativecommons.org/licenses/by/4.0/).

1. Introduction

Hajek, in 1998, introduced *BL*-algebra in order to provide a general framework for formalizing statements of a fuzzy nature [1]. *BL*-algebra is the algebraic structure arising from the continuous triangular norms, and it has certain logical axioms similarly to Boolean algebras or *MV*-algebras from classical logic or Lukasiewicz logic, respectively. In addition, every *MV*-algebra is a *BL*-algebra, whereas the converse is not always true. Thus, the class of *MV*-algebra is a subset of the class of *BL*-algebra, and this is the main reason that we selected *BL*-algebra to investigate the code generated by it. Moreover, *Hajeck* showed that every *BL*-algebra with an involutory complement is *MV*-algebra.

In the twentieth century, there is a problem in engineering about the transmutation of information. Shannon [2] in 1948 and Hamming [3] in 1950 provided some frameworks to solve the problem. Their idea was developed and, as a consequence, the electronic information could be transmitted throughout the noisy channel and stored by minimum errors, and coding theory was born. Because the electronic information is a string of zeros and ones, it uses a finite field as the alphabet set. Thus, coding theory can use different areas of mathematics such as linear algebra, finite geometry, lattices, and combinatorics, especially when the alphabet set is generalized to different types of fields. Coding theory can be viewed not only as a part of computer science and engineering but also as a part of pure mathematics, as the mathematicians were interested in the fundamental aspects of this concept.

Application of coding theory in ordered algebraic structures was initiated by Jun and Song in 2011 [4]. They introduced the notion of BCK-valued functions and established binary block codes by using the notion of BCK-valued functions. After that, in 2014, Borumand et al. [5] and in 2015, Flaut [6] presented some relationships between BCK-algebras and the related binary block codes. They proved that every BCK-algebra determines a binary block code. Gilani [7] studied some properties of the codes generated by BCK-functions in an arbitrary BCK-algebras. Details about the fundamental relations in an arbitrary BCK-algebra and related generated code, namely BCK-code, was investigated by Bordbar in [8]. During the last few years, binary block codes generated by different types of ordered algebraic structures were studied, for instance, codewords in a binary block code generated by a UP-valued function investigated by Chinram and Iampan in 2021 [9], and in 2015, Mostafa et al. [10] applied coding theory to KU-algebras and gave some connection between binary block codes and KU-algebras.

In this paper, we investigate the code generated by a BL-algebra. Our motivation is to study the properties of a code generated by one ordered algebraic structure. We begin with a discussion of the ordered algebraic structure known as BL-algebra as it is an extended algebraic structure and has some other ordered structures as its subsets, such as MV-algebra. Moreover, by using the order in a BL-algebra, we define a new order in a generated code and give the code an algebraic ordered structure. The defined order among the codewords can be useful in decoding and will be our future work. For our goal, in Section 3 we define a BL-function on an arbitrary nonempty set P based on BL-algebra L and by using the BL-function, l-functions of it and l-subsets of P for $l \in L$ are investigated. In addition, properties of BL-function and its l-functions and l-subsets of P that we need for generated code are considered. In Section 4, a binary equivalence relation $\approx$ defines on a BL-algebra L, and by using this relation and BL-function, the code C based on L is generated. Finally, we study an order among the codewords of C that gives the code C an ordered algebraic structure. In Theorem 3, we show that the BL-algebra L with its order and the code C based on L with respect to defined order have the same structures.

2. Preliminaries

A *BL-algebra* is a structure $\mathcal{L} := (L, \wedge, \vee, \odot, \rightarrow, 0, 1)$ such that $(L, \wedge, \vee, 0, 1)$ is a bounded lattice, $(L, \odot, 1)$ is an abelian monoid, i.e., $\odot$ is commutative and associative, and the following conditions hold for all $x, y, z \in L$:

(BL1) $x \odot 1 = x$,
(BL2) $x \odot y \leq z$ if and only if $x \leq y \rightarrow z$,
(BL3) $x \wedge y = x \odot (x \rightarrow y)$,
(BL4) $(x \rightarrow y) \vee (y \rightarrow x) = 1$.

Every BL-algebra $\mathcal{L}$ satisfies the following assertions

$$x = 1 \rightarrow x,\ x \rightarrow x = 1,\ x \rightarrow 1 = 1, \tag{1}$$

$$x \leq y \Leftrightarrow x \rightarrow y = 1, \tag{2}$$

$$x \leq y \rightarrow x, \tag{3}$$

$$x \rightarrow (y \rightarrow z) = y \rightarrow (x \rightarrow z), \tag{4}$$

$$x \leq y \Rightarrow z \rightarrow x \leq z \rightarrow y,\ y \rightarrow z \leq x \rightarrow z \tag{5}$$

for all $x, y, z \in L$.

For more information about BL-algebra, please refer to [1,11].

The alphabets used in coding theory are finite fields with q elements, $GF(q)$. We say that a code is $q - ary$ if its codewords are defined over the $q - ary$ alphabet $GF(q)$. The most commonly used alphabets are binary fields, $GF(2m)$. This article focuses on codes with the familiar alphabet $GF(2)$, which are known as *binary codes*.

Let c be a codeword. Then the *Hamming weight* $w(c)$ of a codeword c is the number of nonzero components in the codeword. The *Hamming distance* between two codewords $d(c_1, c_2)$ is the number of places in which the codewords c_1 and c_2 differ. In other words, $d(c_1, c_2)$ is the Hamming weight of the vector $c_1 - c_2$, representing the component-wise difference of the vectors c_1 and c_2. The minimum (Hamming) distance of a code C is the minimum distance between any two codewords in the code C, that is,

$$d(C) = min\{d(x, y) \mid x \neq y, x, y \in C\}.$$

The notation (n, M, d) is used to represent a code with code length n, a total of M codewords, and minimum distance d. One of the major goals of coding theory is to develop codes that strike a balance between having small n (for fast transmission of messages), large M (to enable transmission of a wide variety of messages), and large d (to detect many errors).

3. BL-Functions on a Nonempty Set P

In this section, the notions of BL-functions on a nonempty set P based on a BL-algebra L, l-functions, and l-subsets of P for an arbitrary element $l \in L$, will be introduced. Some of the properties connected with l-subsets of P and l-functions of a BL-function will be investigated. Throughout this section, unless stated otherwise, $\mathcal{L} := (L, \wedge, \vee, \odot, \rightarrow, 0, 1)$ denotes a BL-algebra. In addition, in this paper, we use the set L for our definitions as a BL-algebra $\mathcal{L} = (L, \wedge, \vee, \odot, \rightarrow, 0, 1)$.

Definition 1. *Let P be a nonempty set and L be a BL-algebra. A mapping $\phi : P \rightarrow L$ is called a BL-function on P based on L and denoted by ϕ_L. If there is no confusion of L, we use ϕ instead of ϕ_L. Moreover, for a BL-function ϕ on P and $l \in L$, define $\phi_l : P \rightarrow \{0,1\}$ for each $p \in P$ as follows:*

$$\phi_l(p) = \begin{cases} 1, & \text{if and only if } \phi(p) \rightarrow l = 1, \\ 0, & \text{otherwise.} \end{cases} \quad (6)$$

The function ϕ_l is called a l-function of ϕ.

Definition 2. *Let P be a nonempty set and L be a BL-algebra. For a BL-function $\phi : P \rightarrow L$ on P and each $l \in L$, the set P_l defined by*

$$P_l := \{p \in P \mid \phi(p) \rightarrow l = 1\}, \quad (7)$$

is called a l-subset of P.

Example 1. *Let $L = \{0, a, b, 1\}$ be a set with the following Cayley tables:*

$\odot$	0	a	b	1
0	0	0	0	0
a	0	a	a	a
b	0	a	b	b
1	0	a	b	1

$\rightarrow$	0	a	b	1
0	1	1	1	1
a	0	1	1	1
b	0	a	1	1
1	0	a	b	1

Then $\mathcal{L} = (L, \wedge, \vee, \odot, \rightarrow, 0, 1)$ is a BL-algebra (see [12]), where $x \wedge y = \min\{x, y\}$ and $x \vee y = \max\{x, y\}$.

(1) For a set $P = \{p_1, p_2, p_3\}$, the function $\phi : P \rightarrow L$ defined by

$$\phi(p_1) = a, \phi(p_2) = b \text{ and } \phi(p_3) = 1,$$

is a BL-function on P, and the l-subsets of P for each $l \in L$ are as follows:

$$P_0 = \emptyset, P_a = \{p_1\}, P_b = \{p_1, p_2\} \text{ and } P_1 = P.$$

In addition, for each $l \in L$, the l-functions of ϕ are as shown in the following table:

ϕ_l	p_1	p_2	p_3
ϕ_0	0	0	0
ϕ_a	1	0	0
ϕ_b	1	1	0
ϕ_1	1	1	1

(2) Let $Q = \{q_1, q_2, q_3, q_4\}$ and define the function $\psi : Q \rightarrow L$ by

$$\psi(q_1) = a, \psi(q_2) = b, \psi(q_3) = 1 \text{ and } \psi(q_4) = 0.$$

Then ψ is a BL-function on Q. For each $l \in L$, the l-subsets of Q are as follows:

$$Q_0 = \{q_4\}, \ Q_a = \{q_1, q_4\}, \ Q_b = \{q_1, q_2, q_4\} \text{ and } Q_1 = Q.$$

In addition, the l-functions of ψ for each $l \in L$ are as shown in the following table:

ψ_l	q_1	q_2	q_3	q_4
ψ_0	0	0	0	1
ψ_a	1	0	0	1
ψ_b	1	1	0	1
ψ_1	1	1	1	1

The following proposition shows the relationship between BL-function ϕ_L on P and its l-functions and l-subsets of P for $l \in L$.

Proposition 1. *Let $\phi : P \to L$ be a BL-function on a nonempty set P based on L, where L is a BL-algebra. Then the function ϕ can be described by its l-functions and l-subsets of P, for $l \in L$, as the infimum of the following sets:*

$$(\forall p \in P)(\phi(p) = \inf\{l \in L \mid p \in P_l\}), \tag{8}$$

in other words,

$$(\forall p \in P)(\phi(p) = \inf\{l \in L \mid \phi_l(p) = 1\}). \tag{9}$$

Proof. Let $p \in P$ be an arbitrary element and $\phi(p) = l$. Then using (1),

$$\phi(p) \to l = l \to l = 1.$$

Thus $\phi_l(p) = 1$, which means that $p \in P_l$. Assume that $p \in P_{l'}$, for $l' \in L$. Then $1 = \phi(p) \to l' = l \to l'$. By using (2), we conclude that $l \leq l'$. Because $l' \in \{l \in L \mid p \in P_l\}$, it follows that

$$\phi(p) = \inf\{l \in L \mid p \in P_l\}.$$

The equality (9) is a direct conclusion of (6) and (7). □

Corollary 1. *For a BL-algebra L, if $\phi : P \to L$ is a BL-function on P based on L, then for $p \in P$,*

$$\phi(p) = \inf\{\phi_l(p) \hookrightarrow l \mid l \in L\},$$

where

$$\phi_l(p) \hookrightarrow l = \begin{cases} l & \text{if } p \in P_l, \\ 1 & \text{otherwise.} \end{cases}$$

Proposition 2. *Let L be a BL-algebra and $\phi : P \to L$ be a BL-function on a nonempty set P based on L. Then for elements $l_1, l_2 \in L$ we have the following assertion,*

$$l_1 \leq l_2 \implies P_{l_1} \subseteq P_{l_2} \tag{10}$$

Proof. Assume that $l_1, l_2 \in L$ are arbitrary elements such that $l_1 \leq l_2$. Hence $l_1 \to l_2 = 1$. Moreover let $x \in P_{l_1}$. Then $\phi(x) \to l_1 = 1$, which means that $\phi(x) \leq l_1$. By using (5) we have

$$l_1 \to l_2 \leq \phi(x) \to l_2.$$

Thus using (1), we conclude that

$$(l_1 \to l_2) \to (\phi(x) \to l_2) = 1 \to (\phi(x) \to l_2) = \phi(x) \to l_2 = 1.$$

Therefore, $x \in P_{l_2}$, that is $P_{l_1} \subseteq P_{l_2}$. □

Theorem 1. *Let $\phi : P \to L$ be a BL-function on P. Then*

(i) $(\forall p_1, p_2 \in P)\big(\phi(p_1) \neq \phi(p_2) \Leftrightarrow P_{\phi(p_1)} \neq P_{\phi(p_2)}\big).$

(ii) $(\forall l \in L)(\forall p \in P)\big(p \in P_l \Leftrightarrow P_{\phi(p)} \subseteq P_l\big).$

Proof. For BL-function $\phi : P \to L$, let $p_1, p_2 \in P$ be such that $P_{\phi(p_1)} \neq P_{\phi(p_2)}$. Moreover, suppose that $\phi(p_1) = \phi(p_2)$. If $x \in P_{\phi(p_1)}$, then

$$\phi(x) \to \phi(p_1) = \phi(x) \to \phi(p_2) = 1,$$

which means that $x \in P_{\phi(p_2)}$. Thus $P_{\phi(p_1)} \subseteq P_{\phi(p_2)}$. Similarly, $P_{\phi(p_2)} \subseteq P_{\phi(p_1)}$, and this is a contradiction. Therefore, for all $p_1, p_2 \in P$, if $P_{\phi(p_1)} \neq P_{\phi(p_2)}$, then $\phi(p_1) \neq \phi(p_2)$. This proves the ($\Leftarrow$).

In order to prove ($\Rightarrow$), suppose that $p_1, p_2 \in P$ such that $\phi(p_1) \neq \phi(p_2)$. Then clearly $\phi(p_1) \to \phi(p_2) \neq 1$ or $\phi(p_2) \to \phi(p_1) \neq 1$. Hence,

$$P_{\phi(p_1)} = \{x \in P \mid \phi(x) \to \phi(p_1) = 1\}$$
$$\neq \{x \in P \mid \phi(x) \to \phi(p_2) = 1\}$$
$$= P_{\phi(p_2)}.$$

Therefore, $P_{\phi(p_1)} \neq P_{\phi(p_2)}$.

(ii) Let $l \in L$ and $p \in P$ be such that $p \in P_l$. Then $\phi(p) \to l = 1$, and using Proposition 2,

$$P_{\phi(p)} \subseteq P_l.$$

Conversely, suppose that $P_{\phi(p)} \subseteq P_l$ for $l \in L$ and $p \in P$. Because $\phi(p) \to \phi(p) = 1$, we conclude that $p \in P_{\phi(p)}$. Therefore, $p \in P_l$ and the proof is complete. □

Theorem 1 part (ii) shows that the converse of Proposition 2 is true. Thus, we have the following corollary.

Corollary 2. *Let $\phi : P \to L$ be a BL-function on a nonempty set P based on L, where L is a BL-algebra. Then*

$$(\forall p_1, p_2 \in A)\big(\phi(p_1) \to \phi(p_2) = 1 \Leftrightarrow P_{\phi(p_1)} \subseteq P_{\phi(p_2)}\big). \tag{11}$$

Proposition 3. *Let $\phi : P \to L$ be a BL-function on a nonempty set P and $M \subseteq L$. Put*

$$\alpha = \inf\{m \mid m \in M\}.$$

Then

$$P_\alpha = \cap\{P_m \mid m \in M\} \tag{12}$$

Proof. Note that there exists the infimum of M in L for any $M \subseteq L$. Thus, for the infimum element α of M we have

$$x \in P_\alpha \Leftrightarrow \phi(x) \to \alpha = 1$$
$$\Leftrightarrow (\forall m \in M)(\phi(x) \to m = 1)$$
$$\Leftrightarrow (\forall m \in M)(x \in P_m)$$
$$\Leftrightarrow x \in \cap\{P_m \mid m \in M\}.$$

□

For a BL-algebra L and a BL-function $\phi : P \to L$ on a nonempty set P, define the sets P_L and ϕ_L as follows:

$$P_L := \{P_l \mid l \in L\}, \ \phi_L := \{\phi_l \mid l \in L\}.$$

Then we have the following corollary.

Corollary 3. *If $\phi : P \to L$ is a BL-function on a nonempty set P, then*
(i) $P_{\inf\{l|l\in L\}} = \cap \{P_l \mid l \in L\}$,
(ii) *for $l_1, l_2 \in L$, we have $P_{l_1} \cap P_{l_2} \in P_L$.*

Proposition 4. *Let $\phi : P \to L$ be a BL-function on a nonempty set P. Then P is represented by the union of P_l for $l \in L$, that is,*
$$P = \cup \{P_l \mid l \in L\}. \tag{13}$$

Proof. Obviously, $\cup \{P_l \mid l \in L\} \subseteq P$. Let $p \in P$ and $l \in L$ be such that $\phi(p) = l$. Then by using the definition of l-subset of P, we have $p \in P_l$. Thus,
$$p \in \cup \{P_l \mid l \in L\},$$
which means that $P \subseteq \cup \{P_l \mid l \in L\}$. Therefore, $P = \cup \{P_l \mid l \in L\}$. □

Proposition 5. *Let $\phi : P \to L$ be a BL-function on a nonempty set P and $p \in P$. Then*
$$\cap \{P_l \mid p \in P_l\} \in P_L. \tag{14}$$

Proof. Remember that by using (6) and (7), we conclude that for any $p \in P$,
$$p \in P_l \Leftrightarrow \phi_l(p) = 1.$$

It follows from Proposition 3 that
$$\cap \{P_l \mid p \in P_l\} = \cap \{P_l \mid \phi_l(p) = 1\} = P_{\inf\{l|p \in P_l\}} \in P_l.$$
□

Let $\phi : P \to L$ be a *BL*-function on a nonempty set P and $\approx$ be a binary relation on L defined by
$$(\forall l_1, l_2 \in L)(l_1 \approx l_2 \Leftrightarrow P_{l_1} = P_{l_2}). \tag{15}$$

The binary relation $\approx$ is an equivalence relation on L. Moreover, for an arbitrary element $l \in L$, define the sets $\phi(P)$ and $\{l\}_\leq$ as follows:
$$\phi(P) := \{l \in L \mid \phi(p) = l \text{ for some } p \in P\}$$
$$\{l\}_\leq := \{x \in L \mid x \leq l\} = \{x \in L \mid x \to l = 1\}.$$

The relationships between an equivalence relation $\approx$ and the sets $\phi(P)$ and $\{l\}_\leq$ are described in the following theorem.

Theorem 2. *For a BL-function $\phi : P \to L$ on a nonempty set P and the elements $l_1, l_2 \in L$, we have the following assertion:*
$$l_1 \approx l_2 \Leftrightarrow \phi(P) \cap \{l_1\}_\leq = \phi(P) \cap \{l_2\}_\leq. \tag{16}$$

Proof. Suppose that $l_1, l_2 \in L$. Then
$$\begin{aligned}
l_1 \approx l_2 &\Leftrightarrow P_{l_1} = P_{l_2} \\
&\Leftrightarrow (\forall p \in P)(\phi(p) \to l_1 = 1 \Leftrightarrow \phi(p) \to l_2 = 1) \\
&\Leftrightarrow \{p \in P \mid \phi(p) \in \{l_1\}_\leq\} = \{p \in P \mid \phi(p) \in \{l_2\}_\leq\} \\
&\Leftrightarrow \phi(P) \cap \{l_1\}_\leq = \phi(P) \cap \{l_2\}_\leq.
\end{aligned}$$
□

4. Code Generated by a BL-Algebra

The relation $\approx$ on L that is defined in (15) is an equivalence relation on L. Thus, it provides the partition of L. For any $l \in L$, let $[l]$ denotes an equivalence class containing l, which means that

$$[l] := \{k \in L \mid l \approx k\}.$$

In what follows, a binary block code of length n will be made from an arbitrary finite BL-algebra. In this method n is a natural number; this helps us to generate a binary block code of the desired length n.

For $n \in \mathbb{N}$, let $P = \{p_1, p_2, \ldots, p_n\}$ and L be a finite BL-algebra. Every BL-function $\phi : P \to L$ on P determines a binary block code C of length n in the following way:

Let $l \in L$. Then for $[l]$ the correspondence codeword is $c_l = c_1 c_2 \cdots c_n$ such that for $1 \leq i \leq n$

$$c_i = \phi_l(p_i) \tag{17}$$

where $p_i \in P$. We called C a BL-code based on L and denoted by C_L. If there is no confusion of L, we use C instead of C_L.

During our study of block code generated by an arbitrary BL-algebra, three parameters are important. The first parameter is the code length n. In the BL-code based on L, we can make a code of the desired length n. This can be helpful as we can choose the length in different situations. The second parameter that we consider is the total number of codewords. In this kind of code, the total number of codewords is equal to the total number of distinct equivalence classes of $\approx$ relation. The third parameter is the distance between pairs of codewords in a code. In the following examples, these notations will be explained much more.

Example 2. *Let $L = \{0, a, b, 1\}$ be a set with Cayley tables as follows:*

$\odot$	0	a	b	1
0	0	0	0	0
a	0	0	a	a
b	0	a	b	b
1	0	a	b	1

$\to$	0	a	b	1
0	1	1	1	1
a	a	1	1	1
b	0	a	1	1
1	0	a	b	1

Then $\mathcal{L} := (L, \wedge, \vee, \odot, \to, 0, 1)$ is a BL-algebra. For a set $P = \{p_1, p_2, p_3, p_4\}$, let $\phi : P \to L$ be a BL-function on P given by

$$\phi(p_1) = a, \ \phi(p_2) = 0, \ \phi(p_3) = 1 \ \text{and} \ \phi(p_4) = b$$

Then the l-subsets of P are

$$P_0 = \{p_2\}, P_a = \{p_1, p_2\}, \ P_b = \{p_1, p_2, p_4\} \ \text{and} \ P_1 = P.$$

In addition, the l-functions of ϕ are

ϕ_p	p_1	p_2	p_3	p_4
ϕ_0	0	1	0	0
ϕ_a	1	1	0	0
ϕ_b	1	1	0	1
ϕ_1	1	1	1	1

Clearly, we have four different equivalence classes, which are $[0], [a], [b]$, and $[1]$. Thus the total number of codewords is 4 ($M = 4$). By using (17), we conclude that

$$c_0 = 0100, \ c_a = 1100, \ c_b = 1101, \ c_1 = 1111.$$

Thus the binary block code C of length $n = 4$ and 4 codewords is $C = \{0100, 1100, 1101, 1111\}$. Besides, the minimum distance of C is 1 ($d(C) = 1$). It is clear that the code C is not a linear code because $0000 \notin C$.

Let $c_l = l_1 l_2 \cdots l_n$ and $c_k = k_1 k_2 \cdots k_n$ be two codewords belonging to a binary block-code C of length n. Define an order relationship $\preceq$ on the set of codewords belonging to a binary block-code C of length n as follows:

$$c_l \preceq c_k \Leftrightarrow l_i \leq k_i \text{ for } i = 1, 2, \cdots, n. \tag{18}$$

By using (2) for the BL-algebra L and (18) for the BL-code C based on L, we conclude that the graphs of L concerning the order $\leq$ and the code C with respect to the order $\preceq$ have the same structures. For instance, in Example 2, we have

Therefore, we can have the following theorem.

Theorem 3. *Let $\mathcal{L} := (L, \wedge, \vee, \odot, \rightarrow, 0, 1)$ be a finite BL-algebra and $|L| = n$, where $n \in \mathbb{N}$. Then L determines a block-code C of length n (namely BL-code) such that the graph of L with respect to its order $\leq$ and the graph of BL-code C with respect to the order $\preceq$ have the same structure.*

Proof. Let $\mathcal{L} := (L, \wedge, \vee, \odot, \rightarrow, 0, 1)$ be a finite BL-algebra and $L = \{l_1, l_2, \cdots, l_n\}$. Moreover, suppose that $P = L$. Then P is a nonempty set and $\phi : L \rightarrow L$ defined by $\phi(l_i) = l_i$, for $1 \leq i \leq n$ is a BL-function on L based on L. Suppose that $\frac{L}{\approx}$ be a set of all equivalence classes of the elements of L regarding the equivalence relation $\approx$ defined in (15). That is,

$$\frac{L}{\approx} = \{[l] \mid l \in L\}.$$

Define the mapping $\psi : \frac{L}{\approx} \longrightarrow C$ by

$$\psi([l_i]) = c_{l_i}, \tag{19}$$

whereby using (17), we have

$$c_{l_i} = \phi_{l_i}(l_1)\phi_{l_i}(l_2)\ldots\phi_{l_i}(l_n),$$

for $1 \leq i \leq n$.

Moreover, let $[l_i] = [l_j]$, for $i, j \in \{1, 2, \ldots, n\}$, and $i \neq j$. Then using (7) and (15), we conclude that

$$\begin{aligned}
[l_i] = [l_j] &\Leftrightarrow L_{l_i} = L_{l_j} \\
&\Leftrightarrow \{l \in L \mid \phi(l) \to l_i = 1\} = \{l \in L \mid \phi(l) \to l_j = 1\} \\
&\Leftrightarrow \{l \in L \mid \phi_{l_i}(l) = 1\} = \{l \in L \mid \phi_{l_j}(l) = 1\} \\
&\Leftrightarrow (\forall l \in L)\left(\phi_{l_i}(l) = 1 \Leftrightarrow \phi_{l_j}(l) = 1\right) \\
&\Leftrightarrow (\forall l \in L)\left(\phi_{l_i}(l) = 0 \Leftrightarrow \phi_{l_j}(l) = 0\right) \\
&\Leftrightarrow c_{l_i} = c_{l_j}.
\end{aligned}$$

This means that the function ψ in (19) is well defined and the inverse implications show that the function ψ is one-to-one.

Now suppose that $l_i, l_j \in L$ are such that $l_i \leq l_j$, for $1 \leq i, j \leq n$. Then Proposition 2 shows that $L_{l_i} \subseteq L_{l_j}$. If $l \in L$ and $l \in L_{l_i}$, then $\phi_{l_i}(l) = 1$. Because $L_{l_i} \subseteq L_{l_j}$, $l \in L_{l_j}$ and $\phi_{l_j}(l) = 1$, therefore $\phi_{l_i}(l) \leq \phi_{l_j}(l)$. Thus, in this case $c_{l_i} \leq c_{l_j}$.

If $l \in L$ and $l \notin L_{l_i}$, then $\phi_{l_i}(l) = 0$ and we have two opportunities. The first one is $l \in L_{l_j}$, which means that $\phi_{l_j}(l) = 1$, and the second one is $l \notin L_{l_j}$, which means that $\phi_{l_j}(l) = 0$. In both case, we have $\phi_{l_i}(l) \leq \phi_{l_j}(l)$. Hence, $c_{l_i} \leq c_{l_j}$. Therefore, if $l_i \leq l_j$, then $\psi(l_i) \preceq \psi(l_j)$, that is, ψ preserves the order. Therefore, the figures of $(L, \leq)$ and $(C, \preceq)$ have the same structures. □

Example 3. Let $L = \{0, a, b, c, d, 1\}$ be a set with the following Cayley tables.

$\odot$	1	a	b	c	d	0
1	1	a	b	c	d	0
a	a	b	b	d	0	0
b	b	b	b	0	0	0
c	c	d	0	c	d	0
d	d	0	0	d	0	0
0	0	0	0	0	0	0

$\to$	1	a	b	c	d	0
1	1	a	b	c	d	0
a	1	1	a	c	c	d
b	1	1	1	c	c	c
c	1	a	b	1	a	b
d	1	1	a	1	1	a
0	1	1	1	1	1	1

Then $\mathcal{L} = (L, \wedge, \vee, \odot, \to, 0, 1)$ is a BL-algebra. For a set $P = \{p_i \mid i = 1, 2, \cdots, 6\}$, let $\phi : P \to L$ be a BL-function on a set P given by

$$\phi(p_1) = 0,\ \phi(p_2) = a,\ \phi(p_3) = b,\ \phi(p_4) = c,\ \phi(p_5) = d,\text{ and }\phi(p_6) = 1.$$

Then

ϕ_p	p_1	p_2	p_3	p_4	p_5	p_6
ϕ_0	1	0	0	0	0	0
ϕ_a	1	1	1	0	1	0
ϕ_b	1	0	1	0	0	0
ϕ_c	1	0	0	1	1	0
ϕ_d	1	0	0	0	1	0
ϕ_1	1	1	1	1	1	1

Thus, using the (17) we have

$$c_0 = 100000,\ c_a = 111010,\ c_b = 101000,\ c_c = 100110,\ c_d = 100010,\ c_1 = 111111.$$

Finally, the generated binary block code C based on L is:

$$C = \{100000, 111010, 101000, 100110, 100010, 111111\}.$$

Moreover, the graph of C using the order (18) is as follows:

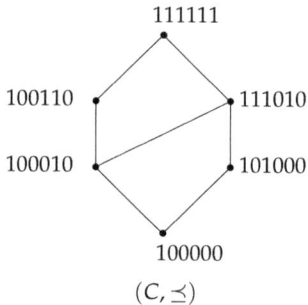

$(C, \preceq)$

which is the same with the graph of L and the order (2), that is,

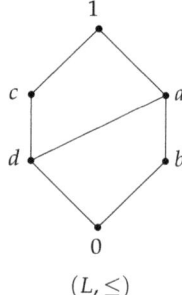

$(L, \leq)$

5. Conclusions

In this paper, we have studied the code generated by a BL-algebra as one of the important classes of ordered algebra. For this goal, the notion of BL-function on a nonempty set P based on BL-algebra L was introduced, and for $l \in L$, l-fuctions of a BL-fuction and l-subsets of P were studied. After investigating some results concerning the BL-functions, our study has focused on a binary equivalence relation $\approx$ on L, and using this relation we define the code C based on L. Finally, we have defined an order between the codewords of C, which gives the code C the ordered structure. Moreover, the graph of C with its order and the graph of L have the same structures.

The results related to BL-code C show that, in general, this code is not linear. For our future work, we will concentrate on some conditions that make this binary block code a linear code. Moreover, using the notations and ideas of this article, the order that we defined between the codewords of the code C based on BL-algebra L can help us to find a new algorithm for decoding the ciphertext. In our future research, we focus on this part of the decoding algorithm.

Author Contributions: Conceptualization, H.B.; methodology, H.B.; investigation, H.B.; writing—original draft preparation, H.B.; writing—review and editing, H.B. All authors have read and agreed to the published version of the manuscript.

Funding: No funding.

Institutional Review Board Statement: Not applicable.

Informed Consent Statement: Not applicable.

Data Availability Statement: Not applicable.

Conflicts of Interest: The author declares no conflict of interest.

References

1. Hájek, P. *Metamathematics of Fuzzy Logic*; Kluwer Academic Publishers: Dordrecht, The Netherlands, 1998.
2. Shannon, C.E. A Mathematical Theory of Communications. *Bell Syst. Tech. J.* **1948**. *27*, 379–423 [CrossRef]
3. Hamming, R.W. Error detecting and error-correcting codes. *Bell Syst. Tech. J.* **1950**, *29*, 147–160. [CrossRef]
4. Jun, Y.B.; Song, S.Z. Codes based on BCK-algebras. *Inf. Sci.* **2011**, *181*, 5102–5109. [CrossRef]
5. Saeid, A.B.; Fatemidokht, H.; Flaut, C.; Rafsanjani, M.K. On codes based on BCK-algebras. *J. Intell. Fuzzy Syst.* **2015**, *29*, 2133–2137. [CrossRef]
6. Flaut, C. BCK-algebras arising from block codes. *J. Intell. Fuzzy Syst.* **2015**, *28*, 1829–1833. [CrossRef]
7. Gilani, A. Codes generates by BCK-functions in BCK-algebras. In Proceedings of the 4th International Conference on Combinatorics, Cryptography, Computer Science and Computing, Tehran City, Iran, 20–21 November 2019.
8. Bordbar, H. BCK codes. *Adv. Comp. Int.* **2022**, *2*, 4. [CrossRef]
9. Chinram, R.; Iampan, A. Codewords generated by UP-valued functions. *AIMS Math.* **2021**, *6*, 4771–4785. [CrossRef]
10. Mostafa, M.; Youssef, B.; Jad, H.A. Coding theory applied to KU-algebras. *J. New Theory* **2015**, *6*, 43–53.
11. Busneag, D.; Piciu, D. On the lattice of deductive systems of a BL-algebra. *Cent. Eur. Sci. J.* **2003**, *2*, 221–237.
12. Hedayati, H.; Jun, Y.B. Chain conditions on fuzzy positive implicative filters of BL-algebras. *Hacet. J. Math. Stat.* **2011**, *40*, 819–828.

Article

A Lower Bound for the Distance Laplacian Spectral Radius of Bipartite Graphs with Given Diameter

Linming Qi [1,*], **Lianying Miao** [2], **Weiliang Zhao** [1] **and Lu Liu** [3]

[1] Department of Fundamental Courses, Zhejiang Industry Polytechnic College, Shaoxing 312000, China; zhaoweiliang@zjipc.edu.cn
[2] School of Mathematics, China University of Mining and Technology, Xuzhou 221116, China; miaolianying@cumt.edu.cn
[3] College of Economics and Management, Shandong University of Science and Technology, Qingdao 266590, China; magic_liu@sdust.edu.cn
* Correspondence: qilinmingxr@163.com or qilinming@zjipc.edu.cn

Abstract: Let G be a connected, undirected and simple graph. The distance Laplacian matrix $\mathcal{L}(G)$ is defined as $\mathcal{L}(G) = diag(Tr) - \mathcal{D}(G)$, where $\mathcal{D}(G)$ denotes the distance matrix of G and $diag(Tr)$ denotes a diagonal matrix of the vertex transmissions. Denote by $\rho_{\mathcal{L}}(G)$ the distance Laplacian spectral radius of G. In this paper, we determine a lower bound of the distance Laplacian spectral radius of the n-vertex bipartite graphs with diameter 4. We characterize the extremal graphs attaining this lower bound.

Keywords: distance Laplacian matrix; spectral radius; diameter

MSC: 05C50

1. Introduction

The distance Laplacian and distance signless Laplacian matrices of a graph G proposed by Aouchiche and Hansen [1] are defined as $\mathcal{L}(G) = diag(Tr) - \mathcal{D}(G)$ and $\mathcal{Q}(G) = diag(Tr) + \mathcal{D}(G)$, respectively. Much attention has been paid to them since they were put forward. Aouchiche et al. [2] described some elementary properties of the distance Laplacian eigenvalues of graphs. Niu et al. [3] determined some extremal graphs minimizing the distance Laplacian spectral radius among bipartite graphs in terms of the matching number and the vertex connectivity, respectively. Nath and Paul [4] focused on the graph whose complement is a tree or a unicyclic graph and considered the second-smallest distance Laplacian eigenvalue. Lin and Zhou [5] determined some extremal graphs among several classes of graphs. Tian et al. [6] proved four conjectures put forward by Aouchiche and Hansen in [2]. One can refer to [7–11] for more details on the distance signless Laplacian spectral radius of graphs.

Although lots of conclusions have been obtained, many more problems remain unsolved. For instance, there are few papers focusing on the distance (signless) Laplacian spectral radius of graphs in terms of diameter, an important parameter of graphs. For adjacency matrices of graphs, several conclusions with respect to the diameter have been derived (e.g., [12–14]). In [12], the authors determined some extremal graphs with small diameter. Generally, the communication network is organized with small diameter to improve the quality of the service on the networks. Motivated by this, in the present paper, we deduce a lower bound of the distance Laplacian spectral radius among bipartite graphs with diameter 4, and we hope that it could be used to address a general case.

This paper is arranged as follows. In Section 2, some elementary notions and lemmas applied in the next parts are presented. In Section 3, the lower bound for the distance Laplacian spectral radius is obtained for bipartite graphs with diameter 4. Moreover, the extremal graph attaining the lower bound is determined.

2. Preliminaries

All graphs considered in this paper are undirected, connected and simple. By $V(G)$, we denote the vertex set of G, and the order of G is $|V(G)|$. Denote by $N_G(u)$ the set of vertices adjacent to u. If $N_G(u) = N_G(v)$ for $u, v \in V(G)$, then they are called twin points. Generally, a subset $S \subset V(G)$ is called a twin point set, if $N_G(u) = N_G(v)$ for any $u, v \in S$. The distance between $u, v \in V(G)$, denoted by $d(u, v)$, is the length of the shortest path between u and v. The diameter of graph G, written as $d(G)$ (d for short), is the maximum distance among all pairs of vertices of G. The chromatic number of G means the least number of colors required to color all the vertices of G such that each pair of adjacent vertices has different colors. The spanning subgraph of G is obtained by deleting some edges from G with order invariable. The transmission $Tr_G(u)$ of a vertex u is referred to as the sum of the distances of u to all other vertices of $V(G)$, i.e., $Tr_G(u) = \sum_{v \in V(G)} d(v, u)$. $Tr_{max}(G)$ means the maximal vertex transmission of G. Let $\mathcal{B}_{n,d}$ be the set of all n-vertex bipartite graphs with diameter d and $\mathcal{C}_{n,k}$ the set of all n-vertex graphs with chromatic number k.

Suppose $V(G) = \{v_1, v_2, \ldots, v_n\}$. The distance matrix $\mathcal{D}(G)$ of G is an $n \times n$ symmetric real matrix with $d(v_i, v_j)$ as the (i, j)-entry. Let the diagonal matrix $diag(Tr)$, called the vertex transmission matrix of G, be

$$diag(Tr) = diag(Tr_G(v_1), Tr_G(v_2), \ldots, Tr_G(v_n)).$$

The largest eigenvalue of the distance Laplacian matrix $\mathcal{L}(G)$ is called the distance Laplacian spectral radius, written as $\rho_\mathcal{L}(G)$. For any matrix M, $\lambda_1(M)$ always denotes the largest eigenvalue of M.

A vector $x = (x_1, x_2, \ldots, x_n)^T$ can be considered as a function defined on $V(G) = \{v_1, v_2, \ldots, v_n\}$, which maps v_i to x_i, i.e., $x(v_i) = x_i$. Thus, for $\mathcal{L}(G)$,

$$x^T \mathcal{L}(G) x = \sum_{\{u,v\} \subseteq V(G)} d(u, v)(x(u) - x(v))^2.$$

It is clear that $\mathbf{1} = (1, 1, \ldots, 1)^T$ is an eigenvector corresponding to the eigenvalue zero of $\mathcal{L}(G)$. Thus, if $x = (x_1, x_2, \ldots, x_n)^T$ is an eigenvector of $\mathcal{L}(G)$ corresponding to a nonzero eigenvalue, then $\sum_{i=1}^n x_i = 0$.

Lemma 1 (Rayleigh's Principal Theorem, p. 29, [15]). *Let A be a symmetric real matrix and u any unit nonzero vector. Then $\lambda_1(A) \geq u^T A u$ with equality if and only if u is the eigenvector corresponding to $\lambda_1(A)$.*

Lemma 2 (Courant-Weyl Inequality, p. 31, [15]). *Let A_1 and A_2 be two symmetric real matrices of order n. Then*

$$\lambda_n(A_2) + \lambda_i(A_1) \leq \lambda_i(A_1 + A_2) \text{ for } 1 \leq i \leq n.$$

Lemma 3 (Interlacing Theorem, p. 30, [15]). *Suppose A is a symmetric real matrix of order n and M a principal submatrix of A with order $s(\leq n)$. Then*

$$\lambda_i(A) \geq \lambda_i(M), \ 1 \leq i \leq s.$$

The next lemma follows from Lemma 3 immediately.

Lemma 4 (Proposition 2.11, [6]). *Let G be an n-vertex graph and M a principal submatrix of $\mathcal{L}(G)$ with order $s \leq n$. Then $\lambda_1(M) \leq \rho_\mathcal{L}(G)$.*

Lemma 5 (Theorem 3.5, [1]). *Suppose $G + e_{uv}$ is the graph obtained from G by adding an edge e_{uv} joining u and v. Then $\rho_\mathcal{L}(G) \geq \rho_\mathcal{L}(G + uv)$.*

3. The Lower Bound of the Distance Laplacian Spectral Radius of Graphs among $\mathcal{B}_{n,4}$

If $G \in \mathcal{B}_{n,d}$, then there exists a partition $\{V_0, V_1, \ldots, V_d\}$ of $V(G)$ such that $|V_0| = 1$ and $d(u,v) = i$ for $u \in V_0$ and $v \in V_i$ $(i = 1, 2, \ldots, d)$.

Lemma 6 (Lemma 2.1, [12]). *Let $G \in \mathcal{B}_{n,d}$ with a vertex partition described as above. Then $G[V_i]$ induces an empty graph (i.e., containing no edge) for each $i \in \{0, 1, \ldots, d\}$.*

Lemma 7. *Let $d \geq 3$ and $G \in \mathcal{B}_{n,d}$. If $d(G+e) < d$ when any edge e is added to G, then $|V_d| = 1$ and the induced subgraph $G[V_{i-1} \cup V_i]$ $(i = 1, 2, \ldots, d)$ is a complete bipartite graph.*

Proof. From Lemma 6, it is clear that $G[V_{i-1} \cup V_i]$ $(i = 1, 2, \ldots, d)$ is a complete bipartite graph. Moreover, let $u \in V_d$ and $v \in V_{d-3}$. Assume, on the contrary, that $|V_d| \geq 2$, then the graph $G + e_{uv} \in \mathcal{B}_{n,d}$, a contradiction. □

Remark 1. *Denote a subset of $\mathcal{B}_{n,d}$ by $\widetilde{\mathcal{B}_{n,d}}$, consisting of all the graphs satisfying Lemma 7. For instance, if $G \in \widetilde{\mathcal{B}_{n,4}}$, then G is of the form shown in Figure 1. Then the partition of $V(G)$ can be written as $V_0 = \{w\}$, $V_1 = \{v_1, \ldots, v_s\}$, $V_2 = \{u_1, \ldots, u_t\}$, $V_3 = \{y_1, \ldots, y_k\}$ and $V_4 = \{z\}$, where $s + t + k + 2 = n$ and $s, t, k \geq 1$.*

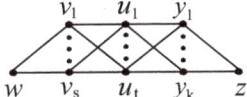

Figure 1. A graph $G \in \widetilde{\mathcal{B}_{n,4}}$.

Before giving the main conclusion of this section, we first investigate the properties of the eigenvector corresponding to $\rho_{\mathcal{L}}(G)$ for $G \in \widetilde{\mathcal{B}_{n,4}}$.

Let $G \in \widetilde{\mathcal{B}_{n,4}}$ and the partition of $V(G)$ be arranged as in Remark 1. Without loss of generality, suppose $|V_3| \geq |V_1| \geq 1$ (i.e., $k \geq s \geq 1$).

Lemma 8. *Let the eigenvector corresponding to $\rho_{\mathcal{L}}(G)$ be x. Then*

$$\begin{cases} x(v_i) = x(v_j) \ (1 \leq i,j \leq s), \\ x(u_i) = x(u_j) \ (1 \leq i,j \leq t), \\ x(y_i) = x(y_j) \ (1 \leq i,j \leq k). \end{cases}$$

Proof. Since the proofs of the three results are parallel, here we only give the first one. As the vertices of V_1 are twin points (if $s > 1$), $d(v, v_i) = d(v, v_j)$ for each $v \in V(G) \setminus \{v_i, v_j\}$, and thus $Tr(v_i) = Tr(v_j) = 2s + t + 2k + 2$. Considering the characteristic equations indexed by v_i and v_j, it is obtained that

$$\begin{cases} \rho_{\mathcal{L}}(G) \cdot x(v_i) = \sum_{v \in V(G)} d(v_i, v)(x(v_i) - x(v)) \\ \rho_{\mathcal{L}}(G) \cdot x(v_j) = \sum_{v \in V(G)} d(v_j, v)(x(v_j) - x(v)). \end{cases}$$

Then it follows that $\rho_{\mathcal{L}}(G) \cdot (x(v_i) - x(v_j)) = (Tr(v_i) + 2)(x(v_i) - x(v_j))$. From Lemma 4, we easily obtain

$$\begin{aligned} \rho_{\mathcal{L}}(G) &\geq Tr_{max} = Tr(w) = s + 2t + 3k + 4 \\ &> 2s + t + 2k + 4 = Tr(v_i) + 2. \end{aligned}$$

Thus, $x(v_i) = x(v_j)$ follows. □

For the eigenvector x in Lemma 8, suppose $x(w) = x_0$, $x(v_i) = x_1$ ($1 \leq i \leq s$), $x(u_i) = x_2$ ($1 \leq i \leq t$), $x(y_i) = x_3$ ($1 \leq i \leq k$) and $x(z) = x_4$. Then x can be written as

$$x = (x_0, \underbrace{x_1, \ldots, x_1}_{s}, \underbrace{x_2, \ldots, x_2}_{t}, \underbrace{x_3, \ldots, x_3}_{k}, x_4)^T.$$

Lemma 9. *Let x be as just described. If $|V_1| = |V_3|$ (i.e., $s = k$), then*

$$\begin{cases} x_0 = -x_4 \neq 0, \\ x_1 = -x_3 \neq 0, \\ x_2 = 0. \end{cases}$$

Proof. Applying Lemma 8, the characteristic equation $\mathcal{L}(G) \cdot x = \rho_\mathcal{L}(G) \cdot x$ can be simplified in the conventional form as follows:

$$\begin{cases} m_0 \cdot x_0 - s \cdot x_1 - 2t \cdot x_2 - 3k \cdot x_3 - 4x_4 = \rho_\mathcal{L}(G) \cdot x_0 \\ -x_0 + m_1 \cdot x_1 - t \cdot x_2 - 2k \cdot x_3 - 3x_4 = \rho_\mathcal{L}(G) \cdot x_1 \\ -2x_0 - s \cdot x_1 + m_2 \cdot x_2 - k \cdot x_3 - 2x_4 = \rho_\mathcal{L}(G) \cdot x_2 \\ -3x_0 - 2s \cdot x_1 - t \cdot x_2 + m_3 \cdot x_3 - x_4 = \rho_\mathcal{L}(G) \cdot x_3 \\ -4x_0 - 3s \cdot x_1 - 2t \cdot x_2 - k \cdot x_3 + m_4 \cdot x_4 = \rho_\mathcal{L}(G) \cdot x_4, \end{cases} \quad (1)$$

where $m_0 = Tr(w)$, $m_1 = Tr(v_i) - 2(s-1)$, $m_2 = Tr(u_i) - 2(t-1)$, $m_3 = Tr(y_i) - 2(k-1)$ and $m_4 = Tr(z)$.

The sum of the first equality and the fifth one gives

$$(n + t + 2k + 2)x_0 - 4(x_0 + sx_1 + tx_2 + kx_3 + x_4) + (n + t + 2s + 2)x_4 = \rho_\mathcal{L}(G) \cdot (x_0 + x_4). \quad (2)$$

Since $s = k$ and $x_0 + sx_1 + tx_2 + kx_3 + x_4 = 0$, we have

$$2n(x_0 + x_4) = \rho_\mathcal{L}(G) \cdot (x_0 + x_4). \quad (3)$$

Take a 2×2 principal submatrix M of $\mathcal{L}(G)$, where $M = \begin{pmatrix} Tr(w) & -4 \\ -4 & Tr(z) \end{pmatrix}$. Note that $Tr(w) = Tr(z) = 2n$ for $s = k$. Then, applying Lemma 4,

$$\rho_\mathcal{L}(G) \geq \lambda_1(M) = Tr(w) + 4 = 2n + 4.$$

Thus, we obtain $x_0 + x_4 = 0$, i.e., $x_0 = -x_4$ from (3). Similarly, from the second and the fourth equalities in (1), it follows that

$$-2(x_0 + x_4) + (n+2)(x_1 + x_3) = \rho_\mathcal{L}(G) \cdot (x_1 + x_3).$$

Since $x_0 + x_4 = 0$ and $\rho_\mathcal{L}(G) \geq 2n + 4 > n + 2$, $x_1 + x_3 = 0$, i.e., $x_1 = -x_3$. The fourth equality in (1) minus the second one indicates that

$$\begin{aligned} 2(x_4 - x_0) &= [\rho_\mathcal{L}(G) - (2s + t + 2k + 4)](x_3 - x_1) \\ &= (\rho_\mathcal{L}(G) - 2n + t)(x_3 - x_1). \end{aligned} \quad (4)$$

If $x_0 = 0$, then $x_4 = -x_0 = 0$. Further, from (4), it follows that $x_1 = -x_3 = 0$ (note that $\rho_\mathcal{L}(G) \geq 2n + 4$). Recalling that $x_0 + sx_1 + tx_2 + kx_3 + x_4 = 0$, we know $x_2 = 0$, and thus x is a zero vector, a contradiction. Hence, $x_0 = -x_4 \neq 0$. Similarly, $x_1 = -x_3 \neq 0$, which implies $x_2 = 0$. The proof is complete. □

For convenience, denote the graph $G \in \widetilde{\mathcal{B}_{n,4}}$ by $G(1, s, t, k, 1)$ with vertex partition shown in Remark 1. We next determine the unique extremal graph minimizing the distance Laplacian spectral radius among $\widetilde{\mathcal{B}_{n,4}}$.

Theorem 1. *The graph $G(1,1,n-4,1,1)$ in Figure 2 is the unique graph with minimum distance Laplacian spectral radius among $\widetilde{\mathcal{B}_{n,4}}$.*

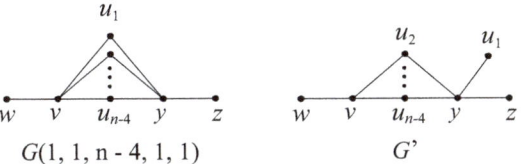

Figure 2. The graph $G(1,1,n-4,1,1) \in \widetilde{\mathcal{B}_{n,4}}$ and graph G'.

Proof. Let $G_0 = G(1,s,t,k,1) \in \widetilde{\mathcal{B}_{n,4}}$ with $s+t+k+2 = n$ and $s,t,k \geq 1$. Without loss of generality, assume that $s \leq k$. We proceed by proving the following three claims, which will imply the conclusion.

Claim 1. If $s \geq 2$ in graph G_0, then let $G_1 = G(1,s-1,t,k+1,1)$. We claim that $\rho_{\mathcal{L}}(G_0) < \rho_{\mathcal{L}}(G_1)$.

In graph G_0, let $V(G_0) = \{V_0, \ldots, V_4\}$ and V_i be expressed as that in Remark 1. Then we easily obtain

$$\begin{cases} Tr(w) = s + 2t + 3k + 4, & Tr(v_i) = Tr(y_i) = 2s + t + 2k + 2, \\ Tr(u_i) = s + 2t + k + 2, & Tr(z) = 3s + 2t + k + 4, \end{cases} \quad (5)$$

and the distance Laplacian matrix of G_0 is

$$\mathcal{L}(G_0) = \begin{pmatrix} Tr(w) & -J_{1\times s} & -2J_{1\times t} & -3J_{1\times k} & -4 \\ -J_{s\times 1} & (Tr(v_i)+2)I_s - 2J_s & -J_{s\times t} & -2J_{s\times k} & -3J_{s\times 1} \\ -2J_{t\times 1} & -J_{t\times s} & (Tr(u_i)+2)I_t - 2J_t & -J_{t\times k} & -2J_{t\times 1} \\ -3J_{k\times 1} & -2J_{k\times s} & -J_{k\times t} & (Tr(y_i)+2)I_k - 2J_k & -J_{k\times 1} \\ -4 & -3J_{1\times s} & -2J_{1\times t} & -J_{1\times k} & Tr(z) \end{pmatrix}.$$

Further, we have

$$|\lambda I_n - \mathcal{L}(G_0)| = (\lambda - Tr(v_i) - 2)^{s-1}(\lambda - Tr(u_i) - 2)^{t-1}(\lambda - Tr(y_i) - 2)^{k-1} \cdot |\lambda I_5 - R(G_0)|, \quad (6)$$

where

$$R(G_0) = \begin{pmatrix} Tr(w) & -s & -2t & -3k & -4 \\ -1 & Tr(v_i) - 2(s-1) & -t & -2k & -3 \\ -2 & -s & Tr(u_i) - 2(t-1) & -k & -2 \\ -3 & -2s & -t & Tr(y_i) - 2(k-1) & -1 \\ -4 & -3s & -2t & -k & Tr(z) \end{pmatrix}. \quad (7)$$

From the above, we say that the largest eigenvalue of $R(G_0)$ is the spectral radius of G_0. In fact, $\lambda_1(R(G_0)) \geq Tr(w)$ from Lemma 3, and $Tr(v_i) + 2$, $Tr(y_i) + 2$ and $Tr(u_i) + 2$ are the eigenvalues of $\mathcal{L}(G_0)$ apart from those of $R(G_0)$ from (6). Furthermore, $Tr(w) > Tr(v_i) + 2 = Tr(y_i) + 2$ and $Tr(w) > Tr(u_i) + 2$ by (5) clearly. Thus, $\lambda_1(R(G_0)) = \rho_{\mathcal{L}}(G_0)$ holds.

For graph G_1, we obtain the matrices $\mathcal{L}(G_1)$ and $R(G_1)$ by substituting $s-1$ and $k+1$ for s and k in $\mathcal{L}(G_0)$ and $R(G_0)$, respectively. Analogously, we have $\lambda_1(R(G_1)) = \rho_{\mathcal{L}}(G_1)$. Denote the characteristic polynomials of $R(G_0)$ and $R(G_1)$ by $\psi_0(\lambda)$ and $\psi_1(\lambda)$, respectively. Next, we are aimed at proving

$$\psi_1(\rho_{\mathcal{L}}(G_0)) < 0. \quad (8)$$

By using MATLAB, we obtain

$$\psi_1(\lambda) - \psi_0(\lambda) = 4(s-k-1)\lambda\left(\lambda^2 - s\lambda - k\lambda - 2n\lambda - 4\lambda + 6n + sn + kn + n^2\right). \quad (9)$$

Let $g(\lambda) = \lambda(\lambda^2 - s\lambda - k\lambda - 2n\lambda - 4\lambda + 6n + sn + kn + n^2)$. Then the derivative of $g(\lambda)$ is
$$g'(\lambda) = 3\lambda^2 - (2s + 2k + 4n + 8)\lambda + n^2 + sn + kn + 6n$$
with symmetry axis $\tilde{\lambda} = \frac{2n+s+k+4}{3}$. Since $s \leq k$, $Tr_{max} = Tr(w) = 2n - s + k$ in graph G_0, and thus $\rho_{\mathcal{L}}(G_0) \geq Tr_{max} = 2n - s + k$ from Lemma 4. By simple calculation, we obtain $\tilde{\lambda} < 2n - s + k$, and since $n \geq s + k + 3$, we have
$$g'(2n - s + k) = 5n^2 + (5k - 11s - 10)n + 5s^2 - 6sk + 8s + k^2 - 8k > 0.$$

We now say that $g(\lambda)$ is strictly increasing for $\lambda \geq 2n - s + k$. Moreover, from $n \geq s + k + 3$, it follows that
$$g(2n - s + k) = (2n - s + k)(n^2 + kn - 3sn - 2n + 4s - 4k - 2sk + 2s^2) > 0.$$

Note that $s - k - 1 < 0$ in (9). Then we have
$$\begin{aligned}\psi_1(\rho_{\mathcal{L}}(G_0)) &= \psi_1(\rho_{\mathcal{L}}(G_0)) - \psi_0(\rho_{\mathcal{L}}(G_0)) < \psi_1(2n - s + k) - \psi_0(2n - s + k) \\ &= 4(s - k - 1) \cdot g(2n - s + k) < 0,\end{aligned}$$
which establishes (8).

Applying (8), we can easily prove that $\rho_{\mathcal{L}}(G_0) < \rho_{\mathcal{L}}(G_1)$. Assume on the contrary that $\rho_{\mathcal{L}}(G_0) > \rho_{\mathcal{L}}(G_1)$ (noting that $\rho_{\mathcal{L}}(G_0) \neq \rho_{\mathcal{L}}(G_1)$ since $\psi_1(\rho_{\mathcal{L}}(G_0)) < 0$ and $\psi_1(\rho_{\mathcal{L}}(G_1)) = 0$). Observing that $\psi_1(\lambda)$ tends to infinity when λ tends to infinity (as the leading coefficient of $\psi_1(\lambda)$ is 1), we can find a sufficiently large $q > \rho_{\mathcal{L}}(G_0)$ such that $\psi_1(q) > 0$. As $\psi_1(\lambda)$ is a continuous function, from $\psi_1(\rho_{\mathcal{L}}(G_0)) < 0$ and $\psi_1(q) > 0$, it follows that $\psi_1(p) = 0$ for a positive number p between $\rho_{\mathcal{L}}(G_0)$ and q, which is a contradiction to the fact that $\rho_{\mathcal{L}}(G_1)$ is the largest root of $\psi_1(\lambda) = 0$. Therefore, $\rho_{\mathcal{L}}(G_0) < \rho_{\mathcal{L}}(G_1)$.

Claim 2. Assume $G_2 = G(1, s - 1, t + 2, k - 1, 1)$, where $k \geq s \geq 2$ and $t \geq 1$. Then we claim that $\rho_{\mathcal{L}}(G_0) > \rho_{\mathcal{L}}(G_2)$.

Let the unit eigenvector corresponding to $\rho_{\mathcal{L}}(G_2)$ be
$$x = (x_0, \underbrace{x_1, \ldots, x_1}_{s-1}, \underbrace{x_2, \ldots, x_2}_{t+2}, \underbrace{x_3, \ldots, x_3}_{k-1}, x_4)^T.$$

By Rayleigh's principle,
$$\begin{aligned}\rho_{\mathcal{L}}(G_0) - \rho_{\mathcal{L}}(G_2) &\geq x^T \cdot (\mathcal{L}(G_0) - \mathcal{L}(G_2)) \cdot x \\ &= \sum_{\{u,v\} \subseteq V(G_0)} (d_{G_0}(u,v) - d_{G_2}(u,v))(x(u) - x(v))^2 \\ &= 2(s - 1)(x_1 - x_2)^2 + 2(k - 1)(x_3 - x_2)^2 \geq 0.\end{aligned} \quad (10)$$

Next, we show that $\rho_{\mathcal{L}}(G_0) - \rho_{\mathcal{L}}(G_2) > 0$. First, if $s = k$, then from Lemma 9, it follows that $x_1 = -x_3 \neq 0$ and $x_2 = 0$, and thus
$$\rho_{\mathcal{L}}(G_0) - \rho_{\mathcal{L}}(G_2) \geq 4(s - 1)x_1^2 > 0.$$

On the other hand, suppose $s < k$ and $2(s - 1)(x_1 - x_2)^2 + 2(k - 1)(x_3 - x_2)^2 = 0$ in (10). Then $x_1 = x_2 = x_3$. Substitute $t + 2$ for t in (4), and then $x_0 = x_4$ follows by applying $x_1 = x_3$. In addition, by replacing s, k and t with $s - 1$, $k - 1$ and $t + 2$ in (2), respectively, it gives
$$\rho_{\mathcal{L}}(G_2) \cdot x_0 = 2n \cdot x_0,$$

and hence $x_0 = 0$ for the reason that $\rho_{\mathcal{L}}(G_2) \geq Tr_{G_2}(w) = 2n + k - s > 2n$. Recalling that $x_0 + (s-1)x_1 + (t+2)x_2 + (k-1)x_3 + x_4 = 0$, we have $x_1 = x_2 = x_3 = 0$, and then eigenvector x is the zero vector, a contradiction. Hence, if $s < k$, then

$$2(s-1)(x_1 - x_2)^2 + 2(k-1)(x_3 - x_2)^2 > 0.$$

In summary, $\rho_{\mathcal{L}}(G_0) - \rho_{\mathcal{L}}(G_2) > 0$ holds.

Claim 3. If $s = k - 1(\geq 1)$ in graph G_0, then let $G_3 = G(1, s, t+1, k-1, 1)$. We claim that $\rho_{\mathcal{L}}(G_0) > \rho_{\mathcal{L}}(G_3)$.

Let the unit eigenvector corresponding to $\rho_{\mathcal{L}}(G_3)$ be

$$x = (x_0, \underbrace{x_1, \ldots, x_1}_{s}, \underbrace{x_2, \ldots, x_2}_{t+1}, \underbrace{x_3, \ldots, x_3}_{k-1=s}, x_4)^T.$$

Then by Rayleigh's principle and Lemma 9,

$$\begin{aligned}
\rho_{\mathcal{L}}(G) - \rho_{\mathcal{L}}(G_3) &\geq x^T \cdot (\mathcal{L}(G) - \mathcal{L}(G_3)) \cdot x \\
&= \sum_{\{u,v\} \subseteq V(G)} (d_G(u,v) - d_{G_3}(u,v))(x(u) - x(v))^2 \\
&= (x_0 - x_2)^2 + s(x_1 - x_2) - t(x_2 - x_2)^2 + s(x_2 - x_3)^2 - (x_2 - x_4)^2 \\
&= 2s \cdot x_1^2 > 0.
\end{aligned}$$

Now we are in a position to complete the proof of the theorem.

For graph $G_0 = G(1, s, t, k, 1) \in \widetilde{\mathcal{B}_{n,4}}$, suppose $k \geq s$. If $k \geq 2$ and $(k - s) \equiv 0 \pmod 2$, then by Claim 1,

$$\rho_{\mathcal{L}}(G_0) \geq \rho_{\mathcal{L}}(G(1, s + \frac{k-s}{2}, t, k - \frac{k-s}{2}, 1)) = \rho_{\mathcal{L}}(G(1, \frac{s+k}{2}, t, \frac{s+k}{2}, 1))$$

with equality if and only if $s = k$. Furthermore, from Claim 2, we have

$$\rho_{\mathcal{L}}(G(1, \frac{s+k}{2}, t, \frac{s+k}{2}, 1)) > \rho_{\mathcal{L}}(G(1, 1, n-4, 1, 1)).$$

On the other side, if $k \geq 3$ and $(k - s) \equiv 1 \pmod 2$, then by Claim 1,

$$\rho_{\mathcal{L}}(G_0) \geq \rho_{\mathcal{L}}(G(1, s + \frac{k-s-1}{2}, t, k - \frac{k-s-1}{2}, 1)) = \rho_{\mathcal{L}}(G(1, \frac{s+k-1}{2}, t, \frac{s+k+1}{2}, 1)),$$

with equality if and only if $k - s = 1$. Moreover, from Claim 3,

$$\rho_{\mathcal{L}}(G(1, \frac{s+k-1}{2}, t, \frac{s+k+1}{2}, 1)) > \rho_{\mathcal{L}}(G(1, \frac{s+k-1}{2}, t+1, \frac{s+k-1}{2}, 1)).$$

Finally, from Claim 2, it follows that

$$\rho_{\mathcal{L}}(G(1, \frac{s+k-1}{2}, t+1, \frac{s+k-1}{2}, 1)) > \rho_{\mathcal{L}}(G(1, 1, n-4, 1, 1)).$$

For the case of $s = 1$ and $k = 2$, from Claim 3, it is straightforward that

$$\rho_{\mathcal{L}}(G(1, 1, n-5, 2, 1)) > \rho_{\mathcal{L}}(G(1, 1, n-4, 1, 1)).$$

This completes the proof. □

From Theorem 1 and Lemma 5, we indicate that if $G \in \mathcal{B}_{n,d}$ is not a spanning subgraph of $G(1,1,n-4,1,1)$, then $\rho_{\mathcal{L}}(G) > \rho_{\mathcal{L}}(G(1,1,n-4,1,1))$. In addition, if $s = k = 1$ and $t = n-4$ in (7), then we obtain $\lambda_1(R_0) = 4 + \frac{1}{2}(3n + \sqrt{n^2 + 16})$ by using MATLAB, i.e.,

$$\rho_{\mathcal{L}}(G(1,1,n-4,1,1)) = 4 + \frac{1}{2}(3n + \sqrt{n^2 + 16}).$$

Thus, we have the following theorem.

Theorem 2. *Let $G \in \mathcal{B}_{n,d}$. Then $\rho_{\mathcal{L}}(G) \geq 4 + \frac{1}{2}(3n + \sqrt{n^2 + 16})$ with equality if and only if $G = G(1,1,n-4,1,1)$.*

Proof. Denote the graph $G(1,1,n-4,1,1) - e_{vu_1}$ by G' (see Figure 2). First, we show that $\rho_{\mathcal{L}}(G') > \rho_{\mathcal{L}}(G(1,1,n-4,1,1))$. Take a 2×2 principal submatrix $M = \begin{pmatrix} Tr(w) & -4 \\ -4 & Tr(z) \end{pmatrix} = \begin{pmatrix} 2n+2 & -4 \\ -4 & 2n \end{pmatrix}$ from $\mathcal{L}(G')$. By simple calculation, $\lambda_1(M) > 2n + 5 > 4 + \frac{1}{2}(3n + \sqrt{n^2 + 16})$. From Lemma 4, it follows that

$$\rho_{\mathcal{L}}(G') \geq \lambda_1(M) > 2n + 5 > 4 + \frac{1}{2}(3n + \sqrt{n^2 + 16}) = \rho_{\mathcal{L}}(G(1,1,n-4,1,1)).$$

Thus, we say that for any spanning subgraph $H \neq G(1,1,n-4,1,1)$ of $G(1,1,n-4,1,1)$, $\rho_{\mathcal{L}}(H) > \rho_{\mathcal{L}}(G(1,1,n-4,1,1))$ from Lemma 5.

Hence, now, it is clear from Theorem 1 and the above result that for any graph $G \in \mathcal{B}_{n,d}$,

$$\rho_{\mathcal{L}}(G) \geq \rho_{\mathcal{L}}(G(1,1,n-4,1,1)) = 4 + \frac{1}{2}(3n + \sqrt{n^2 + 16})$$

with equality if and only if $G = G(1,1,n-4,1,1)$. □

4. Conclusions

In this paper, a lower bound of the distance Laplacian spectral radius of the n-vertex bipartite graphs with diameter 4 is obtained. The method used here is helpful for solving the general case and we conjecture that the graph $G(1,\ldots,1,n-d,1,\ldots,1)$ is the unique one minimizing the distance Laplacian spectral radius among n-vertex bipartite graphs with even diameter $d \geq 4$.

Author Contributions: Conceptualization, L.Q. and L.M.; methodology, W.Z.; formal analysis, L.L.; writing—original draft preparation, L.Q.; writing—review and editing, W.Z. and L.L.; supervision, L.M.; project administration, L.M.; funding acquisition, L.M. All authors have read and agreed to the published version of the manuscript.

Funding: This work was supported by the 2021 Visiting Scholar "Teacher Professional Development Project" in the University of Zhejiang Provincial Department of Education (Grant No. FX2021169).

Institutional Review Board Statement: Not applicable.

Informed Consent Statement: Not applicable.

Data Availability Statement: Not applicable.

Conflicts of Interest: The authors declare no conflict of interest.

References

1. Aouchiche, M.; Hansen, P. Two Laplacians for the distance matrix of a graph. *Linear Algebra Appl.* **2013**, *439*, 21–33. [CrossRef]
2. Aouchiche, M.; Hansen, P. Montréal, Some properties of the distance Laplacian eigenvalues of a graph. *Czechoslov. Math. J.* **2014**, *64*, 751–761. [CrossRef]
3. Niu, A.; Fan, D.; Wang, G. On the distance Laplacian spectral radius of bipartite graphs. *Discret. Appl. Math.* **2015**, *186*, 207–213. [CrossRef]

4. Nath, M.; Paul, S. On the distance Laplacian spectra of graphs. *Linear Algebra Appl.* **2014**, *460*, 97–110. [CrossRef]
5. Lin, H.; Zhou, B. On the distance Laplacian spectral radius of graphs. *Linear Algebra Appl.* **2015**, *475*, 265–275. [CrossRef]
6. Tian, F.; Wong, D. Jianling Rou, Proof for four conjectures about the distance Laplacian and distance signless Laplacian eigenvalues of a graph. *Linear Algebra Appl.* **2015**, *471*, 10–20. [CrossRef]
7. Aouchiche, M.; Hansen, P. On the distance signless Laplacian of a graph. *Linear Multilinear Algebra* **2016**, *64*, 1113–1123. [CrossRef]
8. Xing, R.; Zhou, B. On the distance and distance signless Laplacian spectral radii of bicyclic graphs. *Linear Algebra Appl.* **2013**, *439*, 3955–3963. [CrossRef]
9. Xing, R.; Zhou, B.; Li, J. On the distance signless Laplacian spectral radius of graphs. *Linear Multilinear Algebra* **2014**, *62*, 1377–1387. [CrossRef]
10. Das, K.C. Proof of conjectures on the distance signless Laplacian eigenvalues of graphs. *Linear Algebra Appl.* **2015**, *467*, 100–115 [CrossRef]
11. Lin, H.; Lu, X. Bounds on the distance signless Laplacian spectral radius in terms of clique number. *Linear Multilinear Algebra* **2015**, *63*, 1750–1759. [CrossRef]
12. Zhai, M.; Liu, R.; Shu, J. On the spectral radius of bipartite graphs with given diameter. *Linear Algebra Appl.* **2009**, *430*, 1165–1170. [CrossRef]
13. van Dam, E.R. Graphs with given diameter maximizing the spectral radius. *Linear Algebra Appl.* **2007**, *426*, 454–457. [CrossRef]
14. van Dam, E.R.; Kooij, R.E. The minimal spectral radius of graphs with a given diameter. *Linear Algebra Appl.* **2007**, *423*, 408–419. [CrossRef]
15. Brouwer, A.E.; Haemers, W.H. *Spectra of Graphs*; Springer: New York, NY, USA, 2012.

Article

The Extendability of Cayley Graphs Generated by Transposition Trees

Yongde Feng [1,2], Yanting Xie [1], Fengxia Liu [2] and Shoujun Xu [1,*]

[1] School of Mathematics and Statistics, Gansu Center for Applied Mathematics, Lanzhou University, Lanzhou 730000, China; ydfengxj@163.com (Y.F.); ls_xieyt@lzu.edu.cn (Y.X.)
[2] College of Mathematics and Systems Science, Xinjiang University, Urumqi 830046, China; xjulfx@163.com
* Correspondence: shjxu@lzu.edu.cn

Abstract: A connected graph Γ is k-extendable for a positive integer k if every matching M of size k can be extended to a perfect matching. The extendability number of Γ is the maximum k such that Γ is k-extendable. In this paper, we prove that Cayley graphs generated by transposition trees on $\{1, 2, \ldots, n\}$ are $(n-2)$-extendable and determine that the extendability number is $n-2$ for an integer $n \geq 3$.

Keywords: extendability; cayley graphs; transposition trees; bubble-sort graphs; star graphs

MSC: 05C25; 05C70

1. Introduction

Cayley graphs on a group and a generating set have been an important class of graphs in the study of interconnection networks for parallel and distributed computing [1–6]. Some recent results about topological properties and routing problems on the networks based on Cayley graphs on the symmetric groups with the set of transpositions as the generating sets, including two special classes, the star graphs [5] and bubble-sort graphs [1], can be found in [6–9].

Throughout this paper, we consider finite, simple connected graph. Let Γ be a graph with vertex set $V(\Gamma)$ and edge set $E(\Gamma)$. A graph H is a subgraph of Γ if $V(H) \subseteq V(\Gamma)$ and $E(H) \subseteq E(\Gamma)$. The induced subgraph $\Gamma[C]$ is the subgraph of Γ with vertex set C and edge set $\{uv | u, v \in C, uv \in E(\Gamma)\}$. Let G be a group, S a subset of G such that the identity element does not belong to S and $S = S^{-1}$, where $S^{-1} = \{\tau^{-1} | \tau \in S\}$. The *Cayley graph* Γ, denoted by $\Gamma = \mathrm{Cay}(G, S)$, is the graph whose vertex set $V(\Gamma) = G$ and u, v are adjacent if and only if $u^{-1}v \in S$. It's known that Γ is connected if and only if S is a generating set of G. Furthermore, obviously, all Cayley graphs are vertex-transitive (see [10]).

We denote $\mathfrak{S}_n$ as the symmetric group on n letters (set of all permutations on $\{1, 2, \ldots, n\}$). Now let us restrict S to be a subset of transpositions on $\{1, 2, \ldots, n\}$. Clearly all Cayley graphs $\mathrm{Cay}(\mathfrak{S}_n, S)$ are $|S|$-regular bipartite graphs. The *transposition generating graph* of S, denoted by $T(S)$, is the graph with vertex set $\{1, 2, \ldots, n\}$ and two vertices s and t are adjacent if and only if the transposition (st) is in S. If $T(S)$ is a tree, it is called *transposition trees*.

An edge set $M \subseteq E(\Gamma)$ is called a *matching* of Γ if no two of them share an end-vertex. Moreover, a matching of Γ is said to be *perfect* if it covers all vertices of Γ. A connected graph Γ having at least $2k+2$ vertices is said to be k-*extendable*, introduced by Plummer [11], if each matching M of k edges is contained in a perfect matching of Γ. Any k-extendable graph is $(k-1)$-extendable, but the converse is not true [11]. The *extendability number* of Γ, denoted by $ext(\Gamma)$, is the maximum k such that Γ is k-extendable. Plummer [11,12] studied the relationship between n-extendability and other graph properties. For more research results related to matching extendability, one can refer to [13–17]. Yu et al. [18]

classified the 2-extendable Cayley graphs of finite abelian groups. Chen et al. [19] classified the 2-extendable Cayley graphs of dihedral groups. Recently, Gao et al. [20] characterize the 2-extendable quasi-abelian Cayley graphs. Their research is focused on 2-extendability of some Cayley graphs; in this paper, we focus on the general extendability, i.e., $(n-2)$-extendability of Cayley graphs generated by transposition trees.

We proceed as follows. In Section 2, we provide preliminaries and previous related results on Cayley graphs. In Section 3, we give our main results: show that all Cayley graphs generated by transposition trees are $(n-2)$-extendable and then determine their extendability numbers are $n-2$.

2. Preliminaries

In this section, we shall give some definitions and known results which will be used in this paper.

Denote by $\mathfrak{S}_n$ the group of all permutations on $[n] = \{1, 2, \ldots, n\}$. Obviously, $|\mathfrak{S}_n| = n!$. For convenience, we use $\mathbf{x} = x_1 x_2 \ldots x_n$ to denote the permutation $\begin{pmatrix} 1 & 2 & \ldots & n \\ x_1 & x_2 & \ldots & x_n \end{pmatrix}$ (see [21]); (st) to denote the permutation $\begin{pmatrix} 1 \ldots s \ldots t \ldots n \\ 1 \ldots t \ldots s \ldots n \end{pmatrix}$, which is called a *transposition*. Obviously, $x_1 \ldots x_s \ldots x_t \ldots x_n(st) = x_1 \ldots x_t \ldots x_s \ldots x_n$. The identity permutation $12 \ldots n$ is denoted by $\mathbf{1}$. A permutation of $\mathfrak{S}_n$ is said to be *even* (resp. *odd*) if it can be written as a product of an even (resp. odd) number of transpositions. Let S be a subset of transpositions. Clearly, the Cayley graph $\text{Cay}(\mathfrak{S}_n, S)$ is a bipartite graph with one partite set containing the vertices corresponding to odd permutations and the other partite set containing the vertices corresponding to even permutations.

To better describe a transposition set S as the generating set, we use a simple way to depict S via a graph. The *transposition generating graph* $T(S)$ is the graph with vertex set $[n]$ and two vertices s and t are adjacent if and only if $(st) \in S$. If $T(S)$ is a tree, it is called *transposition trees*, we denote by $\mathbb{T}_n$ the set of Cayley graphs $\text{Cay}(\mathfrak{S}_n, S)$ generated by transposition trees. For any graph $\mathcal{T}_n(S) = \text{Cay}(\mathfrak{S}_n, S) \in \mathbb{T}_n$, $\mathbf{x} = x_1 x_2 \ldots x_n$ is adjacent to $\mathbf{y} = y_1 y_2 \ldots y_n$ if and only if for $(st) \in S$, $x_s = y_t$, $x_t = y_s$ and $x_k = y_k$ for $k \neq s, t$, that is $\mathbf{y} = \mathbf{x}(st)$. In this case, we say that the edge $e = \mathbf{xy}$ is an (st)-edge and denote $g(e) = (st)$, which is the edge e corresponding to transposition. Let $E_{st} = \{e \in E(\mathcal{T}_n(S)) | e \text{ is an } (st)\text{-edge}\}$. Obviously, for every transposition $(st) \in S$, E_{st} is a perfect matching of $\mathcal{T}_n(S)$. We have the following propositions about Cayley graphs generated by transpositions:

Proposition 1 ([10], p. 52). *Let $\Gamma = \text{Cay}(\mathfrak{S}_n, S)$ be a Cayley graph generated by transpositions. Then, Γ is connected if and only if $T(S)$ is connected.*

Proposition 2 ([22]). *Let S and S' be two sets of transpositions on $[n]$. Then, $\text{Cay}(\mathfrak{S}_n, S)$ and $\text{Cay}(\mathfrak{S}_n, S')$ are isomorphic if and only if $T(S)$ and $T(S')$ are isomorphic.*

In all Cayley graphs $\mathbb{T}_n$, there are two classes which are most important, when $T(S)$ is isomorphic to the star $K_{1,n-1}$ and the path P_n. If $T(S) \cong K_{1,n-1}$, $\text{Cay}(\mathfrak{S}_n, S)$ is called n-dimensional star graph and denoted by ST_n. If $T(S) \cong P_n$, $\text{Cay}(\mathfrak{S}_n, S)$ is called n-dimensional bubble-sort graph and denoted by BS_n. The star graph and the bubble-sort graph are illustrated in Figures 1 and 2 for the case $n = 4$. Both ST_n and BS_n are connected bipartite $(n-1)$-regular graph of order $n!$. When $n = 3$, $\mathcal{T}_3(S) \cong ST_3 \cong BS_3 \cong C_6$; $n = 4$, up to isomorphism, there are exactly two different graphs ST_4 and BS_4 (see [23]).

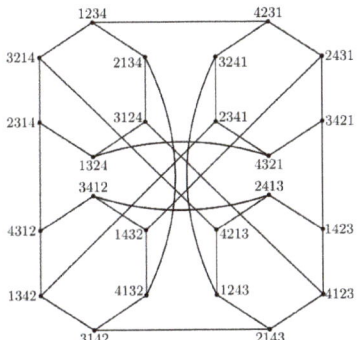

Figure 1. The star graph $ST_4 = \text{Cay}(\mathfrak{S}_4, \{(12), (13), (14)\})$.

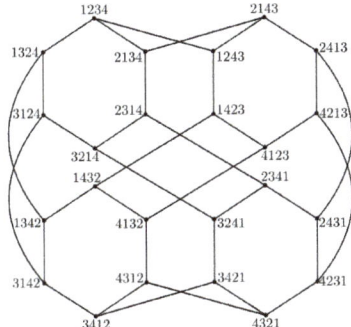

Figure 2. The Bubble-sort graph $BS_4 = \text{Cay}(\mathfrak{S}_4, \{(12), (23), (34)\})$.

Let $\mathbf{x} = x_1 x_2 \ldots x_n$ be a vertex of $\mathcal{T}_n(S)$. We say that x_i is the *i*-th coordinate of $\mathbf{x}$, denoted by $(\mathbf{x})_i$. It is easy to see that the Cayley graph $\mathcal{T}_n(S)$ has the following proposition:

Proposition 3 ([23,24])**.** *Let $T(S)$ be a transposition tree of order n, j one of its leaf and $\mathcal{T}_n^{\{i\}}(S)$ $(1 \leq i \leq n)$ the subgraph of $\mathcal{T}_n(S)$ induced by those vertices $\mathbf{x}$ with $(\mathbf{x})_j = i$. Then, $\mathcal{T}_n(S)$ consists of n vertex-disjoint subgraphs: $\mathcal{T}_n^{\{1\}}(S), \mathcal{T}_n^{\{2\}}(S), \ldots, \mathcal{T}_n^{\{n\}}(S)$; each isomorphic to another Cayley graph $\mathcal{T}_{n-1}(S') = \text{Cay}(\mathfrak{S}_{n-1}, S')$ with $S' = S \setminus \tau$, where τ is the transposition corresponding to the edge incident to the leaf j.*

Readers can refer to [10,21] for the terminology and notation not defined in this paper.

3. Main Results

First, we will give some useful lemmas.

The *Cartesian product* $\Gamma_1 \square \Gamma_2$ of graphs Γ_1 and Γ_2 is a graph with vertex set $V(\Gamma_1) \times V(\Gamma_2)$. Two vertices (u, v) and (u', v') are adjacent in $\Gamma_1 \square \Gamma_2$ if either $u = u'$ and $vv' \in E(\Gamma_2)$ or $uu' \in E(\Gamma_1)$ and $v = v'$. Clearly $\Gamma_1 \square \Gamma_2 = \Gamma_2 \square \Gamma_1$.

Lemma 1. *Let T be a labeled tree of order n, e any edge of T, and T_1, T_2 two components of $T - e$, where $|V(T_1)| = r$. Furthermore, let S (S^-, S_1, S_2, respectively) be the transposition set on $\{1, 2, \ldots, n\}$ satisfying $T(S) = T$ ($T(S^-) = T - e$, $T(S_1) = T_1$, $T(S_2) = T_2$. Then, $\text{Cay}(\mathfrak{S}_n, S^-)$ has $\binom{n}{r}$ components and each component is isomorphic to $\text{Cay}(\mathfrak{S}_r, S_1) \square \text{Cay}(\mathfrak{S}_{n-r}, S_2)$.*

Proof. Without loss of generality, we can assume $r \leq \lfloor \frac{n}{2} \rfloor$.

When $r = 1$, T_1 is an isolated vertex, e is a pendant edge and $S_1 = \emptyset$. Then, $\text{Cay}(\mathfrak{S}_1, S_1) \square \text{Cay}(\mathfrak{S}_{n-1}, S_2) = \text{Cay}(\mathfrak{S}_{n-1}, S_2)$. The lemma is true, following from Proposition 3.

When $r \geq 2$, we relabel T as follows: Relabel the vertices of T_1 as $\{1, 2, \ldots, r\}$ and the vertices of T_2 as $\{r+1, r+2, \ldots, n\}$. Let S', S'^{-}, S'_1, S'_2 be the corresponding transposition sets. Obviously, $S'^{-} = S'_1 \cup S'_2$. By Proposition 2, we know that $\text{Cay}(\mathfrak{S}_n, S) \cong \text{Cay}(\mathfrak{S}_n, S')$, $\text{Cay}(\mathfrak{S}_n, S^{-}) \cong \text{Cay}(\mathfrak{S}_n, S'^{-})$, and so on. Thus, we only need to prove the corresponding result on S', S'^{-}, S'_1 and S'_2. Since $T - e$ is disconnected, $\text{Cay}(\mathfrak{S}_n, S'^{-})$ is also disconnected by Proposition 1. Let Γ_1 be the component of $\text{Cay}(\mathfrak{S}_n, S'^{-})$ containing the identity element 1. Since T_1 and T_2 are connected, S'_1 generates $\mathfrak{S}_r$ and S'_2 generates $\mathfrak{S}_{n-r}$ (let $\mathfrak{S}_{n-r}$ be symmetric group on $\{r+1, r+2, \ldots, n\}$). Then, the vertices in Γ_1 can be represented as $\mathbf{v} = x_1 x_2 \ldots x_r x_{r+1} \ldots x_n$, where $x_1 x_2 \ldots x_r$ is a permutation on $\{1, 2, \ldots, r\}$ and $x_{r+1} \ldots x_n$ is a permutation on $\{r+1, r+2, \ldots, n\}$. Furthermore, let $\mathbf{v} = x_1 x_2 \ldots x_r x_{r+1} \ldots x_n$ and $\mathbf{v}' = x'_1 x'_2 \ldots x'_r x'_{r+1} \ldots x'_n$ be two vertices in Γ_1. Then, $\mathbf{v}$ and $\mathbf{v}'$ are adjacent if and only if for $j, k \leq r$ and $(jk) \in S'_1$, $x_k = x'_j$, $x_j = x'_k$ and $x_l = x'_l$ for other digits, or, for $j, k \geq r+1$ and $(jk) \in S'_2$, $x_k = x'_j$, $x_j = x'_k$ and $x_l = x'_l$ for other digits. Thus, $\Gamma_1 \cong \text{Cay}(\mathfrak{S}_r, S'_1) \square \text{Cay}(\mathfrak{S}_{n-r}, S'_2)$ and $|V(\Gamma_1)| = r!(n-r)!$. Since $\text{Cay}(\mathfrak{S}_n, S'^{-})$ is vertex-transitive, all components of $\text{Cay}(\mathfrak{S}_n, S'^{-})$ are isomorphic and there exist $\frac{n!}{r!(n-r)!} = \binom{n}{r}$ components in it. □

We need to consider the extendability of the Cartesian product when we investigate the extendability of $\mathcal{T}_n(S)$. The following lemmas are used several times in the proof of our theorem.

Lemma 2 ([25,26]). *If Γ is a k-extendable graph, then $\Gamma \square K_2$ is $(k+1)$-extendable.*

Lemma 3 ([25]). *If Γ_1 and Γ_2 are k-extendable and l-extendable graphs, respectively, then their Cartesian product $\Gamma_1 \square \Gamma_2$ is $(k + l + 1)$-extendable.*

Lemma 4 ([27]). *A bipartite Cayley graph is 2-extendable if and only if it is not a cycle.*

In order to prove the main result, we need other definitions and notations. The *symmetric difference* of two sets A and B is defined as the set $A \triangle B = (A - B) \cup (B - A)$. Let Γ be a connected graph. If $e = uv \in E(\Gamma)$, denote $V(e) = \{u, v\}$ and $E(v) = \{e | V(e) \cap \{v\} \neq \emptyset\}$.

Let $\mathbf{x}$ be a permutation of $[n]$. The smallest positive integer k for which $\mathbf{x}^k$ is the identity permutation, this number k is called the *order* of $\mathbf{x}$, denoted by $o(\mathbf{x}) = k$. $fix(\mathbf{x})$ denotes the set of points in $[n]$ fixed by $\mathbf{x}$ (see [10]). Let $\overline{fix(\mathbf{x})} = [n] - fix(\mathbf{x})$. As we know, there is another way of writing the permutation as products of disjoint cycles which are commutative (see [21]). For example, if $\mathbf{x} \in \mathfrak{S}_9$, $\mathbf{x} = 324, 158, 967$, then $\mathbf{x} = (134)(68)(79) = (68)(134)(79)$, and further $fix(\mathbf{x}) = \{2, 5\}$, $\overline{fix(\mathbf{x})} = \{1, 3, 4, 6, 7, 8, 9\}$, $|\overline{fix(\mathbf{x})}| = 7$. We say that $\mathbf{x}$ is a type of $(m_1 m_2 m_3)(m_4 m_5)(m_6 m_7)$ permutation. Clearly $\mathbf{x}^6 = \mathbf{1}$ and $o(\mathbf{x}) = 6$.

Theorem 1. *Any Cayley graph $\mathcal{T}_n(S) \in \mathbb{T}_n$ is $(n-2)$-extendable for any integer $n \geq 3$.*

Proof. We prove the theorem by induction on n. For $n = 3$, the $\mathcal{T}_3(S)$ is 6-cycle, which is 1-extendable. For $n = 4$, the $\mathcal{T}_4(S)$ is a 3-regular bipartite Cayley graph, which is not a cycle. $\mathcal{T}_4(S)$ is 2-extendable by Lemma 4.

Now we assume the statement is true for all integers smaller than n ($n \geq 5$). Let S be a subset of transpositions on $[n]$. The transposition generating graph $T(S)$ is a tree. We will show that any matching M of size $(n-2)$ can be extended to a perfect matching of $\mathcal{T}_n(S)$.

Let M be a matching with $(n-2)$ edges. There are $(n-1)$ classes of edges in $\mathcal{T}_n(S)$ because of $|S| = n - 1$. We may suppose that $E_{s_4 t_4} \cap M = \emptyset$. Let $S^{-} = S \setminus (s_4 t_4)$. By Lemma 1, $\text{Cay}(\mathfrak{S}_n, S^{-})$ has $\binom{n}{r}$ connected components and each component is isomorphic to Cay($\mathfrak{S}_r, S_1$)□Cay($\mathfrak{S}_{n-r}, S_2$). We may assume $1 \leq r \leq \lfloor \frac{n}{2} \rfloor$ by the symmetry of Cartesian product. For the convenience, we denote the components of $\mathcal{T}_n(S) \setminus E_{s_4 t_4} = \text{Cay}(\mathfrak{S}_n, S^-)$ by $\mathcal{C}_i (i = 1, 2, \ldots, l)$, where $l = \binom{n}{r}$.

Claim 1. $\mathcal{C}_i$ is $(n-3)$-extendable.

If $r = 1$, the transposition $(s_4 t_4)$ corresponding to the edge is a leaf of $T(S)$, $\mathcal{C}_i \cong \mathcal{T}_{n-1}(S')$ by Proposition 3, where $S' = S^- = S \setminus (s_4 t_4)$, $\mathcal{C}_i$ is $(n-3)$-extendable by the inductive hypothesis.

If $r = 2$, $\mathcal{T}_2(S) = K_2$, $\mathcal{C}_i \cong K_2 \square \text{Cay}(\mathfrak{S}_{n-2}, S_2) = K_2 \square \mathcal{T}_{n-2}(S_2)$. $\mathcal{T}_{n-2}(S_2)$ is $(n-4)$-extendable by the inductive hypothesis. $\mathcal{C}_i$ is $(n-3)$-extendable by Lemma 2.

If $r \geq 3$, by the inductive hypothesis Cay($\mathfrak{S}_r, S_1$) $\cong \mathcal{T}_r(S_1)$ is $(r-2)$-extendable and Cay($\mathfrak{S}_{n-r}, S_2$) $\cong \mathcal{T}_{n-r}(S_2)$ is $(n-r-2)$-extendable. Hence, Cay($\mathfrak{S}_r, S_1$)□Cay($\mathfrak{S}_{n-r}, S_2$) is $(n-3)$-extendable by Lemma 3. We get the Claim.

Let $J = \{i | E(\mathcal{C}_i) \cap M \neq \varnothing\}$. If $|J| \geq 2$, then $|E(\mathcal{C}_i) \cap M| \leq n - 3$. When $i \in J$, each edge set $E(\mathcal{C}_i) \cap M$ can be extended to a perfect matching of $\mathcal{C}_i$, which is defined by $M(\mathcal{C}_i)$. Clearly, $M \subset \bigcup_{i \in J} M(\mathcal{C}_i)$. When $i \notin J$, let $M(\mathcal{C}_i)$ be an arbitrary perfect matching of $\mathcal{C}_i$. Then,

$$\bigcup_{i=1}^{l} M(\mathcal{C}_i) = \left(\bigcup_{i \in J} M(\mathcal{C}_i) \right) \cup \left(\bigcup_{i \notin J} M(\mathcal{C}_i) \right)$$

is a perfect matching of Cay($\mathfrak{S}_n, S^-$), which is also a perfect matching of $\mathcal{T}_n(S)$.

When $|J| = 1$, without loss of generality, we assume that $M \subset E(\mathcal{C}_1)$ and $\mathcal{C}_1$ contains the identity permutation **1**. If M can be extended to a perfect matching of $\mathcal{C}_1$, we are done. Suppose that M cannot be extended to a perfect matching of $\mathcal{C}_1$. Let $e_2 = v_1 v_2$ be an edge in M. $M \setminus e_2$ can be extended to a perfect matching of $\mathcal{C}_1$ (since $|M \setminus e_2| = n - 3$), which is denoted by $M'(\mathcal{C}_1)$. Let $E(v_1) \cap M'(\mathcal{C}_1) = e_1$, $E(v_2) \cap M'(\mathcal{C}_1) = e_3$, $V(e_1) = \{v_0, v_1\}$, $V(e_3) = \{v_2, v_3\}$ and $e_4 = E(v_3) \cap E_{s_4 t_4}$. By the transitivity of $\mathcal{C}_1$ and without loss of generality, we can assume that $v_0 = \mathbf{1}$. Let $o(g(e_1)g(e_2)g(e_3)g(e_4)) = a$, $v_i = \prod_{j=1}^{i} g(e_j)$, and $e_{4b+1} \in E_{s_1 t_1}$, $e_{4b+2} \in E_{s_2 t_2}$, $e_{4b+3} \in E_{s_3 t_3}$, $e_{4b+4} \in E_{s_4 t_4}$ ($b = 0, \ldots, a-1$), where $\{(s_1 t_1), (s_2 t_2), (s_3 t_3), (s_4 t_4)\} \subset S$. It is easy to see $g(e_2) \neq g(e_i)$ ($i = 1, 3$), $g(e_4) \neq g(e_i)$ ($i = 1, 2, 3$), $g(e_1)g(e_2)g(e_3) \neq g(e_4)$, $v_3 = g(e_1)g(e_2)g(e_3)$ is an odd permutation and $v_4 = g(e_1)g(e_2)g(e_3)g(e_4)$ is an even permutation. The cardinality of $\overline{fix(v_3)}$ can only be 2, 4, 5 and 6. We discuss these four cases one by one in order to prove that M can be extended to a perfect matching of $\mathcal{T}_n(S)$.

Case 1. $|\overline{fix(v_3)}| = 2$.

In this case, v_3 is a transposition and $o(v_3) = 2$. There are two subcases for the order of v_4.

Subcase 1.1. v_4 is a type of $(m_1 m_2)(m_3 m_4)$ permutation.

We have $o(v_4) = 2$, $(v_4)^2 = \mathbf{1}$. Note that $v_i = \prod_{j=1}^{i} g(e_j)$, where $i \in [8]$. Hence, there is an 8-cycle $C_8 = v_0 e_1 v_1 e_2 \ldots v_7 e_8 v_8$ ($v_8 = v_0$). The vertex $v_{4b+i} \in V(\mathcal{C}_{b+1})$ ($i = 0, 1, 2, 3$; $b = 0, 1$). We may take a perfect matching $M'(\mathcal{C}_2)$ of $\mathcal{C}_2$ such that $e_5 \in M'(\mathcal{C}_2), e_7 \in M'(\mathcal{C}_2)$ and $e_6 \notin M'(\mathcal{C}_2)$ because of $\mathcal{C}_2 \cong \mathcal{C}_1$. Now we take $M'' = (M'(\mathcal{C}_1) \cup M'(\mathcal{C}_2)) \triangle E(C_8)$. Clearly $M \subset M''$, M'' is a perfect matching of subgraph $\mathcal{T}_n(S)[V(\mathcal{C}_1) \cup V(\mathcal{C}_2)]$. Let $M(\mathcal{C}_i)$ be a perfect matching of $\mathcal{C}_i$ ($i = 3, \ldots, l$). Hence, $\bigcup_{i=3}^{l} M(\mathcal{C}_i) \cup M''$ is a perfect matching of $\mathcal{T}_n(S)$.

Subcase 1.2. v_4 is a type of $(m_1 m_2 m_3)$ permutation.

We have $o(v_4) = 3$, $(v_4)^3 = \mathbf{1}$. Note that $v_i = \prod_{j=1}^{i} g(e_j)$, where $i \in [12]$. Hence, there is a 12-cycle $C_{12} = v_0 e_1 v_1 e_2 \ldots v_{11} e_{12} v_{12}$ ($v_{12} = v_0$). The vertex $v_{4b+i} \in V(\mathcal{C}_{b+1})$ ($i = 0, 1, 2, 3$; $b = 0, 1, 2$). We may take a perfect matching $M'(\mathcal{C}_{b+1})$ of $\mathcal{C}_{b+1}$ such that $e_{4b+1} \in M'(\mathcal{C}_{b+1}), e_{4b+3} \in M'(\mathcal{C}_{b+1})$ and $e_{4b+2} \notin M'(\mathcal{C}_{b+1})$ ($b = 1, 2$) because of $\mathcal{C}_{b+1} \cong \mathcal{C}_1$.

Now we take $M'' = \left(\bigcup_{i=1}^{3} M'(\mathcal{C}_i)\right) \triangle E(C_{12})$. Clearly $M \subset M''$, M'' is a perfect matching of subgraph $\mathcal{T}_n(S)[\bigcup_{i=1}^{3} V(\mathcal{C}_i)]$. Let $M(\mathcal{C}_i)$ be a perfect matching of $\mathcal{C}_i$ ($i = 4, \ldots, l$). Hence, $\bigcup_{i=4}^{l} M(\mathcal{C}_i) \cup M''$ is a perfect matching of $\mathcal{T}_n(S)$.

Case 2. $|\overline{fix(v_3)}| = 4$.

In this case, v_3 is a type of $(m_1 m_2 m_3 m_4)$ permutation and $o(v_3) = 4$. There are two subcases.

Subcase 2.1. v_4 is a type of $(m_1 m_2 m_3 m_4)(m_5 m_6)$ permutation.

We have $o(v_4) = 4$, $(v_4)^4 = 1$. Note that $v_i = \prod_{j=1}^{i} g(e_j)$, where $i \in [16]$. Hence, there is a 16-cycle $C_{16} = v_0 e_1 v_1 e_2 \ldots v_{15} e_{16} v_{16}$ ($v_{16} = v_0$). The vertex $v_{4b+i} \in V(\mathcal{C}_{b+1})$ ($i = 0, 1, 2, 3; b = 0, 1, 2, 3$). We may take a perfect matching $M'(\mathcal{C}_{b+1})$ of $\mathcal{C}_{b+1}$ such that $e_{4b+1} \in M'(\mathcal{C}_{b+1}), e_{4b+3} \in M'(\mathcal{C}_{b+1})$ and $e_{4b+2} \notin M'(\mathcal{C}_{b+1})$ ($b = 1, 2, 3$) because of $\mathcal{C}_{b+1} \cong \mathcal{C}_1$. Now we take $M'' = \left(\bigcup_{i=1}^{4} M'(\mathcal{C}_i)\right) \triangle E(C_{16})$. Clearly $M \subset M''$, M'' is a perfect matching of subgraph $\mathcal{T}_n(S)[\bigcup_{i=1}^{4} V(\mathcal{C}_i)]$. Let $M(\mathcal{C}_i)$ be a perfect matching of $\mathcal{C}_i$ ($i = 5, \ldots, l$). Hence, $\bigcup_{i=5}^{l} M(\mathcal{C}_i) \cup M''$ is a perfect matching of $\mathcal{T}_n(S)$.

Subcase 2.2. v_4 is a type of $(m_1 m_2 m_3 m_4 m_5)$ permutation.

We have $o(v_4) = 5$, $(v_4)^5 = 1$. Note that $v_i = \prod_{j=1}^{i} g(e_j)$, where $i \in [20]$. Hence, there is a 20-cycle $C_{20} = v_0 e_1 v_1 e_2 \ldots v_{19} e_{20} v_{20}$ ($v_{20} = v_0$). The vertex $v_{4b+i} \in V(\mathcal{C}_{b+1})$ ($i = 0, 1, 2, 3; b = 0, 1, 2, 3, 4$). We may take a perfect matching $M'(\mathcal{C}_{b+1})$ of $\mathcal{C}_{b+1}$ such that $e_{4b+1} \in M'(\mathcal{C}_{b+1}), e_{4b+3} \in M'(\mathcal{C}_{b+1})$ and $e_{4b+2} \notin M'(\mathcal{C}_{b+1})$ ($b = 1, 2, 3, 4$) because of $\mathcal{C}_{b+1} \cong \mathcal{C}_1$. Now we take $M'' = \left(\bigcup_{i=1}^{5} M'(\mathcal{C}_i)\right) \triangle E(C_{20})$. Clearly $M \subset M''$, M'' is a perfect matching of subgraph $\mathcal{T}_n(S)[\bigcup_{i=1}^{5} V(\mathcal{C}_i)]$. Let $M(\mathcal{C}_i)$ be a perfect matching of $\mathcal{C}_i$ ($i = 6, \ldots, l$). Hence, $\bigcup_{i=6}^{l} M(\mathcal{C}_i) \cup M''$ is a perfect matching of $\mathcal{T}_n(S)$.

Case 3. $|\overline{fix(v_3)}| = 5$.

In this case, v_3 is a type of $(m_1 m_2 m_3)(m_4 m_5)$ permutation and $o(v_3) = 6$. There are four subcases.

Subcase 3.1. v_4 is a type of $(m_1 m_2 m_3)(m_4 m_5 m_6)$ permutation.

We have $o(v_4) = 3$, $(v_4)^3 = 1$. There is a 12-cycle $C_{12} = v_0 e_1 v_1 e_2 \ldots v_{11} e_{12} v_{12}$ ($v_{12} = v_0$) in subgraph $\mathcal{T}_n(S)[\bigcup_{i=1}^{3} V(\mathcal{C}_i)]$, where $v_{4b+i} \in V(\mathcal{C}_{b+1})$ for $i = 0, 1, 2, 3; b = 0, 1, 2$. The rest of the proof is similar to Subcase 1.2.

Subcase 3.2. v_4 is a type of $(m_1 m_2 m_3 m_4)(m_5 m_6)$ permutation.

We have $o(v_4) = 4$, $(v_4)^4 = 1$. There is a 16-cycle $C_{16} = v_0 e_1 v_1 e_2 \ldots v_{15} e_{16} v_{16}$ ($v_{16} = v_0$) in subgraph $\mathcal{T}_n(S)[\bigcup_{i=1}^{4} V(\mathcal{C}_i)]$, where $v_{4b+i} \in V(\mathcal{C}_{b+1})$ for $i = 0, 1, 2, 3; b = 0, 1, 2, 3$. The rest of the proof is similar to Subcase 2.1.

Subcase 3.3. v_4 is a type of $(m_1 m_2 m_3 m_4 m_5)$ permutation.

We have $o(v_4) = 5$, $(v_4)^5 = 1$. There is a 20-cycle $C_{20} = v_0 e_1 v_1 e_2 \ldots v_{19} e_{20} v_{20}$ ($v_{20} = v_0$) in subgraph $\mathcal{T}_n(S)[\bigcup_{i=1}^{5} V(\mathcal{C}_i)]$, where $v_{4b+i} \in V(\mathcal{C}_{b+1})$ for $i = 0, 1, 2, 3; b = 0, 1, 2, 3, 4$. The rest of the proof is similar to Subcase 2.2.

Subcase 3.4. v_4 is a type of $(m_1 m_2 m_3)(m_4 m_5)(m_6 m_7)$ permutation.

We have $o(v_4) = 6$, $|\overline{fix(v_4)}| = 7$ and $n \geq 7, l = \binom{n}{r} \geq 7$, $(v_4)^6 = 1$. $v_i = \prod_{j=1}^{i} g(e_j)$, where $i \in [24]$. Hence, there is a 24-cycle $C_{24} = v_0 e_1 v_1 e_2 \ldots v_{23} e_{24} v_{24}$ $(v_{24} = v_0)$. The vertex $v_{4b+i} \in V(\mathcal{C}_{b+1})$ $(i = 0,1,2,3; b = 0,1,2,3,4,5)$. We may take a perfect matching $M'(\mathcal{C}_{b+1})$ of $\mathcal{C}_{b+1}$ such that $e_{4b+1} \in M'(\mathcal{C}_{b+1}), e_{4b+3} \in M'(\mathcal{C}_{b+1})$ and $e_{4b+2} \notin M'(\mathcal{C}_{b+1})$ $(b = 1,2,3,4,5)$ because of $\mathcal{C}_{b+1} \cong \mathcal{C}_1$. Now we take $M'' = \left(\bigcup_{i=1}^{6} M'(\mathcal{C}_i)\right) \triangle E(C_{24})$. Clearly, $M \subset M''$, M'' is a perfect matching of subgraph $\mathcal{T}_n(S)[\bigcup_{i=1}^{6} V(\mathcal{C}_i)]$. Let $M(\mathcal{C}_i)$ be a perfect matching of $\mathcal{C}_i$ $(i = 7, \ldots, l)$. Hence, $\bigcup_{i=7}^{l} M(\mathcal{C}_i) \bigcup M''$ is a perfect matching of $\mathcal{T}_n(S)$.

Case 4. $|\overline{fix(v_3)}| = 6$.

In this case, v_3 is a type of $(m_1 m_2)(m_3 m_4)(m_5 m_6)$ permutation and $o(v_3) = 2$. There are three subcases.

Subcase 4.1. v_4 is a type of $(m_1 m_2)(m_3 m_4)(m_5 m_6)(m_7 m_8)$ permutation.

We have $o(v_4) = 2$. There is an 8-cycle $C_8 = v_0 e_1 v_1 e_2 \ldots v_7 e_8 v_8$ $(v_8 = v_0)$ in subgraph $\mathcal{T}_n(S)[\bigcup_{i=1}^{2} V(\mathcal{C}_i)]$, where $v_{4b+i} \in V(\mathcal{C}_{b+1})$ for $i = 0,1,2,3; b = 0,1$. The rest of the proof is similar to Subcase 1.1.

Subcase 4.2. v_4 is a type of $(m_1 m_2 m_3 m_4)(m_5 m_6)$ permutation.

We have $o(v_4) = 4$. There is a 16-cycle $C_{16} = v_0 e_1 v_1 e_2 \ldots v_{15} e_{16} v_{16}$ $(v_{16} = v_0)$ in subgraph $\mathcal{T}_n(S)[\bigcup_{i=1}^{4} V(\mathcal{C}_i)]$, where $v_{4b+i} \in V(\mathcal{C}_{b+1})$ for $i = 0,1,2,3; b = 0,1,2,3$. The rest of the proof is similar to Subcase 2.1.

Subcase 4.3. v_4 is a type of $(m_1 m_2 m_3)(m_4 m_5)(m_6 m_7)$ permutation.

We have $o(v_4) = 6$. There is a 24-cycle $C_{24} = v_0 e_1 v_1 e_2 \ldots v_{23} e_{24} v_{24}$ $(v_{24} = v_0)$ in subgraph $\mathcal{T}_n(S)[\bigcup_{i=1}^{6} V(\mathcal{C}_i)]$, where $v_{4b+i} \in V(\mathcal{C}_{b+1})$ for $i = 0,1,2,3; b = 0,1,2,3,4,5$. The rest of the proof is similar to Subcase 3.4.

In conclusion, any matching M of size $n - 2$ can be extended to a perfect matching of $\mathcal{T}_n(S)$. The proof is complete. □

The extendability number of Γ, denoted by $ext(\Gamma)$, is the maximum k such that Γ is k-extendable. As we know that $\mathcal{T}_n(S) \in \mathbb{T}_n$ is an $(n-1)$-regular bipartite Cayley graph and not $(n-1)$-extendable. We can obtain the extendability number of $\mathcal{T}_n(S)$ by Theorem 1.

Corollary 1. $ext(\mathcal{T}_n(S)) = n - 2$ for $n \geq 3$.

4. Concluding Remarks

In this paper, we prove that Cayley graph $\mathcal{T}_n(S)$ generated by transposition trees on $\{1, 2, \ldots, n\}$ is $(n-2)$-extendable and determine that the extendability number is $n - 2$, which enriches the results on the extendability of Cayley graphs. Here, the transposition generating graph of S is a tree. A natural problem is whether we can generalize transposition trees to general connected graphs which is worth of further investigation. We present a conjecture.

Conjecture 1. Let S be a transposition generating set of the symmetric group $\mathfrak{S}_n$. Then, the Cayley graph $\mathrm{Cay}(\mathfrak{S}_n, S)$ is $(|S| - 1)$-extendable.

Author Contributions: Methodology: Y.F. and S.X.; writing—original draft preparation: Y.F. and Y.X.; writing—review and editing: Y.F., Y.X., F.L. and S.X. All authors have read and agreed to the published version of the manuscript.

Funding: This work was supported by the National Natural Science Foundation of China (No. 11571155, No. 11961067, No. 12071194).

Informed Consent Statement: Not applicable.

Acknowledgments: Many thanks to the anonymous referees for their helpful comments and suggestions.

Conflicts of Interest: The authors declare no conflict of interest.

References

1. Akers, S.B.; Krishnamurthy, B. A group-theoretic model for symmetric interconnection networks. *IEEE Trans. Comput.* **1989**, *38*, 555–566. [CrossRef]
2. Cooperman, G.; Finkelstein, L. New methods for using Cayley graphs in interconnection networks. *Discret. Appl. Math.* **1992**, *37/38*, 95–118. [CrossRef]
3. Heydemann, M.C. Cayley graphs and interconnection networks. In *Graph Symmetry*; Hahn, G., Sabidussi, G., Eds.; NATO ASI Series, Series C; Kluwer: Dordrecht, The Netherlands, 1997; Volume 497, pp. 167–224.
4. Jwo J.S.; Lakshmivarahan S.; Dhall S.K. A new class of interconnection networks based on the alternating group. *Networks* **1993**, *23*, 315–326. [CrossRef]
5. Lakshmivarahan, S.; Jwo, J.S.; Dhall, S.K. Symmetry in interconnection networks based on Cayley graphs of permutation groups: A survey. *Parallel Comput.* **1993**, *19*, 361–407. [CrossRef]
6. Xu, L.; Zhou, S.; Yang, W. Component connectivity of Cayley graphs generated by transposition trees. *Int. J. Parallel Emergent Distrib. Syst.* **2020**, *35*, 103–110. [CrossRef]
7. Chang N.; Cheng E.; Hsieh, S. Conditional diagnosability of cayley graphs generated by transposition trees under the PMC model. *ACM Tran. Design Autom. Electron. Syst.* **2015**, *20*, 1–16. [CrossRef]
8. Wang, M.; Guo, Y.; Wang, S. The 1-good-neighbour diagnosability of cayley graphs generated by transposition trees under the PMC model and MM* model. *Int. J. Comput. Math.* **2016**, *94*, 620–631. [CrossRef]
9. Zhao, S.L.; Chang, J.M.; Hao, R.X. Reliability assessment of the Caylay graph generated by trees. *Discret. Appl. Math.* **2020**, *287*, 10–14. [CrossRef]
10. Godsil, C.; Royle, G. *Algebraic Graph Theory*; Graduate Texts in Mathematics; Springer: New York, NY, USA, 2001; Volume 207.
11. Plummer, M.D. On n-extendable graphs. *Discret. Math.* **1980**, *31*, 201–210. [CrossRef]
12. Plummer, M.D. Matching extension and the genus of a graph. *J. Comb. Theory Ser. B* **1988**, *44*, 329–337. [CrossRef]
13. Cioabă, S.M.; Koolen, J.H.; Li, W. Max-cut and extendability of matchings in distance-regular graphs. *Eur. J. Comb.* **2017**, *62*, 232–244. [CrossRef]
14. Hackfeld, J.; Koster, A.M.C.A. The matching extension problem in general graphs is co-NP-complete. *J. Comb. Optim.* **2018**, *35*, 853–859. [CrossRef]
15. Zhang, W. Matching Extendability and Connectivity of Regular Graphs from Eigenvalues. *Graphs Comb.* **2020**, *36*, 93–108. [CrossRef]
16. Aldred, R.E.L.; Plummer, M.D. Matching extension in prism graphs. *Discret. Appl. Math.* **2017**, *221*, 25–32. [CrossRef]
17. Aldred, R.E.L.; Plummer, M.D. Extendability and Criticality in Matching Theory. *Graphs Comb.* **2020**, *36*, 573–589. [CrossRef]
18. Chan, O.; Chen, C.C.; Yu, Q.L. On 2-extendable abelian Cayley graphs. *Discret. Math.* **1995**, *146*, 19–32. [CrossRef]
19. Chen, C.C.; Liu, J.; Yu, Q.L. On the classification of 2-extendable Cayley graphs on dihedral groups. *Australas. J. Comb.* **1992**, *6*, 209–219.
20. Gao, X.; Li, Q.L.; Wang, J.W.; Wang, W.Z. On 2-Extendable Quasi-abelian Cayley Graphs. *Bull. Malays. Math. Sci. Soc.* **2016**, *39*, 43–57. [CrossRef]
21. Bóna, M. *A Walk through Combinatorics: An Introduction to Enumeration and Graph Theory*; World Scientific Publishing Co. Inc.: Singapore, 2011.
22. Delorme, C. *Isomorphism of Transposition Graphs*; L.R.I., Orsay: Gif-sur-Yvette, France, 1997.
23. Cheng, E.; Lipták, L. Linearly many faults in Cayley graphs generated by transposition trees. *Inf. Sci.* **2007**, *177*, 4877–4882. [CrossRef]
24. Lin, C.K.; Tan, J.M.; Hsu, L.H; Cheng, E.; Lipták, L. Conditional diagnosability of Cayley graphs generated by transposition trees under the comparison diagnosis model. *J. Interconnect. Netw.* **2008**, *9*, 83–97. [CrossRef]
25. Györi, E.; Plummer, M.D. The cartesian product of a k-extendable and an l-extendable graph is $(k+l+1)$-extendable. *Discret. Math.* **1992**, *101*, 87–96. [CrossRef]
26. Liu, J.; Yu, Q.L. Matching extensions and products of graphs: In *Quo Vadis, Graph Theory?—A Source Book for Challenges and Directions*; John Gimbel, J.W.K., Quintas, L.V., Eds.; Elsevier: Amsterdam, The Netherlands, 1993; Volume 55, pp. 191–200.
27. Li, Q.L.; Gao, X. Classification of 2-extendable bipartite and cubic non-bipartite vertex-transitive graphs. *arXiv* **2016**, arXiv:1612.02988v1.

Article

Parity Properties of Configurations

Michal Staš

Faculty of Electrical Engineering and Informatics, Technical University of Košice, 042 00 Košice, Slovakia; michal.stas@tuke.sk

Abstract: In the paper, the crossing number of the join product $G^* + D_n$ for the disconnected graph G^* consisting of two components isomorphic to K_2 and K_3 is given, where D_n consists of n isolated vertices. Presented proofs are completed with the help of the graph of configurations that is a graphical representation of minimum numbers of crossings between two different subgraphs whose edges do not cross the edges of G^*. For the first time, multiple symmetry between configurations are presented as parity properties. We also determine crossing numbers of join products of G^* with paths P_n and cycles C_n on n vertices by adding new edges joining vertices of D_n.

Keywords: graph; join product; crossing number; configuration; parity properties; path; cycle

MSC: 05C10; 05C38

1. Introduction

The issue of reducing the number of crossings on edges of simple graphs is interesting in a lot of areas. Probably one of the most popular areas is the implementation of the VLSI layout because it caused a significant revolution in circuit design and thus had a strong effect on parallel calculations. Crossing numbers have also been studied to improve the readability of hierarchical structures and automated graphs. The visualized graph should be easy to read and understand. For the sake of clarity of graphic drawings, some reduction of an edge crossing is probably the most important. Note that examining number of crossings of simple graphs is an NP-complete problem by Garey and Johnson [1].

The crossing number $\mathrm{cr}(G)$ of a simple graph G with the vertex set $V(G)$ and the edge set $E(G)$ is the minimum possible number of edge crossings in a drawing of G in the plane (for the definition of a drawing see Klešč [2]). One can easily verify that a drawing with the minimum number of crossings (an optimal drawing) is always a good drawing, meaning that no two edges cross more than once, no edge crosses itself, and also no two edges incident with the same vertex cross. Let D be a good drawing of the graph G. We denote the number of crossings in D by $\mathrm{cr}_D(G)$. Let G_i and G_j be edge-disjoint subgraphs of G. We denote the number of crossings between edges of G_i and edges of G_j by $\mathrm{cr}_D(G_i, G_j)$, and the number of crossings among edges of G_i in D by $\mathrm{cr}_D(G_i)$. For any three mutually edge-disjoint subgraphs G_i, G_j, and G_k of G by [2], the following equations hold:

$$\mathrm{cr}_D(G_i \cup G_j) = \mathrm{cr}_D(G_i) + \mathrm{cr}_D(G_j) + \mathrm{cr}_D(G_i, G_j),$$

$$\mathrm{cr}_D(G_i \cup G_j, G_k) = \mathrm{cr}_D(G_i, G_k) + \mathrm{cr}_D(G_j, G_k).$$

Throughout this paper, some parts of proofs will be based on Kleitman's result [3] on crossing numbers for some complete bipartite graphs $K_{m,n}$ on $m + n$ vertices with a partition $V(K_{m,n}) = V_1 \cup V_2$ and $V_1 \cap V_2 = \emptyset$ containing an edge between every pair of vertices from V_1 and V_2 of sizes m and n, respectively. He showed that

$$\mathrm{cr}(K_{m,n}) = \left\lfloor \frac{m}{2} \right\rfloor \left\lfloor \frac{m-1}{2} \right\rfloor \left\lfloor \frac{n}{2} \right\rfloor \left\lfloor \frac{n-1}{2} \right\rfloor, \quad \text{if} \quad \min\{m, n\} \leq 6. \tag{1}$$

For an overview of several exact values of crossing numbers for specific graphs or some families of graphs, see Clancy [4]. The main goal of this survey is to summarize all such published results for crossing numbers along with references also in an effort to give priority to the author who published the first result. Chapter 4 is devoted to the issue of crossing numbers of join product with all simple graphs of order at most six mainly due to unknown values of $cr(K_{m,n})$ for both m, n more than six in (1). The join product of two graphs G_i and G_j, denoted $G_i + G_j$, is obtained from vertex-disjoint copies of G_i and G_j by adding all edges between $V(G_i)$ and $V(G_j)$. For $|V(G_i)| = m$ and $|V(G_j)| = n$, the edge set of $G_i + G_j$ is the union of the disjoint edge sets of the graphs G_i, G_j, and the complete bipartite graph $K_{m,n}$. Let D_n denote the discrete graph (sometimes called empty graph) on n vertices, and let K_n be the complete graph on n vertices. The exact values for crossing numbers of $G + D_n$ for all graphs G of order at most four are given by Klešč and Schrötter [5], and also for a lot of connected graphs G of order five and six [2,6–24]. Note that $cr(G + D_n)$ are known only for some disconnected graphs G, and so the purpose of this paper is to extend known results concerning this topic to new disconnected graphs [25–28].

Let $G^* = (V(G^*), E(G^*))$ be the disconnected graph of order five consisting of two components isomorphic to the complete graphs K_2 and K_3, respectively, and let also $V(G^*) = \{v_1, v_2, \ldots, v_5\}$. We cannot determine the crossing number of the join product $G^* + D_n$ by a similar technique like in [2,18] because $|E(G^*)| < |V(G^*)|$. From the topological point of view, number of crossings of any drawing D of $G^* + D_n$ placed on surface of the sphere does not matter which of regions is unbounded, but on how many times edges of the graph G^* could be crossed by a subgraph T^i in D. This representation of T^i best describes idea of a configuration utilizing some cyclic permutation on pre-numbered vertices of G^*.

Theorem 1. $cr(G^* + D_1) = 0$ and $cr(G^* + D_n) = n^2 - 2n + \lfloor \frac{n}{2} \rfloor$ for $n \geq 2$, i.e., $cr(G^* + D_n) = 4\lfloor \frac{n}{2} \rfloor \lfloor \frac{n-1}{2} \rfloor + \lfloor \frac{n}{2} \rfloor$ for n even and $cr(G^* + D_n) = 4\lfloor \frac{n}{2} \rfloor \lfloor \frac{n-1}{2} \rfloor + \lfloor \frac{n}{2} \rfloor - 1$ for n odd at least 3.

All subcases of the proof of Theorem 2 will be clearer if a graph of configurations $\mathcal{G}_D$ is used as a graphical representation of minimum numbers of crossings between two different subgraphs. Moreover, in the case of our symmetric graph G^*, the graph $\mathcal{G}_D$ can be linked to parity properties of configurations. Our proof of the main Theorem 2 is therefore an inevitable combination of topological analysis of existing configurations with their parity properties. The color resolution of weighted edges in $\mathcal{G}_D$ will also serve us for a simpler description of existence of its possible subgraphs in the examined drawing D of $G^* + D_n$. Software COGA [29] should be also very helpful in certain parts of presented proofs mainly due to possibility of generating all cyclic permutations of five elements and counting of their subsequent interchanges of adjacent elements.

The obtained crossing number of the join product $G^* + D_n$ is in very special form which is caused by a completely different behavior for n even and odd number. The paper concludes by giving crossing numbers of $G^* + P_n$ and $G^* + C_n$ with same values in Corollaries 3 and 4, respectively, that is something unique in the crossing number theory.

2. Cyclic Permutations and Corresponding Configurations

The join product $G^* + D_n$ (sometimes used notation $G^* + nK_1$) consists of one copy of the graph G^* and n vertices $t_1, \ldots, t_n$, and any vertex t_i is adjacent to every vertex of the graph G^*. We denote the subgraph induced by five edges incident with the fixed vertex t_i by T^i, which yields that

$$G^* + D_n = G^* \cup K_{5,n} = G^* \cup \left(\bigcup_{i=1}^{n} T^i \right). \qquad (2)$$

We consider a good drawing D of $G^* + D_n$. By the rotation $\text{rot}_D(t_i)$ of a vertex t_i in D we understand the cyclic permutation that records the (cyclic) counterclockwise order in which edges leave t_i, as defined by Hernández-Vélez et al. [30] or Woodall [31]. We use the notation (12345) if the counter-clockwise order of edges incident with the fixed vertex t_i is $t_i v_1, t_i v_2, t_i v_3, t_i v_4$, and $t_i v_5$. We recall that rotation is a cyclic permutation. By $\overline{\text{rot}}_D(t_i)$, we understand the inverse permutation of $\text{rot}_D(t_i)$. In the given drawing D, it is highly desirable to separate n subgraphs T^i into three mutually disjoint subsets depending on how many times edges of G^* could be crossed by T^i in D. Let us denote by R_D and S_D the set of subgraphs for which $\text{cr}_D(G^*, T^i) = 0$ and $\text{cr}_D(G^*, T^i) = 1$, respectively. Edges of G^* are crossed by each remaining subgraph T^i at least twice in D.

First, note that if D is a drawing of the join product $G^* + D_n$ with the empty set R_D, then $\sum_{i=1}^{n} \text{cr}_D(G^*, T^i) \geq n$ enforces at least $n^2 - 2n + \lfloor \frac{n}{2} \rfloor$ crossings in D provided by

$$\text{cr}_D(G^* + D_n) \geq \text{cr}_D(K_{5,n}) + \text{cr}_D(G^*, K_{5,n}) \geq 4 \left\lfloor \frac{n}{2} \right\rfloor \left\lfloor \frac{n-1}{2} \right\rfloor + n \geq n^2 - 2n + \left\lfloor \frac{n}{2} \right\rfloor.$$

Based on this argument, we will only consider drawings of the graph G^* for which there is a possibility to obtain a subgraph $T^i \in R_D$. Moreover, let F^i denote the subgraph $G^* \cup T^i$ for any $T^i \in R_D$, which yields that each such subgraph F^i is represented by its $\text{rot}_D(t_i)$.

Let us discuss all possible subdrawings of G^* induced by D. As edges of its subgraph isomorphic to K_3 do not cross each other, it is obvious there are only two such possible drawings of G^* presented in Figure 1.

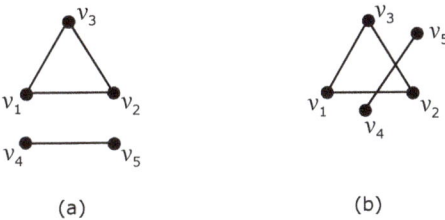

(a)   (b)

Figure 1. Two possible non isomorphic drawings of the graph G^*. (**a**): the planar drawing of G^*; (**b**): the drawing of G^* with two crossings among edges.

Assume there is a good drawing D of $G^* + D_n$ with planar subdrawing of the graph G^* induced by D and also the vertex notation of G^* in such a way as shown in Figure 1a. Our aim is to list all possible rotations $\text{rot}_D(t_i)$ which can appear in D if edges of G^* are not crossed by T^i. Since there is only one subdrawing of $F^i \setminus \{v_4, v_5\}$ represented by the rotation (132), there are three possibilities to obtain the subdrawing of F^i without the edge $v_4 v_5$ depending on in which region both edges $t_i v_4$ and $t_i v_5$ are placed. Of course, there are two next ways how to place the corresponding two edges together with the edge $v_4 v_5$ for each mentioned case. These $3 \times 2 = 6$ possibilities under our consideration can be denoted by $\mathcal{A}_k$, for $k = 1, \ldots, 6$. We will call them by the *configurations* of corresponding subdrawings of the subgraph $G^* \cup T^i$ in D and suppose their drawings as shown in Figure 2.

In the rest of the paper, we present a cyclic permutation by the permutation with 1 in the first position. Thus, the configurations $\mathcal{A}_1, \mathcal{A}_2, \mathcal{A}_3, \mathcal{A}_4, \mathcal{A}_5$, and $\mathcal{A}_6$ are represented by the cyclic permutations (13245), (13254), (14532), (15432), (13452), and (13542), respectively. Clearly, in a fixed drawing of the graph $G^* + D_n$, some configurations from $\mathcal{M} = \{\mathcal{A}_1, \mathcal{A}_2, \mathcal{A}_3, \mathcal{A}_4, \mathcal{A}_5, \mathcal{A}_6\}$ need not appear. We denote by $\mathcal{M}_D$ the set of all configurations that exist in the drawing D belonging to the set $\mathcal{M}$.

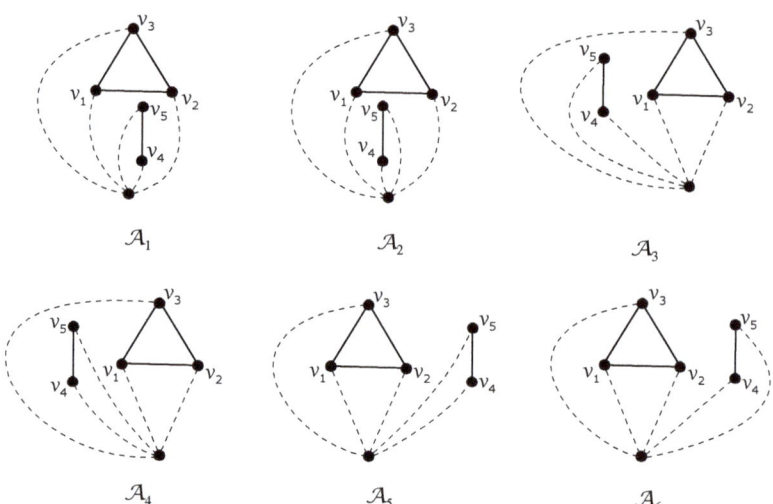

Figure 2. Drawings of six possible configurations $\mathcal{A}_k$ of subgraph $F^i = G^* \cup T^i$ for $T^i \in R_D$.

Let $\mathcal{X}, \mathcal{Y}$ be two configurations from $\mathcal{M}_D$ (not necessary distinct). We denote the number of edge crossings between two different subgraphs T^i and T^j with $\mathrm{conf}(F^i) = \mathcal{X}$ and $\mathrm{conf}(F^j) = \mathcal{Y}$ in D by $\mathrm{cr}_D(\mathcal{X}, \mathcal{Y})$. Finally, let $\mathrm{cr}(\mathcal{X}, \mathcal{Y}) = \min\{\mathrm{cr}_D(\mathcal{X}, \mathcal{Y})\}$ among all good drawings of $G^* + D_n$ with the planar subdrawing of G^* induced by D given in Figure 1a and with $\mathcal{X}, \mathcal{Y} \in \mathcal{M}_D$. Our aim shall be to establish $\mathrm{cr}(\mathcal{X}, \mathcal{Y})$ for all pairs $\mathcal{X}, \mathcal{Y} \in \mathcal{M}$. In particular, the configurations $\mathcal{A}_1$ and $\mathcal{A}_4$ are represented by the cyclic permutations (13245) and (15432), respectively. Each subgraph T^j with $\mathrm{conf}(F^j) = \mathcal{A}_4$ crosses edges of each T^i with $\mathrm{conf}(F^i) = \mathcal{A}_1$ at least once provided by the minimum number of interchanges of adjacent elements of (13245) required to produce $\overline{(15432)} = (12345)$ is one, i.e., $\mathrm{cr}(\mathcal{A}_1, \mathcal{A}_4) \geq 1$. For more details see also Woodall [31]. The same reason gives $\mathrm{cr}(\mathcal{A}_1, \mathcal{A}_2) \geq 3$, $\mathrm{cr}(\mathcal{A}_1, \mathcal{A}_3) \geq 2$, $\mathrm{cr}(\mathcal{A}_1, \mathcal{A}_5) \geq 2$, $\mathrm{cr}(\mathcal{A}_1, \mathcal{A}_6) \geq 1$, $\mathrm{cr}(\mathcal{A}_2, \mathcal{A}_3) \geq 1$, $\mathrm{cr}(\mathcal{A}_2, \mathcal{A}_4) \geq 2$, $\mathrm{cr}(\mathcal{A}_2, \mathcal{A}_5) \geq 1$, $\mathrm{cr}(\mathcal{A}_2, \mathcal{A}_6) \geq 2$, $\mathrm{cr}(\mathcal{A}_3, \mathcal{A}_4) \geq 3$, $\mathrm{cr}(\mathcal{A}_3, \mathcal{A}_5) \geq 2$, $\mathrm{cr}(\mathcal{A}_3, \mathcal{A}_6) \geq 1$, $\mathrm{cr}(\mathcal{A}_4, \mathcal{A}_5) \geq 1$, $\mathrm{cr}(\mathcal{A}_4, \mathcal{A}_6) \geq 2$, and $\mathrm{cr}(\mathcal{A}_5, \mathcal{A}_6) \geq 3$. Clearly, also $\mathrm{cr}(\mathcal{A}_k, \mathcal{A}_k) \geq 4$ for any $k = 1, \ldots, 6$. The lower bounds obtained for number of crossings between two configurations from $\mathcal{M}$ are summarized in the symmetric Table 1 (here, $\mathrm{conf}(F^i) = \mathcal{A}_k$ and $\mathrm{conf}(F^j) = \mathcal{A}_l$ with $k, l \in \{1, \ldots, 6\}$). Note that these values cannot be increased, i.e., the lower bounds can be achieved in some subdrawings of $G^* \cup T^i \cup T^j$ for $T^i, T^j \in R_D$ with desired configurations.

Table 1. The minimum number of crossings between two different subgraphs T^i and T^j such that $\mathrm{conf}(F^i) = \mathcal{A}_k$ and $\mathrm{conf}(F^j) = \mathcal{A}_l$, where the achieved values are color-coded. Namely, the values 1, 2, 3, and 4 will correspond to green, blue, brown, and black, respectively.

-	$\mathcal{A}_1$	$\mathcal{A}_2$	$\mathcal{A}_3$	$\mathcal{A}_4$	$\mathcal{A}_5$	$\mathcal{A}_6$
$\mathcal{A}_1$	4	3	2	1	2	1
$\mathcal{A}_2$	3	4	1	2	1	2
$\mathcal{A}_3$	2	1	4	3	2	1
$\mathcal{A}_4$	1	2	3	4	1	2
$\mathcal{A}_5$	2	1	2	1	4	3
$\mathcal{A}_6$	1	2	1	2	3	4

Further, due to symmetry of mentioned configurations, let us define two functions

$$\pi_1 : \{1, 2, 3\} \to \{1, 2, 3\}, \text{ with } \pi_1(1) = 3, \pi_1(2) = 1, \text{ and } \pi_1(3) = 2,$$

$$\pi_2 : \{4,5\} \to \{4,5\}, \text{ with } \pi_2(4) = 5, \text{ and } \pi_2(5) = 4.$$

Let $\Pi_1, \Pi_2 : \mathcal{M} \to \mathcal{M}$ be the functions obtained by applying π_1 and π_2 on corresponding cyclic permutations of configurations in $\mathcal{M}$, respectively. Thus, we have

$$\Pi_1(\mathcal{A}_1) = \mathcal{A}_3, \ \Pi_1(\mathcal{A}_3) = \mathcal{A}_5, \ \Pi_1(\mathcal{A}_5) = \mathcal{A}_1, \ \Pi_1(\mathcal{A}_2) = \mathcal{A}_4,$$

$$\Pi_1(\mathcal{A}_4) = \mathcal{A}_6, \ \Pi_1(\mathcal{A}_6) = \mathcal{A}_2, \ \Pi_2(\mathcal{A}_1) = \mathcal{A}_2, \ \Pi_2(\mathcal{A}_2) = \mathcal{A}_1,$$

$$\Pi_2(\mathcal{A}_3) = \mathcal{A}_4, \ \Pi_2(\mathcal{A}_4) = \mathcal{A}_3, \ \Pi_2(\mathcal{A}_5) = \mathcal{A}_6, \ \Pi_2(\mathcal{A}_6) = \mathcal{A}_5.$$

Therefore it is not difficult to show that values in rows of Table 1 can be obtained by successive application of the mentioned transformations Π_1 and Π_2.

3. The Graph of Configurations and Parity Properties

Low possible number of crossings between two different subgraphs from the nonempty set R_D is one of main problems in determining $\text{cr}(G^* + D_n)$, and graph of configurations as a graphical representation of Table 1 is going by useful tool in our research. This idea of representation was first introduced in [26].

Let D be a good drawing of $G^* + D_n$ with the planar subdrawing of G^* induced by D given in Figure 1a, and let $\mathcal{M}_D$ be nonempty set of all configurations that exist in D belonging to $\mathcal{M} = \{\mathcal{A}_1, \mathcal{A}_2, \mathcal{A}_3, \mathcal{A}_4, \mathcal{A}_5, \mathcal{A}_6\}$. A graph of configurations $\mathcal{G}_D$ is an ordered triple (V_D, E_D, w_D), where V_D is the set of vertices, E_D is the set of edges formed by all unordered pairs of two vertices (not necessary distinct), and a weight function $w : E_D \to \mathbb{N}$ that associates with each edge of E_D an unordered pair of two vertices of V_D. The vertex $a_k \in V_D$ if the corresponding configuration $\mathcal{A}_k \in \mathcal{M}_D$ for some $k \in \{1, \ldots, 6\}$. The edge $e = a_k a_l \in E_D$ if a_k and a_l are two vertices of $\mathcal{G}_D$. Finally, $w_D(e) = m \in \mathbb{N}$ for the edge $e = a_k a_l$, if m is associated lower bound between two configurations $\mathcal{A}_k$ and $\mathcal{A}_l$ in Table 1. Based on that $\mathcal{G}_D$ is an undirected edge-weighted graph without multiple edges uniquely determined by D and is also subgraph of $\mathcal{G}$ induced by V_D if we define $\mathcal{G} = (V, E, w)$ in the same way over $\mathcal{M}$. The graph $\mathcal{G} = (V, E, w)$ corresponds to the edge-weighted complete graph K_6 in Figure 3, and thus will follow all subcases in the proof of the main Theorem 2 more clearly. In the rest of Figure 3, let any loop of the mentioned graph $\mathcal{G}$ be presented by circle around vertex with respect to weight 4.

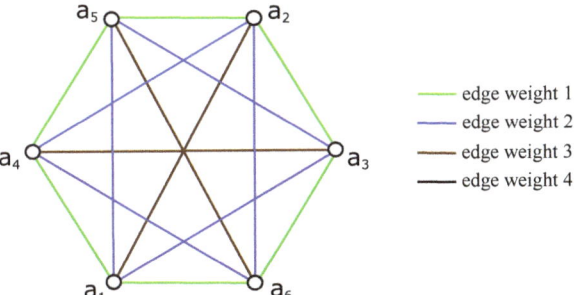

Figure 3. Representation of lower bounds of Table 1 by the graph $\mathcal{G} = (V, E, w)$.

Let α_i denote the number of all subgraphs $T^j \in R_D$ with the configuration $\mathcal{A}_i \in \mathcal{M}_D$ of $F^j = G^* \cup T^j$ for each $i = 1, \ldots, 6$. So, if we denote by $I_o = \{1, 3, 5\}$ and $I_e = \{2, 4, 6\}$, then $\sum_{i \in I_o \cup I_e} \alpha_i = |R_D|$. Moreover, for a better understanding, we get for all $i \in I_o \cup I_e$: $\alpha_i > 0$ if and only if there is a subgraph $T^j \in R_D$ with the configuration $\mathcal{A}_i \in \mathcal{M}_D$ of $F^j = G^* \cup T^j$ if and only if $a_i \in V_D$ in the graph $\mathcal{G}_D$.

Now, let us assume the configurations $\mathcal{A}_1$ of F^i, $\mathcal{A}_4$ of F^j, and $\mathcal{A}_6$ of F^k. The reader can easily find a subdrawing of $G^* \cup T^i \cup T^j \cup T^k$ in which $\text{cr}_D(T^i, T^j) = 1$, $\text{cr}_D(T^i, T^k) = 1$, and $\text{cr}_D(T^j, T^k) = 2$, i.e., $\text{cr}_D(T^i \cup T^j \cup T^k) = 4 = \text{cr}(K_{5,3})$. Further, there is a possibility to add another subgraph T^l that crosses edges of the graph $T^i \cup T^j \cup T^k$ four times. We have to emphasize that the vertex t_l must be placed in the triangular region with three vertices of G^* on its boundary (in the subdrawing of $G^* \cup T^i \cup T^j \cup T^k$), i.e., $T^l \notin R_D \cup S_D$ and the subgraph $F^l = G^* \cup T^l$ is represented by $\mathrm{rot}_D(t_l) = (12435)$. Clearly, the number of adding crossings cannot be smaller than 4 according to the well-known fact that $\mathrm{cr}(K_{5,4}) = 8$. This situation suggests one natural problem which requires the following definition of a new number β_1. If $\alpha_1 > 0$, $\alpha_4 > 0$, and $\alpha_6 > 0$, then let us denote by β_1 the number of subgraphs $T^l \notin R_D \cup S_D$ with $\mathrm{rot}_D(t_l) = (12435)$. It is obvious that any subgraph $T^l \notin R_D \cup S_D$ satisfies the condition $\mathrm{cr}_D(G^* \cup T^i \cup T^j \cup T^k, T^l) \geq 2 + 4 = 6$ with the configurations $\mathcal{A}_1$ of F^i, $\mathcal{A}_4$ of F^j, and $\mathcal{A}_6$ of F^k, and the number of $T^l \notin R_D \cup S_D$ that cross the graph $G^* \cup T^i \cup T^j \cup T^k$ exactly six times is at most β_1. Due to symmetry of some configurations, it is appropriate to use the transform functions Π_1, Π_2 defined above and by the similar way, we can also define the numbers β_i for any $i = 2, \ldots, 6$. Thus, if $\alpha_2 > 0$, $\alpha_3 > 0$, and $\alpha_5 > 0$ or $\alpha_3 > 0$, $\alpha_2 > 0$, and $\alpha_6 > 0$ or $\alpha_4 > 0$, $\alpha_1 > 0$, and $\alpha_5 > 0$ or $\alpha_5 > 0$, $\alpha_2 > 0$, and $\alpha_4 > 0$ or $\alpha_6 > 0$, $\alpha_1 > 0$, and $\alpha_3 > 0$, then let us denote by β_2 or β_3 or β_4 or β_5 or β_6 the number of subgraphs $T^l \notin R_D \cup S_D$ represented by the rotation (12534) or (14253) or (15243) or (15234) or (14235), respectively. The importance of the values β_i will be presented in the proof of the main Theorem 2 as parity properties (6) and (7).

4. The Crossing Number of $G^* + D_n$

A drawing D of $G^* + D_n$ is said to be antipode-free if $\mathrm{cr}_D(T^i, T^j) \geq 1$ for any two different vertices t_i and t_j. In the proof of Theorem 2, the following statements related to some restricted subdrawings of the graph $G^* + D_n$ are required.

Lemma 1. *Let D be a good and antipode-free drawing of $G^* + D_n$, $n > 1$, with the vertex notation of the graph G^* in such a way as shown in Figure 1a. For $T^i \in R_D$, let $\mathcal{A}_k \in \mathcal{M}_D$ be a configuration of the corresponding subgraph $F^i = G^* \cup T^i$ for some $k \in \{1, \ldots, 6\}$. If there is a $T^j \in S_D$ such that $\mathrm{cr}_D(T^i, T^j) = 1$, then all possible $\mathrm{rot}_D(t_j)$ are given in Table 2.*

Table 2. The corresponding rotations of t_j, for $T^i \in R_D$, $F^i = G^* \cup T^i$ and $T^j \in S_D$ satisfying the restriction $\mathrm{cr}_D(T^i, T^j) = 1$.

$\mathrm{conf}(F^i)$	$\mathrm{rot}_D(t_j)$	$\mathrm{conf}(F^i)$	$\mathrm{rot}_D(t_j)$	$\mathrm{conf}(F^i)$	$\mathrm{rot}_D(t_j)$
$\mathcal{A}_1$	(14523)	$\mathcal{A}_3$	(12345)	$\mathcal{A}_5$	(12453)
$\mathcal{A}_2$	(15423)	$\mathcal{A}_4$	(12354)	$\mathcal{A}_6$	(12543)

Proof. Assume the configuration $\mathcal{A}_1$ of the subgraph $F^i = G^* \cup T^i$ for some $T^i \in R_D$, i.e., $\mathrm{rot}_D(t_i) = (13245)$. The subdrawing of F^i induced by D contains just five regions with t_i on their boundaries, see Figure 2. If there is a $T^j \in S_D$ such that $\mathrm{cr}_D(T^i, T^j) = 1$, then the vertex t_j must be placed in the region with the four vertices v_1, v_2, v_4, and v_5 of G^* on its boundary. Besides that only the edge $v_1 v_2$ of G^* can be crossed by $t_j v_3$, and $\mathrm{cr}_D(T^i, T^j) = 1$ is fulfilling for T^j with $\mathrm{rot}_D(t_j) = (14523)$ if $t_j v_4$ crosses $t_i v_5$. The same idea also force that the rotations of the vertex t_j are (15423), (12345), (12354), (12453), and (12543) for the remaining configurations $\mathcal{A}_2$, $\mathcal{A}_3$, $\mathcal{A}_4$, $\mathcal{A}_5$, and $\mathcal{A}_6$ of F^i, respectively. □

Corollary 1. *Let D be a good and antipode-free drawing of $G^* + D_n$, for $n > 3$, with the vertex notation of the graph G^* in such a way as shown in Figure 1a. If T^i, T^j, and $T^k \in R_D$ are three different subgraphs with $\mathrm{cr}_D(T^i, T^j) = 1$, $\mathrm{cr}_D(T^i, T^k) = 1$ and such that F^i, F^j, and F^k have three mutually different configurations from any of the sets $\{\mathcal{A}_1, \mathcal{A}_4, \mathcal{A}_6\}$, $\{\mathcal{A}_2, \mathcal{A}_3, \mathcal{A}_5\}$, $\{\mathcal{A}_3, \mathcal{A}_2, \mathcal{A}_6\}$, $\{\mathcal{A}_1, \mathcal{A}_4, \mathcal{A}_5\}$, $\{\mathcal{A}_2, \mathcal{A}_4, \mathcal{A}_5\}$, and $\{\mathcal{A}_1, \mathcal{A}_3, \mathcal{A}_6\}$, then*

$$\mathrm{cr}_D(T^i \cup T^j \cup T^k, T^l) \geq 6 \qquad \text{for any } T^l \in S_D,$$

i.e.,
$$\mathrm{cr}_D(G^* \cup T^i \cup T^j \cup T^k, T^l) \geq 7 \qquad \text{for any } T^l \in S_D.$$

Proof. Let us assume the configurations $\mathcal{A}_1$ of F^i, $\mathcal{A}_4$ of F^j, and $\mathcal{A}_6$ of F^k with respect to the restrictions $\mathrm{cr}_D(T^i, T^j) = \mathrm{cr}_D(T^i, T^k) = 1$ and recall that they are represented by the cyclic permutations $\mathrm{rot}_D(t_i) = (13245)$, $\mathrm{rot}_D(t_j) = (15432)$, and $\mathrm{rot}_D(t_k) = (13542)$. If there is a subgraph $T^l \in S_D$ with $\mathrm{cr}_D(T^i, T^l) = 1$, then the subgraph F^l can be represented only by $\mathrm{rot}_D(t_l) = (14523)$, where the edge $t_l v_3$ crosses $v_1 v_2$ of G^* and either $t_l v_4$ or $t_l v_5$ crosses corresponding edge of T^i. Any such subgraph T^l must cross edges of both subgraphs T^j and T^k at least twice because the minimum number of interchanges of adjacent elements of (14523) required to produce $\overline{(15432)} = (12345)$ and $\overline{(13542)} = (12453)$ is two. Clearly, if $\mathrm{cr}_D(T^j, T^l) > 2$ or $\mathrm{cr}_D(T^k, T^l) > 2$, we obtain the desired result $\mathrm{cr}_D(T^i \cup T^j \cup T^k, T^l) \geq 1 + 3 + 2 = 6$. Further, if $\mathrm{cr}_D(T^j, T^l) = 2$ and $\mathrm{cr}_D(T^k, T^l) = 2$, then the edge $t_l v_5$ is crossed by $t_l v_4$ in $D(T^i \cup T^j \cup T^l)$ and also $t_i v_4$ by $t_l v_5$ in $D(T^i \cup T^k \cup T^l)$, respectively. However, then $\mathrm{cr}_D(T^i, T^l) \geq 2$, which contradicts the fact that $\mathrm{cr}_D(T^i, T^l) = 1$ in $D(T^i \cup T^j \cup T^k \cup T^l)$.

If there is a $T^l \in S_D$ with $\mathrm{cr}_D(T^j, T^l) = 1$, then the subgraph F^l is represented only by the cyclic permutation (12354). Using same properties as in the previous subcase, we have $\mathrm{cr}_D(T^i, T^l) \geq 2$ and $\mathrm{cr}_D(T^k, T^l) \geq 3$. This in turn implies that $\mathrm{cr}_D(T^i \cup T^j \cup T^k, T^l) \geq 2 + 1 + 3 = 6$. Of course, we can apply the same idea for the case of $\mathrm{cr}_D(T^k, T^l) = 1$.

To finish the proof, let us consider a subgraph $T^l \in S_D$ with $\mathrm{cr}_D(T^i, T^l) = 2$, $\mathrm{cr}_D(T^j, T^l) = 2$, and $\mathrm{cr}_D(T^k, T^l) = 2$. This enforces that the minimum number of interchanges of adjacent elements of $\mathrm{rot}_D(t_l)$ required to produce $\overline{(13245)} = (15423)$, $\overline{(15432)} = (12345)$, and $\overline{(13542)} = (12453)$ must be exactly two. However, it is not difficult to show that such cyclic permutation does not exist. Similar arguments can be applied for remaining five cases (or using the transformations Π_1 and Π_2), and the proof is complete. □

Corollary 2. *Let D be a good and antipode-free drawing of $G^* + D_n$, for $n > 3$, with the vertex notation of the graph G^* in such a way as shown in Figure 1a. If T^i, T^j, and $T^k \in R_D$ are three different subgraphs such that F^i, F^j, and F^k have three mutually different configurations from any of the sets $\{\mathcal{A}_1, \mathcal{A}_3, \mathcal{A}_5\}$ and $\{\mathcal{A}_2, \mathcal{A}_4, \mathcal{A}_6\}$, then*
$$\mathrm{cr}_D(T^i \cup T^j \cup T^k, T^l) \geq 5 \qquad \text{for any } T^l \in S_D,$$
i.e.,
$$\mathrm{cr}_D(G^* \cup T^i \cup T^j \cup T^k, T^l) \geq 6 \qquad \text{for any } T^l \in S_D.$$

Proof. Let us assume the configurations $\mathcal{A}_1$ of F^i, $\mathcal{A}_3$ of F^j, and $\mathcal{A}_5$ of F^k. If there is a subgraph $T^l \in S_D$ with $\mathrm{cr}_D(T^i, T^l) = 1$, then the subgraph F^l can be represented only by the cyclic permutations (14523). Uniqueness of all rotations in Table 2 confirms that $\mathrm{cr}_D(T^j, T^l) \geq 2$ and $\mathrm{cr}_D(T^k, T^l) \geq 2$. Hence, $\mathrm{cr}_D(T^i \cup T^j \cup T^k, T^l) \geq 1 + 2 + 2 = 5$, and the similar way can be applied for the case if $\mathrm{cr}_D(T^j, T^l) = 1$ or $\mathrm{cr}_D(T^k, T^l) = 1$ with $T^l \in S_D$. It remains to consider the case where $\mathrm{cr}_D(T^i, T^l) \geq 2$, $\mathrm{cr}_D(T^j, T^l) \geq 2$, and $\mathrm{cr}_D(T^k, T^l) \geq 2$, which yields that $\mathrm{cr}_D(T^i \cup T^j \cup T^k, T^l) \geq 2 + 2 + 2 = 6$ clearly holds for any such T^l, as claimed. The proof proceeds in the similar way for the second triple of configurations $\{\mathcal{A}_2, \mathcal{A}_4, \mathcal{A}_6\}$, and this completes the proof. □

Lemma 2. $\mathrm{cr}(G^* + D_2) = 1$.

Proof. If we consider the configurations $\mathcal{A}_2$ of F^i and $\mathcal{A}_3$ of F^j, then one can easily find a subdrawing of $T^i \cup T^j$ in which $\mathrm{cr}_D(T^i, T^j) = 1$, i.e., $\mathrm{cr}(G^* + D_2) \leq 1$. The graph $G^* + D_2$ contains a subgraph that is a subdivision of the complete graph K_5 and it is well-known by Guy [32] that $\mathrm{cr}(K_5) = 1$. As $\mathrm{cr}(G^* + D_2) \geq \mathrm{cr}(K_5) = 1$, the proof of Lemma 2 is complete. □

Theorem 2. $cr(G^* + D_1) = 0$ and $cr(G^* + D_n) = n^2 - 2n + \lfloor \frac{n}{2} \rfloor$ for $n \geq 2$, i.e., $cr(G^* + D_n) = 4\lfloor \frac{n}{2} \rfloor \lfloor \frac{n-1}{2} \rfloor + \lfloor \frac{n}{2} \rfloor$ for n even and $cr(G^* + D_n) = 4\lfloor \frac{n}{2} \rfloor \lfloor \frac{n-1}{2} \rfloor + \lfloor \frac{n}{2} \rfloor - 1$ for n odd at least 3.

Proof. The graph $G^* + D_1$ is planar, hence $cr(G^* + D_1) = 0$. For $n \geq 2$, both special drawings in Figures 4 and 5 produce $n^2 - 2n + \lfloor \frac{n}{2} \rfloor$ crossings, and so $cr(G^* + D_n) \leq n^2 - 2n + \lfloor \frac{n}{2} \rfloor$. The opposite inequality can be proved by induction on n, and the result holds for $n = 2$ by Lemma 2. For some $n \geq 3$, suppose a drawing D of $G^* + D_n$ with

$$cr_D(G^* + D_n) < n^2 - 2n + \left\lfloor \frac{n}{2} \right\rfloor \tag{3}$$

and that

$$cr(G^* + D_m) = m^2 - 2m + \left\lfloor \frac{m}{2} \right\rfloor \quad \text{for any integer } 2 \leq m < n. \tag{4}$$

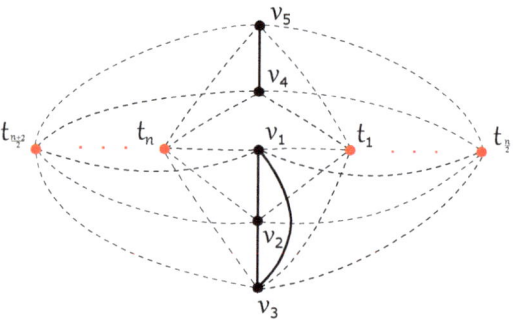

Figure 4. The good drawing of $G^* + D_n$ with $4\lfloor \frac{n}{2} \rfloor \lfloor \frac{n-1}{2} \rfloor + \lfloor \frac{n}{2} \rfloor$ crossings for n even, $n \geq 2$.

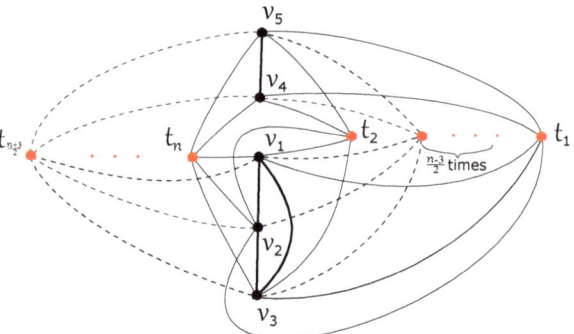

Figure 5. The good drawing of $G^* + D_n$ with $4\lfloor \frac{n}{2} \rfloor \lfloor \frac{n-1}{2} \rfloor + \lfloor \frac{n}{2} \rfloor - 1$ crossings for n odd, $n \geq 3$, where three subgraphs T^1, T^2, and T^n are fixed.

Let us first show that D must be antipode-free. Suppose that, without loss of generality, $cr_D(T^{n-1}, T^n) = 0$. If at least one of T^{n-1} and T^n, say T^n, does not cross G^*, it is not difficult to verify in Figure 1 that $\{T^{n-1}, T^n\} \not\subseteq R_D$, i.e., $cr_D(G^*, T^{n-1} \cup T^n) \geq 1$. By (1), we already know that $cr_D(K_{5,3}) \geq 4$, which yields that edges of the subgraph $T^{n-1} \cup T^n$ must be crossed at least four times by each other T^k. So, by fixing the subgraph $T^{n-1} \cup T^n$ in D, we have

$$cr_D(G^* + D_{n-2}) + cr_D(T^{n-1} \cup T^n) + cr_D(K_{5,n-2}, T^{n-1} \cup T^n) + cr_D(G^*, T^{n-1} \cup T^n)$$

$$\geq (n-2)^2 - 2(n-2) + \left\lfloor \frac{n-2}{2} \right\rfloor + 0 + 4(n-2) + 1 = n^2 - 2n + \left\lfloor \frac{n}{2} \right\rfloor$$

The obtained crossing number contradicts the assumption (3) and confirms that the considered drawing D is antipode-free. For easier reading, if $r = |R_D|$ and $s = |S_D|$, then again (3) together with $cr_D(K_{5,n}) \geq 4\lfloor \frac{n}{2} \rfloor \lfloor \frac{n-1}{2} \rfloor$ using (1) imply the following inequality with respect to possible edge crossings of G^* in D:

$$cr_D(G^*) + s + 2(n - r - s) < \left\lfloor \frac{n}{2} \right\rfloor. \tag{5}$$

The inequality (5) forces more than $\lceil \frac{n}{2} \rceil$ subgraphs T^i by which edges of G^* are not crossed, that is, $r \geq \lceil \frac{n}{2} \rceil + 1 \geq 3$ and $s < \lfloor \frac{n}{2} \rfloor$. Of course, if n is odd then previous inequalities could be strengthened, but this is not necessary in the following process of obtaining a contradiction with number of crossings in D. Moreover, if $n = 3$ then $r = 3$, and $cr_D(G^* + D_3) \geq cr(K_{5,3}) = 4$ with the assumption (3) enforce n at least four.

Case 1: $cr_D(G^*) = 0$ and choose the vertex notation of the graph G^* in such a way as shown in Figure 1a. In this case, we deal with configurations from the nonempty set $\mathcal{M}_D$. As the set R_D is nonempty, recall that

$$\sum_{i \in I_o} \alpha_i + \sum_{i \in I_e} \alpha_i = r \geq 3.$$

Let us first suppose that either $\alpha_1 + \alpha_3 + \alpha_5 = 0$ or $\alpha_2 + \alpha_4 + \alpha_6 = 0$. For the rest of the proof we may therefore assume that $\alpha_2 + \alpha_4 + \alpha_6 = 0$, that is, $\alpha_1 + \alpha_3 + \alpha_5 > 0$. Since $\mathcal{G}_D$ is the subgraph of $\mathcal{G}$ induced by V_D with respect to weights 2 of all its edges (without possible loops), three possible subcases presented in Figure 6 may occur:

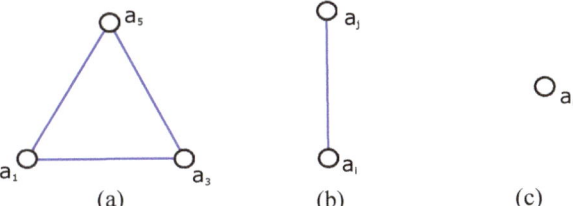

Figure 6. Three possible components of the graph $\mathcal{G}_D$ if $\alpha_2 = \alpha_4 = \alpha_6 = 0$. (**a**): $\alpha_i > 0$ for each $i \in I_o$; (**b**): $\alpha_i > 0$ and $\alpha_j > 0$ for exactly two different $i, j \in I_o$; (**c**): $\alpha_i > 0$ for only one $i \in I_o$.

(a) $\alpha_i > 0$ for each $i \in I_o$. Let us assume three subgraphs T^{n-2}, T^{n-1}, $T^n \in R_D$ such that F^{n-2}, F^{n-1} and F^n have three mutually different configurations from the set $\mathcal{M}_D = \{\mathcal{A}_1, \mathcal{A}_3, \mathcal{A}_5\}$. Then, $cr_D(T^{n-2} \cup T^{n-1} \cup T^n, T^i) \geq 4 + 2 + 2 = 8$ holds for any other $T^i \in R_D$ by summing values in corresponding three rows of Table 1, and $cr_D(G^* \cup T^{n-2} \cup T^{n-1} \cup T^n, T^i) \geq 6$ is true by Corollary 2 for any $T^i \in S_D$. Then, by fixing the graph $G^* \cup T^{n-2} \cup T^{n-1} \cup T^n$

$$cr_D(G^* + D_n) \geq 4 \left\lfloor \frac{n-3}{2} \right\rfloor \left\lfloor \frac{n-4}{2} \right\rfloor + 8(r - 3) + 6s + 7(n - r - s) + 6$$

$$= 4 \left\lfloor \frac{n-3}{2} \right\rfloor \left\lfloor \frac{n-4}{2} \right\rfloor + 7n + r - s - 18 \geq 4 \left\lfloor \frac{n-3}{2} \right\rfloor \left\lfloor \frac{n-4}{2} \right\rfloor$$

$$+ 7n + \left(\left\lceil \frac{n}{2} \right\rceil + 1 \right) + \left(1 - \left\lfloor \frac{n}{2} \right\rfloor \right) - 18 \geq n^2 - 2n + \left\lfloor \frac{n}{2} \right\rfloor.$$

(b) Assuming that $\alpha_i > 0$ for exactly two $i \in I_o$, without lost of generality, let us consider two different subgraphs T^{n-1}, $T^n \in R_D$ such that F^{n-1} and F^n have configurations $\mathcal{A}_1$ and $\mathcal{A}_3$, respectively. As $\mathcal{M}_D = \{\mathcal{A}_1, \mathcal{A}_3\}$, we have $cr_D(T^{n-1} \cup T^n, T^i) \geq 4 + 2 = 6$ for any $T^i \in R_D$, $i \neq n - 1, n$. Therewith, the antipode-free property of D forces that, $cr_D(T^{n-1} \cup T^n, T^i) \geq 2$ trivially holds for any subgraph T^i with $i \neq n - 1, n$. Hence, by fixing the graph $G^* \cup T^{n-1} \cup T^n$

$$\mathrm{cr}_D(G^* + D_n) \geq 4\left\lfloor\frac{n-2}{2}\right\rfloor\left\lfloor\frac{n-3}{2}\right\rfloor + 6(r-2) + 3s + 4(n-r-s) + 2$$

$$= 4\left\lfloor\frac{n-2}{2}\right\rfloor\left\lfloor\frac{n-3}{2}\right\rfloor + 4n + 2r - s - 10 \geq 4\left\lfloor\frac{n-2}{2}\right\rfloor\left\lfloor\frac{n-3}{2}\right\rfloor$$

$$+ 4n + 2\left(\left\lceil\frac{n}{2}\right\rceil + 1\right) + \left(1 - \left\lfloor\frac{n}{2}\right\rfloor\right) - 10 \geq n^2 - 2n + \left\lfloor\frac{n}{2}\right\rfloor.$$

(c) $\alpha_i > 0$ for only one $i \in I_o$. As $\mathcal{M}_D = \{\mathcal{A}_i\}$, in the rest of the paper, we may consider $T^n \in R_D$ with the configuration $\mathcal{A}_1$ of F^n. Then edges of each other subgraph $T^j \in R_D$ cross at least four times edges of T^n provided by $\mathrm{rot}_D(t_n) = \mathrm{rot}_D(t_j)$. Thus, by fixing the graph $G^* \cup T^n$

$$\mathrm{cr}_D(G^* + D_n) \geq 4\left\lfloor\frac{n-1}{2}\right\rfloor\left\lfloor\frac{n-2}{2}\right\rfloor + 4(r-1) + 2s + 3(n-r-s) + 0$$

$$= 4\left\lfloor\frac{n-1}{2}\right\rfloor\left\lfloor\frac{n-2}{2}\right\rfloor + 3n + r - s - 4 \geq 4\left\lfloor\frac{n-1}{2}\right\rfloor\left\lfloor\frac{n-2}{2}\right\rfloor$$

$$+ 3n + \left(\left\lceil\frac{n}{2}\right\rceil + 1\right) + \left(1 - \left\lfloor\frac{n}{2}\right\rfloor\right) - 4 \geq n^2 - 2n + \left\lfloor\frac{n}{2}\right\rfloor.$$

All three subcases contradict the assumption (3). In addition, let us suppose that $\alpha_1 + \alpha_3 + \alpha_5 > 0$ and $\alpha_2 + \alpha_4 + \alpha_6 > 0$. Remark that the subgraph $\mathcal{G}_D$ can be either connected (consisting of a single component) or also disconnected with several components. Now, we are able to discuss over remaining possible components of $\mathcal{G}_D$ in the following subcases:

1. There are no two adjacent edges with weights 1 in the subgraph $\mathcal{G}_D$, that is, there are four possibilities presented in Figure 7.

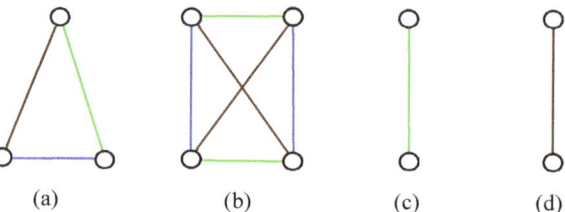

Figure 7. Four possible components of the subgraph $\mathcal{G}_D$ in which there are no two adjacent edges with weights 1. Green, blue, brown, and black correspond to the values 1, 2, 3, and 4, respectively. (**a**): the complete graph K_3 with edge weights 1, 2, and 3; (**b**): the complete graph K_4 with edge weights 1, 1, 2, 2, 3, and 3; (**c**): the complete graph K_2 with edge weight 1; (**d**): the complete graph K_2 with edge weight 3.

- $w_D(a_i a_j) = 1$ for some $i \in I_o, j \in I_e$, i.e., there are three cases mentioned in Figure 7a–c. Let us consider two subgraphs $T^{n-1}, T^n \in R_D$ such that F^{n-1}, F^n have different configurations from $\{\mathcal{A}_i, \mathcal{A}_j\}$, where i, j are associated indexes. Using weights of edges in the considered component of $\mathcal{G}_D$, one can easily verify that edges of the graph $T^{n-1} \cup T^n$ are crossed at least five times by edges of any another subgraph $T^k \in R_D$. Moreover, since the minimum number of interchanges of adjacent elements of $\mathrm{rot}_D(t_n)$ required to produce $\mathrm{rot}_D(t_{n-1})$ is three, any subgraph T^k with $k \neq n-1, n$ crosses edges of $T^{n-1} \cup T^n$ at least thrice. Thus, by fixing the graph $G^* \cup T^{n-1} \cup T^n$

$$\operatorname{cr}_D(G^* + D_n) \geq 4 \left\lfloor \frac{n-2}{2} \right\rfloor \left\lfloor \frac{n-3}{2} \right\rfloor + 5(r-2) + 4s + 5(n-r-s) + 1$$

$$= 4 \left\lfloor \frac{n-2}{2} \right\rfloor \left\lfloor \frac{n-3}{2} \right\rfloor + 5n - s - 9 \geq 4 \left\lfloor \frac{n-2}{2} \right\rfloor \left\lfloor \frac{n-3}{2} \right\rfloor$$

$$+ 5n + \left(1 - \left\lfloor \frac{n}{2} \right\rfloor\right) - 9 \geq n^2 - 2n + \left\lfloor \frac{n}{2} \right\rfloor.$$

- $w_D(a_i a_j) > 1$ for all $i \in I_o$, $j \in I_e$, i.e., there is only one case mentioned in Figure 7d. Let us again consider two subgraphs $T^{n-1}, T^n \in R_D$ such that F^{n-1}, F^n have different configurations from $\{\mathcal{A}_i, \mathcal{A}_j\}$, where i, j are associated indexes. Then, $\operatorname{cr}_D(T^{n-1} \cup T^n, T^k) \geq 7$ holds by summing edge-weights 4 and 3 for any other $T^k \in R_D$. Hence, by fixing the graph $G^* \cup T^{n-1} \cup T^n$

$$\operatorname{cr}_D(G^* + D_n) \geq 4 \left\lfloor \frac{n-2}{2} \right\rfloor \left\lfloor \frac{n-3}{2} \right\rfloor + 7(r-2) + 3s + 4(n-r-s) + 3$$

$$= 4 \left\lfloor \frac{n-2}{2} \right\rfloor \left\lfloor \frac{n-3}{2} \right\rfloor + 4n + 3r - s - 11 \geq 4 \left\lfloor \frac{n-2}{2} \right\rfloor \left\lfloor \frac{n-3}{2} \right\rfloor$$

$$+ 4n + 3\left(\left\lceil \frac{n}{2} \right\rceil + 1\right) + 1 - \left\lfloor \frac{n}{2} \right\rfloor - 11 \geq n^2 - 2n + \left\lfloor \frac{n}{2} \right\rfloor.$$

Both discussed cases again confirm a contradiction with (3) in D, and so, suppose that there are two adjacent edges with weights 1 in the subgraph $\mathcal{G}_D$. Further, only in the case if the number β_j is defined, we claim that the following two properties (6) and (7) must be also fulfilled in D:

$$\beta_j + \sum_{i \in I_o} \alpha_i > \left\lfloor \frac{n}{2} \right\rfloor \quad \text{for some } j \in I_o, \tag{6}$$

$$\beta_j + \sum_{i \in I_e} \alpha_i > \left\lfloor \frac{n}{2} \right\rfloor \quad \text{for some } j \in I_e. \tag{7}$$

For a contradiction, suppose, without loss of generality, that $\beta_1 + \alpha_1 + \alpha_3 + \alpha_5 \leq \left\lfloor \frac{n}{2} \right\rfloor$, that is, $-\alpha_1 - \alpha_3 - \alpha_5 - \beta_1 \geq -\left\lfloor \frac{n}{2} \right\rfloor$. In this case, from the definition of β_1, we have $\alpha_1 > 0$, $\alpha_4 > 0$, and $\alpha_6 > 0$. Thus, in the rest of the paper, let us consider three subgraphs $T^{n-2}, T^{n-1}, T^n \in R_D$ such that F^{n-2}, F^{n-1}, and F^n have configurations $\mathcal{A}_1, \mathcal{A}_4$, and $\mathcal{A}_6$, respectively. Using values in Table 1, one can easily verify that edges of the graph $T^{n-2} \cup T^{n-1} \cup T^n$ are crossed at least six times and seven times by edges of any another subgraph $T^i \in R_D$ with the configuration $\mathcal{A}_1, \mathcal{A}_3, \mathcal{A}_5$ and $\mathcal{A}_2, \mathcal{A}_4, \mathcal{A}_6$ of F^i (of course, if $\mathcal{A}_k \in M_D$ for some $k \in I_o \cup I_e$ in D), respectively. However, from Corollary 1 we get that $\operatorname{cr}_D(G^* \cup T^{n-2} \cup T^{n-1} \cup T^n, T^i) \geq 7$ holds for any $T^i \in S_D$ provided by we can also assume that $\operatorname{cr}_D(T^{n-2}, T^{n-1}) = 1$ and $\operatorname{cr}_D(T^{n-2}, T^n) = 1$ due to the congruence property (If $\operatorname{rot}_D(t_x)$ and $\operatorname{rot}_D(t_y)$ are two cyclic permutations of odd length, and $Q(\operatorname{rot}_D(t_x), \operatorname{rot}_D(t_y))$ denotes the minimum number of interchanges of adjacent elements of $\operatorname{rot}_D(t_x)$ required to produce the inverse cyclic permutation of $\operatorname{rot}_D(t_y)$, then $\operatorname{cr}_D(T^x, T^y) = Q(\operatorname{rot}_D(t_x), \operatorname{rot}_D(t_y)) + 2z$ for some nonnegative integer z, for more see Woodall [31]). Hence, by fixing the graph $G^* \cup T^{n-2} \cup T^{n-1} \cup T^n$

$$\operatorname{cr}_D(G^* + D_n) \geq 4 \left\lfloor \frac{n-3}{2} \right\rfloor \left\lfloor \frac{n-4}{2} \right\rfloor + 6(\alpha_1 + \alpha_3 + \alpha_5 - 1) + 7(\alpha_2 + \alpha_4 + \alpha_6 - 2) + 7s$$

$$+ 6\beta_1 + 7(n - r - s - \beta_1) + 4 = 4 \left\lfloor \frac{n-3}{2} \right\rfloor \left\lfloor \frac{n-4}{2} \right\rfloor + 7n - \alpha_1 - \alpha_3 - \alpha_5 - \beta_1 - 16$$

$$\geq 4 \left\lfloor \frac{n-3}{2} \right\rfloor \left\lfloor \frac{n-4}{2} \right\rfloor + 7n - \left\lfloor \frac{n}{2} \right\rfloor - 16 \geq n^2 - 2n + \left\lfloor \frac{n}{2} \right\rfloor.$$

The obtained crossing number also contradicts the assumption (3) of D and confirms that both parity properties (6) and (7) must be fulfilled in D.

2. There are two adjacent edges with weights 1 in the subgraph $\mathcal{G}_D$, that is, there are five possibilities presented in Figure 8.

   (a) Let the graph $\mathcal{G}_D$ consist of one component in such a way as shown in Figure 8a. Without lost of generality, let us assume that a_2, a_3, a_6 are vertices of the considered path on three vertices with weight 1 of both edges. In this case, it is obvious that $\alpha_2 + \alpha_3 + \alpha_6 = r$. Since the number β_3 can be defined, the property (6) forces $\beta_3 + \alpha_3 > \lfloor \frac{n}{2} \rfloor$. Further, let us also assume that $T^n \in R_D$ with the configuration $\mathcal{A}_3$ of F^n. Then, by fixing the graph $G^* \cup T^n$

   $$\mathrm{cr}_D(G^* + D_n) \geq 4 \left\lfloor \frac{n-1}{2} \right\rfloor \left\lfloor \frac{n-2}{2} \right\rfloor + 4(\alpha_3 - 1) + 1(\alpha_2 + \alpha_6) + 4\beta_3 + 2s$$

   $$+ 3(n - r - s - \beta_3) = 4 \left\lfloor \frac{n-1}{2} \right\rfloor \left\lfloor \frac{n-2}{2} \right\rfloor + 3n + (\alpha_3 + \beta_3 - \alpha_2 - \alpha_6)$$

   $$-(s + \alpha_2 + \alpha_6) - 4 \geq 4 \left\lfloor \frac{n-1}{2} \right\rfloor \left\lfloor \frac{n-2}{2} \right\rfloor + 3n + 0 - \left\lfloor \frac{n}{2} \right\rfloor - 4 \geq n^2 - 2n + \left\lfloor \frac{n}{2} \right\rfloor.$$

   (b) Let the graph $\mathcal{G}_D$ consist of one component in such a way as shown in Figure 8b. Without lost of generality, let us assume that a_2, a_3, a_6 are vertices of the considered path on three vertices with weight 1 of both edges and let a_2, a_4, a_6 be vertices of the 3-cycle with respect to weight 2 of all its edges. In this case, it is obvious that $\alpha_2 + \alpha_3 + \alpha_4 + \alpha_6 = r$. The property (6) enforces again $\beta_3 + \alpha_3 > \lfloor \frac{n}{2} \rfloor$ because the number β_3 can be defined. Further, if $T^n \in R_D$ is assumed with the configuration $\mathcal{A}_4$ of F^n, then by fixing the graph $G^* \cup T^n$

   $$\mathrm{cr}_D(G^* + D_n) \geq 4 \left\lfloor \frac{n-1}{2} \right\rfloor \left\lfloor \frac{n-2}{2} \right\rfloor + 3(\alpha_3 + \beta_3) + 2(n - \alpha_3 - \beta_3 - 1)$$

   $$\geq 4 \left\lfloor \frac{n-1}{2} \right\rfloor \left\lfloor \frac{n-2}{2} \right\rfloor + 2n + \left(\left\lfloor \frac{n}{2} \right\rfloor + 1 \right) - 2 \geq n^2 - 2n + \left\lfloor \frac{n}{2} \right\rfloor.$$

   (c) Let the graph $\mathcal{G}_D$ consist of one component in such a way as shown in Figure 8c–e. Let us take a maximal path P_k on k vertices as the subgraph of $\mathcal{G}_D$ with weights 1 on all its edges. If a_i and a_j are two inner vertices of P_k with $i + 1 \equiv j \pmod{2}$ for which the numbers β_i and β_j satisfy the parity properties (6) and (7), then addition of both inequalities thus obtained enforces a contradiction

   $$n \geq \beta_i + \beta_j + r \geq 2 \left(\left\lfloor \frac{n}{2} \right\rfloor + 1 \right).$$

   The obtained contradictions in all three cases complete the proof for the planar subdrawing of G^* induced by D given in Figure 1a.

   Case 2. $\mathrm{cr}_D(G^*) = 2$ and choose the vertex notation of the graph G^* presented as in Figure 1b. Since the set R_D is nonempty and there is only one subdrawing of a subgraph $F^i = G^* \cup T^i$ for all $T^i \in R_D$ represented by the rotation (13524), the subgraph T^i is crossed at least four times by edges of each subgraph $T^j \in R_D$ with $j \neq i$. Hence, by fixing the graph $G^* \cup T^i$

   $$\mathrm{cr}_D(G^* + D_n) \geq 4 \left\lfloor \frac{n-1}{2} \right\rfloor \left\lfloor \frac{n-2}{2} \right\rfloor + 4(r - 1) + 2(n - r) + 2 = 4 \left\lfloor \frac{n-1}{2} \right\rfloor \left\lfloor \frac{n-2}{2} \right\rfloor$$

   $$+ 2n + 2r - 2 \geq 4 \left\lfloor \frac{n-1}{2} \right\rfloor \left\lfloor \frac{n-2}{2} \right\rfloor + 2n + 2 \left(\left\lceil \frac{n}{2} \right\rceil + 1 \right) - 2 \geq n^2 - 2n + \left\lfloor \frac{n}{2} \right\rfloor.$$

   For all these mentioned cases, it turned out that there is no drawing of the graph $G^* + D_n$ with fewer than $n^2 - 2n + \lfloor \frac{n}{2} \rfloor$ crossings, and the proof of Theorem 2 is complete. □

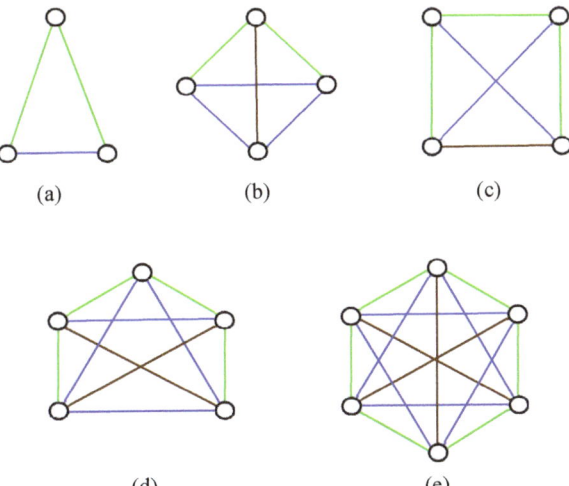

Figure 8. Five possible components of the subgraph $\mathcal{G}_D$ in which there are two adjacent edges with weights 1. Green, blue, brown, and black correspond to the values 1, 2, 3, and 4, respectively. (**a**): the complete graph K_3 with edge weights 1, 1, and 2; (**b**): the complete graph K_4 with edge weights 1, 1, 2, 2, 2, and 3; (**c**): the complete graph K_4 with edge weights 1, 1, 1, 2, 2, and 3; (**d**): the complete graph K_5; (**e**): the complete graph K_6.

5. Conclusions

Into both drawings in Figures 4 and 5, we could add $n-1$ or n edges forming paths P_n, $n \geq 2$ or cycles C_n, $n \geq 3$ on vertices of D_n with no crossing, respectively. Thus, the following surprising results are obvious.

Corollary 3. $\mathrm{cr}(G^* + P_n) = n^2 - 2n + \lfloor \frac{n}{2} \rfloor$ for $n \geq 2$, i.e., $\mathrm{cr}(G^* + P_n) = 4 \lfloor \frac{n}{2} \rfloor \lfloor \frac{n-1}{2} \rfloor + \lfloor \frac{n}{2} \rfloor$ for n even and $\mathrm{cr}(G^* + P_n) = 4 \lfloor \frac{n}{2} \rfloor \lfloor \frac{n-1}{2} \rfloor + \lfloor \frac{n}{2} \rfloor - 1$ for n odd.

Corollary 4. $\mathrm{cr}(G^* + C_n) = n^2 - 2n + \lfloor \frac{n}{2} \rfloor$ for $n \geq 3$, i.e., $\mathrm{cr}(G^* + C_n) = 4 \lfloor \frac{n}{2} \rfloor \lfloor \frac{n-1}{2} \rfloor + \lfloor \frac{n}{2} \rfloor$ for n even and $\mathrm{cr}(G^* + C_n) = 4 \lfloor \frac{n}{2} \rfloor \lfloor \frac{n-1}{2} \rfloor + \lfloor \frac{n}{2} \rfloor - 1$ for n odd.

These results extend already known results of join products of graphs on at most six vertices with paths and cycles, see [2,5,18,20,26,33–41].

Funding: This research received no external funding.

Institutional Review Board Statement: Not applicable.

Informed Consent Statement: Not applicable.

Data Availability Statement: Not applicable.

Conflicts of Interest: The author declares no conflict of interest.

References

1. Garey, M.R.; Johnson, D.S. Crossing number is NP-complete. *SIAM J. Algebraic. Discret. Methods* **1983**, *4*, 312–316. [CrossRef]
2. Klešč, M. The crossing numbers of join of the special graph on six vertices with path and cycle. *Discret. Math.* **2010**, *310*, 1475–1481. [CrossRef]
3. Kleitman, D.J. The crossing number of $K_{5,n}$. *J. Comb. Theory* **1970**, *9*, 315–323. [CrossRef]
4. Clancy, K.; Haythorpe, M.; Newcombe, A. A survey of graphs with known or bounded crossing numbers. *Australas. J. Comb.* **2020**, *78*, 209–296.
5. Klešč, M.; Schrötter, Š. The crossing numbers of join products of paths with graphs of order four. *Discuss. Math. Graph Theory* **2011**, *31*, 321–331. [CrossRef]

6. Asano, K. The crossing number of $K_{1,3,n}$ and $K_{2,3,n}$. *J. Graph Theory* **1986**, *10*, 1–8. [CrossRef]
7. Berežný, Š.; Staš, M. On the crossing number of join of the wheel on six vertices with the discrete graph. *Carpathian J. Math.* **2020**, *36*, 381–390. [CrossRef]
8. Ding, Z.; Huang, Y. The crossing numbers of join of some graphs with n isolated vertices. *Discuss. Math. Graph Theory* **2018**, *38*, 899–909. [CrossRef]
9. Ho, P.T. The Crossing Number of $K_{2,4,n}$. *Ars Comb.* **2013**, *109*, 527–537.
10. Ho, P.T. The crossing number of $K_{1,1,3,n}$. *Ars Comb.* **2011**, *99*, 461–471.
11. Ho, P.T. The crossing number of $K_{1,m,n}$. *Discret. Math.* **2008**, *308*, 5996–6002. [CrossRef]
12. Ho, P.T. The crossing number of $K_{2,2,2,n}$. *Far East J. Appl. Math.* **2008**, *30*, 43–69.
13. Ho, P.T. On the crossing number of some complete multipartite graphs. *Utilitas Math.* **2009**, *79*, 125–143.
14. Huang, Y.; Zhao, T. The crossing number of $K_{1,4,n}$. *Discret. Math.* **2008**, *308*, 1634–1638. [CrossRef]
15. Klešč, M. On the crossing numbers of products of stars and graphs of order five. *Graphs Comb.* **2001**, *17*, 289–294. [CrossRef]
16. Klešč, M. On the Crossing Numbers of Cartesian Products of Stars and Graphs on Five Vertices. In *Combinatorial Algorithms*; LNCS; Springer: Berlin/Heidelberg, Germany, 2009; Volume 5874, pp. 324–333.
17. Klešč, M.; Draženská, E. The crossing numbers of products of the graph $K_{2,2,2}$ with stars. *Carpathian J. Math.* **2008**, *24*, 327–331.
18. Klešč, M.; Schrötter, Š. The crossing numbers of join of paths and cycles with two graphs of order five. In *Lecture Notes in Computer Science: Mathematical Modeling and Computational Science*; Springer: Berlin/Heidelberg, Germany, 2012; Volume 7125, pp. 160–167.
19. Klešč, M.; Schrötter, Š. On the crossing numbers of cartesian products of stars and graphs of order six. *Discuss. Math. Graph Theory* **2013**, *33*, 583–597. [CrossRef]
20. Klešč, M.; Valo, M. Minimum crossings in join of graphs with paths and cycles. *Acta Elec. Inf.* **2012**, *12*, 32–37.
21. Mei, H.; Huang, Y. The Crossing Number of $K_{1,5,n}$. *Int. J. Math. Combin.* **2007**, *1*, 33–44.
22. Staš, M. Determining Crossing Numbers of the Join Products of Two Specific Graphs of Order Six With the Discrete Graph. *Filomat* **2020**, *34*, 2829–2846. [CrossRef]
23. Wang, Y.; Huang, Y. The crossing number of Cartesian product of 5-wheel with any tree. *Discuss. Math. Graph Theory* **2021**, *41*, 183–197.
24. Wang, J.; Zhang, L.; Huang, Y. On the crossing number of the Cartesian product of a 6-vertex graph with S_n. *Ars Combin.* **2013**, *109*, 257–266.
25. Klešč, M.; Staš, M.; Petrillová, J. The crossing numbers of join of special disconnected graph on five vertices with discrete graphs. *Graphs Comb.* **2022**, *38*, 35. [CrossRef]
26. Staš, M. Determining crossing number of join of the discrete graph with two symmetric graphs of order five. *Symmetry* **2019**, *11*, 123. [CrossRef]
27. Staš, M. On the crossing number of join product of the discrete graph with special graphs of order five. *Electron. J. Graph Theory Appl.* **2020**, *8*, 339–351. [CrossRef]
28. Staš, M. On the crossing numbers of join products of five graphs of order six with the discrete graph. *Opusc. Math.* **2020**, *40*, 383–397. [CrossRef]
29. Berežný, Š.; Buša, J., Jr. Algorithm of the Cyclic-Order Graph Program (Implementation and Usage). *J. Math. Model. Geom.* **2019**, *7*, 1–8. [CrossRef]
30. Hernández-Vélez, C.; Medina, C.; Salazar, G. The optimal drawing of $K_{5,n}$. *Electron. J. Comb.* **2014**, *21*, 29.
31. Woodall, D.R. Cyclic-order graphs and Zarankiewicz's crossing number conjecture. *J. Graph Theory* **1993**, *17*, 657–671. [CrossRef]
32. Guy, R.K. Crossing numbers of graphs. *Graph Theory Appl.* **1972**, *303*, 111–124.
33. Draženská, E. On the crossing number of join of graph of order six with path. In Proceedings of the CJS 2019: 22nd Czech-Japan Seminar on Data Analysis and Decision Making, Nový Světlov, Czechia, 25–28 September 2019; pp. 41–48.
34. Draženská, E. Crossing numbers of join product of several graphs on 6 vertices with path using cyclic permutation. In Proceedings of the MME 2019: Proceedings of the 37th International Conference, České Budějovice, Czechia, 11–13 September 2019; pp. 457–463.
35. Klešč, M. The join of graphs and crossing numbers. *Electron. Notes Discret. Math.* **2007**, *28*, 349–355. [CrossRef]
36. Klešč, M. The crossing numbers of join of cycles with graphs of order four. In Proceedings of the Aplimat 2019: 18th Conference on Applied Mathematics, Bratislava, Slovakia, 5–7 February 2019; pp. 634–641.
37. Klešč, M; Kravecová, D.; Petrillová, J. The crossing numbers of join of special graphs. *Electr. Eng. Inform.* **2011**, *2*, 522–527.
38. Klešč, M.; Petrillová, J.; Valo, M. On the crossing numbers of Cartesian products of wheels and trees. *Discuss. Math. Graph Theory* **2017**, *37*, 399–413. [CrossRef]
39. Ouyang, Z.; Wang, J.; Huang, Y. The crossing number of join of the generalized Petersen graph $P(3,1)$ with path and cycle. *Discuss. Math. Graph Theory* **2018**, *38*, 351–370.
40. Staš, M. The crossing numbers of join products of paths and cycles with four graphs of order five. *Mathematics* **2021**, *9*, 1277. [CrossRef]
41. Staš, M.; Valiska, J. On the crossing numbers of join products of $W_4 + P_n$ and $W_4 + C_n$. *Opusc. Math.* **2021**, *41*, 95–112. [CrossRef]

Article

State Machines and Hypergroups

Gerasimos G. Massouros [1,*] and Christos G. Massouros [2]

[1] School of Social Sciences, Hellenic Open University, Aristotelous 18, GR 26335 Patra, Greece
[2] Core Department, National and Kapodistrian University of Athens, Euripus Campus, GR 34400 Euboia, Greece; chrmas@uoa.gr or ch.massouros@gmail.com
* Correspondence: germasouros@gmail.com or masouros.gerasimos@ac.eap.gr

Abstract: State machines are a type of mathematical modeling tool that is commonly used to investigate how a system interacts with its surroundings. The system is thought to be made up of discrete states that change in response to external inputs. The state machines whose environment is a two-element magma are investigated in this study, focusing on the case when the magma is a group or a hypergroup. It is shown that state machines in any two-element magma can only have up to three states. In particular, the quasi-automata and quasi-multiautomata state machines are described and enumerated.

Keywords: hypergroup; magma; state machines; automata

MSC: 20N20; 18B20; 68Q70

1. Introduction

The state machines and the hypergroups are mathematical achievements of the twentieth century. The state machines are mathematical models which are mostly used for the study of actual physical or behavioral processes. Their roots can be traced back to mathematical logic and they are the primary and major components of Computer Theory. Alan Mathison Turing (1912–1954) developed his theoretical universal-algorithm machine to address the question of whether an algorithm for providing proofs whenever they do exist can be found and he discovered that some tasks which this abstract machine is expected to be able to perform are impossible even for it. The usefulness of the state machines quickly began to spread in other sciences as well. For example, Warren Sturgis McCulloch (1898–1969) and Walter Harry Pitts (1923–1969) created a mathematical model in neuroscience [1]. The model they constructed for a "neural net" was a state machine of the same nature as Turing's. Stephen Cole Kleene (1909–1994) later elaborated their model [2], while Noam Chomsky created mathematical models in linguistics for the description of languages [3,4]. The rapid development of technology in the twentieth century has made it possible to materialize such theoretical machines by creating the computers. This development fulfilled the timeless dream of mankind, to create machines like the Antikythera mechanism of the Hellenistic era (the earliest known analog computer, dated back to the second century BC [5–7]), and the mechanical calculating devices created by Blaise Pascal (1623–1662), by Gottfried Wilhelm von Leibniz (1646–1716), by Charles Babbade (1792–1871) and his co-worker Ada Augusta (1815–1852), the daughter of poet Lord Byron, all of which were as powerful as their respective technologies would allow.

The basic building blocks of a state machine are their internal qualities which are named internal states. The internal states are reacting to certain changes in their environment and this reaction causes state transitions. In the general case, it does not matter what the states and the environment of a state machine really are.

For example, in biology, we can consider the state of a cell in its environment, which consists of certain chemical and physical conditions, such as PH, temperature, light and so on. When these chemicals do not remain within a range of concentrations and/or the physical conditions exceed a threshold value, the cell changes its behavior, e.g., a plant cell performs photosynthesis under the "input" of sunlight.

Also, a population group or a business can move to a new state when the economic and social changes in its environment cross a certain threshold.

Respectively in an electronic system which involves various electrical components, the flow of electric current can activate some of these components and the system can switch from one state to another. In fact, as in the case of some circuits with flip-flops, the new state depends on both, the input pulse and the previous state of the circuit.

So, the same environmental change (input) can shift the system into different states, depending on its previous state (condition).

A lot of ink has been shed on the research and study of the behavior of such systems, both living and non-living. In this paper, we approach this issue from a different perspective, i.e., we will address the question of how many different state machines can "survive" or can be acceptable in a given environment. Undoubtedly, the interaction between a system and its environment can be vastly complicated and it still remains an area that needs to be understood, if we wish to be in a position to predict the behaviour of the system and perceive the level of the environmental constraints it can endure as well as their impact on it.

We will attempt this approach with the tools of abstract Algebra, which is the most extensively used branch of Mathematics in the study of the state machines. So, in this paper, the environment is a set equipped with a rule of synthesis such that the result of the synthesis of any two elements is one or more elements. In Éléments de Mathématique, Algèbre [8], Nicolas Bourbaki used the Greek word *magma*, which comes from the verb μάσσω (=knead), to indicate such a set, while in [9] this notion was generalized so as to include more structures. Here, we use the magma of two elements, which is the environment of the binary state machines, like the 0–1 environment which is used in the digital technology. The interaction of the environment on the states is modeled algebraically via the different types of the associativity.

Sometimes, algebraic tools had been developed before even the relevant questions were asked and this is one of the most fascinating aspects of mathematics: to give the answers long before the rest of the world realizes why they should ask the questions.

During the 20th century, Algebra itself faced, and on most occasions overcame, many difficult and serious challenges that were deriving from various mathematical and not only areas, such as the Theory of Equations, Geometry, Topology, Quantum Mechanics, Chemistry, etc. Also, the 20th century put an end to the paradise of determinism. David Hilbert's (1862–1943) visions collapsed under Kurt Gödel's (1906–1978) work. Quantum Mechanics, the uncertainty principle of Werner Heisenberg (1901–1976), the axiomatic foundation of probability by Andrey Kolmogorov (1903–1987) made uncertainty inherent in science and brought into existence its mathematics. In this direction, Frédéric Marty (1911–1940) in a series of three papers [10–12] introduced the hypergroup in Algebra and gave some of its initial properties. His untimely death during World War II, while serving as a French Army officer, did not allow him to write more papers. However, the aforementioned three were enough to bring into being the Hypercompositional Algebra. The fundamental notion of the Hypercompositional Algebra is the hypercomposition, that is, a law of synthesis of any two elements, which yields a set of elements instead of a single element only.

The introduction of Hypercompositional Algebra into Computer Theory occurred in the G.G. Massouros Ph.D. thesis [5], under the supervision of J. Mittas (1921–2012). There followed more papers by the same author and Ch. Massouros, e.g., [13–21], as well as other researchers such as J. Chvalina [22–28], L. Chvalinová [22], Š. Hošková-Mayerová [24,25], M. Novák [26–32], S. Křehlík [26,27,29–31,33], M.M. Zahedi [34], M. Ghorani [34,35], D. Heidari and S. Doostali [36], R.A. Borzooei et al. [37]. In relation to this subject, there are applications of the Hypercompositional Algebra in graph theory, artificial intelligence, cryptography, sensor networks and many more that, indicatively, can be found in [38–60]. Also, results in the above areas, up to the date it was published, can be found in P. Corsini and V. Leoreanu's book [61].

The following section presents the necessary preliminary notions for the self-sufficiency of the paper. The third section contains a study of operators' and hyperoperators' actions and the way they define hypercompositional structures in the set on which they operate. The fourth paragraph focuses on the detailed study of the magma of two elements, which is the environment of the binary state machines. The next (fifth) paragraph is dedicated to the analysis of the binary state machines, when their environment has the structure of a group or a hypergroup. The quasi-automata and quasi-multiautomata that can exist in such an environment are studied and enumerated.

2. Preliminaries

The fundamental notion of the Hypercompositional Algebra is the hypercomposition, that is, a law of synthesis which yields a set of elements instead of a single element, when applied on any two elements. More specifically, we have the definitions [9]:

Definition 1. *Let E be a non-void set. A mapping from $E \times E$ into E is called a composition on E, while a mapping from $E \times E$ into the power set $P(E)$ of E is called a hypercomposition on E. A hypercomposition is called partial, if $ab = \varnothing$, for some a, b in E. A set enriched with a composition or a hypercomposition is called a magma.*

The above definition, which was introduced in [9], extends the definition of the magma given by Nicolas Bourbaki [8] in order to include laws of synthesis which are hypercompositions on a set E.

Let $(E, \perp)$ be a magma. For any two non-void subsets X, Y of E, $X \perp Y = \{x \perp y \in E \mid x \in X, y \in Y\}$, if $\perp$ is a composition and $X \perp Y = \bigcup_{x \in X, y \in Y} (x \perp y)$, if $\perp$ is a hypercomposition.

If X or Y is empty, then $X \perp Y$ is empty. If $a \in E$, we usually write $a \perp Y$ instead of $\{a\} \perp Y$ and $X \perp a$ instead of $X \perp \{a\}$. In general, the singleton $\{a\}$ is identified with its member a. Sometimes it is convenient to use the relational notation $A \approx B$ to assert that subsets A and B have a non-void intersection. Then, as the singleton $\{a\}$ is identified with its member a, the notation $a \approx A$ or $A \approx a$ is used as a substitute for $a \in A$. The relation $\approx$ may be considered as a weak generalization of the equality, since, if A and B are singletons and $A \approx B$, then $A = B$. Thus, $a \approx b \perp c$ means $a = b \perp c$, if the synthesis is a composition and $a \in b \perp c$, if the synthesis is a hypercomposition. This notation is extensively used when it is not necessary to distinguish between a composition or a hypercomposition with respect to a law of synthesis.

Definition 2. *A law of synthesis $(x, y) \to x \perp y$ on a set E is called associative if the property,*

$$(x \perp y) \perp z = x \perp (y \perp z)$$

is valid, for all elements x, y, z in E, while it is called reproductive if for all elements x in E the equality

$$x \perp E = E \perp x = E$$

holds.

Definition 3. *An associative magma is called a semigroup if the law of synthesis on the magma is a composition, while it is called a semihypergroup if the law of synthesis is a non-partial hypercomposition.*

Definition 4. *A reproductive magma is called a quasigroup if the law of synthesis on the magma is a composition, while it is called a quasihypergroup if the law of synthesis is a non-partial hypercomposition.*

Definition 5. *An associative and reproductive magma is called a group if the law of synthesis on the magma is a composition, while it is called a hypergroup if the law of synthesis is a hypercomposition.*

The above Definition 5 appeared in [9], which also contains a detailed presentation of the fundamental properties that derive from the axioms of the associativity and the reproductivity in groups and hypergroups. Among other things, in [9] it is proved that:

Theorem 1. *If G is a group, then:*
i.   *there exists an element $e \in G$ such that $ea = a = ae$ for all $a \in G$*
ii. *for each element $a \in G$ there exists an element $a' \in G$ such that $a'a = e = aa'$*

Theorem 2. *If H is a hypergroup, then:*

$$ab \neq \varnothing \text{ for all } a, b \in H$$

Thus, the hypercomposition in a hypergroup cannot be partial. In this paper, we will consider only non-partial hypercompositions.

It is very common in the bibliography to enrich a magma with the axiom of associativity. Besides, another equality that can be valid in the successive synthesis of the magma's elements is the *inverted associativity*. Recall that a composition or a hypercomposition on a non-void set E is called *left inverted associative* if

$$(a \perp b) \perp c = (c \perp b) \perp a, \text{ for every } a, b, c \in E,$$

while it is called right inverted associative if

$$a \perp (b \perp c) = c \perp (b \perp a), \text{ for every } a, b, c \in E.$$

The notion of the inverted associativity was initially conceived by Kazim and Naseeruddin [62]. A magma equipped with a left inverted associativity is called *left almost semigroup* if the law of synthesis is a composition, while it is called *left almost semihypergroup* if the law of synthesis is a hypercomposition. The terminology is analogous for the right inverted associative magma.

Definition 6. *A reproductive magma which satisfies the axiom of the left inverted associativity is called a left almost-group (LA-group) when the law of synthesis on the magma is a composition, while it is called a left almost-hypergroup (LA-hypergroup) when the law of synthesis is a hypercomposition. A reproductive, right inverted associative magma is called a right almost-group (RA-group) or a right almost-hypergroup (RA-hypergroup) when the law of synthesis is a composition or a hypercomposition respectively.*

Apparently, if the law of synthesis is commutative, then the almost left or almost right groups and hypergroups are groups and hypergroups, respectively. However, it is possible for both associativity and inverted associativity to be valid in a magma. Such cases can be found in the examples of [63], which presents a detailed study of the left/right almost-hypergroups. For the quasi-canonical LA-hypergroups, see [64].

Every law of synthesis in a magma induces two new laws of synthesis. If the law of synthesis is written multiplicatively, then the two induced laws are:

$$a/b = \{x \in E \mid a \approx xb\} \text{ and } b\backslash a = \{x \in E \mid a \approx bx\}$$

Thus $x \approx a/b$ if and only if $a \approx xb$ and $x \approx b\backslash a$ if and only if $a \approx bx$. In the case of a multiplicative magma, the two induced laws are called *inverse laws* and they are named *right division* and *left division*, respectively. If the magma is commutative, it is obvious that the right and left divisions coincide.

Directly connected to the induced laws of synthesis is the transposition axiom, which was firstly introduced by W. Prenowitz (1906–2000) for the study of geometry with the tools of Hypercompositional Algebra (e.g., [65]) and afterwards it was generalized by J. Jantosciak (1942–2017) in [66].

Definition 7. *A magma E is called a transposition magma if it satisfies the axiom:*

$$b\backslash a \approx c/d \text{ implies } ad \approx bc, \text{ for all } a,b,c,d \in E$$

It is obvious that in a transposition magma the following implication

$$a\backslash b \approx d/c \Rightarrow ad \approx bc, \text{ for all } a,b,c,d \in E$$

is valid as well. In [9], the above implications reversed and so we have the two reverse transposition axioms:

Weak reverse transposition axiom:

$$ad \approx bc \text{ implies } b\backslash a \approx c/d \text{ or } a\backslash b \approx d/c, \text{ for all } a,b,c,d \in E$$

Strong reverse transposition axiom:

$$ad \approx bc \text{ implies } b\backslash a \approx c/d \text{ and } a\backslash b \approx d/c, \text{ for all } a,b,c,d \in E$$

However, the following property also applies:

$$ad \approx bc \Leftrightarrow a\backslash b \approx d/c \text{ or } b\backslash a \approx c/d, \text{ for all } a,b,c,d \in E$$

This axiom was named **bilateral transposition** axiom [9].

Special notation: In the following pages, in addition to the typical algebraic notations, we are using Krasner's notation for the complement and difference. So, we denote by $A \cdot\cdot B$ the set of elements that are in the set A, but not in the set B.

3. Action of a Magma on a Set

Let E and S be two non-empty sets. A mapping of E into the set S^S of the mappings of S into itself is called an *action* of E on S. Let $a \to \delta_a$ be an action of E on S. The mapping δ of $S \times E$ to S such that $\delta(s,a) = \delta_a(s)$ is an external law of composition on S, with E being the operating set. δ is called the *law of right action* of E on S. The *law of left action* of E on S is defined in a similar way. The element $\delta_a(s)$ is also called the *transform* of s under a. It is usually denoted by a right (resp. left) multiplicative notation sa (resp. as). The elements of E are called *operators*.

A mapping $\widehat{\delta}$ of $S \times E$ to the power set $P(S)$ of S is an external law of hypercomposition on S. Then, the elements of E are called *hyperoperators* [67]. If $a \in E$ is a hyperoperator, then the multiplicative notation sa (resp. as) signifies an element of $P(S)$, that is, $sa \subseteq S$ (resp. $as \subseteq S$).

A subset T of S is called *stable* under the action $a \to \delta_a$ of E on S if $\delta_a(T) \subseteq T$ for all $a \in E$. The intersection of a family of stable subsets of S under a given action is a stable subset of S as well. Therefore, if X is any subset of S, there exists a smallest stable subset of S that contains it. This subset is said to be *generated* by X and it consists of the elements $(\delta_{a_1} \circ \delta_{a_2} \circ \cdots \circ \delta_{a_n})(x)$, where $x \in X$, $n > 0$, $a_i \in E$ for all i.

Definition 8. *An element s_2 of S is called connected to an element s_1 of S if there exists an element a of E such that $\delta_a(s_1) = s_2$.*

It must be mentioned that s_2 being connected to s_1 does not necessarily imply that s_1 is connected to s_2. If s_2 is connected to s_1, there may be a sequence $a_1, a_2, \ldots, a_n$ of elements of E such that $(\delta_{a_1} \circ \delta_{a_2} \circ \cdots \circ \delta_{a_n})(s_1) = s_2$. Thus, via the notion of the connected elements, a hypercomposition can be defined on S, as follows:

$$s_1 + s_2 = \begin{cases} \{s \in S \mid s = s_1 a \text{ and } s_2 = sb, \text{ with } a, b \in E\}, & \text{if } s_2 \text{ is connected to } s_1 \\ \{s_1, s_2\}, \text{ if } s_2 \text{ is not connected to } s_1 \end{cases}$$

Proposition 1. *If the set of the operators E over a non-void set S is a unitary magma, then $(S, +)$ becomes a hypergroup.*

Proof. Since E is a unitary magma, the result of the hypercomposition always contains the two participating elements, thus $s + S = S + s = S$ for all $s \in S$ and so the reproductive axiom is valid. Moreover, the associativity holds. Indeed, if s_1, s_2 and s_3 are not connected to each other, then

$$s_1 + (s_2 + s_3) = (s_1 + s_2) + s_3 = \{s_1, s_2, s_3\}$$

Next, suppose that s_2 and s_3 are connected to s_1. Also let s_3 be connected to s_2. Then:

$$\begin{aligned}(s_1 + s_2) + s_3 &= \{q \in S \mid q = s_1 a \text{ and } s_2 = (s_1 a)b, \text{ with } a, b \in E\} + s_3 = \\ &= \left\{ \begin{array}{c} s \in S \mid s = (s_1 a)c, \; s_2 = (s_1 a)b, \text{ and} \\ s_3 = ((s_1 a)c)d \text{ with } a, b, c, d \in E \end{array} \right\} = \\ &= s_1 + s_3 \end{aligned}$$

and

$$\begin{aligned}s_1 + (s_2 + s_3) &= s_1 + \{q \in S \mid q = s_2 a \text{ and } s_3 = (s_2 a)b, \text{ with } a, b \in E\} = \\ &= \left\{ \begin{array}{c} s \in S \mid s = s_1 c \text{ and } (s_1 c)d = s_2 a, \; (s_2 a)k = s_3 \text{ or} \\ (s_1 c)l = s_3 \text{ with } a, c, d, k, l \in E \end{array} \right\} = \\ &= s_1 + s_3 \end{aligned}$$

Similar is the proof of all the other cases and hence the proposition. □

Corollary 1. *The set of vertices of a directed graph is endowed with the structure of the hypergroup if the result of the hypercomposition of two vertices v_i and v_j is the set of the vertices which appear in all the possible paths that connect v_i to v_j, or $\{v_i, v_j\}$, if there do not exist any connecting paths from vertex v_i to vertex v_j.*

If E is a magma, an equivalence relation ξ on E is called a *congruence relation* if

$$(a, b) \in \xi, \; (c, d) \in \xi \text{ implies } \begin{cases} [\{y\} \times bd] \cap \xi \neq \varnothing & \text{for all } y \in ac \\ \text{and} \\ [ac \times \{z\}] \cap \xi \neq \varnothing & \text{for all } z \in bd \end{cases}$$

When the law of synthesis in the magma is a composition, then ac and bd are singletons and the above definition is simplified to:

$$(a, b) \in \xi, \; (c, d) \in \xi \text{ implies } (ac, bd) \in \xi$$

The set E/ξ of all equivalence classes defined on E by ξ becomes an associative magma if we define

$$\xi_a \cdot \xi_b = \{\xi_c \mid c \approx ab\} \text{ for all } \xi_a, \xi_b \in E/\xi$$

Proposition 2. *Every congruence relation ξ on a magma E is a normal equivalence relation, and therefore the set E/ξ becomes a magma under the law of synthesis*

$$C_x \cdot C_y = \{C_z \mid z \approx xy\}$$

where C_x is the class of an arbitrary element $x \in E$.

Proof. Since ξ is a congruence relation, for each $x, y \in E$ it holds:

$$z' \in C_x \cdot C_y \Rightarrow$$
$$\Rightarrow (\exists (x', y')) \in C_x \times C_y) \, [z \approx x'y'] \Rightarrow (\exists z \approx xy)[z' \xi z] \Rightarrow z' \in C_z \Rightarrow$$
$$\Rightarrow C_x \cdot C_y \subseteq \bigcup_{z \approx xy} C_z$$

Conversely now:

$$z'' \in \bigcup_{z \approx xy} C_z \Rightarrow$$
$$\Rightarrow (\exists z \approx xy)[z'' \xi z] \Rightarrow (\exists (x'', y'') \in E^2)[x'' \xi x \wedge y'' \xi y \wedge z'' \approx x''y''] \Rightarrow$$
$$\Rightarrow z'' \in C_x \cdot C_y \Rightarrow \bigcup_{z \approx xy} C_z \subseteq C_x \cdot C_y$$

Thus, $C_x \cdot C_y = \bigcup_{z \approx xy} C_z$, and so the quotient set E/ξ, enriched with the law of synthesis $C_x \cdot C_y = \{C_z \mid z \approx xy\}$, is a magma. Obviously, if the law of synthesis is a composition, then the previous equality is simplified to $C_x \cdot C_y = C_{xy}$. □

Corollary 2. *Every congruence relation ξ on a hypergroup E is a normal equivalence relation, and therefore the quotient E/ξ becomes a hypergroup under the hypercomposition $C_x \cdot C_y = \{C_z \mid z \approx xy\}$. If E is a group, then E/ξ is a group as well, under the composition*

$$C_x \cdot C_y = C_{xy}$$

Proposition 3. *If E is a transposition magma and ξ is a congruence relation on E, then E/ξ is a transposition magma.*

Proof. Suppose that for some elements C_x, C_y, C_z, C_w, of the quotient set E/ξ it holds that $C_y \backslash C_x \approx C_z / C_w$. Then there exist elements x', y', z', w' belonging to C_x, C_y, C_z, C_w, respectively, such that $y' \backslash x' \approx z'/w'$. Since the transposition axiom is valid in E, it derives that $x'w' \approx y'z'$. Therefore, $C_x \cdot C_w \approx C_y \cdot C_z$ and hence the proposition. □

Definition 9. *A state machine M is a triplet (S, E, δ) where S and E are sets and δ is mapping of $S \times E$ to S.*

The set S describes the internal qualities of the system. The elements of S are called *internal states* of M. If S is finite, then M is called a *finite state machine*. The set E describes the environmental inputs that can affect the system. The mapping δ describes the environmental influences on the internal qualities of the system and it is called a *state transition function*. Such a system is obviously quite general and can be used in a variety of cases. From the mathematical standpoint, a state machine is a set with operators and the fact that we can successfully approach, describe and examine such systems via algebraic tools and techniques is one of the most impressive and remarkable achievements of modern algebra.

Example 1. *State machines can be depicted by the so-called transition diagrams. Thus, if $S = \{s_1\}$ and E is a finite set, then the relevant state machine is illustrated with the transition diagram presented in Figure 1:*

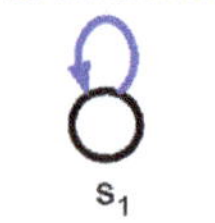

for all α in E

○↻
s_1

Figure 1. Transition diagram for a state machine with one state.

Moreover, if $S = \{s_1, s_2\}$ and $E = \{a\}$, we could have any of the following four state machines that are shown in Figure 2:

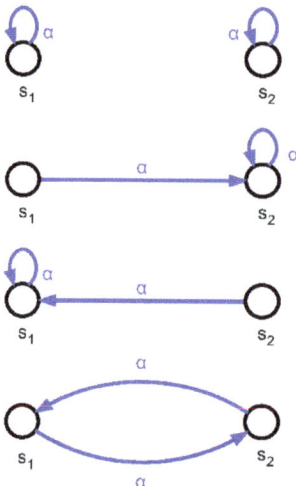

Figure 2. Transition diagrams for a state machine with two states and a singleton as the set of the operators.

Another way of specifying a state machine is by writing out the transition function δ in tabular form, thus creating the so-called **state transition table**. For example, the state transition table of the last in the above figure state machine is the following one:

δ	a
s_1	s_2
s	s_1

Suppose that $a \in E$ is applied to the state $s \in S$ of a state machine M. Then the machine moves to state $\delta(s,a) = \delta_a(s) = sa$. Next, if another input, say $b \in E$, is applied to the machine, the resultant state is:

$$\delta(\delta_a(s), b) = \delta_b(\delta_a(s)) = \delta_b(sa) = (sa)b$$

Since $\delta_b(\delta_a(s)) = (\delta_b \circ \delta_a)(s)$, if E is a magma, we say that the state transition function satisfies the associativity if $\delta_b(\delta_a(s)) = \delta_c(s)$, $c \approx ab$. If the law of synthesis is a composition, then the associativity is of the form $\delta_b(\delta_a(s)) = \delta_{ab}(s)$ or equivalently $(sa)b = s(ab)$ and it is named *mixed associativity*, while if the law of synthesis is a hypercomposition, the associativity is fulfilled if $(sa)b \in s(ab)$ and it is called *generalized mixed associativity* [26,29,68].

Definition 10. *A state machine $M = (S, E, \delta)$ is called quasi-automaton if E is a magma and the state transition function satisfies the mixed associativity, i.e., $(sa)b = s(ab)$ for any pair $a, b \in E$ and any state $s \in S$.*

Definition 11. *A state machine $M = (S, E, \delta)$ is called quasi-multiautomaton if E is a magma and the state transition function satisfies the generalized mixed associativity, i.e., $(sa)b \in s(ab)$ for any pair $a, b \in E$ and any state $s \in S$.*

A detailed presentation of the terminology, based on the historical development of the area, can be found in the well-written paper [29]. The above definitions are in line with [29]. Obviously, every quasi-automaton is a quasi-multiautomaton. Apparently, quasi-multiautomata which are not quasi-automata can only exist when E is a hypercompositional magma. On the contrary, quasi-automata exist when either the magma is endowed with a composition or when it is endowed with a hypercomposition. A special case of quasi-automata occurs when E is a free semigroup or a free monoid instead of an arbitrary magma. In this case, computer theory tends to use the term *"word"* for the elements of E, the term *"letter"* for the elements of its generating set Σ and the term *concatenation of words* for the law of synthesis in E. Moreover, the free semigroup generated by Σ is denoted by Σ^+ and the corresponding free monoid by Σ^*. Also, the quasi-automaton is denoted by $M = (S, E, \delta)$.

Proposition 4. *Let (S, E, δ) be a quasi-automaton and $\sim$ a binary relation on the magma, defined by*

$$a \sim b \Leftrightarrow \delta_a = \delta_b, \text{ where } a, b \in E$$

Then $\sim$ is a congruence relation on E and the magma $E/\sim$ has the same algebraic structure as E.

Proof. This relation is easily seen to be an equivalence relation. Next, let $a \sim b$ and $c \sim d$. From $a \sim b$, it follows that $\delta_a(s) = \delta_b(s)$ for all $s \in S$. Next, since $c \sim d$, the following sequence of equivalent statements holds:

$$\delta_c(\delta_a(s)) = \delta_c(\delta_b(s)); \ \delta_c(\delta_a(s)) = \delta_d(\delta_b(s)); \ \bigcup_{y \approx ca} \delta_y(s) = \bigcup_{z \approx db} \delta_z(s);$$

Therefore, $\sim$ is a congruence relation on E. Next, it is easy to see that the magma $E/\sim$ is of the same type as E, that is, if E is a semigroup, semi-hypergroup, hypergroup, group, etc., then $E/\sim$ has the same algebraic structure, respectively. □

Now, if $M = (S, E, \delta)$ is a quasi-automaton, then the semigroup $E = \Sigma^+/\sim$ or the monoid $E = \Sigma^*/\sim$ can be constructed with the use of Proposition 4. In many cases, it is more convenient to study this semigroup rather than the original machine M. However, if we do not want to lose sight of the set of states, we consider the machine $M = (S, E, \delta)$. Each element of E is an equivalence class of Σ^+ or Σ^*, which acts on S as follows: $s[a] = \delta_a(s)$, where $s \in S$ and $a \in \Sigma^+$ or $a \in \Sigma^*$.

4. The Magma of 2 Elements

In this section, we will proceed to a detailed study and classification of the two-element magma, which is the binary state machines' environment.

While there exists only one single element magma which is a group and also a LA/RA-group, there exist $3^4 = 81$ magmas with 2 elements. These magmas can be constructed, classified and enumerated, with the techniques and methods which are developed in [69–71] and [63]. In the following propositions, these magmas are presented via their Cayley tables. Note that, in a Cayley table, the entry in the row headed by x and the column headed by y is the synthesis $x \cdot y$.

i. **Associative Magmas**

Proposition 5. *There exist 6 semigroups which are classified into 2 classes with 2 isomorphic semigroups each, and into 2 single-member classes, which are presented below via their Cayley tables. Moreover, since SG_1 and SG_2 are commutative, they satisfy both the left and the right inverted associativity.*

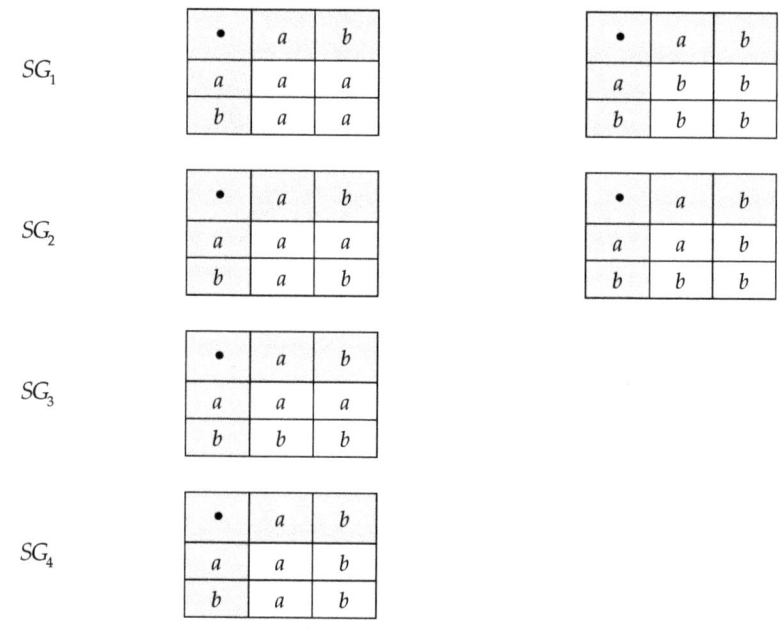

SG_1

·	a	b
a	a	a
b	a	a

·	a	b
a	b	b
b	b	b

SG_2

·	a	b
a	a	a
b	a	b

·	a	b
a	a	b
b	b	b

SG_3

·	a	b
a	a	a
b	b	b

SG_4

·	a	b
a	a	b
b	a	b

Remark 1. When $a = 0$ and $b = 1$ in the two isomorphic semigroups SG_2, we get the two well-known operations of the Boolean algebra:

x	y	x y
0	0	0
0	1	0
1	0	0
1	1	1

x	y	x+y
0	0	0
0	1	1
1	0	1
1	1	1

Proposition 6. *There exist 10 semihypergroups which are classified into 5 classes with 2 isomorphic semihypergroups each. These are displayed below in the form of Cayley tables. Moreover, since SH_1 is commutative, it satisfies both the left and the right inverted associativity.*

SH_1

·	a	b
a	a	a
b	a	{a,b}

·	a	b
a	{a,b}	b
b	b	b

SH_2

·	a	b
a	a	a
b	{a,b}	b

·	a	b
a	a	{a,b}
b	b	b

SH_3

•	a	b
a	a	a
b	{a,b}	{a,b}

•	a	b
a	{a,b}	{a,b}
b	b	b

SH_4

•	a	b
a	a	b
b	{a,b}	b

•	a	b
a	a	{a,b}
b	a	b

SH_5

•	a	b
a	a	{a,b}
b	a	{a,b}

•	a	b
a	{a,b}	b
b	{a,b}	b

ii. **Reproductive Magmas**

Proposition 7. *There exist 21 quasihypergroups which are classified into 9 two-member classes and 3 single-member classes, which are presented below via their Cayley tables.*

QH_1

•	a	b
a	a	b
b	{a,b}	a

•	a	b
a	b	{a,b}
b	a	b

QH_2

•	a	b
a	a	{a,b}
b	b	a

•	a	b
a	b	a
b	{a,b}	b

QH_3

•	a	b
a	a	{a,b}
b	{a,b}	a

•	a	b
a	b	{a,b}
b	{a,b}	b

QH_4

•	a	b
a	b	a
b	a	{a,b}

•	a	b
a	{a,b}	b
b	b	a

QH_5

•	a	b
a	b	a
b	{a,b}	{a,b}

•	a	b
a	{a,b}	{a,b}
b	b	a

QH_6

•	a	b
a	b	{a,b}
b	a	{a,b}

•	a	b
a	{a,b}	b
b	{a,b}	a

QH_7

•	a	b
a	{a,b}	a
b	a	{a,b}

•	a	b
a	{a,b}	b
b	b	{a,b}

QH_8

•	a	b
a	{a,b}	a
b	{a,b}	{a,b}

•	a	b
a	{a,b}	{a,b}
b	b	{a,b}

QH_9

•	a	b
a	{a,b}	b
b	{a,b}	{a,b}

•	a	b
a	{a,b}	{a,b}
b	a	{a,b}

QH_{10}

•	a	b
a	b	{a,b}
b	{a,b}	a

QH_{11}

•	a	b
a	{a,b}	b
b	a	{a,b}

QH_{12}

•	a	b
a	{a,b}	a
b	b	{a,b}

iii. Associative and Reproductive Magmas

Proposition 8. *There exists one class with 2 isomorphic commutative groups.*

G_1

•	a	b
a	a	b
b	b	a

•	a	b
a	b	a
b	a	b

Proposition 9. *There exist 12 hypergroups which are classified into 5 classes with 2 isomorphic hypergroups each and into 2 single-member classes, which are presented below via their Cayley tables. $H_3 - H_7$ are commutative.*

H_1

•	a	b
a	a	b
b	{a,b}	{a,b}

•	a	b
a	{a,b}	{a,b}
b	a	b

H_2

·	a	b
a	a	{a,b}
b	b	{a,b}

·	a	b
a	{a,b}	a
b	{a,b}	b

H_3

·	a	b
a	a	b
b	b	{a,b}

·	a	b
a	{a,b}	a
b	a	b

H_4

·	a	b
a	a	{a,b}
b	{a,b}	{a,b}

·	a	b
a	{a,b}	{a,b}
b	{a,b}	b

H_5

·	a	b
a	b	{a,b}
b	{a,b}	{a,b}

·	a	b
a	{a,b}	{a,b}
b	{a,b}	a

H_6

·	a	b
a	a	{a,b}
b	{a,b}	b

H_7

·	a	b
a	{a,b}	{a,b}
b	{a,b}	{a,b}

H_6 is the two-element *B-hypergroup*. B(inary)-hypergroups came into being during the study of formal languages and automata with the use of hypercompositional algebra [5,13,16,17]. The free monoid of the words generated by an alphabet Σ can be endowed with the B-hypergroup structure, and so become a join hyperringoid [21,72–74], which is named *linguistic hyperringoid* [14,21,73,74]. If the B-hypergroup is fortified with a strong identity [31], which is necessary for the theory of formal languages and automata [14,21], then the join hyperring comes into being [72–74]. H_7 is the two-element *total hypergroup*.

Proposition 10. *All the two-element hypergroups are transposition hypergroups.*

Proof. The Cayley tables of the induced hypercompositions for the seven two-element hypergroups are presented below. For the classes with two elements, we chose the first hypergroup for the presentation of the induced hypercomposition. Observe that the hypergroups H_3, H_4, H_5, H_6 and H_7 are commutative; therefore, the two induced hypercompositions coincide, and so there is only one Cayley table corresponding to each one of them. As mentioned above, in the Cayley tables, the entry in the row headed by x and the column headed by y is the synthesis x/y or $y \backslash x$ respectively.

H_1

/	a	b
a	{a,b}	b
b	b	{a,b}

\	a	b
a	a	{a,b}
b	b	{a,b}

H_2

/	a	b
a	a	{a,b}
b	b	{a,b}

\	a	b
a	{a,b}	b
b	b	{a,b}

H_3

/	a	b
a	a	b
b	b	{a,b}

H_4

/	a	b
a	{a,b}	{a,b}
b	b	{a,b}

H_5

/	a	b
a	b	{a,b}
b	{a,b}	{a,b}

H_6

/	a	b
a	{a,b}	a
b	b	{a,b}

H_7

/	a	b
a	{a,b}	{a,b}
b	{a,b}	{a,b}

The verification of the transposition axiom gives the rest. □

So, according to Proposition 10, there do not exist non-transposition hypergroups with cardinality 1 or 2. However, as shown in the following example, there exist non-transposition hypergroups if their cardinality is greater than or equal to 3.

Example 2. *The hypercomposition on hypergroup H_6 can be written in the following two ways:*

$$a \cdot b = \begin{cases} \{a,b\}, & \text{if } a \neq b \\ a, & \text{if } a = b \end{cases} \quad (1)$$

and

$$a \cdot b = \begin{cases} H, & \text{if } a \neq b \\ a, & \text{if } a = b \end{cases} \quad (2)$$

If $H = \{a,b\}$, then the above two formulas give the same hypercomposition, but if $cardH \geq 3$, then they produce two different hypergroups. The first one, which is the B-hypergroup, satisfies the transposition axiom (see [13] for the proof), while the second one does not. Indeed, the induced hypercomposition of (2) is:

$$a \cdot b = \begin{cases} H \cdot\cdot \{b\}, & \text{if } a \neq b \\ a, & \text{if } a = b \end{cases}$$

Next, if $a \neq b$, we have:

$$a/b \cap b/a = [H \cdot\cdot \{b\}] \cap [H \cdot\cdot \{a\}] \neq \emptyset$$

while

$$aa \cap bb = \{a\} \cap \{b\} = \emptyset.$$

Moreover, the verification of the reverse transposition axiom for hypergroups H_1–H_7 leads to the following result:

Proposition 11. *All the two-element hypergroups satisfy the strong reverse transposition axiom.*

A consequence of Propositions 10 and 11 is the following Theorem:

Theorem 3. *All the two-element hypergroups satisfy the bilateral transposition axiom.*

In [9], following the observation that the quasicanonical hypergroups, the canonical hypergroups, and of course, the groups and the abelian groups satisfy the bilateral transposition axiom, the question arose: *Do there exist other hypergroups satisfying the bilateral transposition axiom apart from the quasicanonical and the canonical ones?* The above Theorem 3 gives the affirmative answer to this question.

iv. **Magmas with inverted associativity**

Proposition 12. *There exists only one class with 2 isomorphic left almost-semihypergroups, as per the following Cayley tables.*

$LA - S_1$

·	a	b
a	b	b
b	{a,b}	{a,b}

·	a	b
a	{a,b}	{a,b}
b	a	a

Proposition 13. *In addition to the five non-isomorphic commutative hypergroups (and hence left and right almost-hypergroups) that are mentioned in Proposition 9, there also exist two non-isomorphic left almost-hypergroups which are classified into one two-member class and into one single-member class, as per the following Cayley tables.*

$LA - H_1$

·	a	b
a	{a,b}	b
b	{a,b}	{a,b}

·	a	b
a	{a,b}	{a,b}
b	a	{a,b}

$LA - H_2$

·	a	b
a	{a,b}	b
b	a	{a,b}

Proposition 14. *There exists only one class with 2 isomorphic right almost-semihypergroups, as per the following Cayley tables.*

$RA-S_1$

•	a	b
a	b	{a,b}
b	b	{a,b}

•	a	b
a	{a,b}	a
b	{a,b}	a

Proposition 15. *In addition to the five non-isomorphic commutative hypergroups (and hence left and right almost-hypergroups) that are mentioned in Proposition 9, there also exist two non-isomorphic right almost-hypergroups which are classified into one two-member class and into one single-member class, as they are presented in the following Cayley tables.*

$RA-H_1$

•	a	b
a	{a,b}	a
b	{a,b}	{a,b}

•	a	b
a	{a,b}	{a,b}
b	b	{a,b}

$RA-H_2$

•	a	b
a	{a,b}	a
b	b	{a,b}

v. **Rigid Magmas**

The remaining 26 magmas are classified into 12 two-member classes and into two single-member classes. The law of synthesis on the 2 magmas of the single-member classes is a composition. The same goes for the magmas in three of the twelve two-member classes.

Proposition 16. *There exist only two non-isomorphic groupoids of two elements, which are presented in the following Cayley tables.*

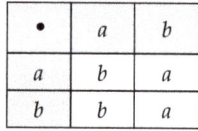

•	a	b
a	b	a
b	b	a

•	a	b
a	b	b
b	a	a

Definition 12. *A magma is called rigid if its group of automorphisms is of order 1.*

As it is shown in [75,76], there exist 21 rigid hypergroupoids whose classification is described in Theorem 4 of [76]. The following Theorem 4 applies to the two-element magmas:

Theorem 4. *There exist 9 rigid magmas of two elements, classified as follows:*
i. *2 non-commutative groupoids, which do not satisfy the transposition axiom;*
ii. *2 non-commutative transposition semigroups;*
iii. *1 commutative quasi-hypergroup, which does not satisfy the transposition axiom;*
iv. *1 LA-hypergroup, which does not satisfy the transposition axiom;*
v. *1 RA-hypergroup, which does not satisfy the transposition axiom;*
vi. *2 hypergroups, which satisfy both the left and right invert associativity.*

Proof.

i. Let us consider the first groupoid of Proposition 15. Then, the two induced hypercompositions are given in the following Cayley tables:

/	a	b
a	∅	{a,b}
b	{a,b}	∅

\	a	b
a	b	b
b	a	a

Next, we have that $a/b \cap b\backslash a = \{a,b\} \cap \{b\} \neq \emptyset$, while $ab \cap ba = \{a\} \cap \{b\} = \emptyset$. Therefore, the transposition axiom is not valid. Analogous is the proof for the second groupoid.

ii. Let us consider the semigroup SG_3. Then, the two induced hypercompositions are the following ones:

/	a	b
a	a	b
b	a	b

\	a	b
a	{a,b}	∅
b	∅	{a,b}

The verification of the transposition axiom, according to the above Cayley tables, proves its validity. The same goes for the case of SG_4.

iii. Since QH_{10} is commutative, the two induced hypercompositions coincide and so we have:
$$a/b = b/a = \{a,b\}, \ a/a = b, \ b/b = a$$
Next, $a/b \cap b/a = \{a,b\} \neq \emptyset$ but $aa \cap bb = \emptyset$. Therefore, the transposition axiom is not valid.

iv. The induced hypercompositions on the LA–H_2 are:

/	a	b
a	{a,b}	b
b	a	{a,b}

\	a	b
a	a	{a,b}
b	{a,b}	b

Since the implication
$$b/a \cap a\backslash b = \{a\} \cap \{a,b\} \neq \emptyset \Rightarrow ab \cap ba = \emptyset$$
holds, the transposition axiom is not valid.

v. It is true as it is the dual of iv.

vi. The two hypergroups are H_6 and H_7. H_6 is a B-hypergroup. As it is well known, the B-hypergroups are join hypergroups (see [13] for the proof), that is, commutative hypergroups which satisfy the transposition axiom. H_7 is the two-element total hypergroup and total hypergroups are join hypergroups as well [5,9,13]. □

5. Binary State Machines

Let $M = (S,E,\delta)$ be a state machine. Two states, s, t, are called *connected* if there exists a sequence of inputs which causes S to leave state s and go into state t, that is, if there exists a sequence $a_1, a_2, \ldots, a_n$ of elements of E such that $(\delta_{a_1} \circ \delta_{a_2} \circ \cdots \circ \delta_{a_n})(s) = t$. The states s, t are called *isolated to each other* if neither s is connected to t, nor t to s. A state machine M is called *connected* if its undirected graph is connected, while it is called *strongly connected* if every ordered pair (s,t) of states in S is connected.

Proposition 17. *Suppose that the connected state machine $M = (S,E,\delta)$ is a quasi-multiautomaton. Then, for every pair (s,t) of states, there exists one element of E which connects them, i.e., $\delta_a(s) = t$ for some $a \in E$.*

Proof. Let $a_1, a_2, \ldots, a_n$ be a sequence of elements of E such that $(\delta_{a_1} \circ \delta_{a_2} \circ \cdots \circ \delta_{a_n})(s) = t$. If $n = 1$ the proposition is obvious. Let $n > 1$. Then

$$(\delta_{a_1} \circ \delta_{a_2} \circ \cdots \circ \delta_{a_n})(s) = \delta_{a_1}(\delta_{a_2}(\cdots \delta_{a_{n-1}}(\delta_{a_n}(s)))) \in$$
$$\in \bigcup_{a \in a_1 a_2 \ldots a_n} \delta_a(s) = \{\delta(s,a) \mid a \in a_1 a_2 \ldots a_n\}$$

Hence, there exists $a \in a_1 a_2 \ldots a_n$ such that $t = \delta(s, a)$. □

Theorem 5. *If the magma E in a quasi-multiautomaton $M = (S, E, \delta)$ has n elements, then the set S cannot have more than $n + 1$ states.*

Proof. As per Proposition 17, for every pair of states (s, t), there exists one element of E which connects them, i.e., there exists $a \in E$ such that $\delta(s, a) = t$. Therefore, if $cardE = n$ e.g., if $E = \{a_1, a_2, \ldots, a_n\}$, then, for each state $s \in S$, there exist at most n states connected with s, which are the $s_i = \delta(s, a_i)$, $1 \leq i \leq n$. Next, if some state $s_k \in S$ yields $\delta(s_k, a_j) = t$, then, since $s_k = \delta(s, a_k)$, it holds that

$$t = \delta(\delta(s, a_k), a_j) \in \{\delta(s, a) \mid a \in a_k a_j\} \subseteq S$$

Hence, $S = \{s, s_1, s_2, \ldots, s_n\}$ and so the Theorem. □

Definition 13. *If the magma E of a state machine $M = (S, E, \delta)$ has 2 elements, then M is called a binary state machine.*

Corollary 3. *The set S of the states of a binary quasi-multiautomaton $M = (S, E, \delta)$ cannot have more than 3 elements.*

Theorem 6. *There exists 1 binary state machine with 1 state, 16 binary state machines with 2 states and 729 with 3 states.*

Proof. The state transition table of a state machine with 1 state is:

δ	a	b
s_1	s_1	s_1

The state transition table of a state machine with 2 states is:

δ	a	b
s_1	x	y
s_2	z	w

where $x, y, z, w \in \{s_1, s_2\}$. Hence, there exist $2^4 = 16$ different to each other binary state machines with 2 states.

Moreover, the state transition table of the state machines with 3 states is:

δ	a	b
s_1	x	y
s_2	z	w
s_3	u	v

where $x, y, z, w, u, v \in \{s_1, s_2, s_3\}$. Therefore, there are $3^6 = 729$ different to each other binary state machines with three states. □

Definition 14. *Two state machines $M_1 = (S_1, E_1, \delta_1)$ and $M_2 = (S_2, E_2, \delta_2)$ are isomorphic if E_1 and E_2 are isomorphic and there exists a one-to-one mapping f from S_1 onto S_2 such that*

$$f(\delta_1(s,a)) = \delta_2(f(s),a).$$

Theorem 7. *There exist 10 isomorphic binary state machines with 2 states, which are classified into 6 two-element classes and into 4 single-element classes, as presented in Figure 3:*

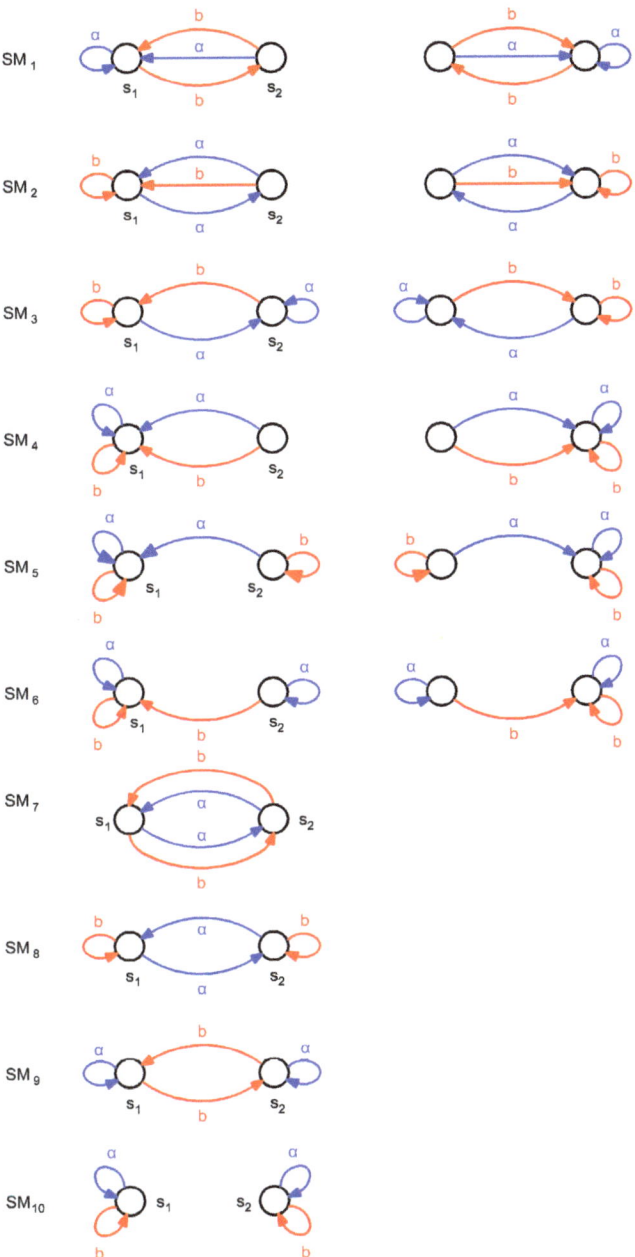

Figure 3. Binary state machines with two states.

Next, we will find which state machines are quasi-multiautomata or quasi-automata when their magma is a group or a hypergroup. In the proofs of the following propositions, we use the first member of each state machine class as the representative of the entire class. Similarly, the representative of every class from the group or the hypergroups will be its first member.

Proposition 18. *If E is the group G_1 (Proposition 8), then the state machines SM_4, SM_9, SM_{10} are quasi-automata.*

Proof. The verification of the axioms shows that SM_4, SM_9, SM_{10} are quasi-automata. The rest state machines do not satisfy the mixed associativity. Indeed:
For SM_1 it holds: $(s_2 a)b = s_1 b = s_2$ while $s_2(ab) = s_2 b = s_1$.
For SM_2 it holds: $(s_1 b)b = s_1 b = s_1$ while $s_1(bb) = s_1 a = s_2$.
For SM_3 it holds: $(s_2 b)a = s_1 a = s_2$ while $s_2(ba) = s_2 b = s_1$.
For SM_5 it holds: $(s_2 a)b = s_1 b = s_1$ while $s_2(ab) = s_2 b = s_2$.
For SM_6 it holds: $(s_2 b)b = s_1 b = s_1$ while $s_2(bb) = s_2 a = s_2$.
For SM_7 it holds: $(s_1 a)a = s_2 a = s_1$ while $s_1(aa) = s_1 a = s_2$.
For SM_8 it holds: $(s_1 b)a = s_1 a = s_2$ while $s_1(ba) = s_1 b = s_1$. □

Proposition 19. *If E is the hypergroup H_1 (Proposition 9), then the state machines SM_3, SM_6, SM_9 are quasi-multiautomata and the state machines SM_4, SM_{10} are quasi-automata.*

Proof. The state machines SM_3, SM_6, SM_9 satisfy the generalized mixed associativity and hence they are quasi-multiautomata. Indicatively, for SM_3 we have:

$$(s_1 b)a = s_1 a = s_2 \text{ and } s_1(ba) = s_1\{a,b\} = \{s_1 a, s_1 b\} = \{s_2, s_1\}, \text{ thus } (s_1 b)a \in s_1(ba)$$

The state machines SM_4 and SM_{10} satisfy the mixed associativity and so they are quasi-automata. Indicatively, for SM_4 we have:

$$(s_2 b)a = s_1 a = s_1 \text{ and } s_2(ba) = s_2\{a,b\} = \{s_2 a, s_2 b\} = s_1, \text{ thus } (s_2 b)a = s_2(ba)$$

The rest state machines do not satisfy any associativity condition. Indeed:
For SM_1 it holds: $(s_2 a)b = s_1 b = s_2$ while $s_2(ab) = s_2 b = s_1$.
For SM_2 it holds: $(s_2 a)a = s_1 a = s_2$ while $s_2(aa) = s_2 a = s_1$.
For SM_5 it holds: $(s_2 a)b = s_1 b = s_1$ while $s_2(ab) = s_2 b = s_2$.
For SM_7 it holds: $(s_1 a)a = s_2 a = s_1$ while $s_1(aa) = s_1 a = s_2$.
For SM_8 it holds: $(s_1 a)b = s_2 b = s_2$ while $s_1(ab) = s_1 b = s_1$. □

Proposition 20. *If E is the hypergroup H_2 (Proposition 9), then the state machines SM_6, SM_9 are quasi-multiautomata and the state machines SM_4, SM_{10} are quasi-automata.*

Proof. The state machines SM_6, SM_9 satisfy the generalized mixed associativity, and therefore they are quasi-multiautomata, while SM_4 and SM_{10} satisfy the mixed associativity and so they are quasi-automata. The rest state machines do not satisfy any associativity condition. Indeed:
For SM_1 it holds: $(s_2 a)b = s_1 b = s_2$ while $s_2(ab) = s_2\{a,b\} = \{s_2 a, s_2 b\} = s_1$.
For SM_2 it holds: $(s_2 a)a = s_1 a = s_2$ while $s_2(aa) = s_2 a = s_1$.
For SM_3 it holds: $(s_2 b)a = s_1 a = s_2$ while $s_2(ba) = s_2 b = s_1$.
For SM_5 it holds: $(s_2 b)a = s_2 a = s_1$ while $s_2(ba) = s_2 b = s_2$.
For SM_7 it holds: $(s_1 a)a = s_2 a = s_1$ while $s_1(aa) = s_1 a = s_2$.
For SM_8 it holds: $(s_1 a)a = s_2 a = s_1$ while $s_1(aa) = s_1 a = s_2$. □

Proposition 21. *If E is the hypergroup H_3 (Proposition 9), then the state machines SM_6, SM_9 are quasi-multiautomata and the state machines SM_4, SM_{10} are quasi-automata.*

Proof. The state machines SM_6, SM_9 satisfy the generalized mixed associativity, and therefore they are quasi-multiautomata, while SM_4 and SM_{10} satisfy the mixed associativity and therefore they are quasi-automata. The rest state machines do not satisfy any associativity condition. Indeed:

For SM_1 it holds: $(s_2a)b = s_1b = s_2$ while $s_2(ab) = s_2b = s_1$.
For SM_2 it holds: $(s_2a)a = s_1a = s_2$ while $s_2(aa) = s_2a = s_1$.
For SM_3 it holds: $(s_2b)a = s_1a = s_2$ while $s_2(ba) = s_2b = s_1$.
For SM_5 it holds: $(s_2a)b = s_1b = s_1$ while $s_2(ab) = s_2b = s_2$.
For SM_7 it holds: $(s_1a)a = s_2a = s_1$ while $s_1(aa) = s_1a = s_2$.
For SM_8 it holds: $(s_1b)a = s_1a = s_2$ while $s_1(ba) = s_1b = s_1$. □

Proposition 22. *If E is the hypergroup H_4 (Proposition 9), then the state machines SM_3, SM_5, SM_6, SM_9 are quasi-multiautomata and the state machines SM_4, SM_{10} are quasi-automata.*

Proof. The state machines SM_3, SM_5, SM_6, SM_9 satisfy the generalized mixed associativity, and therefore they are quasi-multiautomata, while SM_4 and SM_{10} satisfy the mixed associativity and therefore they are quasi-automata. The rest state machines do not satisfy any associativity condition. Indeed:

For SM_1 it holds: $(s_2a)b = s_1b = s_2$ while $s_2(ab) = s_2\{a,b\} = \{s_2a, s_2b\} = s_1$.
For SM_2 it holds: $(s_1a)a = s_2a = s_1$ while $s_1(aa) = s_1a = s_2$.
For SM_7 it holds: $(s_1a)b = s_2b = s_1$ while $s_1(ab) = s_1\{a,b\} = \{s_1a, s_1b\} = s_2$.
For SM_8 it holds: $(s_1a)a = s_2a = s_1$ while $s_1(aa) = s_1a = s_2$. □

Proposition 23. *If E is the hypergroup H_5 (Proposition 9), then the state machine SM_8 is a quasi-multiautomaton and the state machines SM_4, SM_{10} are quasi-automata.*

Proof. The state machine SM_8 satisfies the generalized mixed associativity, and therefore it is a quasi-multiautomaton, while SM_4 and SM_{10} satisfy the mixed associativity and therefore they are quasi-automata. The rest state machines do not satisfy any associativity condition. Indeed:

For SM_1 it holds: $(s_1a)a = s_1a = s_1$ while $s_1(aa) = s_1b = s_2$.
For SM_2 it holds: $(s_2a)a = s_1a = s_2$ while $s_2(aa) = s_2b = s_1$.
For SM_3 it holds: $(s_1a)a = s_2a = s_2$ while $s_1(aa) = s_1b = s_1$.
For SM_5 it holds: $(s_2a)a = s_1a = s_1$ while $s_2(aa) = s_2b = s_2$.
For SM_6 it holds: $(s_2a)a = s_2a = s_2$ while $s_2(aa) = s_2b = s_1$.
For SM_7 it holds: $(s_1a)a = s_2a = s_1$ while $s_1(aa) = s_1b = s_2$.
For SM_9 it holds: $(s_1a)a = s_1a = s_1$ while $s_1(aa) = s_1b = s_2$. □

Proposition 24. *If E is the hypergroup H_6 (Proposition 9), then the state machines SM_3, SM_5, SM_6 are quasi-multiautomata and the state machines SM_4, SM_{10} are quasi-automata.*

Proof. The state machines SM_3, SM_5, SM_6 satisfy the generalized mixed associativity, and therefore they are quasi-multiautomata, while SM_4 and SM_{10} satisfy the mixed associativity and therefore they are quasi-automata. For the rest state machines, observe that they satisfy none of the associativity conditions. Indeed:

For SM_1 it holds: $(s_2b)b = s_1b = s_2$ while $s_2(bb) = s_2b = s_1$.
For SM_2 it holds: $(s_2a)a = s_1a = s_2$ while $s_2(aa) = s_2a = s_1$.
For SM_7 it holds: $(s_1a)a = s_2a = s_1$ while $s_1(aa) = s_1a = s_2$.
For SM_8 it holds: $(s_1a)a = s_2a = s_1$ while $s_1(aa) = s_1a = s_2$.
For SM_9 it holds: $(s_1b)b = s_1b = s_2$ while $s_2(bb) = s_2b = s_1$. □

Proposition 25. *If E is the hypergroup H_7 (Proposition 9), then the state machines SM_3, SM_5, SM_6, SM_8, SM_9 are quasi-multiautomata and the state machines SM_4, SM_{10} are quasi-automata.*

Proof. The state machines SM_3, SM_5, SM_6, SM_8, SM_9 satisfy the generalized mixed associativity, and therefore they are quasi-multiautomata, while SM_4 and SM_{10} satisfy the mixed associativity and therefore they are quasi-automata. The remaining state machines satisfy none of the associativity conditions. Indeed:

For SM_1 it holds: $(s_2a)b = s_1b = s_2$ while $s_2(ab) = s_2\{a,b\} = \{s_2a, s_2b\} = s_1$.
For SM_2 it holds: $(s_2a)a = s_1a = s_2$ while $s_2(aa) = s_2\{a,b\} = \{s_2a, s_2b\} = s_1$.
For SM_7 it holds: $(s_2a)a = s_1a = s_2$ while $s_1(aa) = s_2\{a,b\} = \{s_2a, s_2b\} = s_1$. □

6. Conclusions and Open Problems

This paper approaches the state machines from within their environment in which they can "survive", i.e., exist and operate. They are given an extended definition which derives from the consideration that the environment in which they can exist is an algebraic magma in the sense of [9], where the initial definition of magma by N. Bourbaki [8] is generalized for the purpose of incorporating algebraic structures endowed with hypercompositional laws. Hence, a state machine M is defined as a triplet (S,E,δ), where S is the set of the states of the machine, E is the set of the environmental inputs to the machine and δ is a mapping of $S \times E$ to S, which describes the interaction between each state and its environment. Our study focuses on the binary state machines, where E is a two-element magma and our results can be summarized in Table 1:

Table 1. Classification of the magmas with two elements.

	Total Number	Isomorphism Classes	Classes with 1 Member (Rigid)	Classes with 2 Members
binary magmas	81	45	9	36
binary magmas with composition	16	10	4	6
binary magmas with hypercomposition	65	35	5	30
non-reproductive semigroups	6	4	2	2
non-reproductive semihypergroups	10	5		5
non-reproductive LA-semihypergroups	2	1		1
non-reproductive RA-semihypergroups	2	1		1
non-associative quasihypergroups	21	11	3	9
groups	2	1		1
hypergroups/transposition hypergroups	12	7	2	5
LA-hypergroups	3	2	1	1
RA-hypergroups	3	2	1	1

With regard to the classification given in this table, we note that according to the above Theorem 3, the two-element hypergroups are not just transposition hypergroups (row 10), but bilateral transposition hypergroups. Hence, the hitherto open question which was asked in [9] is answered affirmatively.

Of all the structures that appear in the above table, this paper presents the state machines with two states whose environment is a two-element group or two-element hypergroups. The results are presented in Table 2:

Table 2. Binary state machines with two states.

Group/Hypergroup	Non-Isomorphic Quasiautomata	Non-Isomorphic Quasi-Multiautomata
G_1	3	
H_1	2	3
H_2	2	2
H_3	2	2
H_4	2	4
H_5	2	1
H_6	2	3
H_7	2	5

According to Corollary 3, only quasiautomata and quasi-multiautomata with up to three states can operate in the environment of a two-element magma. The description of the three state binary machines, as well as the investigation of the state machines which correspond to algebraic structures of E that are other than groups or hypergroups, still remain open problems. This question becomes more complicated in the instances when E is enriched with two laws of synthesis, as it happens in a hyperringoid [74]. It is worth mentioning here that E is a hyperringoid in specific state machines like the automata, where the environment is defined via an alphabet [16,17,21].

All of the above refer to deterministic state machines. However, the state transition function can be a mapping from $S \times E$ to the power set $P(S)$ of S, defining thus the non-deterministic state machines (see also [13,14,67]). This consideration broadens the margins of the study as there can exist state transition functions for which $\delta(s,a)$ is not necessarily just a single element. It can be more than one element and it can also be none, as it is possible for $\delta(s,a)$ to be equal to the empty set.

Author Contributions: G.G.M. and C.G.M. contributed equally to this work. All authors have read and agreed to the published version of the manuscript.

Funding: This research received no external funding. The APC was funded by the MDPI journal *Mathematics*.

Institutional Review Board Statement: Not applicable.

Informed Consent Statement: Not applicable.

Conflicts of Interest: The authors declare no conflict of interest.

References

1. McCulloch, W.S.; Pitts, W. A logical calculus of the ideas immanent in nervous activity. *Bull. Math. Biophys.* **1943**, *5*, 115–133. [CrossRef]
2. Kleene, S.C. *Representation of Events in Nerve Nets and Finite Automata. Automata Studies*; Princeton University Press: Princeton, NJ, USA, 1956; pp. 3–42.
3. Chomsky, N. Three models for the description of language. *IRE Trans. Inf. Theory* **1956**, *2*, 113–124. [CrossRef]
4. Chomsky, N. On certain formal properties of grammars. *Inf. Control.* **1959**, *2*, 137–167. [CrossRef]
5. Massouros, G.G. Automata—Languages and Hypercompositional Structures. Ph.D. Thesis, National Technical University of Athens, Athens, Greece, 1993.
6. Grillo, F. Hero of Alexandria's *Automata*: A Critical Edition and Translation, Including a Commentary on Book One. Ph.D. Thesis, University of Glasgow, Glasgow, Scotland, 2019.
7. Freeth, T.; Higgon, D.; Dacanalis, A.; MacDonald, L.; Georgakopoulou, M.; Wojcik, A. A Model of the Cosmos in the ancient Greek Antikythera Mechanism. *Sci. Rep.* **2021**, *11*, 5821. [CrossRef] [PubMed]
8. Bourbaki, N. *Éléments de Mathématique, Algèbre*; Hermann: Paris, France, 1971.
9. Massouros, C.; Massouros, G. An Overview of the Foundations of the Hypergroup Theory. *Mathematics* **2021**, *9*, 1014. [CrossRef]
10. Marty, F. *Sur une Généralisation de la Notion de Groupe*. Huitième Congrès des Mathématiciens Scandinaves Stockholm; 1934; pp. 45–49.
11. Marty, F. Rôle de la notion de hypergroupe dans l' étude de groupes non abéliens. *C. R. Acad. Sci.* **1935**, *201*, 636–638.
12. Marty, F. Sur les groupes et hypergroupes attachés à une fraction rationelle. *Ann. L' Ecole Norm.* **1936**, *3*, 83–123.
13. Massouros, G.G.; Mittas, I.D. Languages—Automata and hypercompositional structures, Algebraic Hyperstructures and Applications. In Proceedings of the 4th International Congress, Xanthi, Greece, 27–30 June 1990; World Scientific: Singapore, 1991; pp. 137–147.
14. Massouros, G.G. Automata and hypermoduloids, Algebraic Hyperstructures and Applications. In Proceedings of the 5th International Congress, Iasi, Romania, 4–10 July 1993; Hadronic Press: Palm Harbor, FL, USA, 1994; pp. 251–265.
15. Massouros, G.G. An automaton during its operation, Algebraic Hyperstructures and Applications. In Proceedings of the 5th International Congress, Iasi, Romania, 4–10 July 1993; Hadronic Press: Palm Harbor, FL, USA, 1994; pp. 267–276.
16. Massouros, G.G. Hypercompositional structures in the theory of languages and automata. *Sci. Ann. Cuza Univ.* **1994**, *3*, 65–73.
17. Massouros, G.G. Hypercompositional structures from the computer theory. *Ratio Math.* **1999**, *13*, 37–42.
18. Massouros, G.G. On the attached hypergroups of the order of an automaton. *J. Discret. Math. Sci. Cryptogr.* **2003**, *6*, 207–215. [CrossRef]
19. Massouros, C.G.; Massouros, G.G. Hypergroups associated with graphs and automata. *AIP Conf. Proc.* **2009**, *1168*, 164–167. [CrossRef]
20. Massouros, C.G. On path hypercompositions in graphs and automata. *MATEC Web Conf.* **2016**, *41*, 05003. [CrossRef]

21. Massouros, G.G.; Massouros, C.G. Hypercompositional algebra, computer science and geometry. *Mathematics* **2020**, *8*, 1338. [CrossRef]
22. Chvalina, J.; Chvalinová, L. State hypergroups of Automata. *Acta Math. Inform. Univ. Ostrav.* **1996**, *4*, 105–120.
23. Chvalina, J. Infinite multiautomata with phase hypergroups of various operators. In Proceedings of the 10th International Congress on Algebraic Hyperstructures and Applications, Brno, Czech Republic, 3–9 September 2008; Hošková, Š., Ed.; University of Defense: Brno, Czech Republic, 2009; pp. 57–69.
24. Hošková, Š.; Chvalina, J.; Raskova, P. Multiautomata with input centralizer alphabet formed by first order partial differential operators. In Proceedings of the 5 Konference o Matematice a Fyzice na Vysokch Kolch Technickch, Brno, Czech Republic, 2007; pp. 99–107, ISBN 978-80-7231-274-0.
25. Chvalina, J.; Hošková, Š. Multiautomata formed by first order partial differential operators. *J. Appl. Math.* **2008**, *1*, 423–430.
26. Chvalina, J.; Křehlík, S.; Novák, M. Cartesian composition and the problem of generalizing the MAC condition to quasi-multiautomata. *An. St. Univ. Ovidius Constanta* **2016**, *24*, 79–100. [CrossRef]
27. Chvalina, J.; Novák, M.; Křehlík, S. Hyperstructure generalizations of quasi-automata induced by modelling functions and signal processing. *AIP Conf. Proc.* **2019**, *2116*, 310006. [CrossRef]
28. Chvalina, J.; Novák, M.; Smetana, B.; Staněk, D. Sequences of Groups, Hypergroups and Automata of Linear Ordinary Differential Operators. *Mathematics* **2021**, *9*, 319. [CrossRef]
29. Novák, M.; Křehlík, S.; Staněk, D. n-ary Cartesian composition of automata. *Soft Comput.* **2019**, *24*, 1837–1849. [CrossRef]
30. Novák, M.; Křehlík, S.; Ovaliadis, K. Elements of Hyperstructure Theory in UWSN Design and Data Aggregation. *Symmetry* **2019**, *11*, 734. [CrossRef]
31. Křehlík, S.; Novák, M.; Vyroubalová, J. From Automata to Multiautomata via Theory of Hypercompositional Structures. *Mathematics* **2022**, *10*, 1. [CrossRef]
32. Novák, M. Some remarks on constructions of strongly connected multiautomata with the input semihypergroup being a centralizer of certain transformation operators. *J. Appl. Math.* **2008**, *I*, 65–72.
33. Křehlík, S. n-Ary Cartesian Composition of Multiautomata with Internal Link for Autonomous Control of Lane Shifting. *Mathematics* **2020**, *8*, 835. [CrossRef]
34. Ghorani, M.; Zahedi, M.M. Some hypergroups induced by tree automata. *Aust. J. Basic Appl. Sci.* **2012**, *6*, 680–692.
35. Ghorani, M. State hyperstructures of tree automata based on lattice-valued logic. *RAIRO—Theor. Inf. Appl.* **2018**, *52*, 23–42. [CrossRef]
36. Heidari, D.; Doostali, S. The application of hypergroups in symbolic executions and finite automata. *Soft Comput.* **2021**, *25*, 7247–7256. [CrossRef]
37. Borzooei, R.A.; Varasteh, H.R.; Hasankhani, A. F-Multiautomata on Join Spaces Induced by Differential Operators. *Appl. Math.* **2014**, *5*, 1386–1391. [CrossRef]
38. Nieminen, J. Join space graphs. *J. Geom.* **1988**, *33*, 99–103. [CrossRef]
39. Nieminen, J. Chordal graphs and join spaces. *J. Geom.* **1989**, *34*, 146–151. [CrossRef]
40. Corsini, P. Graphs and Join Spaces. *J. Comb. Inf. Syst. Sci.* **1991**, *16*, 313–318.
41. Corsini, P. Hypergraphs and hypergroups. *Algebr. Univ.* **1996**, *35*, 548–555. [CrossRef]
42. Rosenberg, I.G. Hypergroupes induced by paths of a direct graph. *Ital. J. Pure Appl. Math.* **1998**, *4*, 133–142.
43. Corsini, P.; Leoreanu-Fotea, V.; Iranmanesh, A. On the sequence of hypergroups and membership functions determined by a hypergraph. *J. Mult. Valued Log. Soft Comput.* **2008**, *14*, 565–577.
44. Chvalina, J.; Hošková-Mayerová, Š.; Nezhad, A.D. General actions of hyperstructures and some applications. *An. St. Univ. Ovidius Constanta* **2013**, *21*, 59–82. [CrossRef]
45. Farshi, M.; Lo Faro, G.; Mirvakili, S. Hypergraphs and hypergroups based on a special relation. *Commun. Algebra* **2014**, *42*, 3395–3406. [CrossRef]
46. De Salvo, M.; Lo Faro, G. Wrapping graphs and partial semi-hypergroups. *J. Inf. Optim. Sci.* **1997**, *18*, 157–166. [CrossRef]
47. De Salvo, M.; Fasino, D.; Freni, D.; Lo Faro, G. Fully simple semihypergroups, transitive digraphs, and sequence A000712. *J. Algebra* **2014**, *415*, 65–87. [CrossRef]
48. Polat, N. On bipartite graphs whose interval space is a closed join space. *J. Geom.* **2017**, *108*, 719–741. [CrossRef]
49. Chowdhury, G. Syntactic Semihypergroup. *Glob. J. Pure Appl. Math.* **2017**, *13*, 1103–1115.
50. Kalampakas, A.; Triantafyllou, N.; Ksystra, K.; Stefaneas, P. A Formal Representation of Video Content with the Picture Hyperoperation. In *Algebraic Modeling of Topological and Computational Structures and Applications*; Springer: Cham, Switzerland, 2015; Volume 219. [CrossRef]
51. Nikkhah, A.; Davvaz, B.; Mirvakili, S. Hypergroups Constructed from Hypergraphs. *Filomat* **2018**, *32*, 3487–3494. [CrossRef]
52. Hošková-Mayerová, Š.; Maturo, A. Decision-making process using hyperstructures and fuzzy structures in social sciences. *Stud. Fuzz. Soft Comput.* **2018**, *357*, 103–111.
53. Heidari, D.; Cristea, I. Breakable semihypergroups. *Symmetry* **2019**, *11*, 100. [CrossRef]
54. Iranmanesh, M.; Jafarpour, M.; Cristea, I. The non-commuting graph of a non-central hypergroup. *Open Math.* **2019**, *17*, 1035–1044. [CrossRef]
55. Cristea, I.; Kocijan, J.; Novák, M. Introduction to dependence relations and their links to algebraic hyperstructures. *Mathematics* **2019**, *7*, 885. [CrossRef]

56. Maturo, F.; Ventre, V.; Longo, A. On Consistency and Incoherence in Analytical Hierarchy Process and Intertemporal Choices Models. In *Models and Theories in Social Systems. Studies in Systems, Decision and Control*; Springer: Cham, Switzerland, 2019; Volume 179. [CrossRef]
57. Hamidi, M.; Borumand Saied, A. Creating and computing graphs from hypergraphs. *Krag. J. Math.* **2019**, *43*, 139–164.
58. Shamsi, K.; Ameri, R.; Mirvakili, S. Cayley graph associated to a semihypergroup. *Algebraic Struct. Appl.* **2020**, *7*, 29–49.
59. Kankaras, M. Reducibility in Corsini hypergroups. *An. St. Univ. Ovidius Constanta* **2021**, *29*, 93–109. [CrossRef]
60. Al-Tahan, M.; Davvaz, B. Hypergroups defined on hypergraphs and their regular relations. *Krag. J. Math.* **2022**, *46*, 487–498. [CrossRef]
61. Corsini, P.; Leoreanu, V. *Applications of Hyperstructures Theory*; Kluwer Academic Publishers: Berlin, Germany, 2003.
62. Kazim, M.A.; Naseeruddin, M. On almost semigroups. *Port. Math.* **1977**, *36*, 41–47.
63. Massouros, C.; Yaqoob, N. On the theory of left/right almost groups and hypergroups with their relevant enumerations. *Mathematics* **2021**, *9*, 1828. [CrossRef]
64. Yaqoob, N.; Cristea, I.; Gulistan, M. Left almost polygroups. *Ital. J. Pure Appl. Math.* **2018**, *39*, 465–474.
65. Prenowitz, W. A Contemporary Approach to Classical Geometry. *Math. Assoc. Am.* **1961**, *68*, 1–67. [CrossRef]
66. Jantosciak, J. Transposition hypergroups, Noncommutative Join Spaces. *J. Algebra* **1997**, *187*, 97–119. [CrossRef]
67. Massouros, C.G.; Massouros, G.G. Operators and Hyperoperators acting on Hypergroups. *AIP Conf. Proc.* **2008**, *1048*, 380–383. [CrossRef]
68. Hošková, Š.; Chvalina, J. Discrete transformation hypergroups and transformation hypergroups with phase tolerance space. *Discret. Math.* **2008**, *308*, 4133–4143. [CrossRef]
69. Tsitouras, C.G.; Massouros, C.G. On enumeration of hypergroups of order 3. *Comput. Math. Appl.* **2010**, *59*, 519–523. [CrossRef]
70. Massouros, C.G.; Tsitouras, C.G. Enumeration of hypercompositional structures defined by binary relations. *Ital. J. Pure Appl. Math.* **2011**, *28*, 43–54.
71. Tsitouras, C.G.; Massouros, C.G. Enumeration of Rosenberg type hypercompositional structures defined by binary relations. *Eur. J. Comb.* **2012**, *33*, 1777–1786. [CrossRef]
72. Massouros, C.G.; Massouros, G.G. On join hyperrings, Algebraic Hyperstructures and Applications. In Proceedings of the 10th International Congress, Brno, Czech Republic, 3–9 September 2008; pp. 203–215.
73. Massouros, G.G.; Massouros, C.G. Homomorphic relations on Hyperringoids and Join Hyperrings. *Ratio Mat.* **1999**, *13*, 61–70.
74. Massouros, G.G. The Hyperringoid. *Mult. Valued Log.* **1998**, *3*, 217–234.
75. Bayon, R.; Lygeros, N. Advanced results in enumeration of hyperstructures. *J. Algebra* **2008**, *320*, 821–835. [CrossRef]
76. Massouros, C.G. On the enumeration of rigid hypercompositional structures. *AIP Conf. Proc.* **2014**, *1648*, 740005. [CrossRef]

Article

Upper and Lower Bounds for the Spectral Radius of Generalized Reciprocal Distance Matrix of a Graph

Yuzheng Ma [1], Yubin Gao [2,*] and Yanling Shao [2]

1. School of Data Science and Technology, North University of China, Taiyuan 030051, China; b20210705@st.nuc.edu.cn
2. School of Mathematical Sciences, North University of China, Taiyuan 030051, China; ylshao@nuc.edu.cn
* Correspondence: ybgao@nuc.edu.cn

Abstract: For a connected graph G on n vertices, recall that the reciprocal distance signless Laplacian matrix of G is defined to be $RQ(G) = RT(G) + RD(G)$, where $RD(G)$ is the reciprocal distance matrix, $RT(G) = diag(RT_1, RT_2, \ldots, RT_n)$ and RT_i is the reciprocal distance degree of vertex v_i. In 2022, generalized reciprocal distance matrix, which is defined by $RD_\alpha(G) = \alpha RT(G) + (1-\alpha)RD(G), \alpha \in [0,1]$, was introduced. In this paper, we give some bounds on the spectral radius of $RD_\alpha(G)$ and characterize its extremal graph. In addition, we also give the generalized reciprocal distance spectral radius of line graph $L(G)$.

Keywords: graph; generalized reciprocal distance matrix; reciprocal distance signless Laplacian matrix; spectral radius

MSC: 05C50; 05C12; 15A18

1. Introduction

In this paper, all graphs considered are finite, simple, and connected. Let G be such a graph with vertex set $V(G) = \{v_1, v_2, \ldots, v_n\}$ and edge set $E(G)$, where $|V(G)| = n$ and $|E(G)| = m$. Let d_{v_i} denote the degree of vertex v_i, which is simply written as d_i. $N(v_i)$ denote the neighbor set of v_i. The distance between vertices v_i and v_j in G is the length of the shortest path connecting v_i to v_j, which is denoted as $d(v_i, v_j)$. We use the notation d_{ij} instead of $d(v_i, v_j)$. The diameter of G, denoted by $diam(G)$, is the maximum distance between any pair of vertices of G. The Harary matrix of G, which is also called the reciprocal distance matrix, is an $n \times n$ matrix defined as [1]

$$RD_{i,j} = \begin{cases} \frac{1}{d(v_i,v_j)}, & \text{if } i \neq j, \\ 0, & \text{if } i = j. \end{cases}$$

Henceforth, we consider $i \neq j$ for $d(v_i, v_j)$.

The transmission of vertex v_i, denoted by $Tr_G(v_i)$ or Tr_i, is defined to be the sum of the distances from v_i to all vertices in G, that is, $Tr_G(v_i) = Tr_i = \sum_{u \in V(G)} d(u, v_i)$. A graph G is said to be k-transmission regular graph if $Tr_G(v) = k$ for each $v \in V(G)$. Transmission of a vertex v is also called the distance degree or the first distance degree of v.

Definition 1. *Let G be a graph with $V(G) = \{v_1, v_2, \ldots, v_n\}$. The reciprocal distance degree of a vertex v, denoted by $RTr_G(v)$, is given by*

$$RTr_G(v) = \sum_{u \in V(G), u \neq v} \frac{1}{d(u,v)}.$$

Let $RT(G)$ be the $n \times n$ diagonal matrix defined by $RT_{i,i} = RTr_G(v_i)$.

Sometimes we use the notation RT_i instead of $RTr_G(v_i)$ for $i = 1, \ldots, n$.

Definition 2. *A graph G is called a k-reciprocal distance degree regular graph if $RT_i = k$ for all $i \in \{1, 2, \ldots, n\}$.*

The Harary index of a graph G, denoted by $H(G)$, is defined in [1] as

$$H(G) = \frac{1}{2}\sum_{i=1}^n \sum_{j=1}^n RD_{i,j} = \frac{1}{2}\sum_{u,v \in V(G), u \neq v} \frac{1}{d(u,v)}.$$

Clearly,

$$H(G) = \frac{1}{2}\sum_{i=1}^n RT_i.$$

In [2], Bapat and Panda defined the reciprocal distance Laplacian matrix as $RL(G) = RT(G) - RD(G)$. It was proved that, given a connected graph G of order n, the spectral radius of its reciprocal distance Laplacian matrix $\rho(RL(G)) \leq n$ if and only if its complement graph, denoted by $\overline{G}$, is disconnected. In [3], Alhevaz et al. defined the reciprocal distance signless Laplacian matrix as $RQ(G) = RT(G) + RD(G)$. Recently, the lower and upper bounds of the spectral radius of the reciprocal distance matrices and reciprocal distance signless Laplacian matrices of graphs were given in [3–6], respectively.

In [7], the author, using the convex linear combinations of the matrices $RT(G)$ and $RD(G)$, introduces a new matrix, that is generalized reciprocal distance matrix, denoted by $RD_\alpha(G)$, which is defined by

$$RD_\alpha(G) = \alpha RT(G) + (1-\alpha)RD(G), \ 0 \leq \alpha \leq 1.$$

Since $RD_0(G) = RD(G)$, $RD_{\frac{1}{2}}(G) = \frac{1}{2}RQ(G)$ and $RD_1(G) = RT(G)$, then $RD_{\frac{1}{2}}(G)$ and $RQ(G)$ have the same spectral properties. To this extent these matrices $RD(G)$, $RT(G)$, and $RQ(G)$ may be understood from a completely new perspective, and some interesting topics arise. For the these matrices $RD(G)$, $RT(G)$, and $RQ(G)$, some spectral extremal graphs with fixed structure parameters have been characterized in [8,9]. It is natural to ask whether these results can be generalized to $RD_\alpha(G)$.

Since $RD_\alpha(G)$ is real symmetric matrics, we can denoted $\lambda_1(RD_\alpha(G)) \geq \lambda_2(RD_\alpha(G)) \geq \cdots \geq \lambda_n(RD_\alpha(G))$ to the eigenvalues of $RD_\alpha(G)$. The maximum eigenvalue $\lambda_1(RD_\alpha(G))$ is called the spectral radius of the matrix $RD_\alpha(G)$, denoted by $\rho(RD_\alpha(G))$.

This paper is organized as follows. In Section 2, we give some definitions, notations, and lemmas of generalized reciprocal distance matrix. In Section 3, we give the upper and lower bounds of the spectral radius of the generalized reciprocal distance matrix $RD_\alpha(G)$ by using the reciprocal distance degree and the second reciprocal distance degree. In Section 4, we give the bounds of the spectral radius of the generalized reciprocal distance matrix of $L(G)$, where $L(G)$ is the line graph of graph G.

2. Lemmas

In this section, we give some definitions, notations, and lemmas to prepare for subsequent proofs.

Definition 3. *Let G be a graph with $V(G) = \{v_1, v_2, \ldots, v_n\}$, the reciprocal distance matrix $RD(G)$ and the reciprocal distance degree sequence $\{RT_1, RT_2, \ldots, RT_n\}$. Then the second reciprocal distance degree of a vertex v_i, denoted by T_i, is given by*

$$T_i = \sum_{j=1, j\neq i}^n \frac{1}{d_{i,j}}RT_j.$$

Definition 4. A graph G is called a pseudo k-reciprocal distance degree regular graph if $\frac{T_i}{RT_i} = k$ for all $i \in \{1, 2, \ldots, n\}$.

Definition 5. The Frobenius norm of an $n \times n$ matrix $M = (m_{i,j})$ is

$$\|M\|_F = \sqrt{\sum_{i=1}^{n}\sum_{j=1}^{n} |m_{i,j}|^2}.$$

We recall that, if M is a normal matrix then $\|M\|_F^2 = \sum_{i=1}^{n} |\lambda_i(M)|^2$ where $\lambda_1(M), \ldots, \lambda_n(M)$ are the eigenvalues of M. In particular, $\|RD_\alpha(G)\|_F^2 = \sum_{i=1}^{n} |\lambda_i(RD_\alpha(G))|^2$.

Lemma 1 ([6]). *Let G be a graph of order n with reciprocal distance degree sequence $\{RT_1, RT_2, \ldots, RT_n\}$ and second reciprocal distance degree sequence $\{T_1, T_2, \ldots, T_n\}$. Then*

$$T_1 + T_2 + \cdots + T_n = RT_1^2 + RT_2^2 + \cdots + RT_n^2.$$

Lemma 2 (Perron–Frobenius theorem [10]). *If A is a non-negative matrix of order n, then its spectral radius $\rho(A)$ is an eigenvalue of A and it has an associated non-negative eigenvector. Furthermore, if A is irreducible, then $\rho(G)$ is a simple eigenvalue of A with an associated positive eigenvector.*

Lemma 3 ([7]). *Let G be a graph with $n \geq 2$ vertices and Harary index $H(G)$. Then*

$$\rho(RD_\alpha(G)) \geq \frac{2H(G)}{n}.$$

The equality holds if and only if G is a reciprocal distance degree regular graph.

Lemma 4 ([11]). *Let $A = (a_{i,j})$ be an $n \times n$ nonnegative matrix with spectral radius $\rho(A)$ and row sums $S_1(A), S_2(A), \ldots, S_n(A)$. Then,*

$$\min_{1 \leq i \leq n} S_i(A) \leq \rho(A) \leq \max_{1 \leq i \leq n} S_i(A).$$

Moreover, if A is an irreducible matrix, then equality holds on either side (and hence both sides) of the equality if and only if all row sums of A are all equal.

Lemma 5 ([6]). *Let G be a graph on n vertices. Let RT_{max} and RT_{min} be the maximum and the minimum reciprocal distance degree of G, respectively. Then, for any $v_i \in V(G)$,*

$$2H(G) + (RT_{max} - 1)RT_i - (n-1)RT_{max} \leq T_i \leq 2H(G) + (RT_{min} - 1)RT_i - (n-1)RT_{min}.$$

Lemma 6 (Cauchy alternating theorem [12]). *Let A be a real symmetric matrix of order n and B be a principal submatrix of order m of A. Suppose A has eigenvalues $\lambda_1 \geq \lambda_2 \geq \cdots \geq \lambda_n$, and B has eigenvalues $\beta_1 \geq \beta_2 \geq \cdots \geq \beta_m$. Then, for all $k = 1, 2, \ldots, m$, $\lambda_{n-m+k} \leq \beta_k \leq \lambda_k$.*

Lemma 7. *Let G be a graph on $n \geq 2$ vertices with $0 \leq \alpha < 1$. The G has exactly two distinct generalized reciprocal distance eigenvalues if and only if G is a complete graph. In particular, $\rho(RD_\alpha(K_n)) = n-1$ and $\lambda_i(RD_\alpha(K_n)) = \alpha n - 1$ for $i = 2, 3, \ldots, n$.*

Proof. Let $n \geq 2$. Clearly, the spectrum of the generalized reciprocal distance matrix of the complete graph K_n is $\{n-1, (\alpha n - 1)^{[n-1]}\}$.

Let G be a graph with generalized reciprocal distance matrix $RD_\alpha(G)$. If G has exactly two distinct RD_α-eigenvalues, then $\lambda_1(RD_\alpha(G)) > \lambda_2(RD_\alpha(G))$. Since G is a connected graph and $RD_\alpha(G)$ is an irreducible matrix. Then, from Lemma 2, $\lambda_1(RD_\alpha(G)) = \rho(RD_\alpha(G))$ is the greatest and simple eigenvalue of $RD_\alpha(G)$. Thus, the algebraic multiplicity of $\lambda_2(RD_\alpha(G))$ is $n-1$, i.e.,

$$\lambda_2(RD_\alpha(G)) = \lambda_3(RD_\alpha(G)) = \cdots = \lambda_n(RD_\alpha(G)). \tag{1}$$

Now, to prove that $G = K_n$, we show that the diameter of G is 1. That is, we prove that G does not contain an shortest path P_k, for $k \geq 3$.

We suppose that G contains an induced shortest path P_k, $k \geq 3$. Let B be the principal submatrix of $RD_\alpha(G)$ indexed by the vertices in P_k. Then by Lemma 6, we have

$$\lambda_i(RD_\alpha(G)) \geq \lambda_i(B) \geq \lambda_{i+n-k}(RD_\alpha(G)), i = 1, 2, \ldots, k.$$

Using the equalities given in (1), we obtain $\lambda_2(RD_\alpha(G)) \geq \lambda_2(B) \geq \lambda_3(B) \geq \cdots \geq \lambda_k(B) \geq \lambda_p(RD_\alpha(G)) = \lambda_2(RD_\alpha(G))$. Thus, for $k \geq 3$, the matrix $B = (RD_\alpha(P_k))$ has at most two different eigenvalues. By definition, we can get the generalized reciprocal distance matrix of P_3, that is

$$RD_\alpha(P_3) = \begin{bmatrix} \frac{3}{2}\alpha & 1-\alpha & \frac{1}{2}(1-\alpha) \\ 1-\alpha & 2(1-\alpha) & 1-\alpha \\ \frac{1}{2}(1-\alpha) & 1-\alpha & \frac{3}{2}\alpha \end{bmatrix}.$$

Using the software Maple 18, it is easy to calculate that the generalized reciprocal distance spectrum of the path of order 3 is $\{\frac{3}{2}\alpha + \frac{1}{4} + \frac{1}{4}\sqrt{36\alpha^2 - 68\alpha + 33}, \frac{3}{2}\alpha + \frac{1}{4} - \frac{1}{4}\sqrt{36\alpha^2 - 68\alpha + 33}, 2\alpha - \frac{1}{2}\}$, this is false.

Therefore, G does not have two vertices at distance two or more. Then, $G = K_n$. □

Lemma 8 ([13]). *If $x_1 \geq x_2 \geq \cdots \geq x_m$ are real numbers such that $\sum_{i=1}^{m} x_i = 0$, then*

$$x_1 \leq \sqrt{\frac{m-1}{m} \sum_{i=1}^{m} x_i^2}.$$

The equality holds if and only if $x_2 = x_3 = \cdots = x_m = -\frac{x_1}{m-1}$.

Lemma 9 (Rayleigh quotient theorem [14]). *let M be a real symmetric matrix of order n whose eigenvalues are $\lambda_1 \geq \lambda_2 \geq \ldots \geq \lambda_n$. Then, for any n-dimensional nonzero column vector x,*

$$\lambda_1 \geq \frac{x^T M x}{x^T x} \geq \lambda_n.$$

Lemma 10 ([15]). *If $diam(G) \leq 2$ and if none of the three graphs F_1, F_2, and F_3 depicted in Figure 1 are induced subgraphs of G, then $diam(L(G)) \leq 2$.*

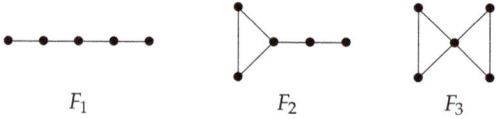

Figure 1. Graphs F_1, F_2, T_3 in Lemma 10.

3. Bounds of $\rho(RD_\alpha(G))$ of Graphs

In this section, we find bounds of the spectral radius of generalizes reciprocal distance matrix in terms of parameters associated with the structure of the graph.

Let **e** be the n-dimensional vector of ones.

Theorem 1. Let G be a graph with reciprocal distance degree sequence $\{RT_1, RT_2, \ldots, RT_n\}$. Then

$$\rho(RD_\alpha(G)) \geq \sqrt{\frac{RT_1^2 + RT_2^2 + \cdots + RT_n^2}{n}}.$$

The equality holds if and only if G is a reciprocal distance degree regular graph.

Proof. Let $\mathbf{x} = [x_1, x_2, \ldots, x_n]^T$ be the unit positive Perron eigenvector of $RD_\alpha(G)$ corresponding to $\rho(RD_\alpha(G))$. We take the unit vector $\mathbf{y} = \frac{1}{\sqrt{n}}\mathbf{e}$. Then, we have

$$\rho(RD_\alpha(G)) = \sqrt{\rho^2(RD_\alpha(G))} = \sqrt{\mathbf{x}^T(RD_\alpha(G))^2\mathbf{x}} \geq \sqrt{\mathbf{y}^T(RD_\alpha(G))^2\mathbf{y}}. \quad (2)$$

Since $(RD_\alpha(G))\mathbf{y} = \frac{1}{\sqrt{n}}[RT_1, RT_2, \ldots, RT_n]^T$, we obtain

$$\mathbf{y}^T(RD_\alpha(G))^2\mathbf{y} = \frac{RT_1^2 + RT_2^2 + \cdots + RT_n^2}{n}.$$

Therefore,

$$\rho(RD_\alpha(G)) \geq \sqrt{\frac{RT_1^2 + RT_2^2 + \cdots + RT_n^2}{n}}.$$

Now, assume that the equality holds. By Equation (2), we have that $\mathbf{y}$ is the positive eigenvector corresponding to $\rho(RD_\alpha(G))$. From $RD_\alpha(G)\mathbf{y} = \rho(RD_\alpha(G))\mathbf{y}$, we obtain that $RT_i = \rho(RD_\alpha(G))$, for $i = 1, 2, \ldots, n$. Therefore, graph G is a reciprocal distance degree regular graph.

Conversely, if G is a reciprocal distance degree regular graph, then $RT_1 = RT_2 = \cdots = RT_n = k$. From Lemma 2, $k = \rho(RD_\alpha(G))$. So

$$\rho(RD_\alpha(G)) = k = \sqrt{\frac{nk^2}{n}} = \sqrt{\frac{RT_1^2 + RT_2^2 + \cdots + RT_n^2}{n}}.$$

The equality holds. □

Theorem 2. Let G be a graph with reciprocal distance degree sequence $\{RT_1, RT_2, \ldots, RT_n\}$ and second reciprocal distance degree sequence $\{T_1, T_2, \ldots, T_n\}$. Then

$$\rho(RD_\alpha(G)) \geq \sqrt{\frac{(\alpha RT_1^2 + (1-\alpha)T_1)^2 + (\alpha RT_2^2 + (1-\alpha)T_2)^2 + \cdots + (\alpha RT_n^2 + (1-\alpha)T_n)^2}{\sum_{i=1}^{n} RT_i^2}}.$$

The equality holds if and only if G is a pseudo reciprocal distance degree regular graph.

Proof. Using $\mathbf{y} = \frac{1}{\sqrt{\sum_{i=1}^{n} RT_i^2}}[RT_1, RT_2, \ldots, RT_n]^T$, the proof is similar to Theorem 1. □

Remark 1. The lower bound given in Theorem 2 improves the bound given in Theorem 1, and the bound given in Theorem 1 improves the bound given in Lemma 3.

In fact, from Lemma 1, we have $\sum_{i=1}^{n} T_i = \sum_{i=1}^{n} RT_i^2$. By Cauchy–Schwarz inequality

$$n \sum_{i=1}^{n} (\alpha RT_i^2 + (1-\alpha)T_i)^2 \geq (\sum_{i=1}^{n} (\alpha RT_i^2 + (1-\alpha)T_i))^2$$

$$= (\alpha \sum_{i=1}^{n} RT_i^2 + (1-\alpha) \sum_{i=1}^{n} T_i)^2$$

$$= (\alpha \sum_{i=1}^{n} RT_i^2 + (1-\alpha) \sum_{i=1}^{n} RT_i^2)^2$$

$$= (\sum_{i=1}^{n} RT_i^2)^2.$$

Moreover, we recall that, $n \sum_{i=1}^{n} RT_i^2 \geq (\sum_{i=1}^{n} RT_i)^2$. Thus

$$\sqrt{\frac{\sum_{i=1}^{n} (\alpha RT_i^2 + (1-\alpha)T_i)^2}{\sum_{i=1}^{n} RT_i^2}} \geq \sqrt{\frac{(\sum_{i=1}^{n} RT_i^2)^2}{n \sum_{i=1}^{n} RT_i^2}} = \sqrt{\frac{\sum_{i=1}^{n} RT_i^2}{n}}$$

and

$$\sqrt{\frac{\sum_{i=1}^{n} RT_i^2}{n}} \geq \sqrt{\frac{(\sum_{i=1}^{n} RT_i)^2}{n^2}} = \frac{2H(G)}{n}.$$

Theorem 3. *Let G be a graph with reciprocal distance degree sequence $\{RT_1, RT_2, \ldots, RT_n\}$ and second reciprocal distance degree sequence $\{T_1, T_2, \ldots, T_n\}$. Then*

$$\min_{1 \leq i \leq n} \left\{ \sqrt{(1-\alpha)T_i + \alpha(RT_i)^2} \right\} \leq \rho(RD_\alpha(G)) \leq \max_{1 \leq i \leq n} \left\{ \sqrt{(1-\alpha)T_i + \alpha(RT_i)^2} \right\}.$$

Proof. Let $RD_\alpha(G) = (b_{i,j})$. Then $(RD_\alpha(G))^2_{i,j} = \sum_{k=1}^{n} b_{i,k} b_{k,j}$, and the row sum of $(RD_\alpha(G))^2$ should be

$$S_i((RD_\alpha(G))^2) = \sum_{j=1}^{n} \sum_{k=1}^{n} b_{i,k} b_{k,j} = \sum_{k=1}^{n} (b_{i,k} \sum_{j=1}^{n} b_{k,j}) = \sum_{k=1}^{n} (b_{i,k} RT_k).$$

Hence, $S_i((RD_\alpha(G))^2) = (1-\alpha)T_i + \alpha RT_i^2$.

Now, let **x** be the unit Perron vector corresponding to $\rho(RD_\alpha(G))$. Clearly, $RD_\alpha(G)\mathbf{x} = \rho(RD_\alpha(G))\mathbf{x}$ and $(RD_\alpha(G))^2 \mathbf{x} = (\rho(RD_\alpha(G)))^2 \mathbf{x}$. By Lemma 4, we have

$$\min_{1 \leq i \leq n} \left\{ (1-\alpha)T_i + \alpha(RT_i)^2 \right\} \leq (\rho(RD_\alpha(G)))^2 \leq \max_{1 \leq i \leq n} \left\{ (1-\alpha)T_i + \alpha(RT_i)^2 \right\}.$$

Thus

$$\min_{1 \leq i \leq n} \left\{ \sqrt{(1-\alpha)T_i + \alpha(RT_i)^2} \right\} \leq \rho(RD_\alpha(G)) \leq \max_{1 \leq i \leq n} \left\{ \sqrt{(1-\alpha)T_i + \alpha(RT_i)^2} \right\}.$$

□

Theorem 4. *Let G be a graph with n vertices, RT_{max} and T_{max} be the maximum reciprocal distance degree and the maximum second reciprocal distance degree of G, respectively. Then*

$$\rho(RD_\alpha(G)) \leq \frac{\alpha RT_{max} + \sqrt{(\alpha RT_{max})^2 + 4(1-\alpha)T_{max}}}{2}.$$

The equality holds if and only if G is a reciprocal distance degree regular graph.

Proof. Since $RD_\alpha(G) = \alpha RT(G) + (1-\alpha)RD(G), 0 \leq \alpha \leq 1$, it can be obtained by simple calculation

$$S_i(RD_\alpha(G)) = RT_i,$$
$$S_i((RT(G))^2) = S_i(RT(G)RD(G)) = RT_i^2,$$
$$S_i((RD(G))^2) = S_i(RD(G)RT(G)) = T_i.$$

Then

$$\begin{aligned}S_i((RD_\alpha(G))^2) =& S_i(\alpha^2(RT(G))^2 + \alpha(1-\alpha)RT(G)RD(G) \\ & + \alpha(1-\alpha)RD(G)RT(G) + (1-\alpha)^2RD(G)^2) \\ =& S_i(\alpha RT(G)(\alpha RT(G) + (1-\alpha)RD(G))) \\ & + \alpha(1-\alpha)S_i(RD(G)RT(G)) + (1-\alpha)^2S_i(RD(G)^2) \\ =& \alpha RT_i S_i(RD_\alpha(G)) + (1-\alpha)T_i \\ \leq & \alpha RT_{max} S_i(RD_\alpha(G)) + (1-\alpha)T_{max},\end{aligned}$$

that is,

$$S_i((RD_\alpha(G))^2 - \alpha RT_{max}RD_\alpha(G)) \leq (1-\alpha)T_{max}.$$

By Lemma 4,

$$\rho^2(RD_\alpha(G)) - \alpha RT_{max}\rho(RD_\alpha(G)) - (1-\alpha)T_{max} \leq 0.$$

For any vertex v_i, when the inequality is equal, $RT_i = RT_{max}$, $T_i = T_{max}$. That is, G is a reciprocal distance degree regular graph.

On the contrary, when G is a reciprocal distance degree regular graph, the inequality is equal. □

Theorem 5. *Let G be a graph with n vertices, RT_{min} and T_{min} be the minimum reciprocal distance degree and the minmum second reciprocal distance degree of G, respectively. Then*

$$\rho(RD_\alpha(G)) \geq \frac{\alpha RT_{min} + \sqrt{(\alpha RT_{min})^2 + 4(1-\alpha)T_{min}}}{2}.$$

The equality holds if and only if G is a reciprocal distance degree regular graph.

Proof. The method is the same as Theorem 4. □

Theorem 6. *Let G be a graph with reciprocal distance degree sequence $\{RT_1, RT_2, \ldots, RT_n\}$ and second reciprocal distance degree sequence $\{T_1, T_2, \ldots, T_n\}$. Then*

$$\rho(RD_\alpha(G)) \leq \max_{1 \leq i,j \leq n} \left\{ \frac{\alpha(RT_i + RT_j) + \sqrt{\alpha^2(RT_i - RT_j)^2 + 4(1-\alpha)^2 \frac{T_iT_j}{RT_iRT_j}}}{2} \right\}. \quad (3)$$

The equality holds if and only if G is a reciprocal distance degree regular graph.

Proof. Let $\mathbf{x} = (x_1, x_2, \ldots, x_n)$ be the eigenvector corresponding to the eigenvalue $\rho(G)$ of the matrix $RT(G)^{-1}RD_\alpha(G)RT(G)$, $x_s = \max\{x_i | i = 1, 2, \ldots, n\}$, $x_t = \max\{x_i | x_i \neq x_s, i = 1, 2, \ldots, n\}$.

Through simple calculation, the value of the (i,j)-th element of $RT(G)^{-1}RD_\alpha(G)RT(G)$ is

$$\begin{cases} \alpha RT_i, & \text{if } i = j, \\ (1-\alpha)\frac{RT_j}{RT_i}\frac{1}{d_{ij}}, & \text{if } i \neq j. \end{cases}$$

Because

$$RT(G)^{-1}RD_\alpha(G)RT(G)\mathbf{x} = \rho(RD_\alpha(G))\mathbf{x}, \tag{4}$$

row s and t in Equation (4) are

$$\rho(RD_\alpha(G))x_s = \alpha RT_s x_s + (1-\alpha)\sum_{i=1}^n \frac{RT_i}{RT_s}\frac{x_i}{d_{si}}, \tag{5}$$

$$\rho(RD_\alpha(G))x_t = \alpha RT_t x_t + (1-\alpha)\sum_{i=1}^n \frac{RT_i}{RT_t}\frac{x_i}{d_{ti}}. \tag{6}$$

After shifting the item of Equations (5) and (6), we can get

$$(\rho(RD_\alpha(G)) - \alpha RT_s))x_s = (1-\alpha)\sum_{i=1}^n \frac{RT_i}{RT_s}\frac{x_i}{d_{si}}$$
$$\leq (1-\alpha)\frac{x_t}{RT_s}\sum_{i=1}^n RT_i\frac{1}{d_{si}} \tag{7}$$
$$= (1-\alpha)\frac{T_s}{RT_s}x_t,$$

$$(\rho(RD_\alpha(G)) - \alpha RT_t))x_t = (1-\alpha)\sum_{i=1}^n \frac{RT_i}{RT_t}\frac{x_i}{d_{ti}}$$
$$\leq (1-\alpha)\frac{x_s}{RT_t}\sum_{i=1}^n RT_i\frac{1}{d_{ti}} \tag{8}$$
$$= (1-\alpha)\frac{T_t}{RT_t}x_s.$$

Multiply Equation (7) and (8) to simplify $(\rho(RD_\alpha(G)) - \alpha RT_s)(\rho(RD_\alpha(G)) - \alpha RT_t))x_s x_t \leq (1-\alpha)^2 \frac{T_s T_t}{RT_s RT_t} x_t x_s$. Then

$$(\rho(RD_\alpha(G)))^2 - \alpha(RT_s + RT_t)\rho(RD_\alpha(G)) + \alpha^2 RT_s RT_t - (1-\alpha)^2 \frac{T_s T_t}{RT_s RT_t} \leq 0.$$

$$\rho(RD_\alpha(G)) \leq \frac{\alpha(RT_s + RT_t) + \sqrt{\alpha^2(RT_s - RT_t)^2 + 4(1-\alpha)^2 \frac{T_s T_t}{RT_s RT_t}}}{2}.$$

Hence

$$\rho(RD_\alpha(G)) \leq \max_{1 \leq i,j \leq n} \left\{ \frac{\alpha(RT_i + RT_j) + \sqrt{\alpha^2(RT_i - RT_j)^2 + 4(1-\alpha)^2 \frac{T_i T_j}{RT_i RT_j}}}{2} \right\}.$$

Suppose G is a k-reciprocal distance regular graph, $RT_i = k$, $T_i = k^2$, $i = 1, 2, \ldots, n$. According to Lemma 2, $\rho(RD_\alpha(G)) = k$, so Equation (3) holds. On the contrary, if inequality (3) is equal, $x_1 = x_2 = \cdots = x_n$ can be obtained from (7) and (8), that is, $\rho(RD_\alpha(G)) =$

$\alpha RT_1 + (1-\alpha)\frac{T_1}{RT_1} = \alpha RT_2 + (1-\alpha)\frac{T_2}{RT_2} = \cdots = \alpha RT_n + (1-\alpha)\frac{T_n}{RT_n}$, which means that G is a reciprocal distance degree regular graph. □

Theorem 7. *Let G be a graph with reciprocal distance degree sequence $\{RT_1, RT_2, \ldots, RT_n\}$ and second reciprocal distance degree sequence $\{T_1, T_2, \ldots, T_n\}$. Then*

$$\rho(RD_\alpha(G)) \geq \min_{1 \leq i,j \leq n} \left\{ \frac{\alpha(RT_i + RT_j) + \sqrt{\alpha^2(RT_i - RT_j)^2 + 4(1-\alpha)^2 \frac{T_i T_j}{RT_i RT_j}}}{2} \right\}.$$

The equality holds if and only if G is a reciprocal distance degree regular graph.

Proof. The method is the same as Theorem 6. □

Theorem 8. *Let G be a graph of order n and $0 \leq \alpha < 1$, then*

$$\rho(RD_\alpha(G)) \leq \frac{2\alpha H(G)}{n} + \sqrt{\frac{n-1}{n}\left(\|RD_\alpha(G)\|_F^2 - \frac{(2\alpha H(G))^2}{n}\right)}.$$

The equality holds if and only if $G = K_n$.

Proof. We recall that $\sum_{i=1}^n \lambda_i(RD_\alpha(G)) = \alpha \sum_{i=1}^n RT_i = 2\alpha H(G)$, and $\sum_{i=1}^n \lambda_i(RD_\alpha(G))^2 = \|RD_\alpha(G)\|_F^2$. Clearly,

$$\sum_{i=1}^n \left(\lambda_i(RD_\alpha(G)) - \frac{2\alpha H(G)}{n}\right) = 0.$$

By Lemma 8,

$$\rho(RD_\alpha(G)) - \frac{2\alpha H(G)}{n} \leq \sqrt{\frac{n-1}{n}\sum_{i=1}^n \left(\lambda_i(RD_\alpha(G)) - \frac{2\alpha H(G)}{n}\right)^2}, \quad (9)$$

with equality holds if and only if

$$\lambda_2(RD_\alpha(G)) - \frac{2\alpha H(G)}{n} = \cdots = \lambda_n(RD_\alpha(G)) - \frac{2\alpha H(G)}{n} = -\frac{\rho(RD_\alpha(G)) - \frac{2\alpha H(G)}{n}}{n-1}. \quad (10)$$

Since

$$\sum_{i=1}^n \left(\lambda_i(RD_\alpha(G)) - \frac{2\alpha H(G)}{n}\right)^2 = \sum_{i=1}^n (\lambda_i(RD_\alpha(G)))^2 - \frac{4\alpha H(G)}{n}\sum_{i=1}^n \lambda_i(RD_\alpha(G)) + n\left(\frac{2\alpha H(G)}{n}\right)^2$$

$$= \|RD_\alpha(G)\|_F^2 - 2\frac{(2\alpha H(G))^2}{n} + \frac{(2\alpha H(G))^2}{n}$$

$$= \|RD_\alpha(G)\|_F^2 - \frac{(2\alpha H(G))^2}{n}.$$

The upper bound (9) is equivalent to

$$\rho(RD_\alpha(G)) \leq \frac{2\alpha H(G)}{n} + \sqrt{\frac{n-1}{n}\left(\|RD_\alpha(G)\|_F^2 - \frac{(2\alpha H(G))^2}{n}\right)} \quad (11)$$

with the necessary and sufficient condition for the equality given in (10).

Now, suppose that the equality holds. Therefore, the equality condition for (11) can be given in (10), and we obtain that G has only two distinct generalized reciprocal distance eigenvalues. Hence, from Lemma 7, $G = K_n$.

Conversely, from Lemma 7 the generalized reciprocal distance eigenvalues of K_n are $\rho(RD_\alpha(K_n)) = n - 1$ and $\lambda_i(RD_\alpha(G)) = \alpha n - 1$, for $i = 2, 3, \ldots, n$. Then, the equality holds. □

4. Bounds of $\rho(RD_\alpha(G))$ of Line Graph $L(G)$

The line graph $L(G)$ of G is the graph whose vertices correspond to the edges of G, and two vertices of $L(G)$ are adjacent if and only if the corresponding edges of G are adjacent. In this section, we give the bounds of the spectral radius of the generalized reciprocal distance matrix of $L(G)$.

Theorem 9. *Let graph G have n vertices and m edges, and the degree of vertex v_i be recorded as d_i. If $diam(G) \leq 2$ and graphs F_i, $i = 1, 2, 3$ in Lemma 10 are not induced subgraphs of G, then*

$$\rho(RD_\alpha(L(G))) \geq \frac{\frac{1}{2}(m^2 - 3m + \sum\limits_{i=1}^n d_i^2)}{m}.$$

Proof. If $diam(G) \leq 2$, the i-th row element of $RD_\alpha(G)$ is composed of $\{\frac{1}{2}\alpha(n + d_i - 1), (1-\alpha)^{d_i}, \frac{1}{2}(1-\alpha)^{[n-d_i-1]}\}$, which can be obtained from Lemma 9

$$\rho(RD_\alpha(L(G))) \geq \frac{\mathbf{e}^T RD_\alpha(G)\mathbf{e}}{\mathbf{e}^T\mathbf{e}} = \frac{\sum\limits_{i=1}^n \frac{1}{2}(n+d_i-1)}{n} = \frac{\frac{1}{2}(n^2+2m-n)}{n}.$$

Hence, line graph $L(G)$ has $n_1 = m$ vertices and $m_1 = \frac{1}{2}\sum\limits_{i=1}^n d_i^2 - m$ edges. Because graphs F_i, $i = 1, 2, 3$ are not induced subgraphs of G, from Lemma 10, $diam(L(G)) \leq 2$, then

$$\rho(RD_\alpha(L(G))) \geq \frac{\frac{1}{2}(n_1^2 + 2m_1 - n_1)}{n_1}$$

$$= \frac{\frac{1}{2}[m^2 + 2(\frac{1}{2}\sum\limits_{i=1}^n d_i^2 - m) - m]}{m}$$

$$= \frac{\frac{1}{2}(m^2 - 3m + \sum\limits_{i=1}^n d_i^2)}{m}.$$

□

Theorem 10. *Let graph G be r-regular graph with n vertices, and graphs F_i, $i = 1, 2, 3$ be not-induced subgraphs of G. Then*

$$\rho(RD_\alpha(L(G))) \geq \frac{nr}{4} + r - 3.$$

Proof. Let graph G be r-regular graph with n vertices, the number of edges in graph G is $m = \frac{nr}{2}$, $d_i = deg(v_i) = r$. It is proved by Theorem 9. □

Theorem 11. *Let the vertices set and edges set of G be $V(G) = \{v_1, v_2, \ldots, v_n\}$ and $E(G) = \{e_1, e_2, \ldots, e_m\}$, $deg(e_i)$ represent the number of edges adjacent to edge e_i. Then,*

$$\rho(RD_\alpha(L(G))) \leq \max_{1 \leq i \leq m} \left\{\frac{1}{2}(m - deg(e_i) - 1)\right\}.$$

Proof. Let $e = uv$ be an edge of G. Then, the degree of vertex $e \in V(L(G))$ is $deg_{L(G)}(e) = deg_G(u) + deg_G(v) - 2$.

In graph G, if edge $e = uv$ is adjacent to $deg(u) + deg(v) - 2 = deg(e)$, then denoted $|E_e| = m - 1 - deg(e)$ as the number of edges which are not adjacent to edge e. Therefore, in the graph $L(G)$, there are $|E_e|$ vertices, and their distance from vertex e is greater than 1. Thus, the maximum element of generalized reciprocal distances matrix of the corresponding vertices should be $\frac{1}{2}(1 - \alpha)$. We can get

$$S_i(RD_\alpha(L(G))) \leq \frac{1}{2}(1-\alpha)(m - 1 - deg(e_i))$$
$$+ (1-\alpha)deg(e_i) + \alpha(\frac{1}{2}m - \frac{1}{2} + \frac{1}{2}deg(e_i))$$
$$= \frac{1}{2}(m - deg(e_i) - 1).$$

By Lemma 4, $\rho(RD_\alpha(L(G))) \leq \max\limits_{1 \leq i \leq m} \{\frac{1}{2}(m - deg(e_i) - 1)\}$. □

5. Conclusions

In this paper, we find some bounds for the spectral radius of the generalized reciprocal distance matrix of a simple undirected connected graph G, and we also give the generalized reciprocal distance spectral radius of line graph $L(G)$. The graphs for which those bounds are attained are characterized.

Author Contributions: Investigation, Y.M., Y.G. and Y.S.; writing—original draft preparation, Y.M.; writing—review and editing, Y.M., Y.G.; All authors have read and agreed to the published version of the manuscript.

Funding: Research was supported by Shanxi Scholarship Council of China (No. 201901D211227).

Institutional Review Board Statement: Not applicable.

Informed Consent Statement: Not applicable.

Data Availability Statement: All data generated or analyzed during this study are included in this published article.

Acknowledgments: The authors are grateful to the anonymous referees for helpful suggestions and valuable comments, which led to an improvement of the original manuscript.

Conflicts of Interest: The authors declare no conflict of interest.

References

1. Plavšić, D.; Nikolić, S.; Trinajstić, N.; Mihalić, Z. On the Harary index for the characterization of chemical graphs. *J. Math. Chem.* **1993**, *12*, 235–250. [CrossRef]
2. Bapat, R.; Panda, S.K. The spectral radius of the Reciprocal distance Laplacian matrix of a graph. *Bull. Iran. Math. Soc.* **2018**, *44*, 1211–1216. [CrossRef]
3. Alhevaz, A.; Baghipur, M.; Ramane, H.S. Computing the reciprocal distance signless Laplacian eigenvalues and energy of graphs. *Matematiche* **2019**, *74*, 49–73.
4. Das, K.C. Maximum eigenvalue of the reciprocal distance matrix. *J. Math. Chem.* **2010**, *47*, 21–28. [CrossRef]
5. Zhou, B.; Trinajstić, N. Maximum eigenvalues of the reciprocal distance matrix and the reverse Wiener matrix. *Int. J. Quantum Chem.* **2008**, *108*, 858–864. [CrossRef]
6. Medina, L.; Trigo, M. Upper bounds and lower bounds for the spectral radius of Reciprocal Distance, Reciprocal Distance Laplacian and Reciprocal Distance signless Laplacian matrices. *Linear Algebra Appl.* **2021**, *609*, 386–412. [CrossRef]
7. Tian, G.X.; Cheng, M.J.; Cui, S.Y. The generalized reciprocal distance matrix of graphs. *arXiv* **2022**, arXiv: 2204.03787.
8. Baghipur, M.; Ghorbani, M.; Ganie, H.A.; Shang, Y. On the Second-Largest Reciprocal Distance Singless Laplacian Eigenvalue. *Mathematics* **2021**, *9*, 1113–1123. [CrossRef]
9. Alhevaz, A.; Baghipur, M.; Alizadeh, Y.; Pirzada, S. On eigenvalues of the reciprocal distance signless Laplacian matrix of graphs. *Asian-Eur. J. Math.* **2021**, *14*, 2150176. [CrossRef]
10. Varga, R. *Matrix Iterative Analysis*; Springer Sreies in Computational Mathematics; Springer: Berlin/Heidelberg, Germany, 2000.
11. Minc, H. *Nonnegative Matrices*; John Wiley Sons: New York, NY, USA, 1988.
12. Parlett, B.N. *The Symmetric Eigenvalue Problem*; Prentice-Hall: Englewood Cliiffs, NJ, USA, 1980.

13. Rojo, O.; Rojo, H. A decresing sequence of upper bounds on the largest Laplacian eigenvalue of a graph. *Linear Algebra Appl.* **2004**, *318*, 97–116. [CrossRef]
14. Zhang, F. *Matrix Theory Basic Results and Techniques*; Springer: New York, NY, USA, 1999.
15. Ramane, H.S.; Revankar, D.S.; Gutman, I.; Walikar, H.B. Distance spectra and distance energies of iterated line graphs of regular graphs. *Publ. Inst. Math.* **2009**, *85*, 39–46. [CrossRef]

Article

The t-Graphs over Finitely Generated Groups and the Minkowski Metric

Gabriela Diaz-Porto, Ismael Gutierrez * and Armando Torres-Grandisson

Department of Mathematics, Universidad del Norte, Km 5 via a Puerto Colombia, Barranquilla 081007, Colombia
* Correspondence: isgutier@uninorte.edu.co

Abstract: In this paper, we introduce t-graphs defined on finitely generated groups. We study some general aspects of the t-graphs on two-generator groups, emphasizing establishing necessary conditions for their connectedness. In particular, we investigate properties of t-graphs defined on finite dihedral groups.

Keywords: finitely generated groups; finite groups; t-graph; subgraph; connected components; chromatic number

MSC: 20F05; 20E65; 05C12; 05C15

1. Introduction

One of the best-known connections between groups and graph theory was presented by A. Cayley [1]. He gave a group G as a directed graph, where the vertices correspond to elements of G and the edges to multiplication by group generators and their inverses. Such a graph is called a Cayley diagram or Cayley graph of G. It is a central tool in combinatorial and geometric group theory.

Recent works reveal many different ways of associating a graph to a given finite group, most of which were inspired by a question posed by P. Erdös [2]. These differences lie in the adjacency criterion used to relate two group elements constituting the set of vertices of such a graph. Some essential authors in this context are A. Abdollahi [3], A. Ballester-Bolinches et al. [4–8], A. Lucchini [9,10], and D. Hai-Reuven [11], among others.

Our notation will be standard, as in [12] and [13] for groups and graphs. Let $G = \langle g_1, \ldots, g_n \rangle$ be a finitely generated group and suppose now that every element $g \in G$ can be uniquely written as follows

$$g = \prod_{i=1}^{n} g_i^{\epsilon_i}, \tag{1}$$

with $0 \leq \epsilon_i < m_i$, and $1 \leq i \leq n$. The numbers m_i can be, for example, the orders of the corresponding elements in the finite case, but they may also differ from these orders.

To determine a measure of the separation between two elements of G, we introduce the following distance map $d_1 : G \times G \longrightarrow \mathbb{N}_0$, defined by

$$d_1(g,h) = d_1\left(\prod_{i=1}^{n} g_i^{\epsilon_i}, \prod_{i=1}^{n} g_i^{\delta_i}\right) = \sum_{i=1}^{n} |\epsilon_i - \delta_i|. \tag{2}$$

The set G endowed with this distance d is a metric space. Note that d_1 is just the Minkowski l_p metric for $p = 1$ in $\{(\epsilon_1, \ldots, \epsilon_n) \mid 0 \leq \epsilon_i < m_i\}$. This is also called the taxicab distance, Manhattan distance, or grid distance.

G. Diaz-Porto and A. Torres-Grandisson introduced t-graphs using Minkowski's metric in [14,15]. These graphs can be defined by the group G as the underlying set of

vertices and the following adjacency criteria: Let t be an integer number with $1 \leq t \leq n$. We say that $g, h \in G$ are adjacent if and only if $d_1(g, h) = t$.

The simplest example is when G is a finite cyclic group. Let $G = \langle g \rangle$ be a cyclic group with the finite order m. That is, $G = \{1, g, \cdots, g^{m-1}\}$. From (2), we have

$$d_1(g^i, g^j) = |i - j|, \text{ for all } 0 \leq i, j \leq m - 1. \tag{3}$$

This means that in the t-graph of G there exists an edge between g^i and g^j if and only if $|i - j| = t$. Defining on G the following relation

$$g^i \sim g^j \iff i \equiv j \bmod t, \tag{4}$$

where $\sim$ is an equivalence relation, and then we have a partition of G in t classes given by

$$[g^i] := \{g^j \in G \mid j \equiv i \bmod t\}, \tag{5}$$

where $i \in \{0, 1, \ldots, t - 1\}$. Then, the t-graph of a finite cyclic group G can be viewed as the union of t connected components, consisting of path graphs or isolated points. Consequently for $t \geq 2$, the t-graph is non-connected and 2-chromatic. The 1-graph of G is a finite path graph and then connected.

If t is a divisor of the group order m, then it is well known that G has a cyclic subgroup U of order m/t, and the elements of U form a subgraph with m/n vertices, which is a connected component of the t-graph of G.

If $G = \langle g \rangle$ is an infinite cyclic group, then the 1-graph of G is an infinite path graph. This statement follows directly from the definition of the t-graph.

An immediate consequence of the above discussion is that if G is a finite abelian group, say $G = \langle g_1 \rangle \times \cdots \times \langle g_n \rangle$, with $\text{ord}(g_j) = \epsilon_j$. Then, the 1-graph of G is the Cartesian product of n path graphs of lengths ϵ_j, respectively. That is an n-dimensional square grid graph. In general, using the above example, the t-graph of G is the Cartesian product of t components.

In the general case, if G is a direct product of cyclic groups with at least one infinite factor, then the t-graph of G is an infinite rectangular grid graph.

The first thing we can observe is that for a group G different generating systems can give different graphs. For instance, the groups $\mathbb{Z}_4 \times \mathbb{Z}_6$ and $\mathbb{Z}_2 \times \mathbb{Z}_{12}$ are isomorphic, but the graphs associated with the natural generating sets corresponding to these ways to present the group G are different.

On the other hand, if two groups admit generating systems such that every element g can be described as in (1), then it is possible that the corresponding t-graphs are the same, even though the groups are not isomorphic. We can see this in the following example. It is well known that the dihedral group D_n and the quaternion group Q_8 have the subsequent group presentation, respectively,

$$D_n = \langle a, b \mid a^2 = b^n = 1, aba = b^{-1} \rangle, \tag{6}$$

$$Q_8 = \langle a, b \mid a^4 = 1, a^2 = b^2, bab^{-1} = a^{-1} \rangle. \tag{7}$$

Furthermore,

$$\mathbb{Z}_2 \times \mathbb{Z}_4 = \langle a, b \mid a^2 = b^4 = 1, ab = ba \rangle. \tag{8}$$

Note that, in terms of their generators, the elements of D_4, Q_8, and $\mathbb{Z}_2 \times \mathbb{Z}_4$ can be written as follows

$$\{1, a, b, b^2, b^3, ab, ab^2, ab^3\}. \tag{9}$$

This means that the three groups have the same distance table (see Table 1) and, consequently, the same t-graphs for all t.

Table 1. Table of distances of $\mathbb{Z}_2 \times \mathbb{Z}_4$, D_4, and Q_8.

d_1	1	a	b	b^2	b^3	ab	ab^2	ab^3
1	0	1	1	2	3	2	3	4
a	1	0	2	3	4	1	2	3
b	1	2	0	1	2	1	2	3
b^2	2	3	1	0	1	2	1	2
b^3	3	4	2	1	0	3	2	1
ab	2	1	1	2	3	0	1	2
ab^2	3	2	2	1	2	1	0	1
ab^3	4	3	3	2	1	2	1	0

An illustration of the first four t-graphs of these three groups is presented in the following Figure 1.

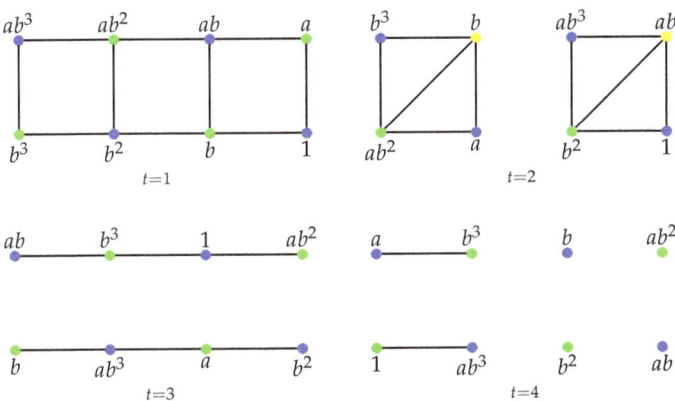

Figure 1. Some t-graphs of $\mathbb{Z}_2 \times \mathbb{Z}_4$, D_4, and Q_8.

Despite being non-isomorphic groups, these groups have precisely the same t-graphs since the metric used to define the adjacency criterion only considers the writing of the group's elements and not how they interact. This leads to the conclusion that any two-generator finite group $G = \langle a, b \rangle$, in which every element can be written in the form $a^i b^j$ with $0 \leq i \leq \text{ord}(a) - 1$ and $0 \leq j \leq \text{ord}(b) - 1$, has the same t-graphs as the group $\mathbb{Z}_{\text{ord}(a)} \times \mathbb{Z}_{\text{ord}(b)}$ since, when considering the form in which its elements are written in terms of the generators, the underlying sets are the same.

Therefore, to study the t-graphs of a finite group G, it is sufficient to consider abelian groups, expressed as products of cyclic groups. Naturally, this implies asking oneself, given an arbitrary group G, how to determine the abelian group with which it will share the same t-graphs. For example, the symmetric group of degree five has the same t-graph as $\mathbb{Z}_2 \times \mathbb{Z}_3 \times \mathbb{Z}_4 \times \mathbb{Z}_5$. In fact, in general, the group $\text{Sym}(n)$ can be factorized in the form $\text{Sym}(n) = \text{Sym}(n-1)\langle (12 \cdots n) \rangle$, and, applying this property inductively, we have that $\text{Sym}(n)$ is generated by the set $\{(12), (123), \cdots, (12 \cdots, n)\}$. In particular, the set

$$\{(12)^i (123)^j (1234)^k (12345)^l \mid i = 0, 1, j = 0, 1, 2, k = 0, 1, 2, 3, l = 0, 1, 2, 3, 4\}$$

is exactly $\text{Sym}(5)$.

On the other hand, this situation brings the possibility of studying t-graphs by defining the adjacency criterion in terms of another metric. This change may imply that the group structure plays a more critical role.

The main goal of this paper is to obtain some characterizations of the *t*-graphs $\mathcal{G}$ associated with the two-generator finite group G that can be expressed in the form

$$G = \langle a, b \rangle = \{a^i b^j \mid 0 \le i \le m, \ 0 \le j \le n\}. \tag{10}$$

where $m \le \text{ord}(a)$ y $n \le \text{ord}(b)$; $n, m \in \mathbb{Z}$. These numbers *m* and *n* depend exclusively on the structure, namely on the group's presentation and the order of G. We determine the number of connected components of $\mathcal{G}$ depending on whether *t* is an even or odd number.

2. Preliminaries on *t*-Graphs

A desirable property of the *t*-graphs is that every subgroup H of a group G naturally results in a subgraph. However, this is, in general, not true. For example, let G be the Klein four-group, say $G = \{1, a, b, ab\}$ and $H = \{1, ab\}$. Concerning their natural generating systems, *ab* and one are not adjacent in the 1-graph of G. Nevertheless, in H, they are adjacent.

Lemma 1. *Let $G = \langle g_1, \ldots, g_n \rangle$ be a finitely generated group and $H \le G$, with $H = \langle h_1, \ldots, h_n \rangle$ and $h_j = g_j^{k_j}$ for some natural numbers k_j. Then, the t-graph of H is a subgraph of the t-graph of G.*

Proof. It follows immediately from the definition of the *t*-graph. □

Lemma 2. *Let $G = \langle g_1, \cdots, g_n \rangle$ be a finitely generated group and suppose now that every element in $g \in G$ can be uniquely written as $g = \prod_{i=1}^{n} g_i^{\epsilon_i}$ with $0 \le \epsilon_i < m_i$ and $1 \le i \le n$. Further, let $H = \langle h_1, \cdots, h_n \rangle$ be a finitely generated group with the same property. If G and H are isomorphic, then the corresponding t-graphs are isomorphic, for all natural numbers t.*

Proof. Let $f: G \longrightarrow H$ be a group isomorphism with $f(g_i) = h_i$, and let $\mathcal{G} = (G, E_1)$ and $\mathcal{H} = (H, E_2)$ be the corresponding *t*-graphs of G and H, respectively. Suppose that $\{x, y\} \in E_1$ with $x = \prod_{i=1}^{n} g_i^{\epsilon_i}$ and $y = \prod_{i=1}^{n} g_i^{\delta_i}$. Then, $d_1(x, y) = t$, and we have

$$d_1(f(x), f(y)) = d_1\left(\prod_{i=1}^n f(g_i)^{\epsilon_i}, \prod_{i=1}^n f(g_i)^{\delta_i}\right) = d_1\left(\prod_{i=1}^n h_i^{\epsilon_i}, \prod_{i=1}^n h_i^{\delta_i}\right)$$
$$= \sum_{i=1}^n |\epsilon_i - \delta_i| = d_1(x, y).$$

It follows that $\{f(x), f(y)\} \in E_2$. □

Remark 1. *Note that the reciprocal of the statement in Lemma 2 is, in general, not true. For example, the t-graphs of the dihedral D_4 and the quaternions group, Q_8 are isomorphic even though $D_4 \not\cong Q_8$.*

To study *t*-graphs in the given context, we can use the spectral theory of graphs, which consists of studying the properties of the Laplacian matrix of a graph, more specifically, its eigenvalues and eigenvectors.

The Laplacian matrix of $\mathcal{G} = (V, E)$ is the $n \times n$ matrix $L = (l_{ij})$ indexed by V, whose (i, j)-entry is defined as follows

$$l_{ij} = \begin{cases} -1 & \text{if } \{v_i, v_j\} \in E \\ \deg(v_i) & \text{if } i = j \\ 0 & \text{otherwise.} \end{cases} \tag{11}$$

To analyze the behavior of the number of connected components $k(\mathcal{G})$ of the *t*-graphs defined on a group G, we use the following theorem, which allows us to realize Tables 2 and 3. A proof of this theorem can be found in [16] (Theorem 7.1).

Theorem 1. *A graph $\mathcal{G}$ has k connected components if and only if the algebraic multiplicity of zero as the Laplacian eigenvalue is k.*

In the following, to study the *t*-graphs associated with a finite group G, we will consider only finite two-generator groups, which can be expressed in the form (10). These numbers m and n depend exclusively on the structure, namely on the group's presentation and the order of G.

Let G be such a group. To observe the behavior of the number of connected components $k(\mathcal{G})$ of a *t*-graph $\mathcal{G}$ determined by a group G, we make use of Theorem 1, with which we were able to make the following tables:

Table 2. Number of connected components of the *t*-graphs on $\mathbb{Z}_n \times \mathbb{Z}_2$.

n\t	1	2	3	4	5	6	7	8	9	10	11	12	13	14	15	16	17	18	19	20
2	1	2	-	-	-	-	-	-	-	-	-	-	-	-	-	-	-	-	-	-
3	1	2	4	-	-	-	-	-	-	-	-	-	-	-	-	-	-	-	-	-
4	1	2	2	6	-	-	-	-	-	-	-	-	-	-	-	-	-	-	-	-
5	1	2	1	4	8	-	-	-	-	-	-	-	-	-	-	-	-	-	-	-
6	1	2	1	2	6	10	-	-	-	-	-	-	-	-	-	-	-	-	-	-
7	1	2	1	2	4	8	12	-	-	-	-	-	-	-	-	-	-	-	-	-
8	1	2	1	2	2	6	10	14	-	-	-	-	-	-	-	-	-	-	-	-
9	1	2	1	2	1	4	8	12	16	-	-	-	-	-	-	-	-	-	-	-
10	1	2	1	2	1	2	6	10	14	18	-	-	-	-	-	-	-	-	-	-
11	1	2	1	2	1	2	4	8	12	16	20	-	-	-	-	-	-	-	-	-
12	1	2	1	2	1	2	2	6	10	14	18	22	-	-	-	-	-	-	-	-
13	1	2	1	2	1	2	1	4	8	12	16	20	24	-	-	-	-	-	-	-
14	1	2	1	2	1	2	1	2	6	10	14	18	22	26	-	-	-	-	-	-
15	1	2	1	2	1	2	1	2	4	8	12	16	20	24	28	-	-	-	-	-
16	1	2	1	2	1	2	1	2	2	6	10	14	18	22	26	30	-	-	-	-
17	1	2	1	2	1	2	1	2	1	4	8	12	16	20	24	28	32	-	-	-
18	1	2	1	2	1	2	1	2	1	2	6	10	14	18	22	26	30	34	-	-
19	1	2	1	2	1	2	1	2	1	2	4	8	12	16	20	24	28	32	36	-
20	1	2	1	2	1	2	1	2	1	2	2	6	10	14	18	22	26	30	34	38

Table 3. Number of connected components of the *t*-graphs on $\mathbb{Z}_n \times \mathbb{Z}_3$.

n\t	1	2	3	4	5	6	7	8	9	10	11	12	13	14	15	16	17	18	19	20
2	1	2	4	-	-	-	-	-	-	-	-	-	-	-	-	-	-	-	-	-
3	1	2	2	7	-	-	-	-	-	-	-	-	-	-	-	-	-	-	-	-
4	1	2	1	4	10	-	-	-	-	-	-	-	-	-	-	-	-	-	-	-
5	1	2	1	3	7	13	-	-	-	-	-	-	-	-	-	-	-	-	-	-
6	1	2	1	2	4	10	16	-	-	-	-	-	-	-	-	-	-	-	-	-
7	1	2	1	2	2	7	13	19	-	-	-	-	-	-	-	-	-	-	-	-
8	1	2	1	2	1	4	10	16	22	-	-	-	-	-	-	-	-	-	-	-
9	1	2	1	2	1	3	7	13	19	25	-	-	-	-	-	-	-	-	-	-
10	1	2	1	2	1	2	4	10	16	22	28	-	-	-	-	-	-	-	-	-
11	1	2	1	2	1	2	2	7	13	19	25	31	-	-	-	-	-	-	-	-
12	1	2	1	2	1	2	1	4	10	16	22	28	34	-	-	-	-	-	-	-
13	1	2	1	2	1	2	1	3	7	13	19	25	31	37	-	-	-	-	-	-
14	1	2	1	2	1	2	1	2	4	10	16	22	28	34	40	-	-	-	-	-
15	1	2	1	2	1	2	1	2	2	7	13	19	25	31	37	43	-	-	-	-
16	1	2	1	2	1	2	1	2	1	4	10	16	22	28	34	40	46	-	-	-
17	1	2	1	2	1	2	1	2	1	3	7	13	19	25	31	37	43	49	-	-
18	1	2	1	2	1	2	1	2	1	2	4	10	16	22	28	34	40	46	52	-
19	1	2	1	2	1	2	1	2	1	2	2	7	13	19	25	31	37	43	49	55
20	1	2	1	2	1	2	1	2	1	2	1	4	10	16	22	28	34	40	46	52

Remark 2. *Note in the previous tables that $k(\mathcal{G})$ has the same value up to a certain value of t where, if t is even, $k(\mathcal{G}) = 2$, and, if t is odd, then $k(\mathcal{G}) = 1$ and, when $t > \lceil \frac{m+n-2}{2} \rceil$, then $k(\mathcal{G})$ has a value with the following possible pattern:*

1. *If $m = 2$ (for example by dihedral groups), then the number of connected components of the t-graph increases by four. We conjecture that $K(G) = 2(2t - n) - 2$.*
2. *If $m = 3$, then the number of connected components starts with seven and so progresses from six to six, if n is odd, and starts at four and progresses from six to six when n is even. We conjecture that $K(G) = 3(2t - n - 2) - 2$.*

This fact leads us to state the first theorem in the next section, which allows us to characterize first the t-graphs associated with two-generator groups in the form (10), concerning the number of connected components.

3. The t-Graph of Some Two-Generator Groups

This section considers the t-graph of a particular case of two-generator groups. Specifically, we suppose that a is an involution and b has an order n. For example, the group G can be the abelian group $\mathbb{Z}_2 \times \mathbb{Z}_n$ or the dihedral group D_n of order n.

Lemma 3. *Let G be a two-generator group in the form (10) with $n, m \geq 2$, and $\mathcal{G}$ is the corresponding t-graph of G. Then, $\mathcal{G}$ has no isolated points if and only if $t \leq \lceil \frac{m+n-2}{2} \rceil$.*

Proof. Let $x = a^i b^j, y = a^k b^l \in G$ with

$$d_1(x,y) = |i - k| + |j - l| = t. \quad (12)$$

Then, $t \in \{0, \ldots, m + n - 2\}$, and suppose $|i - k| = s \in \{0, \ldots, m - 1\}$. This implies that $|j - l| = t - s \in \{0, \ldots, n - 1\}$. Note that if $t - s > n - 1$, the equality (12) is not verified. That is, there is no edge between x and y. Then, in order not to have isolated points, it must be fulfilled that $t - s \leq n - 1$ with $s \in \{0, \ldots, m - 1\}$. Moreover, $t \leq n - 1$. Analogously, it follows that $t \leq m - 1$. Consequently, $2t \leq m + n - 2$, and, therefore, $t \leq \lceil \frac{m+n-2}{2} \rceil$. □

Theorem 2. *Let G be a two-generator group in the form (10) with $n, m \geq 2$, and $\mathcal{G} = (G, E)$ be the corresponding t-graph with $t \leq \lceil \frac{m+n-2}{2} \rceil$.*

1. *If t is an even number, then $k(\mathcal{G}) = 2$.*
2. *If t is an odd number, then $\mathcal{G}$ is connected.*

Proof. From the above lemma, we have that the condition $t \leq \lceil \frac{m+n-2}{2} \rceil$ implies that $\mathcal{G}$ has no isolated points. We now differentiate two possible cases.

1. Let t be an even number. We define $\mathcal{C}_1 = (V_1, E_1)$ and $\mathcal{C}_2 = (V_2, E_2)$, the subgraph of $\mathcal{G}$, as follows:

$$V_1 := \{a^i b^j \mid i + j \equiv 0 \bmod 2\}, \quad (13)$$

$$E_1 := \{\{a^i b^j, a^k b^l\} \mid i + j, k + l \equiv 0 \bmod 2 \wedge |i - k| + |j - l| = t\}, \quad (14)$$

and

$$V_2 := \{a^i b^j \mid i + j \equiv 1 \bmod 2\}, \quad (15)$$

$$E_2 := \{\{a^i b^j, a^k b^l\} \mid i + j, k + l \equiv 1 \bmod 2 \wedge |i - k| + |j - l| = t\}. \quad (16)$$

It is clear that $V_1 \cup V_2 = G$, and then $k(\mathcal{G}) = 2$.

2. Let t be an even number, and $x = a^i b^j \in G$ be arbitrary. If $i + j \equiv 1 \mod 2$, then we consider the sets

$$\{a^k b^l \mid i, k + l \equiv 0 \mod 2, j \equiv 1 \mod 2 \wedge |i - k| + |j - l| = t\} \quad (17)$$
$$\{a^k b^l \mid j, k + l \equiv 0 \mod 2, i \equiv 1 \mod 2 \wedge |i - k| + |j - l| = t\}. \quad (18)$$

Since $\mathcal{G}$ has no isolated points, at least one of these sets is non-empty, and then $\{a^i b^j, a^k b^l\} \in E$.

If $i + j \equiv 0 \mod 2$, then a similar analysis leads to the same conclusion. Then, we have that $\mathcal{G}$ is a connected graph. □

The next theorem shows that the 1-graph associated with a finite dihedral group D_n has a simple structure. It corresponds to a square $(n \times 2)$-grid, as shown in Figure 2 below. Therefore, this graph is bichromatic or bipartite.

Theorem 3. *The 1-graph of D_n is bipartite.*

Proof. From (6), we have that

$$D_n = \{1, b, \cdots, b^{n-1}\} \cup \{ab, \ldots, ab^{n-1}\}. \quad (19)$$

Note that
$$d_1(b^i, b^{i+1}) = d_1(ab^i, ab^{i+1}) = 1, \quad (20)$$

then, the sets $\{1, b, \cdots, b^{n-1}\}$ and $\{a, ab, \cdots, ab^{n-1}\}$ form a bipartition of the vertex set D_n. □

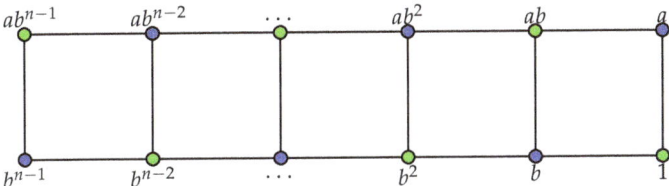

Figure 2. The 1-graph of D_n.

Theorem 2 leads to a complete characterization of the t-graphs associated with D_n. However, before characterizing the t-graphs on dihedral groups, let us first look at some useful lemmas.

Lemma 4. *Let $\mathcal{G} = (D_n, E)$ be the t-graph of D_n. Then,*

$$|E| = \begin{cases} 4(n - t) + 2 & \text{If } t > 1 \\ 3n - 2 & \text{If } t = 1. \end{cases} \quad (21)$$

Proof. Let $x = a^i b^j, y = a^k b^l \in D_n$, then, $0 \leq i, k \leq 1$ and $0 \leq j, l \leq n - 1$. If $d_1(x, y) = |i - k| + |j - l| = t$, then, for $|i - k|$, we have the following cases:

1. If $i = k$, then $|j - l| = t$. Note that there are $n - t$ ways to choose $j, l \in \{0, \ldots, n - 1\}$ such that the absolute value of their difference is t.
2. If $i \neq k$, then $|j - l| = t - 1$. In this case, there are $n - t + 1$ forms to choose $j, l \in \{0, \ldots, n - 1\}$ such that the absolute value of their difference is $t - 1$.

If $t > 1$, then there are $2(n - t) + 2(n - t + 1)$ ways of constructing an edge between two elements of D_n. Therefore, we have that $|E| = 4(n - t) + 2$.

If $t = 1$ then we the same argument we have that $|E| = 3n - 2$. □

Lemma 5. Let $f : D_n \longrightarrow D_n$ be defined as follows

$$f(a^i b^j) = \begin{cases} b^j & \text{If } i = 1 \\ ab^j & \text{If } i = 0. \end{cases} \tag{22}$$

Then, f is an isometry under the Minkowski metric (2). Further, if we restrict f to $U \subset D_n$, we have that U and $f(U)$ are also isometric under the Minkowski metric.

Proof. It is immediate that f is an injective function and $(f \circ f)(x) = x$, for all $x \in D_n$. That is, f is bijective. To prove that f is an isometry, let $a^i b^j, a^k b^l \in D_n$. Then,

1. If $i, k = 1$, then $d_1(f(a^i b^j), f(a^k b^l)) = d_1(b^j, b^l) = d_1(a^i b^j, a^k b^l)$.
2. If $i, k = 0$, then it is similar to the previous case.
3. If $i = 0$ and $k = 1$, then $d_1(f(a^i b^j), f(a^k b^l)) = d_1(ab^j, b^l) = d_1(a^i b^j, a^k b^l)$.
4. If $i = 1$ and $k = 0$, then it is similar to the previous case.

Therefore, f is an isometry on D_n. The other statement is clear. □

Theorem 4. *(Characterization of t-graphs on D_n)*
Let $\mathcal{G} = (D_n, E)$ the t-graph of D_n with $n \geq 2$. We define $r := \lceil \frac{n}{2} \rceil$.

1. If $t \leq r$ and t is an even number, then $k(\mathcal{G}) = 2$, and these connected components are isomorphic.
2. If $t \leq r$ and t is an odd number, then $\mathcal{G}$ is an connected graph.
3. If $t = r + s$, with $1 \leq s \leq n - r$, then the number $K(\mathcal{G})$ of connected components of $\mathcal{G}$ is given by

$$k(\mathcal{G}) = \begin{cases} 4(s-1) + 2 & \text{If } n \text{ is even} \\ 4s & \text{If } n \text{ is odd,} \end{cases} \tag{23}$$

where two of the connected components of $\mathcal{G}$ are an isomorphic path graph.

Proof.

1. It follows from Theorem 2 that $k(\mathcal{G}) = 2$. The connected components of $\mathcal{G}$ are $\mathcal{C}_1 = (V_1, E_1)$ and $\mathcal{C}_2 = (V_2, E_2)$, as in the proof of Theorem 2 (1). It is then sufficient to show that $\mathcal{C}_1 \cong \mathcal{C}_2$. Using the function f defined in Lemma 5, we have for $a^i b^j, a^k b^l \in V_1$ that

$$\{a^i b^j, a^k b^l\} \in E_1 \iff \{f(a^i b^j), f(a^k b^l)\} \in E_2, \tag{24}$$

   which leads to $\mathcal{C}_1 \cong \mathcal{C}_2$.
2. This follows immediately from Theorem 2.
3. We differentiate two cases:

   (a) Suppose t is an even number. The condition $t > r$ implies that $\mathcal{G}$ has isolated points, and then, using Theorem 2, we have that $\mathcal{G}$ has at least two connected components. Let $\mathcal{C}_1 = (V_1, E_1)$ and $\mathcal{C}_2 = (V_2, E_2)$ for the connected components constructed in the proof of Theorem 2 (1).
   We prove first that $|V_1| = |V_2|$. In fact, we have that $|j - l| = t$ or $|j - l| = t - 1$, which implies that

$$j \in \{t - 1, \ldots, n - 1\} \cup \{0, \ldots, n - t\} =: A, \tag{25}$$

   since $l \in \{0, \ldots, n - 1\}$.
   It is clear that $\{t - 1, \ldots, n - 1\} \cap \{0, \ldots, n - t\} = \emptyset$, therefore

$$|A| = 2(n - t) + 2. \tag{26}$$

   On the other hand, it follows immediately that $j \in A$ and $i + j$ are even numbers if and only if $a^i b^j \in V_1$, and then

$$|V_1| = |A| = 2(n-t) + 2. \tag{27}$$

Analogously, $|V_2| = |A|$, and we have $|V_1| = |V_2|$.
To demonstrate that $C_1 \cong C_2$, we consider again the function f defined in Lemma 5. Note that $f(V_1) = V_2$, and, since f is an isometry, we have the statement.
Finally, using Lemma 4, we have that $|E| = 4(n-t) + 2$, and the isomorphy between C_1 and C_2 implies that $|E_1| = |E_2|$. Further, note that the minimum value for $|E_1|$ and $|E_2|$ is $2(n-t) + 1$. This proves that C_1 and C_2 are the unique connected components of $\mathcal{G}$, which are not isolated points, and these are actually isomorphic path graphs.
The number of isolated points of $\mathcal{G}$ is $|D_n| - |V_1| - |V_2| = 2n - 4(n-t) - 4 = -2n + 4t - 4$, and, consequently, $k(\mathcal{G}) = -2n + 4t - 2 = -2n + 4r + 4s - 2$. That is,

- If n is even, then $k(\mathcal{G}) = -2n + 4(\frac{n}{2}) + 4s - 2 = 4(s-1) + 2$.
- If n is odd, then $k(\mathcal{G}) = -2n + 4(\frac{n+1}{2}) + 4s - 2 = 4s$.

(b) Suppose now that t is an odd number. Similar to before, the graph $\mathcal{G}$ has isolated points, and the set

$$\{\{a^i b^j, a^k b^l\} \mid i+j \equiv 0 \bmod 2, k+l \equiv 1 \bmod 2 \land |i-k| + |j-l| = t\}, \tag{28}$$

is a subset of E. Let V' be the set consisting of the non-isolated points of $\mathcal{G}$. Using the same argument as in (a), we obtain

$$|V'| = 2|A| = 4(n-t) + 4. \tag{29}$$

By Lemma 4, we have that $|E| = 4(n-t) + 2$, then, comparing $|V'|$ and $|E|$ excluding the isolated points, it follows that $\mathcal{G}$ cannot be connected.
Let m be an even number such that

$$\begin{cases} 0 \leq m \leq n-t-1 & \text{if } n-t-1 \text{ is even, and} \\ 0 \leq m \leq n-t & \text{if } n-t-1 \text{ is odd,} \end{cases} \tag{30}$$

and consider the subgraph $C_1 = (V_1, E_1)$ of $\mathcal{G}$ with the following edges:

$$\{ab^{t+m-1}, b^m\}, \{b^m, b^{t+m}\}, \{b^{t+m}, ab^{m+1}\}, \{ab^{m+1}, ab^{t+m+1}\}.$$

Then, C_1 is a connected component of $\mathcal{G}$, and, furthermore,

$$|V_1| = 2(n-t) + 2 \land |E_1| = 2(n-t) + 1, \tag{31}$$

whence it is concluded that C_1 is a path graph.
As before, using the function f from Lemma 5, we have that there exists another connected component $C_2 = (f(V_1), E_2)$, isomorphic to C_1. Thus,

$$|E_1| + |E_2| = |E| \land |V_1| + |V_2| = |V'|. \tag{32}$$

This means that C_1 and C_2 are the unique connected components of $\mathcal{G}$, and, analogously to the previous case, we have the same values for $k(\mathcal{G})$.

□

The following corollary is a generalization of Theorem 3.

Corollary 1. *Let G be a two-generator group in the form (10) with $n, m \geq 2$, and t be an odd number. Let further r be defined as in Theorem 2. If $t \leq r$, then $\mathcal{G} = (G, E)$ is a bipartite graph.*

Proof. From Theorem 4, we have that $\mathcal{G}$ is connected. Now, we define the sets V_1 and V_2 as follows

$$V_1 := \{a^i b^j \mid i + j \equiv 0 \bmod 2\} \tag{33}$$
$$V_2 := \{a^i b^j \mid i + j \equiv 1 \bmod 2\} \tag{34}$$

It is immediate to verify that V_1 and V_2 form a bipartition of G, and $\mathcal{G}$ is a bipartite graph. □

An illustration of the previous Corollary is presented in Figure 3.

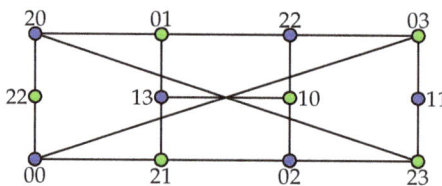

Figure 3. The 3-graph of $G = \mathbb{Z}_3 \times \mathbb{Z}_4$.

Corollary 2. *Let n be an odd number, $n \geq 5$ and $t = \frac{n+1}{2}$.*
1. *If t is odd, then $\mathcal{G} = (D_n, E)$ is a cycle of even length.*
2. *If t is even, then $\mathcal{G} = (D_n, E)$ is non-connected, and it has two isomorphic components, which are two cycles. Furthermore, $\chi(\mathcal{G}) = 3$.*

Proof. These statements follow directly from Theorem 4. Note that $t = r$.
1. From Lemma 4, it follows that

$$|E| = 4(n - (\tfrac{n+1}{2})) + 2 = 2n = |D_n|, \tag{35}$$

and then $\mathcal{G}$ is a cycle of even length.

2. $\mathcal{G}$ has two isomorphic connected components, say $\mathcal{C}_1 = (V_1, E_1)$ and $\mathcal{C}_2 = (V_2, E_2)$. Lemma 4 implies that

$$|E| = 4(n - (\tfrac{n+1}{2})) + 2 = 2n, \tag{36}$$

and it follows that $|E_1| = |E_2| = n$, so $\mathcal{G}$ is constituted by two isomorphic cycles. Finally, note that each component has an odd number of vertices. Then, $\chi(\mathcal{G}) = 3$. □

Corollary 3. *Let n be an even number, $n \geq 2$ and $t = \frac{n}{2} + 1$. Then, the t-graph of D_n consists of two isomorphic paths graphs.*

Proof. Using Theorem 4, and since n is an even number, we have that $r = \frac{n}{2}$, and then $t = r + 1$, and $k(\mathcal{G}) = 2$. The rest is clear. □

Corollary 4. *Let $n \geq 2$ and r be as in Theorem 4. Then, the t-graph of D_n is 2-chromatic if $t \leq r$ and t is an odd number or $t > r$.*

Proof. It follows immediately from Theorem 4 and Corollary 1. □

Corollary 5. *The n-graph of D_n has $2(n-1)$ connected components, and two of these are path graphs with two vertices.*

Proof. Let $\mathcal{G} = (D_n, E)$ be the n-graph of D_n. From Lemma 4, it follows that $|E| = 2$. Note that

$$\{a, b^{n-1}\}, \{ab^{n-1}, 1\} \in E. \tag{37}$$

The other $2n - 4$ elements of D_n are isolated points, and the proof is complete. □

4. Some Questions and Conjectures

Some open questions and conjectures are presented below.

Question 1. *Is it possible to characterize the t-graphs on two-generator groups, when $t > r$ and r is as in Theorem 2?*

Question 2. *Is it possible to generalize a version of Theorem 2 for an n-generator group for n, an arbitrary natural number?*

Question 3. *It is possible to determine in a finite group the existence (or not) of a generating system with the conditions stated for the definition of the t-graphs?*

Conjecture 1. *With respect to Theorem 2, if m is an even number and $t \leq r$, it follows that the two connected components of the t-graph $\mathcal{G}$ are isomorphic.*

Conjecture 2. *Let $n \geq 2$ and r be as in Theorem 4. Then, the t-graph of D_n is 3-chromatic, if $t \leq r$ and t is an even number.*

Conjecture 3. *If $G = \mathbb{Z}_n \times \mathbb{Z}_2$, then $K(G) = 2(2t - n) - 2$.*

Conjecture 4. *If $G = \mathbb{Z}_n \times \mathbb{Z}_3$, then $K(G) = 3(2t - n - 2) - 2$.*

5. Discussion

In the present research, we introduce and investigate the t-graph on a finitely generated group G. It leads to an interesting combinatorial problem. We establish conditions for t to guarantee the existence of isolated points in the t-graph when G is a two-generator group. We also propose an expression to determine the number of the connected components of the t-graph. Other results have to do with the conditions that must be fulfilled for the t-graphs of the dihedral groups to be a path graph or a cycle. Consequently, we can characterize the chromatic number of the t-graph depending exclusively on the parity of t.

Author Contributions: Conceptualization, I.G.; methodology, I.G., G.D.-P. and A.T.-G.; software, A.T.-G.; validation, I.G., G.D.-P. and A.T.-G.; formal analysis, I.G. and A.T.-G. investigation, I.G., G.D.-P. and A.T.-G.; resources, I.G., G.D.-P. and A.T.-G.; writing—original draft preparation, I.G., G.D.-P. and A.T.-G.; writing—review and editing, I.G. All authors have read and agreed to the published version of the manuscript.

Funding: This research received no external funding.

Institutional Review Board Statement: Not applicable.

Informed Consent Statement: Not applicable.

Data Availability Statement: Not applicable.

Conflicts of Interest: The authors declare no conflict of interest.

References

1. Cayley, A. Desiderata and suggestions: No. 2. The Theory of groups: Graphical representation. *Am. J. Math.* **1878**, *1*, 403–405. [CrossRef]
2. Neumann, B.H. A problem of Paul Erdös on groups. *J. Aust. Math. Soc. (Ser. A)* **1976**, *21*, 467–472. [CrossRef]
3. Abdollahi, A.; Zarrin, M. Non-nilpotent graph of a group. *Commun. Algebra* **2010**, *38*, 4390–4403. [CrossRef]
4. Ballester-Bolinches, A.; Cossey, J.; Esteban-Romero, R. On a graph related to permutability in finite groups. *Ann. Mat. Pura Appl.* **2010**, *189*, 567–570. [CrossRef]
5. Ballester-Bolinches, A.; Cossey, J.; Esteban-Romero, R. Graphs and Classes of Finite Groups. *Note Mat.* **2013**, *33*, 89–94.
6. Ballester-Bolinches, A.; Cossey, J. Graphs, partitions and classes of groups. *Monatshefte Math.* **2012**, *166*, 309–318. [CrossRef]

7. Ballester-Bolinches, A.; Cossey, J.; Esteban-Romero, R. A characterization via graphs of the soluble groups in which permutability is transitive. *Algebra Discret. Math.* **2009**, *4*, 10–17.
8. Ballester-Bolinches, A.; Cosme-Llópez, E.; Esteban-Romero, R. Group extensions and graphs. *Expo. Math.* **2016**, *34*, 327–334. [CrossRef]
9. Lucchini, A. The independence graph of a finite group. *Monatshefte Math.* **2020**, *193*, 845–856. [CrossRef]
10. Lucchini, A.; Nemmi, D. The non-$\mathfrak{F}$ graph of a finite group. *Math. Nachrichten* **2021**, *294*, 1912–1921. [CrossRef]
11. Hai-Reuven, D. Non-Solvable Graph of a Finite Group and Solvabilizers. *arXiv* **2013**, arXiv:1307.2924v1.
12. Huppert, B. *Endliche Gruppen I*; Springer: Berlin, Germany, 1967.
13. Sherman-Bennett, M. *On Groups and Their Graphs*; Bard College at Simon's Rock: Great Barrington, MA, USA, 2016.
14. Díaz-Porto, G. Caracterización de *t*-Grafos de Distancia Sobre Grupos Finitamente Generados. Bachelor's Thesis, Universidad del Norte, Barranquilla, Colombia, 2020.
15. Torres-Grandisson, A. Estudio de los *t*-Grafos de Distancia Definidos Sobre Grupos 2-Generados. Bachelor's Thesis, Universidad del Norte, Barranquilla, Colombia, 2021.
16. Nica, B. *A Brief Introduction to Spectral Graph Theory*; EMS Textbooks in Mathematics; European Mathematical Society: Zürich, Switzerland, 2018.

Article
On Some Properties of Addition Signed Cayley Graph $\Sigma_n^{\wedge}$

Obaidullah Wardak [1], Ayushi Dhama [2] and Deepa Sinha [1,*]

[1] Department of Mathematics, South Asian University, New Delhi 110 021, India
[2] Centre for Mathematical Sciences, Banasthali University, Rajasthan 304 022, India
* Correspondence: deepasinha@sau.ac.in

Abstract: We define an addition signed Cayley graph on a unitary addition Cayley graph G_n represented by $\Sigma_n^{\wedge}$, and study several properties such as balancing, clusterability and sign compatibility of the addition signed Cayley graph $\Sigma_n^{\wedge}$. We also study the characterization of canonical consistency of $\Sigma_n^{\wedge}$, for some n.

Keywords: addition signed Cayley graph $\Sigma_n^{\wedge}$

MSC: 05C 22; 05C 75

1. Introduction

We refer to standard books of Harary [1] and West [2] for graph theory. For the signed graphs, we refer to Zaslavsky [3,4]. All the signed graphs considered in this paper are simple, finite and loopless.

For the preliminaries, definition and notation of signed graph S, underlying graph S^u, its negation $\eta(S)$, signed isomorphism and its positive (negative) section, we refer to [5,6].

Some Basic Lemma and Theorems which are used in this paper are stated below as a reference.

Lemma 1 ([7]). *A signed graph in which every chordless cycle is positive is balanced.*

Theorem 1 ([8]). *A signed graph S is clusterable if—and only if—S does not contains a cycle with exactly one negatively charged edge.*

For balancing, clusterability, marking, canonical marking ($\mathcal{C}$-marking), consistency, $\mathcal{C}$-consistency, S consistency, sign compatibility, line signed graph $L(S)$, line signed root graph, ×-line signed graph, ×-line signed root graph and the common-edge signed graph $C_E(S)$ of signed graph, S we refer to [6,9–16].

Addition Signed Cayley Graph $\Sigma_n^{\wedge}$

A *unitary addition Cayley graph* G_n, where $n \in I^+$, I^+ is set of positive integers, is a graph in which the vertex set is a ring of integers modulo n, Z_n. Any two vertices x_1 and x_2 are adjacent in G_n if—and only if—$(x_1 + x_2) \in U_n$, where U_n denotes the unit set.

Unitary addition Cayley graphs for n = 2, 3, 4, 5, 6 and 7 are shown in Figure 1.

The study of unitary Cayley graphs began in order to gain some insight into the graph representation problem (see [17]), and we can extend it to the signed graphs (see [18]). Now, we introduce the definition of an *addition signed Cayley graph* $\Sigma_n^{\wedge}$ as follows:

The *addition signed Cayley graph* $\Sigma_n^{\wedge} = (G_n, \sigma^{\wedge})$ is a signed graph whose underlying graph is a unitary addition Cayley graph G_n, where $n \in I^+$ and for an edge ab of $\Sigma_n^{\wedge}$,

$$\sigma^{\wedge}(ab) = \begin{cases} + & \text{if } a, b \in U_n, \\ - & \text{otherwise.} \end{cases}$$

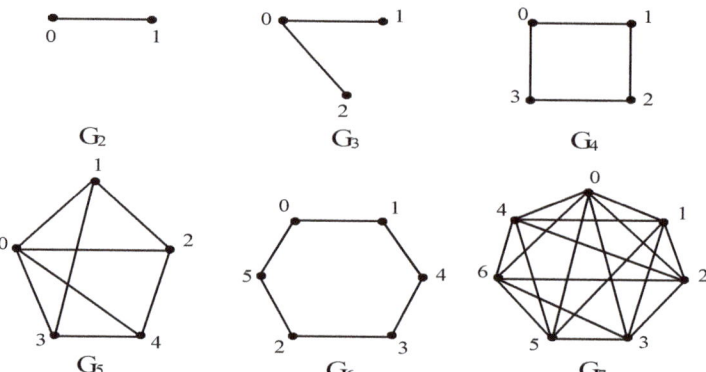

Figure 1. Examples of unitary addition Cayley graphs.

Examples of addition signed Cayley graph for $n = 5, 6$ and 10 can be seen in Figure 2a–c. Throughout the paper, we consider $n \geq 2$.

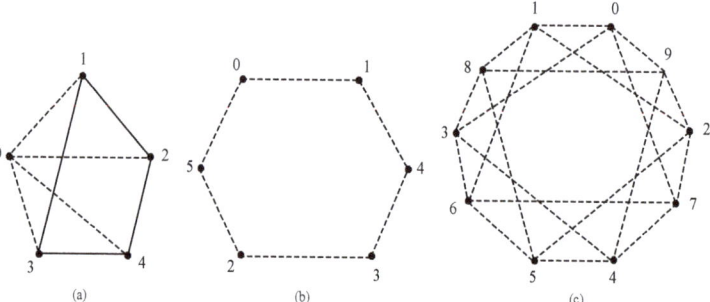

Figure 2. Examples of addition signed Cayley graph $\Sigma_n^{\wedge}$.

2. Some Properties of $\Sigma_n^{\wedge}$

2.1. Balancing in $\Sigma_n^{\wedge}$

The balancing of some derived signed Cayley graphs has been studied in the literature (see [19]). Here, we find out the property of balancing for the addition signed Cayley graph $\Sigma_n^{\wedge}$, for which the following well-known result can be used as a tool.

Theorem 2 ([20])**.** G_n, $n \geq 2$, is bipartite if—and only if—either $n = 3$ or n is even.

Lemma 2. $i \in U_n \Rightarrow (n - i) \in U_n$ and $i \notin U_n \Rightarrow (n - i) \notin U_n$.

Lemma 3. Addition signed Cayley graph $\Sigma_n^{\wedge} = (G_n, \sigma^{\wedge})$, for n even, is an all-negative signed graph.

Proof. Given an addition signed Cayley graph $\Sigma_n^{\wedge} = (G_n, \sigma^{\wedge})$, where n is even. Suppose the conclusion is false. Let there be a positive edge, say ij, in $\Sigma_n^{\wedge}$. By the definition of $\Sigma_n^{\wedge}$, $i, j \in U_n$. Since n is even, U_n consists only of odd numbers. Thus, i and j are odd numbers and their addition $i + j$ is an even number. This shows that $i + j \notin U_n$, i.e., i and j, are not adjacent in $\Sigma_n^{\wedge}$. Thus, we have a contradiction. Hence, if n is even, then $\Sigma_n^{\wedge}$ is all-negative signed graph. □

Sampathkumar [21] gave the famous characterization to prove the balancing in a signed graph, which is as follows:

Theorem 3 (Marking Criterion [21]). *A signed graph $S = (G, \sigma)$ is balanced if—and only if—there exists a marking μ of its vertices such that each edge uv in S satisfies $\sigma(uv) = \mu(u)\mu(v)$.*

Lemma 4. *For the addition signed Cayley graph $\Sigma_n^{\wedge} = (G_n, \sigma^{\wedge})$, $\Sigma_n^{\wedge}$ is a balanced signed graph, if for any prime p, $n = p^a$.*

Proof. $n = p^a$, where p is a prime number. Now, we assign a marking μ to the vertices of $\Sigma_n^{\wedge}$ in such a manner that if $u \in U_n$, then $\mu(u) = +$ and if $u \notin U_n$, then $\mu(u) = -$, $\forall u \in V(\Sigma_n^{\wedge})$. Suppose there is an edge, say ij, in $\Sigma_n^{\wedge}$.
  Case I: Let $\sigma^{\wedge}(ij) = +$. Then, $i, j \in U_n$ and according to the marking $\mu(i) = \mu(j) = +$. Thus, $\sigma^{\wedge}(ij) = \mu(i)\mu(j) = +$.
  Case II: Let $\sigma^{\wedge}(ij) = -$. Then, there are three possibilities:

(a) $i \in U_n, j \notin U_n$.
(b) $i \notin U_n, j \in U_n$.
(c) $i, j \notin U_n$.

Now, for (a) and (b), by marking μ, we get $\mu(j) = -$ and $\mu(i) = +$ or vice versa. Therefore, $\sigma^{\wedge}(ij) = \mu(i)\mu(j) = -$. Now, if $i, j \notin U_n$. Then, i and j are both multiples of p, and then $i + j = kp$, where k is some positive integer and $i + j \notin U_n$. So $ij \notin E(\Sigma_n^{\wedge})$. Thus, condition (c) is not possible. So in every condition we get $\sigma^{\wedge}(ij) = \mu(i)\mu(j)$. Since ij is an arbitrary edge, using Theorem 3, $\Sigma_n^{\wedge}$ is balanced. □

Theorem 4. *The addition signed Cayley graph $\Sigma_n^{\wedge}$ is balanced if—and only if—either n is even or if n has exactly one prime factor, then n is odd.*

Proof. *Necessity*: First, suppose $\Sigma_n^{\wedge}$ is balanced. Now, let $n = p_1^{\alpha_1} p_2^{\alpha_2} \ldots p_m^{\alpha_m}$; $p_1, p_2, \ldots, p_m$ being distinct primes, $p_1 \neq 2$, $p_1 \leq p_2 \leq \ldots \leq p_m$.
  In the unitary addition Cayley graph G_n, $p_1 + 1 \neq k_1 p_i$ for $i = 1, 2, \ldots, m$ and k_1 are some positive integers i.e., $p_1 + 1 \in U_n$, so p_1 is adjacent with one. Now, we claim that p_1 and p_2 are adjacent in G_n. On the contrary, suppose $p_1 p_2$ is not an edge in G_n. Then, $p_1 + p_2 \notin U_n$. Thus, $p_1 + p_2 = k_2 p_i$ for some $i = 1, 2, \ldots, m$ and k_2 are some positive integers. Let $p_1 + p_2$ be a multiple of p_1.

$$p_1 + p_2 = \alpha p_1$$
$$p_2 = \alpha p_1 - p_1$$
$$= (\alpha - 1) p_1$$

for the positive integer α, a contradiction. With the same argument, we can show that $p_1 + p_2$ is not a multiple of p_2. Now, let $p_1 + p_2 = \alpha p_i$, for $i = 3, 4, \ldots, m$. As we know, the addition of two prime factors is always even; $p_1 + p_2$ is even. So, α is even and is at least 2. However, as $p_1 < p_2 < p_i$, $p_1 + p_2$ is always less than any multiple of p_i for $i = 3, 4, \ldots, m$. Thus, $p_1 + p_2 \in U_n$ and $p_1 p_2$ is an edge in G_n. Next, if p_2 is adjacent to 1 in G_n, we get a cycle

$$C = (p_1, p_2, 1, p_1)$$

in $\Sigma_n^{\wedge}$. Clearly, p_1 and p_2 are not in U_n, then by definition of $\Sigma_n^{\wedge}$, C is a negative cycle. Thus, we have a negative cycle in $\Sigma_n^{\wedge}$, implying that $\Sigma_n^{\wedge}$ is not balanced. Now, suppose $p_2 + 1 \notin E(G_n)$, since $p_2 + 1 \notin U_n$. Then, $p_2 + 1 = cp_i$; $i = 1, 2, \ldots, m$, c are positive integers. Clearly,

$$p_2 + 1 = \alpha p_1 \tag{1}$$

α is a positive integer.
  Since $p_2 \notin U_n$, according to Lemma 2, $n - p_2 \notin U_n$. Next, we claim that $n - p_2$ is adjacent to 1 or $n - p_2 + 1 = n - (p_2 - 1) \in U_n$. If $p_2 - 1 \in U_n$, then according to Lemma 2, $n - p_2 + 1 = n - (p_2 - 1) \in U_n$. Suppose $p_2 - 1 \notin U_n$. Then, $p_2 - 1 = \beta p_i$; $i = 1, 2, \ldots, m$,

β are positive integers. Let $p_2 - 1 = \beta p_1$. However, from Equation (1), $p_2 = \alpha p_1 - 1$. This implies

$$p_2 - 1 = \beta p_1$$
$$\alpha p_1 - 1 - 1 = \beta p_1$$
$$\alpha p_1 - 2 = \beta p_1$$
$$\alpha p_1 - \beta p_1 = 2$$
$$(\alpha - \beta) p_1 = 2.$$

This is not possible, as p_1 is at least 3. Thus, $p_2 - 1$ is not a multiple of any of the p_is, whence $p_2 - 1 \in U_n$. Hence, $n - p_2 + 1 = n - (p_2 - 1) \in U_n$, whence $n - p_2$ is adjacent to 1 in G_n. Now, $n - p_2 + p_1 = n - (p_2 - p_1)$. Since $p_1 < p_2 < \cdots p_m$, $p_2 - p_1 \neq k p_i$; $i = 2, 3, \ldots m$, k is a positive integer. Additionally, $p_2 - p_1$ is not a multiple of p_1. This shows that $p_2 - p_1 \in U_n$ and by Lemma 2, $n - (p_2 - p_1) \in U_n$. This shows that $n - p_2$ is adjacent to p_1 in Σ_n. Thus, we get a cycle

$$C' = (p_1, n - p_2, 1, p_1)$$

in Σ_n. Clearly, p_1 and $n - p_2$ do not belong to U_n and $1 \in U_n$. Then, by definition $\Sigma_n^{\wedge}$, we have a cycle C' with three negative edges. Thus, a contradiction. So, by contraposition, necessity is true.

Sufficiency: Let n be even. Then, according to Lemma 3, $\Sigma_n^{\wedge}$ is an all-negative signed graph. Additionally, according to Theorem 2, G_n is a bipartite graph. Hence, $\Sigma_n^{\wedge}$, by Lemma 3 and Theorem 2, is balanced.

Now, let n be odd, with exactly one prime factor. Then, according to Lemma 4, $\Sigma_n^{\wedge}$ is balanced, hence the theorem. □

2.2. Clusterability in $\Sigma_n^{\wedge}$

Theorem 5. *The addition signed Cayley graph $\Sigma_n^{\wedge} = (G_n, \sigma^{\wedge})$ is clusterable.*

Proof. Given an addition signed Cayley graph $\Sigma_n^{\wedge} = (G_n, \sigma^{\wedge})$. Suppose $v \in V(\Sigma_n^{\wedge})$. Define $V^* \subseteq V(\Sigma_n^{\wedge})$, such that $V^* = \{u_i : u_i \in V(\Sigma_n^{\wedge}) \text{ and } \sigma^{\wedge}(vu_i) = +\}$. By the definition of $\Sigma_n^{\wedge}$, clearly u_i and v are in U_n.

If, for i and j, $(i \neq j)$, u_i and u_j are adjacent, then $\sigma^{\wedge}(u_i u_j) = +$. Thus, $U_n \subseteq V^*$. Since $|U_n| = \phi(n)$, $n - \phi(n) = k$ (say) vertices are not in U_n. Thus, only negative edges are incident on these k vertices. Put all these vertices in the k partition $V_1, V_2, \ldots, V_k$, such that each partition contains exactly only one vertex. The clearly induced subgraph $< V^* >$ is all positive. Additionally, no positive edge joins the vertex of V^* with the vertex of any of V_i, for $i = 1, 2, \ldots, k$, and there is no edge xy, such that $\sigma^{\wedge}(xy) = -$ and $x, y \in V^*$. Thus, there exists a partition of the $V(\Sigma_n^{\wedge})$, such that every positive edge has end vertices within the same subset and every negative edge has end vertices in a different subset. Hence, the proof. □

2.3. Sign-Compatibility in $\Sigma_n^{\wedge}$

Theorem 6 ([22]). *A signed graph S is sign compatible if—and only if—S does not contain a sub signed graph isomorphic to either of the two signed graphs. S_1 formed by taking the path $P_4 = (x, u, v, y)$ with both edges xu and vy negative and edge uv positive, and S_2 formed by taking S_1 and identifying the vertices x and y (Figure 3).*

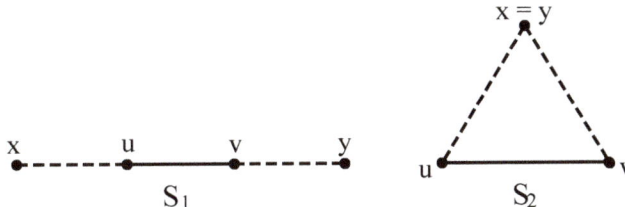

Figure 3. Two forbidden sub signed graphs for a sign-compatible signed graph [13].

Theorem 7. *Addition signed Cayley graph $\Sigma_n^{\wedge}$ is sign compatible if—and only if—n is 3 or even.*

Proof. Let addition signed Cayley graph $\Sigma_n^{\wedge}$ be sign compatible. If possible, suppose the conclusion is not true. Let n be odd but not 3. Now, $01 \in E(\Sigma_n^{\wedge})$. As, $n - 2 + 1 = n - 1 \in U_n$, $1(n - 2) \in E(\Sigma_n^{\wedge})$. Additionally, $n - 2 + 0 = n - 2 \in U_n$. Thus, we have a triangle $(0, 1, n - 2, 0)$ with one positive edge $1(n - 2)$ and two negative edges 01 and $(n - 2)0$, which again contradict Theorem 6. Hence, the condition is necessary.

Next, let n be even. Thus, according to Lemma 3, $\Sigma_n^{\wedge}$, which is all-negative, is trivially sign compatible. If $n = 3$, then $\Sigma_n^{\wedge}$ is P_3, which is trivially sign compatible. □

Acharya and Sinha [23] showed that every line signed graph is sign compatible. Next, we discuss the value of n for which $\Sigma_n^{\wedge}$ is a line signed graph.

Theorem 8. *G_n is a line graph if—and only if—n is equal to 2 or 3 or 4 or 6.*

Proof. *Necessity*: Let G_n be a line graph. Meanwhile, n is not equal to 2, 3, 4 and 6.

Case I: n is prime. It is clear that $n \geq 5$. Here, n is prime, so by the definition of U_n, there are numbers from 1 to $(n - 1)$ in U_n. 0 is connected to every vertex of G_n. The other vertex, $i \neq 0$, in G_n is not connected to only $(n - i)$ by definition. For any $i, j \in V(G_n)$; $i \neq 0, j \neq 0$ there is an induced subgraph in G_n (see Figure 4).

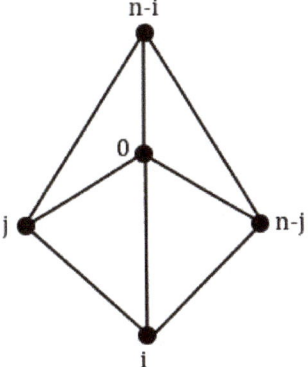

Figure 4. A forbidden subgraph for a line graph in G_n.

Thus, G_n contains forbidden subgraph for a line graph. Thus, G_n is not a line graph.

Case II: n is not prime. 1 is connected to 0 in G_n. Next, 1 is connected to p_1, as $p_1 + 1 \in U_n$, where p_1 is the smallest factor of n. Let $\alpha p_1 = n$, for a positive integer α. Now,

$$1 + (\alpha - 1)p_1 = 1 + \alpha p_1 - p_1$$
$$= 1 + n - p_1$$
$$= n - (p_1 - 1).$$

Since $p_1 - 1 \in U_n$, by Lemma 2, $n - (p_1 - 1) \in U_n$. Thus, 1 and $(a-1)p_1$ are adjacent in G_n. Additionally, 0 is not adjacent to p_1 and $(a-1)p_1$, because their sum is a multiple of p_1. In the same way, p_1 and $(a-1)p_1$ are not connected in G_n because their sum is a multiple of p_1. So, we have an induced subgraph in G_n (see Figure 5). Thus, there is a forbidden subgraph $K_{1,3}$ of a line graph. Additionally, G_n is not a line graph.

Sufficiency: Let $n = 2$ or $n = 3$ or $n = 4$ or $n = 6$. Then, $G_2 \cong L(P_3)$, $G_3 \cong L(P_4)$, $G_4 \cong L(C_4)$ and $G_6 \cong L(C_6)$ (see Figure 6). Hence, the result. □

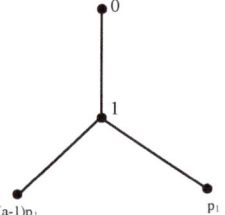

Figure 5. A forbidden subgraph for a line graph in G_n.

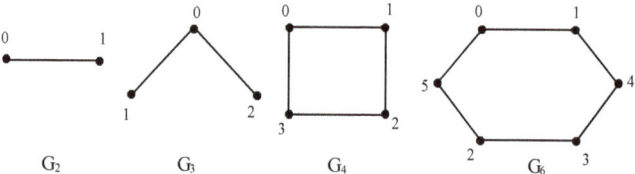

Figure 6. Showing G_2, G_3, G_4 and G_6.

Theorem 9. $\Sigma_n^\wedge$ *is a line signed graph if—and only if—$n = 2$ or $n = 3$ or $n = 4$ or $n = 6$.*

Proof. Necessity: Let, if possible, n be unequal to 2, 3, 4 and 6. Theorem 8 shows that $G_n \not\cong L(G)$, for any graph G. Thus, a contradiction and the condition are necessary.

Sufficiency: Now, suppose $n = 2$ or $n = 3$ or $n = 4$ or $n = 6$. Line signed graphs of an addition signed Cayley graph, for these values of n, are displayed in Figure 7, hence the sufficiency. □

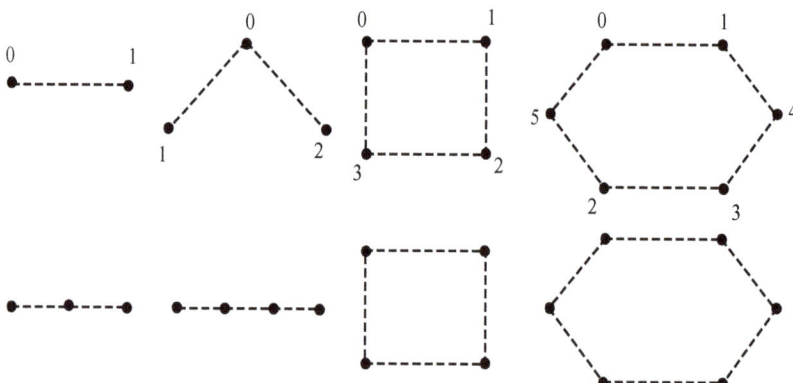

Figure 7. Showing $\Sigma_2^\wedge$, $\Sigma_3^\wedge$, $\Sigma_4^\wedge$ and $\Sigma_6^\wedge$ and its line signed root graphs.

Remark 1. $\Sigma_n^\wedge$ *is a $\times$-line signed graph if—and only if—$n = 2$ or $n = 3$ or $n = 4$ or $n = 6$.*

Proof. Let $\Sigma_n^\wedge$ be a $\times$–line signed graph. We know that the underlying structure for line signed graphs and $\times$–line signed graphs is the same. Thus, the condition comes from Theorem 8.

Next, let $n \in \{2, 3, 4, 6\}$. $\Sigma_2^\wedge, \Sigma_3^\wedge, \Sigma_4^\wedge$ and $\Sigma_6^\wedge$ and its $\times$–line signed root graphs are displayed in Figure 8. From Theorem 4, it is clear that for these values of n, an addition signed Cayley graph is balanced. Additionally, $L_\times(S)$ of any signed graph is always balanced, and its underlying graph is a line graph (see [24]). This result comes from Theorems 4 and 8. □

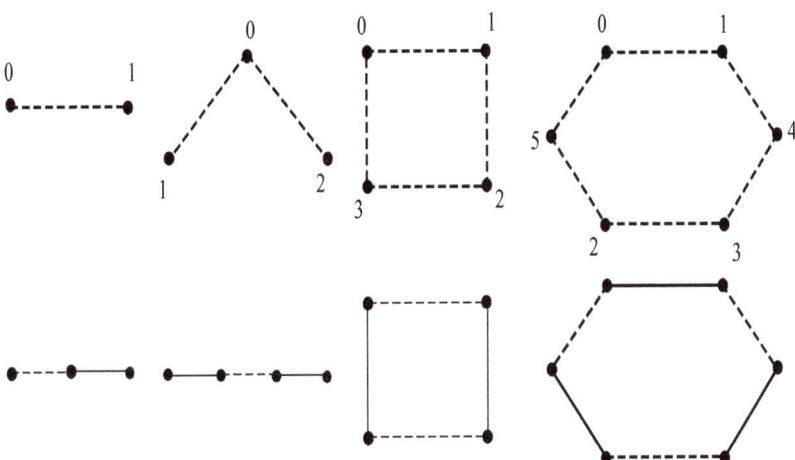

Figure 8. Showing $\Sigma_2^\wedge, \Sigma_3^\wedge, \Sigma_4^\wedge$ and $\Sigma_6^\wedge$ and its $\times$–line signed root graphs.

2.4. C-Consistency of $\Sigma_n^\wedge$

Lemma 5. *For any prime p, $p \neq 2$ and $n = p^\alpha$, the $d^-(2)$ and $d^-(4)$ in $\Sigma_n^\wedge$ is odd.*

Proof. Given a $\Sigma_n^\wedge = (G_n, \sigma^\wedge)$, where $n = p^\alpha$ and p is an odd prime. Since n is odd; 2, $4 \in U_n$. It is obvious that $d^-(2)$ and $d^-(4)$ in $\Sigma_n^\wedge$ appear only when 2 and 4 are adjacent to kp, where k is some positive integer. Now, $(2+4) + cp \neq kp$; positive integers c and k. Additionally, 2 and 4 are connected to all the multiples of p, which are $p^{\alpha-1}$. Therefore $d^-(2)(d^-(4)) = p^{\alpha-1}$ is odd, hence the lemma. □

Theorem 10 ([25]). *Let a, b and m be integers with m positive. The linear congruence $ax \equiv b \pmod{m}$ is soluble if and only if $(a, m) | b$. If x_0 is a solution, there are exactly (a, m) incongruent solutions given by $\{x_0 + tm/(a,m)\}$, where $t = 0, 1, \ldots, (a, m) - 1$.*

Corollary 1. *If $(a, m) = 1$ then the congruence $ax \equiv b \pmod{m}$ has exactly one incongruent solution.*

Lemma 6. *In addition, signed Cayley graph $\Sigma_n^\wedge = (G_n, \sigma^\wedge)$, if $n = p_1^{a_1} p_2^{a_2}$, where p_1 and p_2 are two distinct odd primes, then $d^-(2)(d^-(4)) = odd$.*

Proof. Given that $n = p_1^{a_1} p_2^{a_2}$ in $\Sigma_n^\wedge$, p_1 and p_2 are distinct odd primes. As n is odd, $2 \in U_n$. Now, the negative degree of 2 of $\Sigma_n^\wedge$ appears only when 2 is adjacent with the multiples of p_1 and p_2. Let $A_i = \{cp_i; c \text{ certain positive integers}, i = 1, 2\}$. Then,

$$|A_1| = p_1^{a_1 - 1} p_2^{a_2}$$
$$|A_2| = p_1^{a_1} p_2^{a_2 - 1}$$

and
$$|A_1 \cap A_2| = p_1^{a_1-1} p_2^{a_2-1}$$

Thus, using the inclusion–exclusion principle
$$|A_1 \cup A_2| = p_1^{a_1-1} p_2^{a_2} + p_1^{a_1} p_2^{a_2-1} - p_1^{a_1-1} p_2^{a_2-1}$$

Since $cp_1(p_2) + 2 = p_2(p_1)$, for certain positive integers c and so, $cp_1(p_2)2 \notin E(\Sigma_n^\wedge)$ for those c. Thus, according to Theorem 10, we have

$$p_1 x \equiv -2 \pmod{p_2} \tag{2}$$

and

$$p_2 y \equiv -2 \pmod{p_1} \tag{3}$$

Due to Corollary 1, we have an incongruent solution x_0 (say), which is unique for Equation (2). So, for Equation (2) where $p_1 x + 2 < n$, we have:

$$x_0 + 0(p_2), x_0 + 1(p_2), x_0 + 2(p_2), \ldots, x_0 + (p_1^{a_1-1} p_2^{a_2-1} - 1)(p_2) \tag{4}$$

Thus, Equation (2) has $p_1^{a_1-1} p_2^{a_2-1}$ total solutions. Similarly, the total solutions of Equation (3) are $p_1^{a_1-1} p_2^{a_2-1}$. Hence,

$$d^-(2) = p_1^{a_1-1} p_2^{a_2} + p_1^{a_1} p_2^{a_2-1} - p_1^{a_1-1} p_2^{a_2-1} - p_1^{a_1-1} p_2^{a_2-1} - p_1^{a_1-1} p_2^{a_2-1}$$
$$= p_1^{a_1-1} p_2^{a_2-1}(p_1 + p_2 - 3)$$

p_1 and p_2 are odd primes, which implies $d^-(2)$ is odd. The proof for $d^-(4)$ is analogous. □

Lemma 7. *In $\Sigma_n^\wedge = (G_n, \sigma^\wedge)$, if $n = 3^{a_1} 5^{a_2}$, then $d^-(7) = $ odd.*

Proof. This is easy to prove using the same logic as mentioned in Lemma 6. □

Theorem 11. *Let n have at most two distinct odd prime factors, then $\Sigma_n^\wedge$ is $\mathcal{C}$ consistent if—and only if—n is even or 3.*

Proof. *Necessity*: Let n have, at most, two distinct prime factors and let $\Sigma_n^\wedge$ be $\mathcal{C}$ consistent. If possible, let n be odd but not 3.

Case (a): Let $n \equiv 1 \pmod 3$ or $n \equiv 2 \pmod 3$. As n is odd, $2 \in U_n$. Clearly, 0 is adjacent with $1, 2, n-1$ in $\Sigma_n^\wedge$. Since, $n - 1 + 2 = 1 \in U_n$, $n-1$ and 2 are connected in $\Sigma_n^\wedge$. Since, 3 is not a factor of n, $3 \in U_n$. Now, $2 + 1 = 3 \in U_n$. Hence, 2 and 1 are adjacent in $\Sigma_n^\wedge$. Now, the cycles $Z_1 = (0, 1, 2, 0)$, $Z_2 = (0, 2, n-1, 0)$ have a common chord with end vertices 0 and 2. By Lemma 6,

$$\mu_\sigma(2) = -$$

Since the vertex $0 \notin U_n$, $d(0) = d^-(0) = \phi(n) = $ even. It follows,

$$\mu_\sigma(0) = +.$$

Now, if either Z_1 or Z_2 is not a $\mathcal{C}$-consistent cycle, a contradiction. Thus, Z_1 and Z_2 both cycle are $\mathcal{C}$-consistent. The common chord with end vertices zero and two are oppositely marked, in contradiction with (Theorem 2, [26]).

Case (b): Let $n \equiv 0 \pmod 3$. Then, either $n = 3^{a_1}$ or $n = 3^{a_1} \times p_2^{a_2}$. First, suppose $p_2 \neq 5$. Since, n is odd, $2, 4 \in U_n$. According to Lemma 2, $n - 2 \in U_n$. Clearly, 0 is adjacent to 1, 4 and $n - 2$ in $\Sigma_n^\wedge$. Since, $n - 2 + 4 = n + 2 = 2 \in U_n$, $n - 2$ is adjacent to 4 in $\Sigma_n^\wedge$. Now, for cycle $Z_1 = (0, 1, 4, 0)$, $Z_2 = (0, 4, n - 2, 0)$; Z_1, Z_2 have a common chord with end vertices 0 and 4. According to Lemma 6,

$$\mu_\sigma(4) = -$$

Since the vertex $0 \notin U_n$, $d(0) = d^-(0) = \phi(n) = $ even. It follows,

$$\mu_\sigma(0) = +.$$

Now, if either Z_1 or Z_2 is a cycle which is not $\mathcal{C}$ consistent, a contradiction. Therefore, Z_1 and Z_2 are the cycles which are $\mathcal{C}$-consistent. However, there is a chord whose end vertices 0 and 4 have opposite marking. Here again, we find a contradiction to the (Theorem 2, [26]).

Now, suppose $p_2 = 5$. In this case, we consider two cycles $Z_1 = (0, 1, 7, 0)$ and $Z_2 = (0, 7, 10, 13, 0)$ in $\Sigma_n^\wedge$. For cycles Z_1, Z_2 have a common chord with end vertices 0 and 7, according to Lemma 7,

$$\mu_\sigma(7) = -$$

Since the vertex $0 \notin U_n$, $d(0) = d^-(0) = \phi(n) = $ even. It follows that

$$\mu_\sigma(0) = +.$$

Now, if either Z_1 or Z_2 is a cycle which is not $\mathcal{C}$ consistent, this is a contradiction. Therefore, Z_1 and Z_2 are the cycles which are $\mathcal{C}$ consistent. However, the end vertices 0 and 7 have the opposite marking. Here, we have a contradiction to the (Theorem 2, [26]). Hence, n is either even or $n = 3$.

Sufficiency: Let n be even. According to Lemma 3, $\Sigma_n^\wedge$ is all negative. Additionally, according to Theorem 13, $d(v) = d^-(v) = $ even $\forall v \in V(\Sigma_n^\wedge)$. So, according to canonical marking $\mu_\sigma(v) = +$ $\forall v \in V(\Sigma_n^\wedge)$. So when n is even, $\Sigma_n^\wedge$ is trivially $\mathcal{C}$ consistent. If $n = 3$, then G_3 is a path, which is trivially $\mathcal{C}$-consistent, hence the result. □

3. Balance in Certain Derived Signed Graphs of $\Sigma_n^\wedge$

Theorem 12. $\eta(\Sigma_n^\wedge)$ *is balanced if—and only if—n is 3 or even.*

Proof. Let $\eta(\Sigma_n^\wedge)$ be balanced. If possible, n is odd but not 3, and p is the smallest prime factor of n. Since $n - 2 + 1 = n - 1 \in U_n$, $n - 2$ and 1 are connected in $\Sigma_n^\wedge$. $p + 1 \in U_n$ implies that p and 1 are connected in $\Sigma_n^\wedge$. Additionally, as n is odd, $2 \in U_n$ and $n - 2 \in U_n$. according to Lemma 2. Since, $n - 2 + p = n + (p - 2) = p - 2 \in U_n$, $(n - 2)p \in E(\Sigma_n^\wedge)$. Now, for the cycle $Z = (1, p, n - 2, 1)$ in $\Sigma_n^\wedge$ we have a one positive edge $1(n - 2)$ and two negative edges $1p$ and $p(n - 2)$ in Z. However, in $\eta(\Sigma_n^\wedge)$, there is a cycle $Z' = (1, p, n - 2, 1)$ with one negative edge $1(n - 2)$ and two positive edges $1p$ and $p(n - 2)$. Thus, we have a negative cycle that contradicts the given condition. Therefore, the only possibility is that n is 3 or even.

Conversely, let n be even. $\Sigma_n^\wedge$, according to Lemma 3 is an all-negative signed graph. So $\eta(\Sigma_n^\wedge)$ is balanced and is all positive. $\eta(\Sigma_n^\wedge)$ for $n = 3$ is a tree which is trivially balanced, hence the converse. □

We present the following theorem for the degree of the vertices of G_n (see [20]).

Theorem 13 ([20]). *Let m be any vertex of the unitary addition Cayley graph G_n. Then,*

$$d(m) = \begin{cases} \phi(n) & \text{if } n \text{ is even,} \\ \phi(n) & \text{if } n \text{ is odd and } (m,n) \neq 1, \\ \phi(n) - 1 & \text{if } n \text{ is odd and } (m,n) = 1. \end{cases}$$

Additionally, for a signed graph S, the balance property of $L(S)$ is discussed in ([27], Theorem 4).

Theorem 14. *For an additional signed Cayley graph $\Sigma_n^\wedge = (G_n, \sigma^\wedge)$, its line signed graph $L(\Sigma_n^\wedge)$ is balanced if—and only if—$n \in \{2, 3, 4, 6\}$.*

Proof. Let $L(\Sigma_n^\wedge)$ be balanced and $n \neq 2, 3, 4$ and 6. Now, according to Theorem 13, $d(0) = d^-(0) = \phi(n) = $ even, which implies $d(0) = d^-(0) = \phi(n) \geq 4$. This shows that condition ii (of Theorem 4, [27]) is not satisfied for $\Sigma_n^\wedge$. This is a contradiction. Hence, $n \in \{2, 3, 4, 6\}$. The converse part is easy to prove. □

For a signed graph S, the balance property of $C_E(S)$ is discussed in ([9], Theorem 13).

Theorem 15. *For an additional signed Cayley graph $\Sigma_n^\wedge = (G_n, \sigma^\wedge)$, its common-edge signed graph $C_E(\Sigma_n^\wedge)$ is balanced if—and only if—$n \in \{3, 4, 6\}$.*

Proof. Let $n \notin \{3, 4, 6\}$. It is clear that $0 \notin U_n$. Now, by Theorem 13, $d(0) = d^-(0) = \phi(n) = $ even, which implies $d(0) = d^-(0) = \phi(n) \geq 4$. This shows that condition ii (of Theorem 13, [9]) is not satisfied for $\Sigma_n^\wedge$. Thus, $C_E(\Sigma_n^\wedge)$ is not balanced, which is a contradiction. Hence, $n \in \{3, 4, 6\}$. The converse part is easy to prove. □

Author Contributions: Conceptualization, D.S.; Formal analysis, O.W., D.S. and A.D.; Methodology, O.W.; Supervision, D.S.; Writing—review & editing, O.W. and D.S. All authors have read and agreed to the published version of the manuscript.

Funding: The first author thanks the South Asian University for research grant support. The third author is grateful to DST [MTR/2018/000607] for the support under the Mathematical Research Impact Centric Support (MATRICS).

Data Availability Statement: No data were used to support the findings of the study.

Conflicts of Interest: All the authors declare that they have no conflict of interest regarding the publication of this paper.

References

1. Harary, F. *Graph Theory*; Addison-Wesley Publ. Comp.: Boston, MA, USA, 1969.
2. West, D.B. *Introduction to Graph Theory*; Prentice-Hall of India Pvt. Ltd.: New Delhi, India, 1996.
3. Zaslavsky, T. A mathematical bibliography of signed and gain graphs and allied areas. *Electron. J. Combin.* **2018**, DS8. [CrossRef]
4. Zaslavsky, T. Glossary of signed and gain graphs and allied areas. *Electron. J. Combin.* **1998**, DS9. Available online: https://www.combinatorics.org/files/Surveys/ds9/ds9v1-1998.pdf (accessed on 11 July 2022).
5. Harary, F. On the notion of balance of a signed graph. *Mich. Math.* **1953**, *2*, 143–146. [CrossRef]
6. Sinha, D.; Wardak, O.; Dhama, A. On Some Properties of Signed Cayley Graph Sn. *Mathematics* **2022**, *10*, 2633. [CrossRef]
7. Zaslavsky, T. Signed analogs of bipartite graphs. *Discrete Math.* **1998**, *179*, 205–216. [CrossRef]
8. Davis, J.A. Clustering and structural balance in graphs. *Hum. Relat.* **1967**, *20*, 181–187. [CrossRef]
9. Acharya, M.; Sinha, D. Common-edge sigraphs. *AKCE Int. J. Graphs Comb.* **2006**, *3*, 115–130.
10. Behzad, M.; Chartrand, G.T. Line coloring of signed graphs. *Elem. Math.* **1969**, *24*, 49–52.
11. Gill, M.K. Contribution to Some Topics in Graph Theory and It's Applications. Ph.D. Thesis, Indian Institute of Technology, Bombay, India, 1969.
12. Sharma, P.; Acharya, M. Balanced signed total graphs of commutative ring. *Graphs Combin.* **2016**, *32*, 1585–1597. [CrossRef]
13. Sinha, D. New Frontiers in the Theory of Signed Graphs. Ph.D. Thesis, University of Delhi, New Delhi, India, 2005.
14. Acharya, B.D. A characterization of consistent marked graphs. *Nat. Acad. Sci. Lett. USA* **1983**, *6*, 431–440.
15. Sinha, D.; Acharya, M. Characterization of signed graphs whose iterated line graphs are balanced and S-consistent. *Bull. Malays. Math. Sci. Soc.* **2016**, *39*, 297–306. [CrossRef]

16. Zaslavsky, T. Consistency in the naturally vertex-signed line graph of a signed graph. *Bull. Malays. Math. Sci. Soc.* **2016**, *39*, 307–314. [CrossRef]
17. Evans, A.B.; Fricke, G.H.; Maneri, C.C.; McKee, T.A.; Perkel, M. Representations of graphs modulo n. *J. Graph Theory* **1994**, *18*, 801–815. [CrossRef]
18. Sinha, D.; Dhama, A. Unitary Cayley Meet Signed Graphs. *Electron. Notes Discret. Math.* **2017**, *63*, 425–434. [CrossRef]
19. Sinha, D.; Garg, P. On the unitary Cayley signed graphs. *Electron. J. Comb.* **2011**, *18*, P229. [CrossRef]
20. Sinha, D.; Garg, P.; Singh, A. Some properties of unitary addition Cayley graphs. *Notes Number Theory Discret. Math.* **2011**, *17*, 49–59.
21. Sampathkumar, E. Point-signed and line-signed graphs. *Natl. Acad. Sci. Lett.* **1984**, *7*, 91–93.
22. Sinha, D.; Dhama, A. Sign-Compatibility of common-edge signed graphs and 2-path signed graphs. *Graph Theory Notes N.Y.* **2013**, *65*, 55–61.
23. Acharya, M.; Sinha, D. Characterizations of line sigraphs. *Nat. Acad. Sci. Lett.* **2005**, *28*, 31–34. [CrossRef]
24. Acharya, M. ×-line sigraph of a sigraph. *J. Combin. Math. Combin. Comput.* **2009**, *69*, 103–111.
25. Rose, H.E. *A Course in Number Theory*; Oxford Science Publications; Oxford University Press: Oxford, UK, 1988.
26. Hoede, C. A characterization of consistent marked graphs. *J. Graph Theory* **1992**, *16*, 17–23. [CrossRef]
27. Acharya, M.; Sinha, D. A Characterization of Sigraphs Whose Line Sigraphs and Jump Sigraphs are Switching Equivalent. *Graph Theory Notes N. Y.* **2003**, *44*, 30–34.

Article
Congruence for Lattice Path Models with Filter Restrictions and Long Steps

Dmitry Solovyev [1,2,3]

1 Euler International Mathematical Institute, Pesochnaya nab. 10, 197022 St. Petersburg, Russia
2 Yau Mathematical Sciences Center, Tsinghua University, Jingzhai, Beijing 100084, China
3 Department of Quantum Mechanics, Saint Petersburg State University, 7-9 Universitetskaya Emb., 199034 St. Petersburg, Russia

Abstract: We derive a path counting formula for a two-dimensional lattice path model with filter restrictions in the presence of long steps, source and target points of which are situated near the filters. This solves the problem of finding an explicit formula for multiplicities of modules in tensor product decomposition of $T(1)^{\otimes N}$ for $U_q(sl_2)$ with divided powers, where q is a root of unity. Combinatorial treatment of this problem calls for the definition of congruence of regions in lattice path models, properties of which are explored in this paper.

Keywords: lattice path models; enumerative combinatorics; quantum groups at roots of unity; tensor product decomposition

MSC: 05E10; 16T30; 05A15; 05A19

1. Introduction

Representation theory of Kac–Moody algebras to this day serves as inspiration for numerous combinatorial problems, solutions to which give rise to interesting combinatorial structures. Examples of this can be met in [1–3] and many other well-known works. The problem of tensor power decomposition, in turn, can be considered from the combinatorial perspective as a problem of counting lattice paths in Weyl chambers [4–7]. In this paper, we count paths on the Bratteli diagram [8], reproducing the decomposition of tensor powers of the fundamental module of the quantum group $U_q(sl_2)$ with divided powers, where q is a root of unity ([9–12]), into indecomposable modules. Combinatorial treatment of this problem gives rise to some interesting structures on lattice path models, such as filter restrictions, first introduced in [13], and long steps, which are introduced in the present paper.

In [13], the considered lattice path model was motivated by the problem of finding explicit formulas for multiplicities of indecomposable modules in the decomposition of tensor power of fundamental module $T(1)$ of the small quantum group $u_q(sl_2)$ ([14]). We call this model the auxiliary lattice path model [9]. It consists of the left wall restriction at $x=0$ and filter restrictions located periodically at $x=nl-1$ for $n \in \mathbb{N}$. For $n=1$, the filter restriction is of type 1, and the rest of the values of n filter restrictions are of type 2. Applying periodicity conditions $(M+2l, N) = (M, N)$, $M, N \geq l-1$ to the Bratteli diagram of this model allows one to obtain another lattice path model, recursion for weighted numbers of paths that coincide with recursion for multiplicities of indecomposable $u_q(sl_2)$-modules in the decomposition of $T(1)^{\otimes N}$. Counting weighted numbers of paths descending from $(0,0)$ to (M, N) on this folded Bratteli diagram allows one to obtain desired formula for multiplicity, where M stands for the highest weight of a module, the multiplicity of which is in question, and N stands for the tensor power of $T(1)$. This has been performed in [9].

We found that the auxiliary lattice path model can be modified in a different way, giving results for representation theory of $U_q(sl_2)$, the quantized universal enveloping algebra of sl_2 with divided powers, when q is a root of unity ([15]). Instead of applying periodicity conditions to the auxiliary lattice path model, as in the case of $u_q(sl_2)$, for $U_q(sl_2)$ we consider all filters to be of the 1st type and also allow additional steps from $x = nl - 2$ to $x = (n-2)l - 1$, where $n \geq 3$. Counting weighted numbers of paths descending from $(0,0)$ to (M, N) on the Bratteli diagram of the lattice path model obtained by this modification gives a formula for the multiplicity of $T(M)$ in the decomposition of $T(1)^{\otimes N}$.

The main goal of this paper is to give a more in-depth combinatorial treatment of the auxiliary lattice path model in the presence of long steps and obtain explicit formulas for weighted numbers of paths, descending from $(0,0)$ to (M, N). We explore combinatorial properties of long steps, as well as define boundaries and congruence of regions in lattice path models. Latter is found to be useful for deriving formulas for weighted numbers of paths. For any considered region, weighted numbers of paths at boundary points uniquely define such for the rest of the region by means of recursion. So, for congruent regions in different lattice path models, regions where, roughly speaking, recursion is similar, it is sufficient to prove identities only for boundary points of such regions.

This paper is organized as follows. In Section 2, we introduce the necessary notation. In Section 3, we give background on the auxiliary lattice path model. In Section 4, we introduce the notion of regions in lattice path models, boundary points and congruence of regions. In Section 5, we explore combinatorial properties of long steps in periodically filtered lattice path models and consider the auxiliary lattice path model in the presence of long steps. We do so by means of boundary points and congruence of regions. In Section 6, we modify the auxiliary lattice path model and argue that the recursion for the weighted number of paths in such modified model coincides with the recursion for multiplicities of modules in tensor product decomposition of $T(1)^{\otimes N}$ for $U_q(sl_2)$ with divided powers, where q is a root of unity. In Section 7, we prove formulas for the weighted numbers of descending paths, relevant to this modified model. In Section 8, we conclude this paper with observations for possible future directions or research.

2. Notations

In this paper, we use the notation following [16]. For our purposes of counting multiplicities in tensor power decomposition of $U_q(sl_2)$-module $T(1)$, throughout this paper, we consider the lattice

$$\mathcal{L} = \{(n,m) | n + m = 0 \bmod 2\} \subset \mathbb{Z}^2,$$

and the set of steps $\mathbb{S} = \mathbb{S}_L \cup \mathbb{S}_R$, where

$$\mathbb{S}_R = \{(x,y) \to (x+1, y+1)\}, \quad \mathbb{S}_L = \{(x,y) \to (x-1, y+1)\}.$$

A *lattice path* $\mathcal{P}$ in $\mathcal{L}$ is a sequence $\mathcal{P} = (P_0, P_1, \ldots, P_m)$ of points $P_i = (x_i, y_i)$ in $\mathcal{L}$ with starting point P_0 and the endpoint P_m. The pairs $P_0 \to P_1, P_1 \to P_2 \ldots P_{m-1} \to P_m$ are called steps of $\mathcal{P}$.

Given starting point A and endpoint B, a set $\mathbb{S}$ of steps and a set of restrictions $\mathcal{C}$ we write

$$L(A \to B; \mathbb{S} \mid \mathcal{C})$$

for the set of all lattice paths from A to B that have steps from $\mathbb{S}$ and obey the restrictions from $\mathcal{C}$. We denote the number of paths in this set as

$$|L(A \to B; \mathbb{S} \mid \mathcal{C})|.$$

The set of restrictions $\mathcal{C}$ in lattice path models considered throughout this paper mostly contain wall restrictions and filter restrictions. Left(right) wall restrictions forbid steps in the left(right) direction, reflecting descending paths and preventing them from crossing the 'wall'. Filter restrictions forbid steps in certain directions and provide other steps with non-uniform weights, so paths can cross the 'filter' in one direction, but cannot cross it in the opposite direction. A rigorous definition of these restrictions is given in subsequent sections.

To each step from (x,y) to $(\tilde{x},\tilde{y})$ we assign the weight function $\omega : \mathbb{S} \longrightarrow \mathbb{R}_{>0}$ and use notation $(x,y) \xrightarrow{\omega} (\tilde{x},\tilde{y})$ to denote that the step from (x,y) to $(\tilde{x},\tilde{y})$ has the weight ω. By default, all unrestricted steps from $\mathbb{S}$ will have weight 1 and is denoted by an arrow with no number at the top. The *weight* of a path $\mathcal{P}$ is defined as the product

$$\omega(\mathcal{P}) = \prod_{i=0}^{m-1} \omega(P_i \to P_{i+1}).$$

For the set $L(A \to B; \mathbb{S} \mid \mathcal{C})$ we define the *weighted number of paths* as

$$Z(L(A \to B; \mathbb{S} \mid \mathcal{C})) = \sum_{\mathcal{P}} \omega(\mathcal{P}),$$

where the sum is taken over all paths $\mathcal{P} \in L(A \to B; \mathbb{S} \mid \mathcal{C})$.

3. The Auxiliary Lattice Path Model

In this section, we briefly revise notions and results obtained in [13], relevant for future considerations. It is convenient for us to omit mentioning $\mathbb{S}$ in $L(A \to B; \mathbb{S} \mid \mathcal{C})$. All paths considered below involve steps from set $\mathbb{S}$ unless stated otherwise.

3.1. Unrestricted Paths

Let $L(A \to B)$ be the set of unrestricted paths from A to B on lattice $\mathcal{L}$ with the steps $\mathbb{S}$. An example of such a path is given in Figure 1.

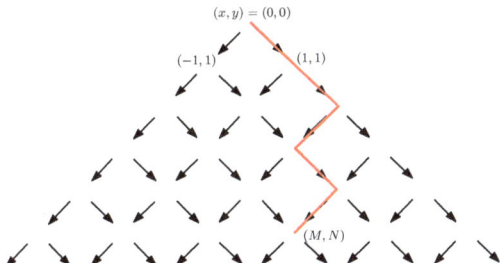

Figure 1. Example of an unrestricted path in $L((0,0) \to (M,N))$ for lattice $\mathcal{L}$ and set of steps $\mathbb{S}$.

Lemma 1. *For a set of unrestricted paths with steps $\mathbb{S}$ we have*

$$|L((0,0) \to (M,N))| = \binom{N}{\frac{N-M}{2}}. \tag{1}$$

3.2. Wall Restrictions

Definition 1. *For lattice paths that start at $(0,0)$ we will say that $\mathcal{W}_d^L$ with $d \leq 0$ is a left wall restriction (relative to $x = 0$) if at points (d,y) paths are allowed to take steps of type $\mathbb{S}_R$ only*

$$\mathcal{W}_d^L = \{(d,y) \to (d+1, y+1) \text{ only}\}.$$

Lemma 2. *The number of paths from $(0,0)$ to (M,N) with the set of steps $\mathbb{S}$ and one wall restriction $\mathcal{W}_a^L$ can be expressed via the number of unrestricted paths as*

$$|L((0,0) \to (M,N) \mid \mathcal{W}_a^L)| = \binom{N}{\frac{N-M}{2}} - \binom{N}{\frac{N-M}{2}+a-1}, \text{ for } M \geq a, \tag{2}$$

We considered the left walls located at $x = 0$. An example of possible steps for paths descending from $(0,0)$ in the presence of this restriction is given in Figure 2.

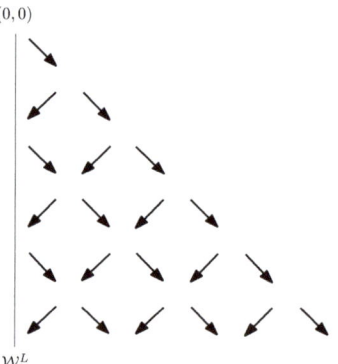

Figure 2. Arrangement of steps for points of $\mathcal{L}$ in presence of restriction $\mathcal{W}_0^L$.

3.3. Filter Restrictions

Definition 2. *For $n \in \mathbb{N}$, we say that there is a filter $\mathcal{F}_d^n$ of type n, located at $x = d$ if at $x = d, d + 1$ only the following steps are allowed:*

$$\mathcal{F}_d^n = \{(d,y) \xrightarrow{n} (d+1, y+1), (d+1, y+1) \to (d+2, y+2), (d+1, y+1) \xrightarrow{2} (d, y+2)\}.$$

The index above the arrow is the weight of the step.

Note that by default, an arrow with no number at the top means that the corresponding step has a weight of 1. An example of possible steps for descending paths in the presence of this restriction is given in Figure 3. We highlighted steps of weight 2 with red instead of an arrow with a superscript 2 for future convenience, as those are the most common for the auxiliary lattice path model and its modifications. We were mostly involved with filters of type 1, so superscripts n were avoided, leaving Bratteli diagrams with black and red arrows, with weights 1 and 2 correspondingly.

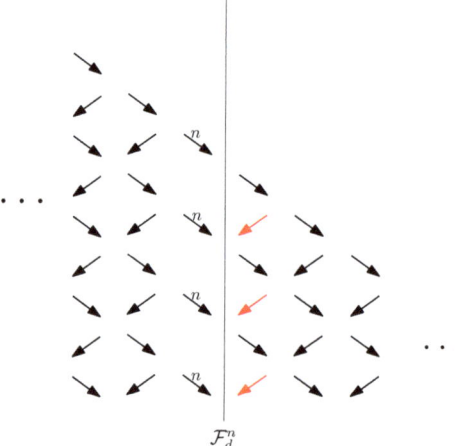

Figure 3. Filter $\mathcal{F}_d^n$. Red arrows correspond to steps $(d+1, y+1) \xrightarrow{2} (d, y+2)$ that has a weight 2. Black arrows with superscript n correspond to steps $(d,y) \xrightarrow{n} (d+1, y+1)$. Other steps have weight 1.

Lemma 3. *The number of lattice paths from $(0,0)$ to (M,N) with steps from $\mathbb{S}$ and filter restriction $\mathcal{F}_d^n$ with $x = d > 0$ and $n \in \mathbb{N}$ is*

$$Z(L_N((0,0) \to (M,N) \mid \mathcal{F}_d^n)) = \binom{N}{\frac{N-M}{2}} - \binom{N}{\frac{N-M}{2}+d}, \text{ for } M < d, \qquad (3)$$

$$Z(L_N((0,0) \to (M,N) \mid \mathcal{F}_d^n)) = n\binom{N}{\frac{N-M}{2}}, \text{ for } M > d. \qquad (4)$$

Proof. The proof is the same as for Lemma 4.8 and Lemma 4.9 in [13]. □

3.4. Counting Paths in the Auxiliary Lattice Path Model

Consider the lattice path model for the set of paths on $\mathcal{L}$ descending from $(0,0)$ to (M,N) with steps $\mathbb{S}$ in the presence of restrictions $\mathcal{W}_0^L, \mathcal{F}_{l-1}^1, \mathcal{F}_{nl-1}^2, n \in \mathbb{N}, n \geq 2$. Such set is denoted as

$$L_N((0,0) \to (M,N); \mathbb{S} \mid \mathcal{W}_0^L, \mathcal{F}_{l-1}^1, \{\mathcal{F}_{nl-1}^2\}_{n=2}^\infty)$$

and such a model is called the auxiliary lattice path model. The main theorem of [13] gives an explicit formula for weighted numbers of paths in the auxiliary lattice path model. Then, in [9], periodicity conditions $(M + 2l, N) = (M, N), M, N \geq l - 1$ were applied, resulting in a folded Brattelli diagram. For such a diagram, recursion on the weighted numbers of paths coincides with recursion on multiplicities of indecomposable $u_q(sl_2)$-modules in tensor product decomposition of $T(1)^{\otimes N}$. Note that due to properties of the category **Rep**$(u_q(sl_2))$, we mostly considered odd values of l; however, the results remain to be true for even values of l as well.

Before coming to modifications of the auxiliary lattice path model relevant to the representation theory of $U_q(sl_2)$ at the roots of unity, we need to slightly tweak it. We are interested in paths descending from $(0,0)$ to (M,N) with steps $\mathbb{S}$ in the presence of restrictions $\mathcal{W}_0^L, \mathcal{F}_{nl-1}^1, n \in \mathbb{N}$, instead of filters of type 2. Such lattice path model is depicted in Figure 4.

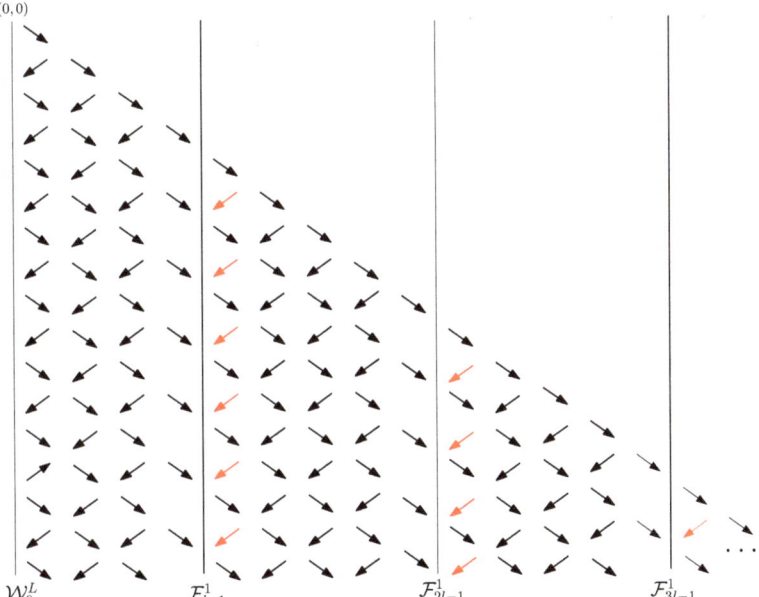

Figure 4. Arrangement of steps for points of $\mathcal{L}$ in the considered, slightly tweaked version of the auxiliary lattice path model. Here, we depict the case, where $l = 5$.

Definition 3. *We denote by multiplicity function in the j-th strip $M^j_{(M,N)}$ the weighted number of paths in set*

$$L_N((0,0) \to (M,N); \mathbb{S} \mid \mathcal{W}_0^L, \{\mathcal{F}^1_{nl-1}\}, n \in \mathbb{N})$$

with the endpoint (M,N) that lies within $(j-1)l - 1 \leq M \leq jl - 2$

$$M^j_{(M,N)} = Z(L_N((0,0) \to (M,N); \mathbb{S} \mid \mathcal{W}_0^L, \{\mathcal{F}^1_{nl-1}\}, n \in \mathbb{N})), \tag{5}$$

where $M \geq 0$ and $j = \left[\frac{M+1}{l} + 1\right]$.

Now consider the version of the main theorem in [13] corresponding to this model.

Theorem 1 ([13]). *The multiplicity function in the j-th strip is given by*

$$M^j_{(M,N)} = \sum_{k=0}^{\left[\frac{N-(j-1)l+1}{4l}\right]} P_j(k) F^{(N)}_{M+4kl} + \sum_{k=0}^{\left[\frac{N-jl}{4l}\right]} P_j(k) F^{(N)}_{M-4kl-2jl} -$$

$$- \sum_{k=0}^{\left[\frac{N-(j+1)l+1}{4l}\right]} Q_j(k) F^{(N)}_{M+2l+4kl} - \sum_{k=0}^{\left[\frac{N-jl-2l}{4l}\right]} Q_j(k) F^{(N)}_{M-4kl-2(j+1)l'}$$

where

$$P_j(k) = \sum_{i=0}^{\left[\frac{j}{2}\right]} \binom{j-2}{2i} \binom{k-i+j-2}{j-2}, \quad Q_j(k) = \sum_{i=0}^{\left[\frac{j}{2}\right]} \binom{j-2}{2i+1} \binom{k-i+j-2}{j-2}, \tag{6}$$

$$F^{(N)}_M = \binom{N}{\frac{N-M}{2}} - \binom{N}{\frac{N-M}{2} - 1}.$$

Proof. The proof is the same as the proof of the main theorem in [13], except that instead of Lemma 4.9 in [13], for the slightly tweaked model one should use Lemma 3. □

From now on, when mentioning the auxiliary lattice path model, we mean its slightly tweaked version. This model will be further modified in subsequent sections. Instead of applying periodicity conditions, as for $u_q(sl_2)$, we enhance this model with long steps, source and target points which are located near filters. As a result, recursion for the weighted numbers of paths on the resultant Bratteli diagram recreates recursion for multiplicities of indecomposable $U_q(sl_2)$-modules in the decomposition of $T(1)^{\otimes N}$.

4. Boundary Points and Congruent Regions

In this section, we consider notions, which are convenient for counting paths in the auxiliary lattice path model in the presence of long steps. We will see, that multiplicities on the boundary of a region uniquely define multiplicities in the rest of the region. For proving identities between multiplicities in two congruent regions, it is sufficient to prove such identities for their boundary points.

Definition 4. *Consider the lattice path model, defined by a set of steps $\mathbb{S}$ and a set of restrictions $\mathcal{C}$ on lattice $\mathcal{L}$. Subset $\mathcal{L}_0 \subset \mathcal{L}$ with steps $\mathbb{S}$ and restrictions $\mathcal{C}$ is called a region of the lattice path model under consideration.*

Intuitively, region $\mathcal{L}_0 \subset \mathcal{L}$ is a restriction of the lattice path model defined by $\mathbb{S}, \mathcal{C}$ on lattice $\mathcal{L}$ to the subset $\mathcal{L}_0$. The word 'restriction' is overused, so we consider regions of lattice path models instead.

Definition 5. Consider $\mathcal{L}_0 \subset \mathcal{L}$ a region of the lattice path model defined by steps $\mathbb{S}$ and restrictions $\mathcal{C}$. Point $B \in \mathcal{L}_0$ is called a boundary point of $\mathcal{L}_0$ if there exists $B' \in \mathcal{L}$, $B' \notin \mathcal{L}_0$ such that step $B' \to B$ is allowed in $\mathcal{L}$ by a set of steps $\mathbb{S}$ and restrictions $\mathcal{C}$. The union of all such points is a boundary of $\mathcal{L}_0$ and is denoted by $\partial \mathcal{L}_0$.

The Definition 5 introduces a notion, reminiscent of the outer boundary in graph theory. Note that boundary points are defined with respect to some lattice path models under consideration. For brevity, we assume that this lattice path model is known from the context, and mentioning it will be mostly omitted.

Example 1. *For a strip in the auxiliary lattice path model, its boundary is in the left filter. It is depicted in Figure 5.*

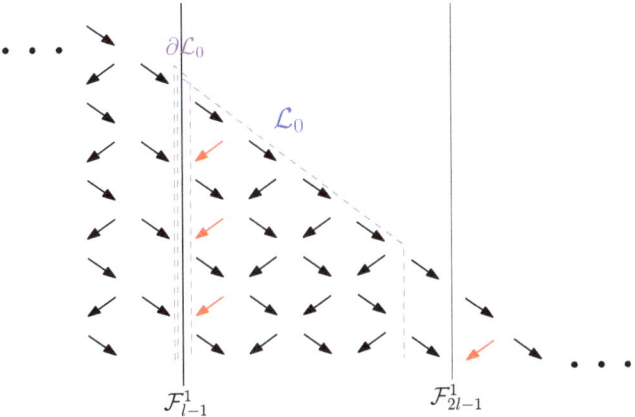

Figure 5. Region $\mathcal{L}_0$, highlighted with blue dashed lines, is a 2nd strip for $l = 5$. Its boundary $\partial \mathcal{L}_0$ is a set of points in the left filter restriction $\mathcal{F}^1_{l-1}$, which is highlighted with purple dashed lines.

Example 2. *Consider region $\mathcal{L}_0$ of the unrestricted lattice path model, as depicted in Figure 6 and highlighted with blue dashed lines. Its boundary is a set of points highlighted with purple dashed lines.*

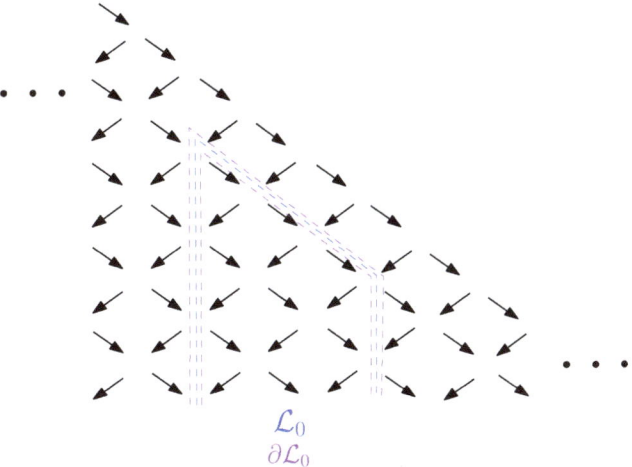

Figure 6. Region $\mathcal{L}_0$ is highlighted with blue dashed lines. Its boundary $\partial \mathcal{L}_0$ is a set of points highlighted with purple dashed lines.

Lemma 4. *Consider region $\mathcal{L}_0$ of a lattice path model defined by $\mathbb{S}, \mathcal{C}$ on lattice $\mathcal{L}$. Weighted numbers of paths $Z(L_N((0,0) \to (M,N); \ldots)$ for $(M,N) \in \mathcal{L}_0$ are uniquely defined by weighted numbers of paths for its boundary points $\partial \mathcal{L}_0$.*

Proof. Suppose weighted numbers of paths for $\partial \mathcal{L}_0$ are known. Suppose that there exists some point $A \in \mathcal{L}_0$, such that its weighted number of paths cannot be expressed in terms of weighted numbers of paths for points in $\partial \mathcal{L}_0$.

The first case is that recursion for a weighted number of paths for A involves some point $A' \in \mathcal{L}_0$, a weighted number of paths for which cannot be expressed in terms of such for points in $\partial \mathcal{L}_0$. In this case, we need to consider A' and recursion on the weighted number of paths for such a point instead of A.

The second case is that recursion for a weighted number of paths for A involves a weighted number of paths for some point $A' \notin \mathcal{L}_0$. Then, $A \in \partial \mathcal{L}_0$ by definition of a boundary point and weighted number of paths for such point is known by the initial supposition of the lemma.

Note that due to the fact that we consider descending paths, M and N, to be finite, the first case can be iterated finitely many times at most. □

Definition 6. *Consider two lattice path models with steps $\mathbb{S}_1, \mathbb{S}_2$ and restrictions $\mathcal{C}_1, \mathcal{C}_2$ defined on lattice $\mathcal{L}$. Subset $\mathcal{L}_1 \subset \mathcal{L}$ is a region in the lattice path model defined by $\mathbb{S}_1, \mathcal{C}_1$. Subset $\mathcal{L}_2 \subset \mathcal{L}$ is a region in the lattice path model defined by $\mathbb{S}_2, \mathcal{C}_2$. Regions $\mathcal{L}_1$ and $\mathcal{L}_2$ are congruent if there exists a translation T in $\mathcal{L}$ such that*

- *$T\mathcal{L}_1 = \mathcal{L}_2$ as sets of points in $\mathcal{L}$*
- *Translation T induces a bijection between steps in $\mathcal{L}_1$ and $\mathcal{L}_2$, meaning that there is a one-to-one correspondence between steps with source and target points related by T, with preservation of weights.*

The second condition can be written down explicitly. Firstly, for each $(M,N) \in \mathcal{L}_1$ and each step $(M,N) \xrightarrow{w} (P,Q)$ in $\mathbb{S}_1$ obeying $\mathcal{C}_1$ such that $(P,Q) \in \mathcal{L}_1$, there is a step $(M',N') \xrightarrow{w} (P',Q')$ in $\mathbb{S}_2$ obeying $\mathcal{C}_2$, where $T(M,N) = (M',N')$, $T(P,Q) = (P',Q')$. Secondly, for each $(M',N') \in \mathcal{L}_2$ and each step $(M',N') \xrightarrow{w} (P',Q')$ in $\mathbb{S}_2$ obeying $\mathcal{C}_2$ such that $(P',Q') \in \mathcal{L}_2$, there is a step $(M,N) \xrightarrow{w} (P,Q)$ in $\mathbb{S}_1$ obeying $\mathcal{C}_1$, where $T^{-1}(M',N') = (M,N)$, $T^{-1}(P',Q') = (M,N)$. To put it simply, if we forget about lattice path models outside $\mathcal{L}_1$ and $\mathcal{L}_2$, these two regions will be indistinguishable. Due to translations in $\mathcal{L}$ being invertible, it is easy to see that congruence defines an equivalence relation.

Now we must prove the main theorem of this subsection.

Theorem 2. *Consider two lattice path models with steps $\mathbb{S}_1, \mathbb{S}_2$ and restrictions $\mathcal{C}_1, \mathcal{C}_2$ defined on lattice $\mathcal{L}$. Region $\mathcal{L}_1$ of the lattice path model defined by $\mathbb{S}_1, \mathcal{C}_1$ is congruent to region $\mathcal{L}_2$ of the lattice path model defined by $\mathbb{S}_2, \mathcal{C}_2$, where $T\mathcal{L}_1 = \mathcal{L}_2$. If equality*

$$Z(L_N((0,0) \to (M,N); \mathbb{S}_1 \mid \mathcal{C}_1)) = Z(L_N((0,0) \to T(M,N); \mathbb{S}_2 \mid \mathcal{C}_2)) \tag{7}$$

holds for all $(M,N) \in \partial \mathcal{L}_1 \cup T^{-1}(\partial \mathcal{L}_2)$, then it holds for all $(M,N) \in \mathcal{L}_1$.

Note, that if $(M,N) \in \partial \mathcal{L}_1$ it does not necessarily follow that $T(M,N) \in \partial \mathcal{L}_2$, due to $\mathcal{C}_1$ and $\mathcal{C}_2$ being different. So, it is natural to ask Formula (7) to hold for $\partial \mathcal{L}_1 \cup T^{-1}(\partial \mathcal{L}_2)$.

Proof. We need to prove that Formula (7) is true for $(M,N) \in \mathcal{L}_1$. The l.h.s. can be uniquely expressed in terms of its values at $\partial \mathcal{L}_1 \cup T^{-1}(\partial \mathcal{L}_2)$, following procedure in Lemma 4. Due to the congruence between $\mathcal{L}_1$ and $\mathcal{L}_2$, recursion for the r.h.s. of (7) coincides with the one for the l.h.s., so we can obtain the same expression on the r.h.s., but with values of weighted numbers of paths for $T(\partial \mathcal{L}_1 \cup T^{-1}(\partial \mathcal{L}_2)) = T(\partial \mathcal{L}_1) \cup \partial \mathcal{L}_2$ instead of $\partial \mathcal{L}_1 \cup T^{-1}(\partial \mathcal{L}_2)$. We can compare the l.h.s. and the r.h.s. term by term, for points related by translation T. All of such terms have the same values due to the initial supposition of the theorem. □

Corollary 1. *Consider lattice path models with steps $\mathbb{S}_1, \mathbb{S}_2, \mathbb{S}_3$ and restrictions $\mathcal{C}_1, \mathcal{C}_2, \mathcal{C}_3$ defined on lattice $\mathcal{L}$. Region $\mathcal{L}_1$ is congruent to $\mathcal{L}_2$ and $\mathcal{L}_3$, where $T_1(M_1, N_1) = (M_2, N_2)$, $T_2(M_1, N_1) = (M_3, N_3)$ for $(M_1, N_1) \in \mathcal{L}_1$. If equality*

$$Z(L_N((0,0) \to (M,N); \mathbb{S}_1 \mid \mathcal{C}_1)) =$$
$$= Z(L_N((0,0) \to T_1(M,N); \mathbb{S}_2 \mid \mathcal{C}_2)) + Z(L_N((0,0) \to T_2(M,N); \mathbb{S}_3 \mid \mathcal{C}_3)) \quad (8)$$

holds for all $(M, N) \in \partial \mathcal{L}_1 \cup T_1^{-1}(\partial \mathcal{L}_2) \cup T_2^{-1}(\partial \mathcal{L}_3)$, then it holds for all $(M, N) \in \mathcal{L}_1$.

Proof. Due to linearity of the r.h.s. of Formula (8), the proof repeats the one of Theorem 2. □

The moral of this section is that for two congruent regions, weighted numbers of paths are defined by values of such at the boundary of the considered regions. For proving identities, it is sufficient to establish equality for weighted numbers of paths at boundary points, while equality for the rest of the region will follow due to the congruence.

5. The Auxiliary Lattice Path Model in the Presence of Long Steps
Long Steps in Lattice Path Models with Filter Restrictions

Long step is a step $(x, y) \xrightarrow{w} (x', y+1)$ in $\mathcal{L}$ such that $|x - x'| > 1$. We denote the sequence of long steps as

$$\mathbb{S}[M_1, M_2] = \{(M_1, M_1 + 2m) \to (M_2, M_1 + 1 + 2m)\}_{m=0}^{\infty},$$

where $x = M_1$ is the source point for the sequence and $x = M_2$ is the target point, $|M_1 - M_2| > 1$. For the purposes of this paper, we are mainly interested in sequences

$$\mathbb{S}(k) \equiv \mathbb{S}[l(k+2) - 2, lk - 1] = \{(l(k+2) - 2, lk - 2 + 2m) \to (lk - 1, lk - 2 + 1 + 2m)\}_{m=0}^{\infty},$$

where $k \in \mathbb{N}$ and $\mathcal{C}$ consists of $\mathcal{F}_{lk-1}^1$ and $\mathcal{F}_{l(k+2)-1}^1$. We need such sequences of long steps for modification of the auxiliary lattice path model, relevant to the representation theory of $U_q(sl_2)$ at roots of unity.

Lemma 5. *Fix $k \in \mathbb{N}$. Let*

$$Z_{(M,N)} \equiv Z(L_N((0,0) \to (M,N)); \mathbb{S} \mid \mathcal{F}_{lk-1}^1, \mathcal{F}_{l(k+2)-1}^1)$$

be the weighted number of lattice paths from $(0,0)$ to (M, N) with filter restrictions $\mathcal{F}_{lk-1}^1$, $\mathcal{F}_{l(k+2)-1}^1$ and set of unrestricted elementary steps $\mathbb{S}$. Let

$$Z'_{(M,N)} \equiv Z(L_N((0,0) \to (M,N)); \mathbb{S} \cup \mathbb{S}(k) \mid \mathcal{F}_{lk-1}^1, \mathcal{F}_{l(k+2)-1}^1)$$

be the weighted number of lattice paths from $(0,0)$ to (M,N) with the same restrictions, with steps $\mathbb{S} \cup \mathbb{S}(k)$. Then for $lk - 1 \leq M \leq l(k+2) - 2$ we have

$$Z'_{(M,N)} = Z_{(M,N)}, \quad \text{if } N \leq M + 2l - 2, \quad (9)$$

$$Z'_{(M,N)} = Z_{(M,N)} + Z_{(M+2l,N)}, \quad \text{if } M + 2l \leq N \leq l(k+4) - 2. \quad (10)$$

Proof. In Figure 7, we depict the setting of the Lemma 5. Long steps do not impact region I, so Formula (9) is true.

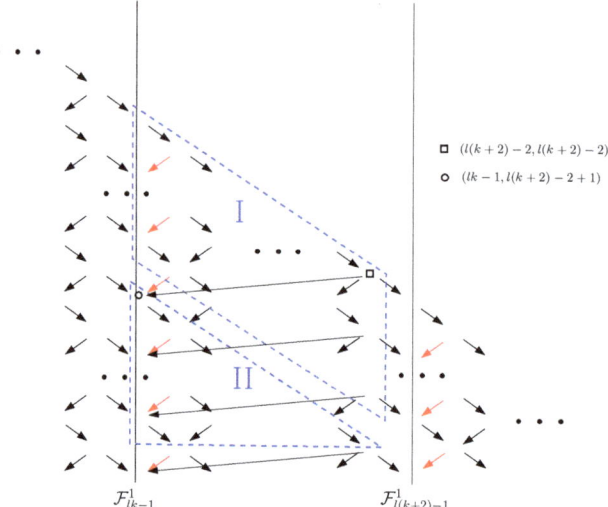

Figure 7. By square and circle we denote points, where long steps first appear. Regions I and II highlighted with blue dashed lines correspond to cases $N \leq M + 2l - 2$, as in (9), and $M + 2l \leq N \leq l(k+4) - 2$, as in (10).

Consider Formula (10). The weighted number of paths in the l.h.s. involves points from region II. Its boundary contains points of the left cathetus of region II, of the form $(lk - 1, N)$ for $l(k+2) - 1 \leq N \leq l(k+4) - 2$, and points of the hypotenuse of the region II, of the form $(lk - 1 + j, l(k+2) - 1 + j)$ for $j = 1, \ldots, 2l - 1$. Denote this set by $\partial \mathcal{L}_{II}$. The r.h.s. of (10) has two terms. The first involves region II, the boundary of which we have already considered. The second term involves points of the region congruent to region II, as they are related by translation $T(M, N) = (M + 2l, N)$, satisfying Definition 6. Its boundary consists of the image of the left cathetus of region II under translation T. Denote this set by $\partial \mathcal{L}'_{II}$. By Corollary 1, it is sufficient to prove Formula (10) for $\partial \mathcal{L}_{II} \cup T^{-1}(\partial \mathcal{L}'_{II}) = \partial \mathcal{L}_{II}$.

We proceed by induction over n, where $N = l(k+2) - 1 + 2n$. For $n = 0$ from recursion we have

$$Z'_{(lk-1,l(k+2)-1)} = Z_{(lk-2,l(k+2)-2)} + 2Z_{(lk,l(k+2)-2)} + Z_{(l(k+2)-2,l(k+2)-2)}, \quad (11)$$

which, taking into account that

$$Z_{(lk-2,l(k+2)-2)} + 2Z_{(lk,l(k+2)-2)} = Z_{(lk-1,l(k+2)-1)},$$

$$Z_{(l(k+2)-2,l(k+2)-2)} = Z_{(l(k+2)-1,l(k+2)-1)},$$

gives us

$$Z'_{(lk-1,l(k+2)-1)} = Z_{(lk-1,l(k+2)-1)} + Z_{(l(k+2)-1,l(k+2)-1)}. \quad (12)$$

We obtained the base of induction.

In a similar manner, it also follows, that Formula (10) is true for boundary points of the hypotenuse of region II. In order to show this, one must consider recursion explicitly and use the fact that

$$Z_{(j,j)} = Z_{(k,k)}, \quad \text{for all } j, k > 0. \quad (13)$$

Now it is sufficient to prove Formula (10) for boundary points, situated in the left cathetus of region II.

Suppose

$$Z'_{(lk-1,l(k+2)-1+2n)} = Z_{(lk-1,l(k+2)-1+2n)} + Z_{(l(k+2)-1,l(k+2)-1+2n)} \quad (14)$$

is true. For the sake of brevity, we rewrite this expression as

$$Z'_{(p,q+2n)} = Z_{(p,q+2n)} + Z_{(q,q+2n)}, \tag{15}$$

where $p = lk - 1$, $q = l(k+2) - 1$, $q = p + 2l$. By Theorem 2, it follows that Formula (10) is true for the region, corresponding to boundary points, covered by the inductive supposition. In particular, this region includes points $(p+j, q+2n+j)$ for $j = 0, \ldots, 2l-1$. Need to prove that

$$Z'_{(p,q+2(n+1))} = Z_{(p,q+2(n+1))} + Z_{(q,q+2(n+1))} \tag{16}$$

Taking into account, that

$$Z'_{(p,q+2n+2)} = Z_{(p-1,q+2n+1)} + 2Z'_{(p+1,q+2n+1)} + Z_{(q-1,q+2n+1)},$$

$$Z_{(p,q+2n+2)} = Z_{(p-1,q+2n+1)} + 2Z_{(p+1,q+2n+1)},$$

$$Z_{(q,q+2n+2)} = Z_{(q-1,q+2n+1)} + 2Z_{(q+1,q+2n+1)},$$

after getting rid of the factors, we obtain

$$Z'_{(p+1,q+2n+1)} = Z_{(p+1,q+2n+1)} + Z_{(q+1,q+2n+1)}. \tag{17}$$

However, this is true from the inductive supposition. □

Note that Formula (10) is not true for greater values of N. Region II indeed can be made into a parallelogram, similar to region I, since the set of boundary points will remain the same. However, the region corresponding to this parallelogram being translated by T contains new boundary points, where (10) does not hold and Corollary 1 cannot be used further, even though these regions are congruent to each other. The formula for greater values of N needs to include some new terms. In this parallelogram-like region, we need to take into account the reflection of paths, induced by the term $Z_{(M+2l,N)}$ in $Z'_{(M,N)}$, from the filter restriction $\mathcal{F}^1_{l(k+2)-1}$. This is achieved by means of the first part of Lemma 3. Now consider the triangular region, which, similarly to region II being below region I, is below the parallelogram-like region considered previously. There, we need to take into account long steps, acting on paths induced by the term $Z_{(M+2l,N)}$, which have descended to $(l(k+2)-2,N)$ and were acted upon by long steps for the second time. This is being conducted in a similar fashion to Corollary 1, where $Z_{(M+2l,N)}$ is assumed to be known from the second part of Lemma 3. This situation for the case of the auxiliary lattice path model in the presence of long steps will be elaborated upon later.

Corollary 2. *Fix $j, k \in \mathbb{N}$, $j \leq k$. Let*

$$Z_{(M,N)} \equiv Z(L_N((0,0) \to (M,N)); \mathbb{S} \mid \mathcal{W}_0^L, \{\mathcal{F}^1_{nl-1}\}_{n=j}^{\infty})$$

be the weighted number of lattice paths from $(0,0)$ to (M,N) with filter restrictions $\{\mathcal{F}^1_{nl-1}\}_{n=j}^{\infty}$ and set of unrestricted elementary steps $\mathbb{S}$. Let

$$Z'_{(M,N)} \equiv Z(L_N((0,0) \to (M,N)); \mathbb{S} \cup \mathbb{S}(k) \mid \mathcal{W}_0^L, \{\mathcal{F}^1_{nl-1}\}_{n=j}^{\infty})$$

be the weighted number of lattice paths from $(0,0)$ to (M,N) with the same restrictions, with steps $\mathbb{S} \cup \mathbb{S}(k)$. Then, for $lk - 1 \leq M \leq l(k+2) - 2$ we have

$$Z'_{(M,N)} = Z_{(M,N)}, \quad \text{if } N \leq M + 2l - 2, \tag{18}$$

$$Z'_{(M,N)} = Z_{(M,N)} + Z_{(M+2l,N)}, \quad \text{if } M + 2l \leq N \leq l(k+4) - 2. \tag{19}$$

Proof. The proof is the same, as for Lemma 5. When proving the inductive step, we still can apply Corollary 1 as region II is still congruent to the one, translated by T. □

Note, that Formula (19), unlike (10), is true for greater values of N, as making region II into a parallelogram-like region will not add new boundary points. The manifestation of this fact is that we do not need to take into account the reflection of paths, as they have already been dealt with in term $Z_{(M+2l,N)}$ due to the periodicity of filter restrictions. So, for such a region Formula (19) holds. However, for the triangular region below the same problem remains.

Consider the auxiliary lattice path model in the presence of the sequence of steps $\mathbb{S}(k)$.

Definition 7. *We denote by multiplicity function in the j-th strip $\tilde{M}^j_{(M,N)}$ the weighted number of paths in set*

$$L_N((0,0) \to (M,N); \mathbb{S} \cup \tilde{\mathbb{S}} \mid \mathcal{W}_0^L, \{\mathcal{F}^1_{nl-1}\}, n \in \mathbb{N})$$

with the endpoint (M,N) that lies within $(j-1)l - 1 \leq M \leq jl - 2$

$$\tilde{M}^j_{(M,N)} = Z(L_N((0,0) \to (M,N); \mathbb{S} \cup \tilde{\mathbb{S}} \mid \mathcal{W}_0^L, \{\mathcal{F}^1_{nl-1}\}, n \in \mathbb{N})), \quad (20)$$

where $\tilde{\mathbb{S}}$ is a set of some additional steps and $M \geq 0$ and $j = \left[\frac{M+1}{l} + 1\right]$.

In this subsection, $\tilde{\mathbb{S}} = \mathbb{S}(k)$ if not stated otherwise.

Lemma 6. *For fixed $k \in \mathbb{N}$*

$$\tilde{M}^{k+1}_{(M,N)} = \sum_{j=0}^{\left[\frac{N-lk+1}{2l}\right]} M^{k+1+2j}_{(M+2jl,N)}, \quad (21)$$

$$\tilde{M}^{k+3}_{(M,N)} = \sum_{j=0}^{\left[\frac{N-l(k+2)+1}{2l}\right]} M^{k+3+2j}_{(M+2jl,N)}, \quad (22)$$

where $\tilde{M}^j_{(M,N)}$ is the multiplicity function for j-th strip in the auxiliary model with steps $\mathbb{S} \cup \tilde{\mathbb{S}}$, $M^j_{(M,N)}$ is the multiplicity function for j-th strip in the auxiliary model with steps $\mathbb{S}$.

Proof. We proceed by induction over n, where $n = \left[\frac{N-l(k+2)+2}{2l}\right]$, first proving (22), then (21). For $n = 0$, Formula (22) follows immediately from the Theorem 2, as long steps do not impact this region. Formula (21) follows from Corollary 2. As was discussed, Formula (19) is true for greater values of N, mainly, it is true for a parallelogram-like region, satisfying $n = 0$. So, we obtained the base of induction.

Suppose, that

$$\tilde{M}^{k+1}_{(M,N)} = \sum_{j=0}^{n+1} M^{k+1+2j}_{(M+2jl,N)} \quad (23)$$

$$\tilde{M}^{k+3}_{(M,N)} = \sum_{j=0}^{n} M^{k+3+2j}_{(M+2jl,N)} \quad (24)$$

is true.

Need to prove the inductive step for (24) first, thus we need to prove (22) for $M + 2ln \leq N \leq M + 2l(n+1)$, where $l(k+2) - 1 \leq M \leq l(k+3) - 2$. Denote this region as $\mathcal{L}_1$. The l.h.s. of (22) is a weighted number of paths, $\partial \mathcal{L}_1$ consists of points $(l(k+2) - 1, N)$ for $l(k+2) - 1 + 2ln \leq N \leq l(k+2) - 1 + 2l(n+2)$ and $(l(k+2) - 1 + j, l(k+2) - 1 + 2ln + j)$ for $j = 0, \ldots, l-1$. We divide the r.h.s. of (22) into two terms. The first corresponds to the sum given by inductive supposition in (24). It is also a weighted number of paths defined for $\mathcal{L}_2 = \mathcal{L}_1$ with the same boundary points $\partial \mathcal{L}_2 = \partial \mathcal{L}_1$, as the l.h.s. of (22). These two regions are congruent, $T_1 = id$. The second is an additional term, which we expect to appear during an inductive step. It is given by

$$M^{k+3+2(n+1)}_{(M+2(n+1)l,N)} = Z(L_N((0,0)) \to (M+2(n+1)l, N)); \mathbb{S} \mid \mathcal{W}^L_0, \{\mathcal{F}^1_{ml-1}\}^\infty_{m=1})$$

for region $l(k+2+2(n+1)) - 1 \leq M \leq l(k+3+2(n+1)) - 2$ and $M \leq N \leq M + 2l$. Denote it by $\mathcal{L}_3$. Its boundary $\partial \mathcal{L}_3$ consists of points $(l(k+2+2(n+1)) - 1, N)$ for $l(k+2+2(n+1)) - 1 \leq N \leq l(k+2+2(n+1)) - 1 + 2l$. This region is an image of $\mathcal{L}_1$ under translation $T_2(M, N) = (M + 2l(n+1), N)$, they are congruent. By Corollary 1, it is sufficient to prove inductive step at points $(l(k+2) - 1, N)$ for $l(k+2) - 1 + 2ln \leq N \leq l(k+2) - 1 + 2l(n+2)$ and points $(l(k+2) - 1 + j, l(k+2) - 1 + 2ln + j)$ for $j = 0, \ldots, l-1$. These are drawn in Figure 8.

Consider points of a form $(l(k+2) - 1, N)$. At n-th iteration we added $M^{k+1+2(n+1)}_{(M+2(n+1)l,N)}$ to $\tilde{M}^{k+1}_{(M,N)}$. This term induces paths, which further descend from $(k+1)$-th strip to boundary points of $(k+3)$-th strip. The region in which induced paths descend is congruent to the region, where paths corresponding to $M^{k+1+2(n+1)}_{(M+2(n+1)l,N)}$ continue to descend to the boundary of $(k+3+2(n+1))$-th strip in the auxiliary lattice path model. This is due to the periodicity of filter restrictions. Here, we can apply Theorem 2 to conclude that the weighted number of induced paths arriving at the boundary of $(k+3)$-th strip is equal to $M^{k+3+2(n+1)}_{(2(n+1)l,N)}$.

Consider points of a form $(l(k+2) - 1 + j, l(k+2) - 1 + 2ln + j)$. For such points, the proof is the same as for the Formula (10) for the hypotenuse of region II.

Now that we proved the inductive step for the boundary of the considered region, by Corollary 1, it follows that

$$\tilde{M}^{k+3}_{(M,N)} = \sum_{j=0}^{n} M^{k+3+2j}_{(M+2jl,N)} + M^{k+3+2(n+1)}_{(M+2(n+1)l,N)} = \sum_{j=0}^{n+1} M^{k+3+2j}_{(M+2jl,N)}, \qquad (25)$$

is true for the whole region, which proves the inductive step for Formula (24).

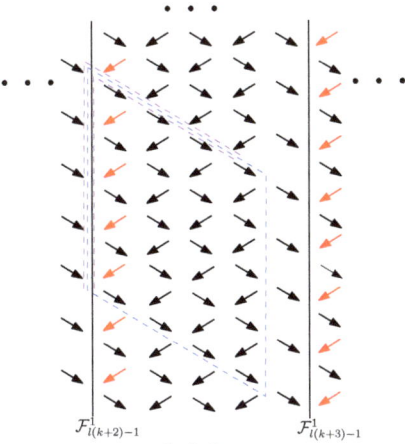

Figure 8. Region $\mathcal{L}_1$, for which it is sufficient to prove (22) consists of points $M + 2ln \leq N \leq M + 2l(n+1)$ where $l(k+2) - 1 \leq M \leq l(k+3) - 2$. It is highlighted with blue dashed lines. Union of boundaries for all terms of the considered expression $\partial \mathcal{L}_1 \cup T_1^{-1}(\partial \mathcal{L}_2) \cup T_2^{-1}(\partial \mathcal{L}_3)$ consists of points $(l(k+2) - 1, N)$ for $l(k+2) - 1 + 2ln \leq N \leq l(k+2) - 1 + 2l(n+2)$ and points $(l(k+2) - 1 + j, l(k+2) - 1 + 2ln + j)$ for $j = 0, \ldots, l-1$. It is highlighted with purple dashed lines. Here, we depict the case, where $l = 5$.

This process for the first iterations is shown in Figure 9.

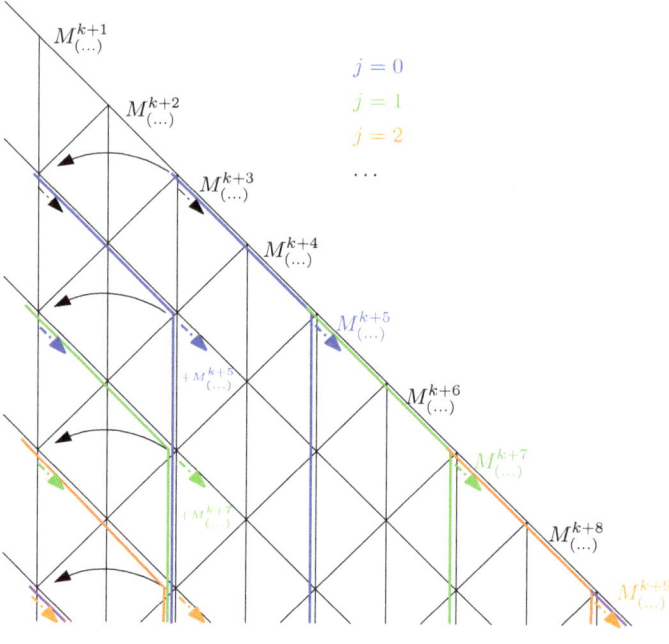

Figure 9. Color emphasizes the number of iterations in the induction. Paths induced at the boundary of $(k+1)$-th strip during $(j-1)$-th iteration descend in the region, highlighted with color, corresponding to j-th iteration. Colored lines outline regions congruent to each other. Dashed colored arrows denote weighted numbers of induced paths, inflicted to $(k+3)$-th strip once they have descended, and their equivalents in strips of the auxiliary lattice path model.

This figure also shows how long steps act on descended paths corresponding to dashed colored arrows, inducing paths at boundary points of $(k+1)$-th strip, highlighted with the dashed arrow of the same color. These induced paths, in turn, descend in the region, highlighted with a color corresponding to the next, $(j+1)$-th iteration. Proving that long steps induce paths at boundary points of $(k+1)$-th strip following this scenario amounts to proving the inductive step for Formula (23).

Now we need to prove the inductive step for Formula (23), which amounts to proving (21) for $lk - 1 \leq M \leq l(k+1) - 2$ and $M + 2ln \leq N \leq M + 2l(n+1)$. It is being conducted in a fashion similar to the proof of (22). Again, we divide the r.h.s. of (21) in two terms. The first one corresponds to the sum given by inductive supposition in (23). The second one is an additional term, which we expect to appear during an inductive step. It is given by

$$M^{k+1+2(n+1)}_{(M+2(n+1)l,N)} = Z(L_N((0,0) \to (M+2(n+1)l, N)); \mathbb{S} \mid \mathcal{W}_0^L, \{\mathcal{F}^1_{ml-1}\}_{m=1}^\infty)$$

for region $l(k + 2(n+1)) - 1 \leq M \leq l(k+1+2(n+1)) - 2$ and $M \leq N \leq M + 2l$. By Corollary 1, it is sufficient to prove inductive step at points $(lk - 1, N)$ for $lk - 1 + 2ln \leq N \leq l(k+1) - 1 + 2l(n+2)$ and points $(lk - 1 + j, lk - 1 + 2ln + j)$ for $j = 0, \ldots, l - 1$. It is shown in Figure 10. This region is the same, as depicted in Figure 8, but translated by $T(M, N) = (M - 2l, N)$.

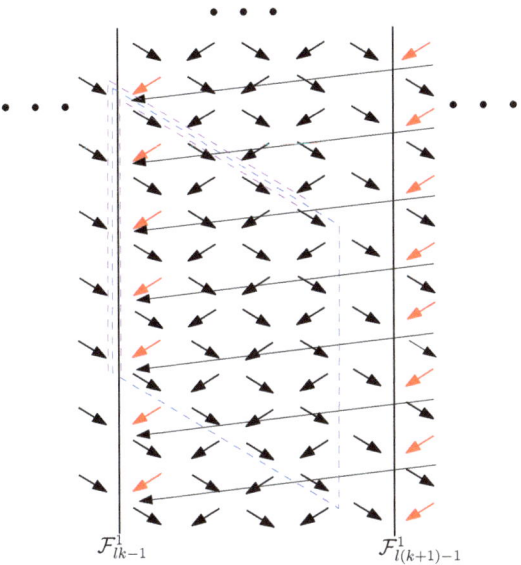

Figure 10. Points $M + 2ln \leq N \leq M + 2l(n+1)$ where $l(k+2) - 1 \leq M \leq l(k+3) - 2$ are highlighted with blue dashed lines. Union of boundaries for all terms of the considered expression consists of points $(l(k+2) - 1, N)$ for $l(k+2) - 1 + 2ln \leq N \leq l(k+2) - 1 + 2l(n+2)$ and points $(l(k+2) - 1 + j, l(k+2) - 1 + 2ln + j)$ for $j = 0, \ldots, l-1$. They are highlighted with purple dashed lines. Here, we depict the case where $l = 5$.

Consider points of a form $(lk - 1, N)$. Above, we have seen that Formula (22) receives term $M^{k+3+2(n+1)}_{(M+2(n+1)l,N)}$ during the inductive step. By inductive supposition (23), it is left to account for the action of long steps, acting on paths, induced by this term. Denote the weighted number of paths, corresponding to this term as

$$Z_{(M,N)} \equiv Z(L_N((-2l(n+1), 0) \to (M, N)); \mathbb{S} \mid \mathcal{W}^L_{-2l(n+1)}, \{\mathcal{F}^1_{ml-1-2l(n+1)}\}_{m=1}^\infty) =$$
$$= M^{k+3+2(n+1)}_{(M+2(n+1)l,N)},$$

where $lk - 1 \leq M \leq l(k+2) - 1$, $M + 2ln \leq N \leq M + 2l(n+1)$. Now, we want to calculate

$$Z'_{(M,N)} \equiv Z(L_N((-2l(n+1), 0) \to (M, N)); \mathbb{S} \cup \tilde{\mathbb{S}} \mid \mathcal{W}^L_{-2l(n+1)}, \{\mathcal{F}^1_{ml-1-2l(n+1)}\}_{m=1}^\infty).$$

From Corollary 2, it is given by

$$Z'_{(M,N)} = Z_{(M,N)} + Z_{(M+2l,N)},$$

where

$$Z_{(M+2l,N)} = M^{k+3+2(n+2)}_{(M+2(n+2)l,N)}.$$

Consider points of a form $(lk - 1 + j, lk - 1 + 2ln + j)$. For such points, the proof is the same as for Formula (10) for the hypotenuse of region II.

Now that we have proven the inductive step for the boundary of the considered region, by Corollary 1, it follows that

$$\tilde{M}^{k+1}_{(M,N)} = \sum_{j=0}^{n+1} M^{k+1+2j}_{(M+2jl,N)} + M^{k+1+2(n+2)}_{(M+2(n+2)l,N)} = \sum_{j=0}^{n+1} M^{k+1+2j}_{(M+2jl,N)}, \qquad (26)$$

is true for the whole region, which proves the inductive step for Formula (23). □

Corollary 3. *For fixed $k \in \mathbb{N}$ and $m \geq k$*

$$\tilde{M}^{m+1}_{(M,N)} = \sum_{j=0}^{[\frac{N-lm+1}{2l}]} M^{m+1+2j}_{(M+2jl,N)}, \quad (27)$$

Proof. The result of Lemma 6 can be extended to other strips in a similar fashion to the proof of (22). Each new term $M^{k+1+2j}_{(M+2jl,N)}$ in $\tilde{M}^{k+1}_{(M,N)}$ induces paths, which further descend from $(k+1)$-th strip to boundary points of each consequent $(k+1+m)$-th strip. The region in which these induced paths descend is congruent to the region, where they would continue to descend in the auxiliary path model due to the periodicity of filter restrictions. Hence, each $\tilde{M}^{k+1+m}_{(M,N)}$ acquires term $M^{k+1+m+2j}_{(M+2jl,N)}$, which proves the statement. □

During this subsection, we introduced long steps and proved lemmas, necessary for counting weighted numbers of paths in modification of the auxiliary lattice path model, relevant for the representation theory of $U_q(sl_2)$ at roots of unity.

6. On Decomposition of $T(1)^{\otimes N}$ for $U_q(sl_2)$ at Roots of Unity

Consider the auxiliary lattice path model with filter restrictions of type 1, in the presence of steps

$$\mathbb{S}_U \equiv \mathbb{S} \cup \left(\bigcup_{k=1}^{\infty} \mathbb{S}(k) \right),$$

and denote it as $\mathcal{L}_U$. The arrangement of steps for points of $\mathcal{L}_U$ is depicted in Figure 11.

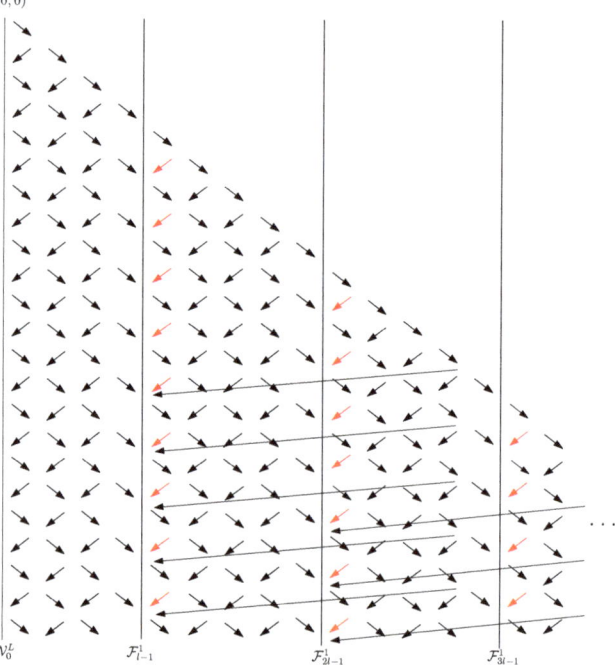

Figure 11. Arrangement of steps for points of the lattice path model $\mathcal{L}_U$. Here, we depict the case, where $l = 5$.

Now, fix $q = e^{\frac{\pi i}{l}}$ and l is odd. Category $\mathbf{Rep}(U_q(sl_2))$ is the category of representations of $U_q(sl_2)$, a quantized universal enveloping algebra of sl_2 with divided powers. Consider tensor product decomposition of a tensor power of fundamental $U_q(sl_2)$-module

$$T(1)^{\otimes N} = \bigoplus_{k=0}^{N} M_{T(k)}^{(l)}(N) T(k), \quad T(1), T(k) \in \mathbf{Rep}(U_q(sl_2)), \tag{28}$$

where $M_{T(k)}^{(l)}(N)$ is the multiplicity of $T(k)$ in tensor product decomposition. We consider the tensor powers of a tilting module, and as a category of tilting modules is closed under finite tensor products, it can be decomposed into a direct sum of tilting modules. The highest weight of $T(k)$ can be written as $k = lk_1 + k_0$. The Grothendieck ring of the category of tilting modules over $U_q(sl_2)$ at odd roots of unity gives the following tensor product rules ([9,17])

$$T(k_0) \otimes T(1) = T(k_0 + 1) \oplus T(k_0 - 1), \quad 0 \le k_0 \le l - 2;$$

$$T(lk_1 + k_0) \otimes T(1) = T(lk_1 + k_0 + 1) \oplus T(lk_1 + k_0 - 1), \quad 1 \le k_0 \le l - 3, k_1 \ge 1;$$

$$T(lk_1 + l - 2) \otimes T(1) = T(l(k_1+1) - 3) \oplus T((k_1+1)l - 1) \oplus T((k_1-1)l - 1), \quad k_1 \ge 1;$$

$$T(lk_1 - 1) \otimes T(1) = T(lk_1), \quad k_1 \ge 1;$$

$$T(lk_1) \otimes T(1) = T(lk_1 + 1) \oplus 2T(lk_1 - 1), \quad k_1 \ge 1.$$

Theorem 3 ([9]). *The multiplicity of the tilting $U_q(sl_2)$-module $T(k)$ in the decomposition of $T(1)^{\otimes N}$ is equal to the weighted number of lattice paths on $\mathcal{L}_U$ connecting $(0,0)$ and (k,N) with weights given by multiplicities of elementary steps $\mathbb{S}_U$.*

Proof. Tensor product rules allow the following recursive description of multiplicities

$$M_{T(0)}^{(l)}(N+1) = M_{T(1)}^{(l)}(N);$$

$$M_{T(lk_1+k_0)}^{(l)}(N+1) = M_{T(lk_1+k_0-1)}^{(l)}(N) + M_{T(lk_1+k_0+1)}^{(l)}(N), \quad 1 \le k_0 \le l-3, k_1 \ge 0;$$

$$M_{T(lk_1-2)}^{(l)}(N+1) = M_{T(lk_1-3)}^{(l)}(N), \quad k_1 \ge 1;$$

$$M_{T(lk_1-1)}^{(l)}(N+1) = M_{T(lk_1-2)}^{(l)}(N) + 2M_{T(lk_1)}^{(l)}(N) + M_{T((k_1+2)l-2)}^{(l)}(N), \quad k_1 \ge 1;$$

$$M_{T(lk_1)}^{(l)}(N+1) = M_{T(lk_1-1)}^{(l)}(N) + M_{T(lk_1+1)}^{(l)}(N), \quad k_1 \ge 1.$$

This recursion coincides with the recursion for weighted numbers of paths descending from $(0,0)$ to (k,N) in lattice path model $\mathcal{L}_U$. The latter is depicted in Figure 11. □

The main goal of the following section is to obtain the explicit formula by combinatorial means, mainly counting lattice paths in modification $\mathcal{L}_U$ of the auxiliary lattice path model.

7. Counting Paths

Consider the lattice path model $\mathcal{L}_U$. From now on, following Definition 7, we denote by multiplicity function in the j-th strip $\tilde{M}_{(M,N)}^j$ the weighted number of paths in set

$$L_N((0,0) \to (M,N); \mathbb{S}_U \mid \mathcal{W}_0^L, \{\mathcal{F}_{nl-1}^1\}, n \in \mathbb{N})$$

with the endpoint (M, N) that lies within $(j-1)l - 1 \leq M \leq jl - 2$

$$\tilde{M}^{j}_{(M,N)} = Z(L_N((0,0) \to (M,N); \mathbb{S}_U \mid \mathcal{W}_0^L, \{\mathcal{F}^1_{nl-1}\}, n \in \mathbb{N})), \quad (29)$$

where

$$\mathbb{S}_U = \mathbb{S} \cup \left(\bigcup_{k=1}^{\infty} \mathbb{S}(k)\right).$$

and $M \geq 0$ and $j = \left[\frac{M+1}{l} + 1\right]$. The main goal of this section is to derive an explicit formula for $\tilde{M}^{j}_{(M,N)}$.

Lemma 7. *For the lattice path model* $\mathcal{L}_U$

$$\tilde{M}^{1}_{(M,N)} = M^{1}_{(M,N)}, \quad (30)$$

and for $k \in \mathbb{N}$,

$$\tilde{M}^{k+1}_{(M,N)} = \sum_{j=0}^{\left[\frac{N-lk+1}{2l}\right]} F^{(k-1+2j)}_{k-1} M^{k+1+2j}_{(M+2jl,N)}, \quad (31)$$

Proof. The formula for the 1st strip follows immediately as long steps have no impact and multiplicity is the same as in the auxiliary lattice path model.

This lemma follows from gradually adding each $\mathbb{S}(k)$ for $k = 1, 2, \ldots$ to the initial set of steps $\mathbb{S}$ and applying results of the Corollary 3 repeatedly. Let us start with $k = 1$. From Corollary 3, it follows that in case of having one series of long steps, for $(m+1)$-th strip we would simply have

$$\tilde{M}^{m+1}_{(M,N)}\big|_{k=1} = \sum_{j=0}^{\left[\frac{N-lm+1}{2l}\right]} M^{m+1+2j}_{(M+2jl,N)}, \quad (32)$$

where $m \in \mathbb{N}$. This is a summation of multiplicities in the auxiliary lattice path model with trivial coefficients. This situation is depicted in Figure 12.

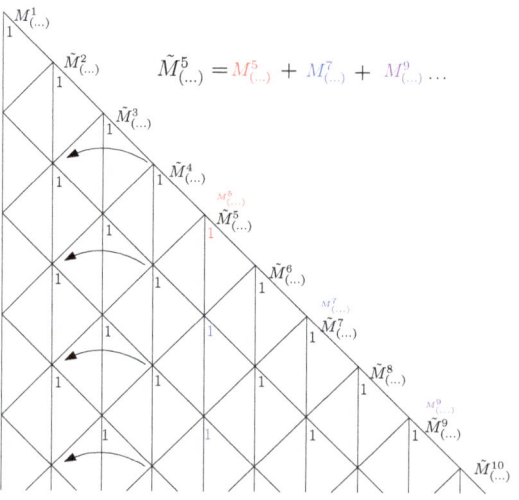

Figure 12. In each strip, triangles with coefficients correspond to terms in Formula (32). Each term is given by the multiplicity of a strip in the auxiliary lattice path model, situated to the far right of the considered triangle. As an example, we show how this mnemonic rule works for $\tilde{M}^{5}_{(M,N)}$.

In each strip, triangles with coefficients correspond to terms in Formula (32). Each term is given by the multiplicity function of a strip in the auxiliary lattice path model, situated to the far right of the considered triangle. The coefficient in a triangle tells us how many terms corresponding to this multiplicity function are in Formula (32). This mnemonic rule comes from considerations in Figure 9. The proofs of Theorem 6 and Corollary 3 define a recursion on the coefficients near multiplicity functions from the auxiliary lattice path model in Formula (32). This recursion is depicted in Figure 13.

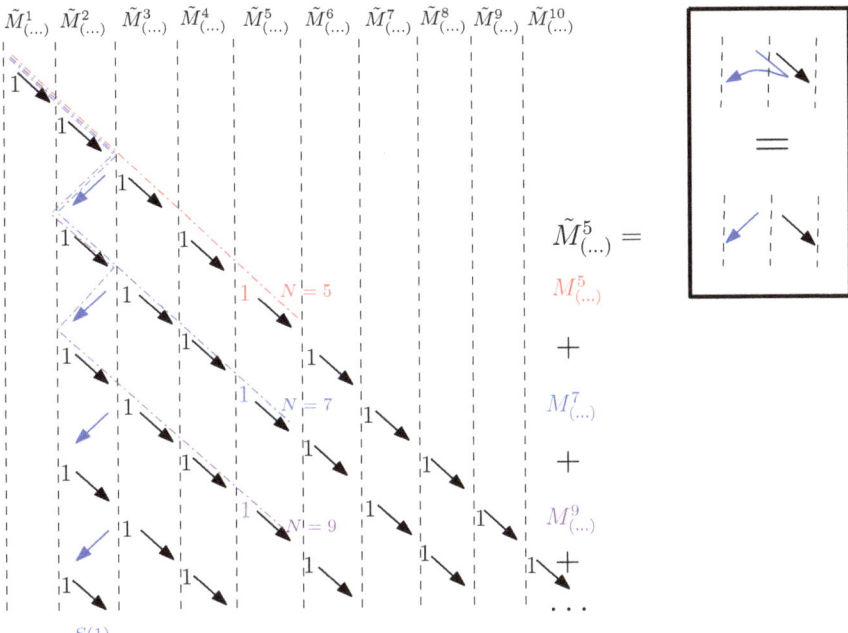

Figure 13. Numbers near vertices of the lattice are the coefficients in Figure 12. Blue arrows denote steps in the recursion, which were added by the long steps $\mathbb{S}(1)$ in the lattice path model. Length of paths N, descending to a considered vertex of a lattice gives the number of the strip in the auxiliary lattice path model, the multiplicity function of which is being added to (32) as a term. As an example, we show formula for $\tilde{M}^5_{(M,N)}|_{k=1}$.

Blue arrows correspond to steps in the recursion on the coefficients, which were added as a consequence of the presence of long steps $\mathbb{S}(1)$. In the black frame, it is noted that although long steps have length $2l$, as shown, for example, in Figure 12, their source and target points belong to two adjacent strips, so when dealing with the coefficients it is convenient to denote blue arrows as in the Figure 13. Long steps, following the idea of the proof of Formula (22) depicted in Figure 9, induce paths that descend further, giving the result as in Corollary 3. Similarly, blue arrows induce paths in the lattice, which descend further, adding new terms in (32).

Note, that without blue arrows we would have obtained a single diagonal path with weighted numbers of paths equal to 1. This situation would give us coefficients as in the formula for multiplicities in the auxiliary lattice path model, meaning that we would have $\tilde{M}^k_{(...)} = M^k_{(...)}$. This is exactly what we would have in case we removed the long steps $\mathbb{S}(1)$ in the lattice path model.

Again, following the idea of the proof of Corollary 3, induced paths descend further to each consequent strip as if they were to continue to descend in the auxiliary lattice path model, so additional terms are dependent on how many strips these induced paths will cross while they descend. In the recursion on the coefficients, it is manifested in the fact that the length of a descending path in Figure 13 gives the number of strips in the auxiliary lattice path model, to which the additional term corresponds.

Now, our main goal is to apply $\mathbb{S}(k)$ for other k. As the considerations above suggest, applying $\mathbb{S}(k)$ for $k = 1, 2, \ldots$ induces other sequences of blue arrows. From Figure 14, we see that the recursion for the coefficients near multiplicity functions is satisfied by Catalan numbers. This proves Formula (31).

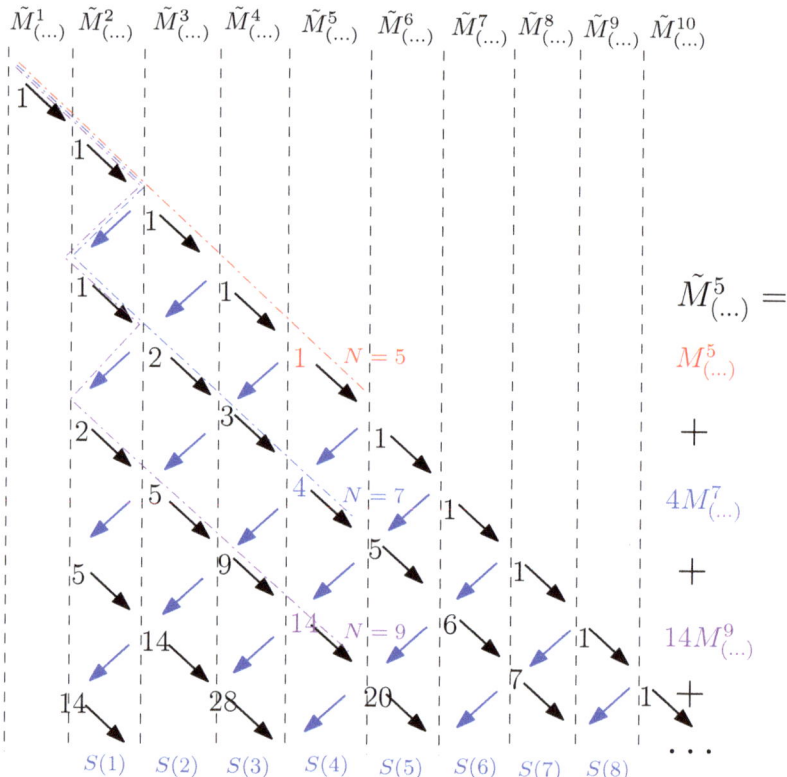

Figure 14. In each strip, numbers in vertices of a lattice correspond to terms in Formula (32). Each term is given by the multiplicity of a strip in the auxiliary lattice path model, the number of which is given by the length of a path, descending to the considered coefficient. As an example, we show formula for $\tilde{M}^5_{(M,N)}$.

□

Note, that action of black arrows on terms in (31) follows from Lemma 3 and the periodicity of filter restrictions. The action of blue arrows on terms in (31) follows from Corollary 1. Now let us prove the main theorem.

Theorem 4. *For $k \in \mathbb{N} \cup \{0\}$ we have*

$$\tilde{M}^{k+1}_{(M,N)} = F^{(N)}_M + \sum_{j=1}^{\left[\frac{N-lk+1}{2l} + \frac{1}{2}\right]} F^{(N)}_{-2lk+M-2jl} + \sum_{j=1}^{\left[\frac{N-lk+1}{2l}\right]} F^{(N)}_{M+2jl} \quad (33)$$

where $lk - 1 \leq M \leq l(k+1) - 2$.

Proof. We proceed by induction over $\left[\frac{N-lk+1}{2l} + \frac{1}{2}\right]$. For $\left[\frac{N-lk+1}{2l} + \frac{1}{2}\right] = 1$ the Formula (33) obviously gives the same result as (31), which is the base of induction.

Suppose, that

$$\tilde{M}_{(M,N)}^{k+1} = F_M^{(N)} + \sum_{j=1}^{n} F_{-2lk+M-2jl}^{(N)} + \sum_{j=1}^{n-1} F_{M+2jl}^{(N)} \qquad (34)$$

is true. We need to prove this statement for $n+1$. It is sufficient to compare coefficients in (33) and (31) near $F_{M+2nl}^{(N)}$ and $F_{-2lk+M-2(n+1)l}^{(N)}$. We focus on the term $F_{M+2nl}^{(N)}$, the rest can be performed in a similar fashion. From the structure of $M_{(M,N)}^k$, given by Theorem 1 and depicted in Figure 15, we have two cases: $n+1$ is odd and $n+1$ is even.

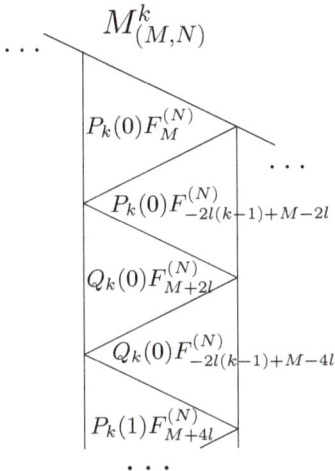

Figure 15. Graphical presentation of terms in the formula for $M_{(M,N)}^k$, given by Theorem 1. Each term is depicted in accordance to the domain of the lattice where it appears for the first time.

For the case of odd $n+1$ the last term in (31) is associated with $P_j(\frac{n}{2})$, for the case of even n it is $Q_j(\frac{n+1}{2})$. We focus on the case of odd $n+1$, the other case can be proven in a similar manner. So, the proof boils down to a combinatorial identity

$$\sum_{\substack{j=0 \\ \text{even } j}}^{n} F_{k-1}^{(k-1+2j)} P_{k+1+2j}\left(\frac{n-j}{2}\right) - \sum_{\substack{j=1 \\ \text{odd } j}}^{n-1} F_{k-1}^{(k-1+2j)} Q_{k+1+2j}\left(\frac{n-1-j}{2}\right) = 1. \qquad (35)$$

From comparing coefficients near $F_{M+2(n-2)l}^{(N)}$ in the inductive supposition (34), we know that

$$\sum_{\substack{j=0 \\ \text{even } j}}^{n-2} F_{k-1}^{(k-1+2j)} P_{k+1+2j}\left(\frac{n-2-j}{2}\right) - \sum_{\substack{j=1 \\ \text{odd } j}}^{n-3} F_{k-1}^{(k-1+2j)} Q_{k+1+2j}\left(\frac{n-3-j}{2}\right) = 1 \qquad (36)$$

is true. Take into account, that

$$P_j(k) - P_j(k-1) = \binom{j+2k-3}{j-3}, \qquad (37)$$

$$Q_j(k) - Q_j(k-1) = \binom{j+2k-2}{j-3}. \qquad (38)$$

Subtracting (36) from (35), we obtain

$$\left(\sum_{\substack{j=0 \\ \text{even } j}}^{n-2} - \sum_{\substack{j=1 \\ \text{odd } j}}^{n-3}\right) F_{k-1}^{(k-1+2j)} \binom{j+n+k-2}{2j+k-2} \tag{39}$$

$$+ F_{k-1}^{(k-1+2n)} P_{k+1+2n}(0) - F_{k-1}^{(k-1+2n-2)} Q_{k+1+2(n-1)}(0) = 0.$$

Taking into account that $P_j(0) = 1$ and $Q_j(0) = j - 2$ and simplifying further, we arrive at

$$\sum_{j=0}^{n} (-1)^j F_{k-1}^{(k-1+2j)} \binom{j+n+k-2}{2j+k-2} = 0. \tag{40}$$

$$\sum_{j=0}^{n} (-1)^j \binom{j+n+k-2}{2j+k-2} \left(\binom{2j+k-1}{j} - \binom{2j+k-1}{j-1}\right) = 0 \tag{41}$$

The last identity follows from the following lemma.

Lemma 8. *For* $n, k \in \mathbb{N}$

$$\sum_{j=0}^{n} (-1)^j \binom{j+n+k-2}{2j+k-2} \binom{2j+k-1}{j} = 2(-1)^n, \tag{42}$$

$$\sum_{j=0}^{n} (-1)^j \binom{j+n+k-2}{2j+k-2} \binom{2j+k-1}{j-1} = 2(-1)^n. \tag{43}$$

Proof. Let us first prove Formula (42). Denote

$$F(n,j) = \frac{(-1)^{j+n}}{2} \binom{j+n+k-2}{2j+k-2} \binom{2j+k-1}{j}.$$

We need to show, that

$$\sum_{j=0}^{n} F(n,j) = 1, \quad \forall n \in \mathbb{N}.$$

For $n = 1$ it is true, which gives us the base of induction. Using Zeilberger's algorithm ([18–20]), we obtain its Wilf–Zeilberger pair

$$G(n,j) = \frac{(-1)^{j+n} j(j+k-1)(k+2n)}{2n(n+1)(j-n-1)(k+2j-1)(k^2+n(n-1)+k(2n-1))}$$

$$\times (1 + k^2 n - 3n^2 + k(n^2 - 3n - 1) + j(2n^2 + 2kn + k - 1)) \binom{j+n+k-2}{2j+k-2} \binom{2j+k-1}{j},$$

for which

$$-F(n+1,j) + F(n,j) = G(n,j+1) - G(n,j) \tag{44}$$

is true. Applying sum over j to both sides and simplifying telescopic sum to the right, we obtain that

$$\sum_{j=0}^{n} F(n+1,j) = \sum_{j=0}^{n} F(n,j) + G(n,0) - G(n,n+1). \tag{45}$$

Taking into account, that $G(n,0) = 0$ and $G(n,n+1) = F(n+1,n+1)$, we have

$$\sum_{j=0}^{n+1} F(n+1,j) = \sum_{j=0}^{n} F(n,j), \tag{46}$$

which, considering inductive supposition, proves Formula (42).

Formula (43) can be proven in a similar fashion, the corresponding Wilf–Zilberger pair is given by

$$G(n,j) = \frac{(-1)^{j+n}(j-1)(j+k)(k+2n)}{2n(n-1)(j-n-1)(k+2j-1)(k^2+n(n-1)+k(2n-1))}$$
$$\times (1+k^2(n-1)+k(n^2-3n)-3n^2+j(2n^2+2kn-k-1))\binom{j+n+k-2}{2j+k-2}\binom{2j+k-1}{j-1}.$$

□

The proof of this lemma concludes the proof of the identity and finishes the proof of the initial theorem. □

8. Conclusions

In this paper, we considered the lattice path model $\mathcal{L}_U$, which is the auxiliary lattice path model in the presence of long steps. Weighted numbers of paths in this model recreate multiplicities of $U_q(sl_2)$-modules in tensor product decomposition of $T(1)^{\otimes N}$, where $U_q(sl_2)$ is a quantum deformation of the universal enveloping algebra of sl_2 with divided powers and q is a root of unity. Explicit formulas for multiplicities of all tilting modules in tensor product decomposition were derived by purely combinatorial means in the main theorem of this paper Theorem 4.

We found that the auxiliary lattice model defined in [13] is of great use for counting multiplicities of modules of differently defined quantum deformations of $U(sl_2)$ at q root of unity. For instance, in [9] we applied periodicity conditions to the auxiliary lattice path model to obtain a folded Bratteli diagram, weighted numbers of paths for which recreate multiplicities of modules in tensor product decomposition of $T(1)^{\otimes N}$, where $T(1)$ is a fundamental module of the small quantum group $u_q(sl_2)$. In this paper, we modified the auxiliary lattice path model by applying long steps to obtain multiplicities for the case of $U_q(sl_2)$ with divided powers in a similar fashion.

The model defined in [13] required analysis of combinatorial properties of filters, which we heavily relied on. In this paper, we introduced long steps and explored their combinatorial properties. In order to derive formulas for weighted numbers of paths in this setting, we also defined boundary points and congruence of regions in lattice path models. The philosophy of congruence is fairly easy to understand. Two different lattice path models can be locally indistinguishable due to coinciding recursions for weighted numbers of paths in these regions. Weighted numbers of paths at boundary points of the considered region uniquely define weighted numbers of paths for the rest of the region by recursion. So, instead of proving identities for the whole region, it is sufficient to prove such only for boundary points of the region. At boundary points, an identity can be represented as a linear combination of weighted numbers of paths from different lattice path models and one needs to take into account boundary points of congruent regions with respect to all these models.

We found that besides applying periodicity conditions to the auxiliary lattice path model, one can take $U_q(sl_2)$, consider its restriction to $u_q^- U_q^0 u_q^+$, where $u_q^\pm$ are subalgebras of the small quantum group $u_q(sl_2)$, generated by F and E, respectively, and U_q^0 is a subalgebra of $U_q(sl_2)$, generated by $K^{\pm 1}$ and $\begin{bmatrix} K;c \\ t \end{bmatrix}$, for $t \geq 0, c \in \mathbb{Z}$. Then, we can restrict $u_q^- U_q^0 u_q^+$ to $u_q(sl_2)$. This procedure defines another modification of the auxiliary lattice path model and, remarkably, gives the same result as with periodicity conditions. The lattice path model corresponding to $u_q^- U_q^0 u_q^+$ will be considered in the upcoming paper.

Considering other possible directions for further research, the following questions remain open:

- Multiplicity formulas for decomposition of tensor powers of fundamental representations of $U_q(sl_n)$ at roots of unity remain out of reach and can be a source of inspiration for other interesting combinatorial constructions. We expect that for $U_q(sl_n)$ derivation of such formulas will rely on similar combinatorial ideas. It is worth mentioning that obtaining such formulas explicitly is of interest for asymptotic representation theory, mainly, for constructing Plancherel measure and possibly obtaining its limit shape in different regimes, including regime when $n \to \infty$ ([7,9,21]).
- In [22], similar lattice path models emerge when studying the Grothendieck ring of the category of tilting modules for $U_q(sl_2)$ in the mixed case: when q is an odd root of unity and the ground field is $\overline{\mathbb{F}_p}$. One can expand the combinatorial analysis presented in this paper to a mixed case.
- In [23], it was shown that $U_q(sl_n)$ at roots of unity is in Schur–Weyl duality with Hecke algebra $nH_N(q)$ on $\otimes^N \mathbb{C}^n$. For the case of $U_q(sl_2)$ at roots of unity, multiplicity formulas should give answers for dimensions of certain representations of Temperley–Lieb algebra $TL_N(q)$ at roots of unity. Dimensions of which representations were obtained is an open question, at least to the knowledge of the author of this paper.

Funding: This research was supported by the grant "Agreement on the provision of a grant in the form of subsidies from the federal budget for implementation of state support for the creation and development of world-class scientific centers, including international world-class centers and scientific centers, carrying out research and development on the priorities of scientific technological development No. 075-15-2019-1619" dated 8th November 2019.

Data Availability Statement: Not applicable.

Acknowledgments: We are grateful to Olga Postnova, Nicolai Reshetikhin, Pavel Nikitin, Fedor Petrov for fruitful discussions. We thank Lacey Lindsay for her help with corrections and support.

Conflicts of Interest: The authors declare no conflict of interest.

References

1. Littelmann, P. Paths and root operators in representation theory. *Ann. Math.* **1995**, *142*, 499–525. [CrossRef]
2. Macdonald, I. *Symmetric Functions and Hall Polynomials*, 2nd ed.; Oxford mathematical Monographs; Clarendon Press: Oxford, UK, 1998.
3. Kashivara, M. Crystalizing the q-analogue of universal enveloping algebras. *Commun. Math. Phys.* **1990**, *133*, 249–260. [CrossRef]
4. Grabiner, D.; Magyar, P. Random walks in Weyl chambers and the decomposition of tensor powers. *J. Algebr. Combin.* **1993**, *2*, 239–260. [CrossRef]
5. Grabiner, D. Random walk in an alcove of an affine Weyl group, and non-colliding random walks on an interval. *J. Comb. Theory Ser. A* **2002**, *97*, 285–306. [CrossRef]
6. Tate, T.; Zelditch, S. Lattice path combinatorics and asymptotics of multiplicities of weights in tensor powers. *J. Funct. Anal.* **2004**, *217*, 402–447. [CrossRef]
7. Postnova, O.; Reshetikhin, N. On multiplicities of irreducibles in large tensor product of representations of simple Lie algebras. *arXiv* **2020**, arXiv:1812.11236.
8. Bratteli, O. Inductive limits of finite dimensional C*-algebras. *Trans. Am. Math. Soc.* **1972**, *171*, 195–234. [CrossRef]
9. Lachowska, A.; Postnova, O; Reshetikhin, N.; Solovyev, D. Tensor Powers of Vector Representation of $U_q(sl_2)$ at Even Roots of Unity, TBA. 2022, in press.
10. Faddeev, L.; Reshetikhin, N.; Takhtajan, L. Quantization of Lie groups and Lie algebras. *Algebr. Anal.* **1988**, *1*, 129–139.
11. Chari, V.; Pressley, A. *A Guide to Quantum Groups*; Cambridge University Press: Cambridge, UK, 1995.
12. Andersen, H. Tensor products of quantized tilting modules. *Comm. Math. Phys.* **1992**, *149*, 149–159. [CrossRef]
13. Postnova, O.; Solovyev, D. Counting filter restricted paths in $\mathbb{Z}^2$ lattice. *arXiv* **2021**, arXiv:2107.09774.
14. Lusztig, G. Finite-dimensional Hopf algebras arising from quantized universal enveloping algebra. *J. Am. Math. Soc.* **1990**, *3*, 257–296.
15. Solovyev, D. Towards counting paths in lattice path models with filter restrictions and long steps. *Zap. Nauch Sem. POMI* **2021**, *509*, 201–215.
16. Krattenthaler, C. Lattice path combinatorics chapter. In *Handbook of Enumerative Combinatorics*; Bona, M., Ed.; Chapman and Hall: London, UK, 2015.
17. Andersen, H.H.; Paradowski, J. Fusion categories arising from semisimpleLie algebras. *Comm. Math. Phys.* **1995**, *169*, 563–588. [CrossRef]

18. Zeilberger, D. The Method of Creative Telescoping. *J. Symb. Comput.* **1991**, *11*, 195–204. [CrossRef]
19. Wilf, H.; Zeilberger, D. Rational Functions Certify Combinatorial Identities. *J. Am. Math. Soc.* **1990**, *3*, 147–158. [CrossRef]
20. Paule, P.; Schorn, M. A Mathematica Version of Zeilberger's Algorithm for Proving Binomial Coefficient Identities. *J. Symb. Comput.* **1994**, *11*, 673–698. [CrossRef]
21. Borodin, A.; Okounkov, A.; Olshanski, G. Asymptotics of Plancherel measures for symmetric groups. *J. Am. Math. Soc.* **2000**, *13*, 481–515. [CrossRef]
22. Sutton, L.; Tubbenhauer, D.; Wedrich, P.; Zhu, J. SL2 tilting modules in the mixed case. *arXiv* **2021**, arXiv:2105.07724.
23. Martin, P.P. On Schur-Weyl duality, A_n Hecke algebras and quantum sl(N) on $\otimes^{n+1}\mathbb{C}^N$. *Int. J. Mod. Phys. A* **1992**, *7*, 645–673. [CrossRef]

Article

Semihypergroup-Based Graph for Modeling International Spread of COVID-*n* in Social Systems

Narjes Firouzkouhi [1], Reza Ameri [2], Abbas Amini [3,4] and Hashem Bordbar [5,*]

[1] Department of Mathematics, Golestan University, Gorgan 15759-49138, Iran
[2] School of Mathematics, Statistic, and Computer Science, University of Tehran, Tehran 79416-55665, Iran
[3] Department of Mechanical Engineering, Australian University-Kuwait, Mishref, Safat 13015, Kuwait
[4] Centre for Infrastructure Engineering, Western Sydney University, Penrith, NSW 2751, Australia
[5] Center for Information Technologies and Applied Mathematics, University of Nova Gorica, 5000 Nova Gorica, Slovenia
* Correspondence: hashem.bordbar@ung.si

Abstract: Graph theoretic techniques have been widely applied to model many types of links in social systems. Also, algebraic hypercompositional structure theory has demonstrated its systematic application in some problems. Influenced by these mathematical notions, a novel semihypergroup-based graph (SBG) of $G = \langle H, E \rangle$ is constructed through the fundamental relation γ_n on H, where semihypergroup H is appointed as the set of vertices and E is addressed as the set of edges on SBG. Indeed, two arbitrary vertices x and y are adjacent if $x\gamma_n y$. The connectivity of graph G is characterized by $x\gamma^* y$, whereby the connected components SBG of G would be exactly the elements of the fundamental group H/γ^*. Based on SBG, some fundamental characteristics of the graph such as complete, regular, Eulerian, isomorphism, and Cartesian products are discussed along with illustrative examples to clarify the relevance between semihypergroup H and its corresponding graph. Furthermore, the notions of geometric space, block, polygonal, and connected components are introduced in terms of the developed SBG. To formulate the links among individuals/countries in the wake of the COVID (coronavirus disease) pandemic, a theoretical SBG methodology is presented to analyze and simplify such social systems. Finally, the developed SBG is used to model the trend diffusion of the viral disease COVID-*n* in social systems (i.e., countries and individuals).

Keywords: graph theory; hypergroup; fundamental relation; social systems; geometric space

MSC: 05C25; 20N20

Citation: Firouzkouhi, N.; Ameri, R.; Amini, A.; Bordbar, H. Semihypergroup-Based Graph for Modeling International Spread of COVID-*n* in Social Systems. *Mathematics* 2022, *10*, 4405. https://doi.org/10.3390/math10234405

Academic Editor: Patrick Solé

Received: 19 September 2022
Accepted: 17 November 2022
Published: 22 November 2022

Publisher's Note: MDPI stays neutral with regard to jurisdictional claims in published maps and institutional affiliations.

Copyright: © 2022 by the authors. Licensee MDPI, Basel, Switzerland. This article is an open access article distributed under the terms and conditions of the Creative Commons Attribution (CC BY) license (https://creativecommons.org/licenses/by/4.0/).

1. Introduction

Graph theory with its systematic structure is applied to different complicated problems such as physical, biological, and social systems. By employing graph theory, social network structures can be modeled and analyzed to provide simplified knowledge of such systems, where nodes (vertices) are users and lines (edges) are the links among users. Graph theory was first proposed by Euler to solve Konigsberg's seven-bridge problem [1]. After that, he established a novel graph structure called an Eulerian graph [2]. The concepts of a complete graph [3] and a bipartite graph was defined along with tree structure and coloring problems [4]. With the integration of graph theory and fuzzy set theory, the notion of fuzzy graph theory was proposed by Kaufmann. Then, this theory was developed by Rosenfeld, where fuzzy relations on fuzzy sets were introduced to improve graph-theoretic concepts (e.g., bridges and trees) [5]. To eliminate new problems in science, especially combinatorics, hypergraph theory was initiated and formulated by Berge [6] as the generalization of graph theory, where the edges are arbitrary subsets of the vertices to effectively analyze and simplify complex relations in various spectra for real-world problems [7].

Algebraic hypercompositional structure theory, with its dynamic multi-valued systems, is enumerated as the extension of a classical algebraic structure. Marty introduced a hyperoperation (hypercomposition) on a nonvoid set H, which is a map from $H \times H$ to the power set $P(H)$ of H, such that with associative property and reproductivity, H would be hypergroup [8]. Then, the hypercompositional structure theory was improved in terms of theory and applications by Corsini et al. [9]. Freni determined a novel characterization of the derived hypergroup via strongly regular equivalence relation γ on a hypergroup H, and a binary operation on the quotient set H/γ^* so that H/γ^* is a group with relation γ^* as a fundamental relation (γ^* is the transitive closure of γ and H/γ^* is the fundamental group) [10,11]. Indeed, a fundamental relation is a powerful gadget for the derivation of universal algebra (group, ring, module, etc.) on algebraic hypercompositional structures as well as fuzzy algebraic hypercompositional structures. The present authors studied and formulated the fundamental relations on the fuzzy hypergroup, fuzzy hyperring, and fuzzy hypermodule, where their fundamental relations have the smallest equivalence relation resulting in their quotients being a group, ring, and module, respectively, [12–14]. In other studies, they appointed the fundamental functor between the category of fuzzy hyperrings (hypermodules) and the category of rings (modules) [15,16].

The relevance between graphs/hypergraphs and hypergroups has been investigated by many scholars such as Corsini [17] and Leoreanu [18]. Farshi et al. studied hypergroups associated with hypergraphs and established a ρ-hypergroup with a given hypergraph by describing a relation ρ which resulted in the fundamental relation of an ρ-hypergroup [19]. Kalampakas et al. surveyed path hypergroupoids, especially commutativity and graph connectivity, along with the directed graph isomorphism classes of C-hypergroupoids [20]. Nikkhah et al. developed hypergroups constructed from hypergraphs using a hyperoperation upon the set of vertice degrees of a hypergraph, where the established hypergroupoid is H_v-group [21]. Recently, the present authors proposed a Caley graph related to a semi-hypergroup (hypergroup) with some important features including the category of Cayley graphs and a functor with an application in social networks [22].

With dynamic and potential applications of graph theory in various fields of science, i.e., computer science, linguistics, physics, chemistry, social sciences, biology, mathematics, bioinformatics, etc., many studies have been conducted [23]. For example, Savinkov et al. analyzed and modeled human lymphatic systems via graph theory [24]. The systematic converter derivation/modeling and advanced control in an emerging/challenging power electronics converter was simulated by graph theory as a powerful mathematical structure [25]. Park et al. indicated important insights from complex travel mobility networks with graph-based spatiotemporal analytics [26]. In another work, an effective transductive learning technique was proposed by employing variational nonlocal graph theory for hyperspectral image classification [27]. Recently, the authors presented a soft hypergraph as the generalization of graph theory with the pragmatic application for modeling global interactions in social media networks [28].

The COVID-19 (coronavirus disease 2019) pandemic is considered the most fatal global health catastrophe to date with its serious negative and destructive impact on human life, i.e., social, economical, and environmental challenges. After its detection, the virus extended globally and caused innumerable death. At present, there is no definitive treatment of clinical antiviral drugs or vaccines against the virus [29]. Almost whole nations attempted to decline the transition of the disease via examination and treating patients, quarantining suspected persons through contact tracing, limiting large gatherings, maintaining complete or partial lockdowns, etc. The impact of COVID-19 on various societies and useful ways for controlling viral disease were investigated in [30].

The principal objective of this study is to establish a novel framework of a graph called SBG using a specific relation of algebraic hypercompositional structures in the context of social systems, i.e., the spread trend of the coronavirus disease among societies and individuals. After the Introduction and the Preliminary sections, in Section 3, we appoint a neoteric graph $G = \langle H, E \rangle$ by applying a fundamental relation γ^* on a semihypergroup H.

The elements of H are vertices and two vertices x and y are adjacent if $x\gamma_n y$, that is, they are considered edges. The connectivity SBG of G is defined as $x\gamma^* y$, where the connected components of G are precisely the elements of the fundamental group H/γ^*. Certain fundamental properties of graph theory such as complete, regular, Eulerian, isomorphism, and Cartesian products are proposed. In addition, elucidatory examples are applied to demonstrate the relationship between semihypergroup (hypergroup) H and its associated graph. The mathematical notions of geometric space, block, polygonal, and connected components are discussed. In the end, in Section 4, the developed SBG is utilized to model the global outbreak of COVID-n in social systems (i.e., individuals as well as countries) (Figure 1).

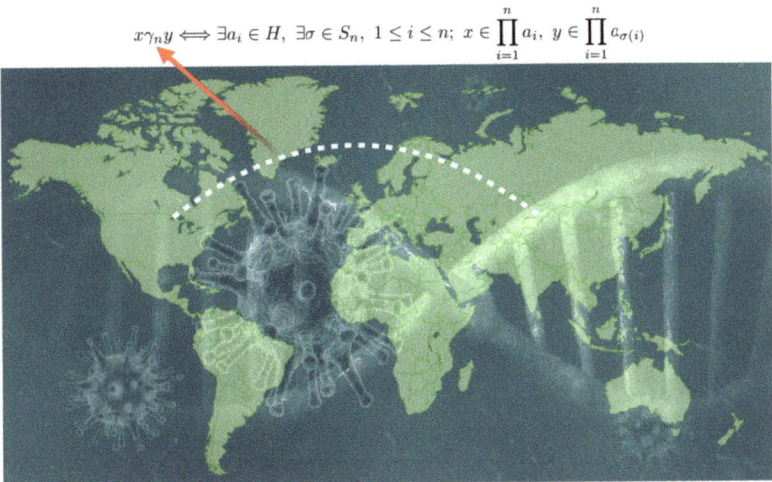

Figure 1. SBG for modeling global spread of COVID-n.

2. Preliminaries

Definition 1. *A hypergroupoid $(L, \diamond)$ is a nonvoid set L with a hyperoperation $\diamond$, which is a map $\diamond : L \times L \to P^*(L)$, where $P^*(L)$ implies the family of all nonvoid subsets of L [9]. Denote $c \diamond d$ as the hyperproduct of c and d for every $c, d \in L$. A hypergroupoid $(L, \diamond)$ is described as a semihypergroup if L has associative property, i.e., $(c \diamond d) \diamond e = c \diamond (d \diamond e)$ for all $c, d, e \in L$. A hypergroup is a semihypergroup along with reproductivity axiom, that is $e \diamond L = L \diamond e = L$ for all $e \in L$. A hypergroupoid $(L, \diamond)$ is called quasihypergroup if the reproductivity property holds. The hypergroup is commutative if $e \diamond f = f \diamond e$ for all $e, f \in L$. A nonvoid subset M of a hypergroup L is a subhypergroup of L if $z \diamond M = M \diamond z = M$ for every $z \in M$.*

Assume E and F are nonvoid subsets of L, hence $E \diamond F = \bigcup_{e \in E, f \in F} e \diamond f$. Moreover, $l \in L$ and $E \subseteq L$, we have $l \diamond E = \bigcup_{e \in E} l \diamond e$. If associativity holds, then we denote the hyperproduct of elements $x_1, \ldots, x_n$ of L by $\prod_{i=1}^{n} x_i := x_1 \diamond x_2 \diamond \ldots \diamond x_n$.

Suppose that $(L, \diamond)$ and $(L', \diamond')$ are two hypergroups. A map $\psi : L \longrightarrow L'$ is determined as a homomorphism if $\psi(k \diamond l) = \psi(k) \diamond' \psi(l)$ for all $k, l \in L$. Furthermore, ψ is named an isomorphism if it is one to one and onto homomorphism written by $L \cong L'$.

The following Definition 2, Proposition 1, Theorem 1, Proposition 2, and Theorem 2 are taken from [31].

Definition 2. *Assume that L is a nonvoid set and σ is a binary relation on L. Consider the following hypercomposition "$\circ$" on L as:*

$$x \circ y = \{z \in L : (x, z) \in \sigma, (z, y) \in \sigma\} \qquad (1)$$

$(L, \circ)$ is a hypergroupoid provided there exists $z \in L$ so that $(x, z) \in \sigma$ and $(z, y) \in \sigma$ for every couple of elements $x, y \in L$.

Denote the hypercompositional structure in Equation (1) by L_σ. The reproductivity property in L_σ is satisfied if and only if $(x, y) \in \sigma$ for all $x, y \in L_\sigma$.

Proposition 1.
- L_σ is a quasihypergroup if and only if $(x, y) \in \sigma$ for all $x, y \in L_\sigma$.
- L_σ is a semihypergroup if and only if $(x, y) \in \sigma$ for all $x, y \in L_\sigma$.

Theorem 1. Let σ be a binary relation on the nonvoid set L. Then, the hypercomposition $x \circ y$ satisfies the reproductivity or associativity only when L_σ is total (i.e., $x \circ y = L_\sigma$).

Each relation σ on finite set $L = \{a_1, a_2, \ldots, a_n\}$ can be represented through a Boolean matrix M_σ with $n \times n$ elements. The Boolean matrix $M_\sigma = (m_{ij})$ is defined as follows:

$$m_{ij} = \begin{cases} 1, & \text{if } (a_i, a_j) \in \sigma \\ 0, & \text{otherwise} \end{cases}$$

In Boolean algebra, we have

$$0 + 1 = 1 + 0 = 1 + 1 = 1, \ 0 + 0 = 0$$
$$0.0 = 0.1 = 1.0 = 0, \ 1.1 = 1$$

L_σ is hypergroupoid if and only if $M_\sigma^2 = S$, where $S = (s_{ij})$ with $s_{ij} = 1$ for all i, j.

Proposition 2.
- L_σ is a quasihypergroup if and only if $M_\sigma = S$.
- L_σ is a semihypergroup if and only if $M_\sigma = S$.

Theorem 2. The only relation σ which results in a quasihypergroup or semihypergroup is the one with $M_\sigma = S$. Additionally, L_σ is the total hypergroup.

It was revealed that with a few lines of the Mathematica program, the results were constructed for the enumeration of the hypergroupoid associated with binary relations of orders 2, 3, 4, and 5 by a significantly simpler procedure [31].

Definition 3. A graph G is a pair $G = (V, E)$, where V is a set of elements described as vertices and E is a set of edges [32]. The two vertices associated with an edge are called endpoints. If $x = y$, then the edge is considered as a loop. A vertex is isolated if it is incident with no edges. The graph G is simple if it has no loops and no two distinct edges have the same pair of ends. The graph G is called null graph when its edges set is empty. Graph H is named a subgraph of graph G if $V(H) \subseteq V(G)$, $E(H) \subseteq E(G)$, and the ends of an edge $e \in E(H)$ are the same as its ends in G. Denote $d(x)$ as the degree of vertex x as well as the number of edges incident with x.

A path in graph G consists of a sequence $x_1, e_1, x_2, e_2, \ldots, e_k, x_k$ that the edges e_i are distinct. Furthermore, if $x_1 = x_k$ then, we call the path a cycle. Consider that $d(x, y)$ is the length of the shortest path between two vertices x and y. Note that $diam(G) = sup\{d(a, b)\}$ for all a and b that are vertices of G, which is called the diameter of graph G. The graph G is connected if there exists a path from vertex x to vertex y, or graph G includes several connected components. A tree is a connected graph that includes no simple cyclic path. Denote k_n as a complete graph, where every pair of vertices is adjacent. An Eulerian circuit is a closed path through a graph applying each edge once and an Eulerian graph is a graph that has this property. Furthermore, graph G is called a Hamiltonian graph if it has a cycle that passes each vertex exactly once. If every vertex has the same degree, the graph is regular, or k-regular if $\forall x \in V, d(x) = k$.

Theorem 3. *A finite graph G without isolated vertices is Eulerian if and only if G is connected and each vertex has an even degree* [32].

Definition 4. *The Cartesian product of two graphs $G_1 = \langle V_1, E_1 \rangle$ and $G_2 = \langle V_2, E_2 \rangle$ is denoted by $G_1 \square G_2$, that is a graph with vertices set $V_1 \times V_2$, where vertices $(t_1, t_2), (w_1, w_2)$ are adjacent if and only if $t_1 = w_1, (t_2, w_2) \in E_2$ or $t_2 = w_2, (t_1, w_1) \in E_1$ for $t_1, w_1 \in V_1$, $t_2, w_2 \in V_2$* [33].

3. Semihypergroup-Based Graph (SBG) Based on Relation γ

Consider an SBG of $G = \langle H; E = (\gamma_n)_{n \in \mathbb{N}} \rangle$, where $(H, \circ)$ is a semihypergroup and γ_n is the relation on H. The order of G is $o(G) = |H|$. The elements of H are represented as vertices and the relations γ_n are appointed as edges. We assign x and y to be adjacent, if $x \gamma_n y$. Clearly, for $n = 1$ and $x \gamma_1 x$, the edge is a loop.

Indeed, γ_n was determined in [10] as follows:

$$x \gamma_n y \iff \exists (a_1, \ldots, a_n) \in H^n, \exists \sigma \in S_n : x \in \prod_{i=1}^{n} a_i, y \in \prod_{i=1}^{n} a_{\sigma(i)} \quad (2)$$

Consider $\gamma_1 = \{(a, a) \mid a \in H\}$. Clearly, the relations γ_n have symmetric property and relation γ has a reflexive and symmetric property for every $n \in \mathbb{N}$, where $\gamma = \bigcup_{n \geq 1} \gamma_n$. Let γ^* be the transitive closure of γ. The class of H/γ^* was addressed as $\gamma^*(z) = \{w \mid z \gamma^* w\}$, for $z, w \in H$. It was proven that for hypergroup H, the relation γ is transitive and γ^* has the smallest strongly regular equivalence property that results H/γ^* is an Abelian group (fundamental group).

Theorem 4. *Assume that H is a hypergroup. Then, for an SBG of $G = \langle H; E = (\gamma_n)_{n \in \mathbb{N}} \rangle$, the following statements hold:*

(i) *A path exists between two vertices x and y of G if and only if $x \gamma^* y$.*
(ii) *The SBG of G is connected if and only if the fundamental group H/γ^* is a singleton, that is $|H/\gamma^*| = 1$.*

Proof. Proof of (i): Consider a path from vertex x to vertex y. Then, there exists a sequence $(a_1, \ldots, a_k) \in H^k$ so that $x = a_1 \gamma_1 a_2 \ldots \gamma_k a_k = y$, that is equal to $x \gamma^* y$. Conversely, if $x \gamma^* y$, then $\exists (a_1, \ldots, a_k) \in H^k$ such that $x = a_1 \gamma_1 a_2 \ldots \gamma_k a_k = y$. Therefore, there exists a path from vertex x to vertex y.

Proof of (ii): By applying (i), for $x, y \in H$, a path exists from vertex x to vertex y if and only if $x \gamma^* y$. Therefore, the SBG of G is a connected graph if and only if $\gamma^* = H \times H$ (i.e., clearly, $\gamma^* \subseteq H \times H$. Furthermore, for all $x, y \in H$, since $(x, y) \in \gamma^*$, then $H \times H \subseteq \gamma^*$). Since $x \gamma^* y$, we have $\gamma^*(x) = \gamma^*(y)$ which means that the fundamental group $H/\gamma^* = \{\gamma^*(x) | x \in H\}$ is a singleton, i.e., $|H/\gamma^*| = 1$. □

Theorem 5. *The connected components SBG of G are precisely the elements of the fundamental group H/γ^*.*

Proof. Let x, y be two vertices SBG of G. By employing Theorem 4, vertex x is connected to vertex y if and only if $x \gamma^* y$. Then, for all $a \in H$, every element of $\gamma^*(a)$ is connected. With the equivalence relation of γ^*, the elements of H/γ^* would be the connected components SBG of G. □

Theorem 6. *Let H be a semihypergroup. If the SBG of $G = \langle H, E \rangle$ is complete, then the relation γ is transitive.*

Proof. Let $x \gamma y$ and $y \gamma z$. For some $n_1, n_2 \in \mathbb{N}$, we have $x \gamma_{n_1} y$ and $y \gamma_{n_2} z$. Since the SBG of G is complete, therefore, for some $n \in \mathbb{N}$, we have $x \gamma_n z$ that yields $x \gamma z$. □

Remark 1. Note that a loop is not considered an edge. If $x\gamma x$, then for every $a_i \in H$, $\exists \sigma \in S_n$ we have $x \in \prod_{i=1}^{n} a_i$ and $x \in \prod_{i=1}^{n} a_{\sigma(i)}$. Hence, $\prod_{i=1}^{n} a_i = \prod_{i=1}^{n} a_{\sigma(i)}$.

Definition 5. Let H be a nonvoid set and let γ^* be the defined relation in Equation (2). Consider the hypercomposition "$\odot$" on H as follows:

$$x \odot y = \{w \in H : (x,w) \in \gamma^*, (w,y) \in \gamma^*\} \qquad (3)$$

We denote the hypercompositional structure $(H, \odot)$ by H_{γ^*}. The H_{γ^*} is a hypergroupoid if $\exists w \in H$ so that $(x,w) \in \gamma^*$ and $(w,y) \in \gamma^*$ for every $x,y \in H$. Since γ^* is transitive, we have $(x,y) \in \gamma^*$ for all $x,y \in H_{\gamma^*}$, then the reproductivity property holds. In fact, for the arbitrary element $x \in H_{\gamma^*}$, the reproductivity axiom $y \in x \odot H_{\gamma^*}$ holds for all $y \in H_{\gamma^*}$, as per the transitive property of γ^*.

Proposition 3.
(i) H_{γ^*} is a semihypergroup if and only if $(x,y) \in \gamma^*$ for all $x,y \in H_{\gamma^*}$.
(ii) H_{γ^*} is a quasihypergroup if and only if $(x,y) \in \gamma^*$ for all $x,y \in H_{\gamma^*}$.
(iii) The SBG of $G = \langle H_{\gamma^*}, E \rangle$ is a connected graph if and only if $(x,y) \in \gamma^*$ for all $x,y \in H_{\gamma^*}$.
(iv) The SBG of $G = \langle H_{\gamma^*}, E \rangle$ is a complete graph if and only if H_{γ^*} is total, i.e., $x \odot y = H_{\gamma^*}$ for all $x,y \in H_{\gamma^*}$.

Proof. Proof of (i): It is derived by applying Proposition 1.

Proof of (ii): With the validity of the reproductivity property, the statement is proven.

Proof of (iii): Since H_{γ^*} is a quasihypergroup and considering part (i), we have H_{γ^*} as a hypergroup. By Theorem 4, we have $x\gamma^*y$ for all $x,y \in H_{\gamma^*}$ if and only if the SBG of G is connected.

Proof of (iv): The statement is attained from Equation (3). □

Example 1. Consider $(H, \circ)$ as a semihypergroup that is given in Table 1.

Table 1. Semihypergroup $(H, \circ)$

$\circ$	0	1	2
0	0	1	2
1	1	{0,2}	1
2	2	1	{0,2}

It is seen that $1 \in 1 \circ 2$, $1 \in 2 \circ 1$, then $1\gamma 1$. Furthermore, we have $0\gamma 0$, $2\gamma 2$ and $0\gamma 2$. The corresponding SBG of G is depicted in Figure 2. Moreover, $H/\gamma^* = \{\{0,2\},1\}$ and $|H/\gamma^*| \neq 1$.

Figure 2. SBG of G.

γ is transitive and the SBG of G is not connected, because vertices 0 and 1 are not adjacent. The SBG of G is not complete, which results in the invalidity of the reverse Theorem 6.

Corollary 1. Let $G = \langle H, E \rangle$ be an SBG, and let H be a semihypergroup. If the SBG of G is complete, then H/γ^* is a singleton, and $diam(G) = 1$.

Proof. By applying Theorems 4 and 6, the relation γ is transitive and H/γ^* is a singleton. Since the SBG of G is complete, then every path from vertex x to vertex y has a maximum length of 1, which means $diam(G) = 1$. □

Proposition 4. *Suppose that H is a hypergroup on the SBG of $G = \langle H, E \rangle$. Then, the degree of vertex x in SBG of G is equal to $|\gamma^*(x)|$.*

Proof. Let H be a hypergroup. By employing Theorem 4 and $\gamma^*(x)$ as an equivalence class of x, the results show that the number of edges incident with vertex x is equal to $|\gamma^*(x)|$. □

Corollary 2. *Let $G = \langle H, E \rangle$ be an SBG, and let H be a hypergroup. Assume that $|\gamma^*(x)| = k$ for all $x \in H$. Then, the SBG of G is a k-regular graph.*

Theorem 7. *Let H_1 be a hypergroup on SBG of $G_1 = \langle H_1, E_1 \rangle$. Let H_2 be a subhypergroup of H_1 on SBG of $G_2 = \langle H_2, E_2 \rangle$. Then, the SBG of G_2 is a sub-SBG of G_1.*

Proof. Assume that H_2 is a subhypergroup of H_1, then $H_2 \subset H_1$. Therefore, the vertices SBG of G_2 is contained in the vertices SBG of G_1 and the edges G_2 is included in the edges of G_1. Then, the SBG of G_2 is a sub-SBG of G_1. □

Theorem 8. *Let H be a hypergroup. The SBG of $G = \langle H, E \rangle$ is Eulerian if and only if $|\gamma^*(z)| = 2k$ for all $z \in H, k \in \mathbb{N}$.*

Proof. Let H be a hypergroup. Then, the relation γ is transitive [9]. By applying Theorem 4, the SBG of G is a connected graph. Additionally, with Proposition 4, $d(z) = |\gamma^*(z)|$, for all $z \in H$ and by Theorem 3, the proof is completed. □

Example 2. *Let $(H, \circ)$ be a hypergroup in [34] (Example 28 (3)).*

The corresponding SBG of G is shown in Figure 3, which is a connected and complete graph. Moreover, $H/\gamma^* = \{\{a, b, c\}\}$ and $|H/\gamma^*| = 1$. Additionally, $|\gamma^*(a)| = |\gamma^*(b)| = |\gamma^*(c)| = 2$, that means $d(a) = d(b) = d(c) = 2$. Furthermore, $diam(G) = 1$ and the SBG of G is a 2-regular and Eulerian graph.

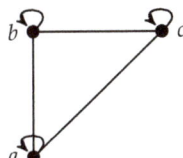

Figure 3. SBG of G.

Definition 6. *The SBG of G is isomorphic to the SBG of G', if there exists a bijection ϕ from the set vertices of G to the set vertices of G', such that $x\gamma_G y \iff \phi(x)\gamma_{G'}\phi(y)$, written by $G \cong G'$.*

Theorem 9. *Let $(H_1, \circ_1)$ and $(H_2, \circ_2)$ be two isomorphic hypergroups and let G_1 and G_2 be two SBGs associated with H_1 and H_2, respectively. Then, the SBG of G_1 and the SBG of G_2 are isomorphisms.*

Proof. Assume H_1 and H_2 are isomorphisms. Then, $|H_1| = |H_2|$ and we have $|G_1| = |G_2|$. Furthermore, if vertex x is connected to vertex y, then $x\gamma^*y$ and we have $\exists (a_1, \ldots, a_n) \in H^n$, $\exists \sigma \in S_n$; $x \in \prod_{i=1}^{n} a_i$, $y \in \prod_{i=1}^{n} a_{\sigma(i)}$. Let $\phi : H_1 \longrightarrow H_2$ be an isomorphism and let $\phi(x) = x'$, $\phi(y) = y'$ and $\phi(a_i) = a_i'$. Furthermore, $\phi(\prod_{i=1}^{n} a_i) = \prod_{i=1}^{n} \phi(a_i) = \prod_{i=1}^{n} a_i'$, which yields x' that is connected to y'. Hence, $G_1 \cong G_2$. □

Example 3. To show that the reverse of Theorem 9 is not satisfied, consider two hypergroups $(H, \circ)$ and $(H', \circ')$ in [34] (Example 16 (3)).

Let $f : H \longrightarrow H'$ with $f(a) = 1$, $f(b) = 1$, $f(c) = 2$. Since $f(a \circ b) = f(H) = \{1, 2\}$ and $f(a) \circ' f(b) = 1 \circ 1 = 1$, means that f is not an isomorphism (i.e., $f(a \circ b) \neq f(a) \circ' f(b)$). The two SBGs are isomorphisms, as depicted in Figure 4.

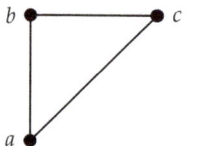 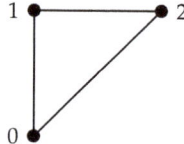

Figure 4. SBGs of G and G' associated with H and H'.

Definition 7. Let $G = \langle H, E \rangle$ and $G' = \langle H', E' \rangle$ be two SBGs, where H and H' are two hypergroups and $E = \{E_1, \ldots, E_m\}$ and $E' = \{E'_1, \ldots, E'_n\}$. Define the Cartesian product $G * G'$ with the vertices set $H \times H'$ and edges set $E_l \times E'_k$ for $1 \leq l \leq m$, $1 \leq k \leq n$.

Example 4. Consider two SBGs in Example 3. By considering $G = \langle H, E \rangle$ and $G' = \langle H', E' \rangle$, the Cartesian product of two SBGs G and G' is depicted in Figure 5. The vertices of $G * G'$ are $H \times H' = \{(a, 0), (a, 1), (a, 2), (b, 0), (b, 1), (b, 2), (c, 0), (c, 1), (c, 2)\}$ and the corresponding edges are $E \times E' = \{[(a, 0), (b, 0)], [(a, 0), (a, 1)], \ldots, [(a, 2), (c, 2)]\}$.

 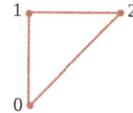

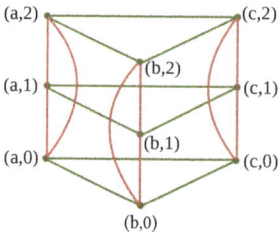

Figure 5. Cartesian product of SBGs G and G'.

Proposition 5. Let $G = \langle H, E \rangle$ and $G' = \langle H', E' \rangle$ be two SBGs and let $(a_1, b_1), (a_2, b_2) \in H \times H'$. Then,

$$(a_1, b_1) \gamma_{G \times G'} (a_2, b_2) \Longleftrightarrow a_1 \gamma_G a_2, \ b_1 \gamma_{G'} b_2.$$

Geometric Concept of SBG

A *geometric space* is a couple (S, V) where S is a nonvoid set and V is the family of a nonvoid subset of S. The elements of S are considered points and the elements of V are represented as blocks. If V covers S, then a *polygonal* of (S, V) is an n-tuple of blocks $(V_1, V_2, \ldots, V_n)$ so that $V_i \cap V_{i+1} \neq \emptyset$, for every $i \in \{1, 2, \ldots, n-1\}$. Introduce the relation $\approx$ on S as follows:

$$x \approx y \Longleftrightarrow \exists (V_1, V_2, \ldots, V_n); \ x \in V_1, \ y \in V_n.$$

If V covers S, then the relation is an equivalence relation. The equivalence class $[x]$ is determined as a *connected component* of x in S [10,11].

According to the SBG of G, we consider a pair $\langle H, E \rangle$ as a *geometric space of SBG*, where H is a semihypergroup (set of vertices) and E is the set of relations γ_n (set of edges) for $n \in \mathbb{N}$ on H. For every $x, y \in H$, we have $xE_iy \iff x\gamma_n y$ with the given relation γ_n as follows:

$$x\gamma_n y \iff \exists (a_1, \ldots, a_n) \in H^n, \exists \sigma \in S_n : x \in \prod_{i=1}^{n} a_i, y \in \prod_{i=1}^{n} a_{\sigma(i)}$$

Take a polygonal SBG of $G = \langle H, E \rangle$ as $(E_1, E_2, \ldots, E_n)$, so that $E_i \cap E_{i+1} \neq \emptyset$ (i.e., $(x, x') \in E_i, (x', x'') \in E_{i+1}$) for $1 \leq i \leq n-1$. By applying the polygonal concept of SBG, the relation $\approx$ is defined as follows:

$$x \approx y \iff \exists E_i, 1 \leq i \leq n; (x, z) \in E_1, (z, y) \in E_n$$

The relation $\approx$ is an equivalence relation. The SBG of G is connected and the equivalence class $[x] = \{y \mid x\gamma^* y\} = |\gamma^*(x)|$, where $[x]$ is a connected component by Theorem 4. Indeed, the connected components SBG of $G = \langle H, E \rangle$ are equivalence classes modulo γ^*. The geometric space $G = \langle H, E \rangle$ is connected if it includes only one connected component, i.e., $H = [x]$, for $x \in H$. Clearly, the relation $\approx$ is the transitive closure of the relation $\gamma = \bigcup_{n \in \mathbb{N}} \gamma_n$. The blocks of the geometric space SBG of $G = \langle H, E \rangle$ using relation γ_n are the constructed sets with permuting finite hyperproducts of distinct finite points (vertices).

4. SBG for Modeling the Spread Trend of COVID-n

SBG can be utilized to model the spread trend of COVID-n by travelers in different countries and on a large scale, involved countries. In this pattern, the vertices represent individuals/countries and edges appoint the relationship among individuals/countries which are based on a fundamental relation.

4.1. Application 1

Let H be the number of individuals. Consider $H = \{\text{Michael, Robert, Emma, Olivia}\}$. Then, the SBG of $G = \langle H, E \rangle$ is determined in the following way:

- Each vertex addresses an individual
- An edge addresses the relationship between two vertices

Define a binary relation "$\circ$" on H as follows:

$$a \circ b = \{x \mid x \text{ get infected to COVID} - n \text{ by person } a \text{ or person } b\}$$

In Table 2, the pair $(H, \circ)$ is a hypergroup.

Table 2. Hypergroup $(H, \circ)$.

$\circ$	Michael = 1	Robert = 2	Emma = 3	Olivia = 4
Michael = 1	1	2	3	4
Robert = 2	2	{1,2}	{3,4}	3
Emma = 3	3	{3,4}	H	{2,3}
Olivia = 4	4	3	{2,3}	{1,4}

The following statements are attained from Table 2:

- Either Robert, Michael, or Emma infected Olivia with COVID.
- Emma is the most infectious the person for the transmission of the coronavirus disease and all members get infected by Emma ($3 \circ 3 = H$).

Consider the relation γ_n as edges for two arbitrary vertices x and y as:

$$x\gamma_n y \iff \exists (a_1,\ldots,a_n) \in H^n, \exists \sigma \in S_n : x \in \prod_{i=1}^n a_i, y \in \prod_{i=1}^n a_{\sigma(i)}$$

Note that $\prod_{i=1}^n a_i$ is regarded as a hyperproduct of distinct elements a_i for $i \in \{1,2,\ldots,n\}$, that is $a_1 \circ a_2 \circ \ldots a_n$. We follow the procedure for all components, i.e.,

$$1\gamma_2 2 \iff 1 \in 2 \circ 2, 2 \in 2 \circ 2$$
$$3\gamma_2 4 \iff 3 \in 2 \circ 3, 4 \in 3 \circ 2$$
$$2\gamma_2 3 \iff 2 \in 3 \circ 4, 3 \in 4 \circ 3$$
$$1\gamma_2 4 \iff 1 \in 4 \circ 4, 4 \in 4 \circ 4$$
$$2\gamma_2 4 \iff 2 \in 3 \circ 3, 4 \in 3 \circ 3$$
$$1\gamma_2 3 \iff 1 \in 3 \circ 3, 3 \in 3 \circ 3$$

This means that $(1,2) \in e_1, (3,4) \in e_2, (2,3) \in e_3, (1,4) \in e_4, (2,4) \in e_5, (1,3) \in e_6$ where, $E = \{e_1, e_2, e_3, e_4, e_5, e_6\}$ are the edges of SBG. The corresponding SBG of G is depicted in Figure 6a and Table 3.

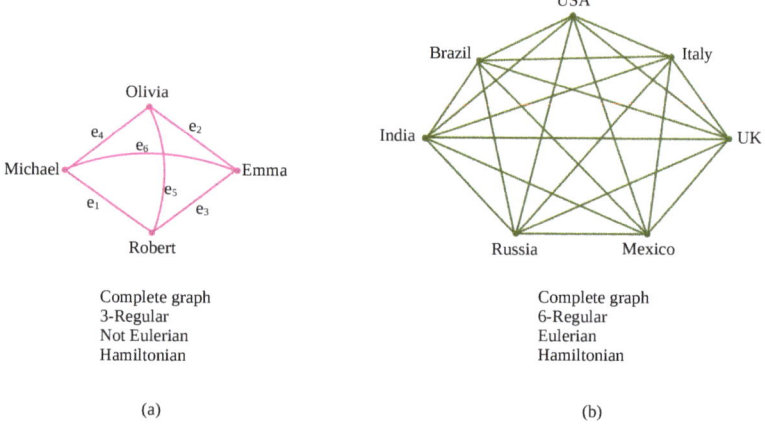

Figure 6. SBGs of G corresponding to (a) Application 1 and (b) Application 2.

Table 3. SBGs of G.

	Michael	Robert	Emma	Olivia
e_1	1	1	0	0
e_2	0	0	1	1
e_3	0	1	1	0
e_4	1	0	0	1
e_5	0	1	0	1
e_6	1	0	1	0

Furthermore, the equivalence class of $[x]$ is considered as the individuals who transmit viral disease COVID to specific person x, that is $[x] = \{y \mid x\gamma^* y\}$, where γ^* is the transitive closure of γ and $\gamma = \bigcup_{n\geq 1} \gamma_n$. Therefore, the class [Michael] = {Robert, Emma, Olivia}, and so on. By applying Proposition 4, the degree of Michael is $|\gamma^*(\text{Michael})| = 3$ and by Corollary 2, the SBG is 3-regular.

4.2. Application 2

Let H be a set of countries with the most reported cases and death in the world. Consider $H = \{\text{USA, Brazil, India, Russia, Mexico, UK, Italy}\}$. Thus, the SBG of $G = \langle H, E \rangle$ is defined as follows

- Every vertex is appointed to a country.
- An edge is appointed to the relationship between two vertices.

Introduce the hyperoperation "$\oplus$" for all $x, y \in H$, as follows:

$x \oplus y = $ The country or set of countries that causes disease outbreak from country x to country y

The couple $(H, \oplus)$ is a hypergroupoid, as given in Table 4.

Table 4. Hypergroupoid $(H, \oplus)$.

$\oplus$	USA = 1	Brazil = 2	India = 3	Russia = 4	Mexico = 5	UK = 6	Italy = 7
USA = 1	{1,2}	2	{1,2,3}	4	{1,2,5}	{1,6,2}	{1,7,2}
Brazil = 2	2	2	2	{2,4}	2	{2,6}	{1,2}
India = 3	{1,2}	{2,3}	3	{2,3,4}	{1,2,3,5}	{1,2,3,6}	{1,2,3,7}
Russia = 4	{1,4}	2	{3,4}	4	{2,4,5}	{2,4,6}	{1,2,4,7}
Mexico = 5	{1,2,5}	2	{2,3,5}	{1,2,4,5}	5	{1,2,5,6}	H
UK = 6	{1,6}	2	{2,3}	{2,4}	{1,2,5,6}	6	{1,2,4,6,7}
Italy = 7	{1,7}	H	{2,3}	{2,4}	{1,2,5,7}	{2,6,7}	7

Consider the relation γ given below:

$$x\gamma_n y \iff \exists (a_1, \ldots, a_n) \in H^n, \exists \sigma \in S_n : x \in \prod_{i=1}^{n} a_i, \; y \in \prod_{i=1}^{n} a_{\sigma(i)}$$

we continue the procedure for all elements of H, according to Table 4, that is

$$1\gamma_2 2 \iff 1 \in 3 \oplus 1,\ 2 \in 1 \oplus 3$$
$$1\gamma_2 3 \iff 1 \in 3 \oplus 5,\ 3 \in 5 \oplus 3$$
$$1\gamma_2 4 \iff 1 \in 4 \oplus 1,\ 4 \in 1 \oplus 4$$
$$1\gamma_2 5 \iff 1 \in 3 \oplus 5,\ 5 \in 5 \oplus 3$$
$$1\gamma_2 6 \iff 1 \in 1 \oplus 6,\ 6 \in 6 \oplus 1$$
$$1\gamma_2 7 \iff 1 \in 1 \oplus 7,\ 7 \in 7 \oplus 1$$
$$2\gamma_2 3 \iff 2 \in 3 \oplus 4,\ 3 \in 4 \oplus 3$$
$$2\gamma_2 4 \iff 2 \in 3 \oplus 4,\ 4 \in 4 \oplus 3$$
$$2\gamma_2 5 \iff 2 \in 4 \oplus 5,\ 5 \in 5 \oplus 4$$
$$2\gamma_2 6 \iff 2 \in 6 \oplus 2,\ 6 \in 2 \oplus 6$$
$$2\gamma_2 7 \iff 2 \in 1 \oplus 7,\ 7 \in 7 \oplus 1$$
$$3\gamma_2 4 \iff 3 \in 3 \oplus 4,\ 4 \in 4 \oplus 3$$
$$3\gamma_2 5 \iff 3 \in 3 \oplus 5,\ 5 \in 5 \oplus 3$$
$$3\gamma_2 6 \iff 3 \in 6 \oplus 3,\ 6 \in 3 \oplus 6$$
$$3\gamma_2 7 \iff 3 \in 7 \oplus 3,\ 7 \in 3 \oplus 7$$
$$4\gamma_2 5 \iff 4 \in 4 \oplus 5,\ 5 \in 5 \oplus 4$$
$$4\gamma_2 6 \iff 4 \in 6 \oplus 7,\ 6 \in 7 \oplus 6$$
$$4\gamma_2 7 \iff 4 \in 6 \oplus 7,\ 7 \in 7 \oplus 6$$
$$5\gamma_2 6 \iff 5 \in 5 \oplus 6,\ 6 \in 6 \oplus 5$$
$$5\gamma_2 7 \iff 5 \in 5 \oplus 7,\ 7 \in 7 \oplus 5$$
$$6\gamma_2 7 \iff 6 \in 6 \oplus 7,\ 7 \in 7 \oplus 6$$

Therefore, $E = \{e_1, \ldots, e_{21}\}$ and the corresponding SBG of G is demonstrated in Figure 6b. By applying Proposition 4, the degree of each vertex is $|\gamma^*(z)| = 6$, and G is complete, and 6-regular. It also has an Eulerian circuit because of connectivity and has an even degree of each vertex; therefore, graph G is Eulerian. The SBG of G is connected and Hamiltonian and the relation γ is transitive.

5. Conclusions

The neoteric structure of a semihypergroup-based graph (SBG) is established using a fundamental relation to advance the mathematical concept of an algebraic hypercompositional structure, namely the hypergroup, in the form of graph theory. Additionally, to model and analyze the links in social systems, the developed SBG approach is recommended to intuitively simplify the complicated procedure. Some significant characteristics of SBG are proposed, including connected, complete, regular, Eulerian, isomorphism, and Cartesian products along with illustrative examples and graphical attitude. As per the engagement of all nations and individuals after the global COVID-n pandemic, the resulting SBG is applied to address the trend of transmission of the coronavirus disease in social systems, particularly countries and individuals. The next phase can be the development of fuzzy SBG and intuitionistic fuzzy SBG with further applicable platforms.

Author Contributions: Conceptualization, N.F. and R.A.; methodology, N.F. and R.A.; software, N.F. and H.B.; validation, A.A. and H.B.; formal analysis, N.F.; investigation, N.F. and R.A.; resources, N.F. and R.A.; data curation, N.F.; writing—original draft preparation, N.F.; writing—review and editing, N.F., R.A., A.A. and H.B.; visualization, N.F. and R.A.; supervision, R.A. and A.A.; project administration, N.F.; funding acquisition, H.B. All authors have read and agreed to the published version of the manuscript.

Funding: This research received no external funding.

Institutional Review Board Statement: Not applicable.

Informed Consent Statement: Not applicable.

Data Availability Statement: Not applicable.

Acknowledgments: We gratefully acknowledge BioRender software, Canada, for assisting in the figure drawing.

Conflicts of Interest: The authors declare no conflict of interests.

References

1. Chen, W.K. *Applied Graph Theory*; Elsevier: Amsterdam, The Netherlands, 2012; Volume 13.
2. Dankelmann, P. Proof of a conjecture on the Wiener index of Eulerian graphs. *Discret. Appl. Math.* **2021**, *301*, 99–108. [CrossRef]
3. Li, Y.; Liu, X.; Zhang, S.; Zhou, S. Perfect state transfer in NEPS of complete graphs. *Discret. Appl. Math.* **2021**, *289*, 98–114. [CrossRef]
4. Dokeroglu, T.; Sevinc, E. Memetic Teaching-Learning-Based Optimization algorithms for large graph coloring problems. *Eng. Appl. Artif. Intell.* **2021**, *102*, 104282. [CrossRef]
5. Rosenfeld, A. Fuzzy graphs. In *Fuzzy Sets and Their Applications to Cognitive and Decision Processes*; Academic Press: Cambridge, MA, USA, 1975; pp. 77–95.
6. Bretto, A. *Hypergraph Theory. An introduction*; Mathematical Engineering; Springer: Cham, Switzerland, 2013.
7. Gerbner, D. A note on the uniformity threshold for Berge hypergraphs. *Eur. J. Comb.* **2022**, *105*, 103561. [CrossRef]
8. Massouros, C.; Massouros, G. An overview of the foundations of the hypergroup theory. *Mathematics* **2021**, *9*, 1014. [CrossRef]
9. Corsini, P.; Leoreanu, V. *Applications of Hyperstructure Theory*; Springer Science and Business Media: Berlin/Heidelberg, Germany, 2013; Volume 5.
10. Freni, D. A new characterization of the derived hypergroup via strongly regular equivalences. *Commun. Algebra* **2002**, *30*, 3977–3989. [CrossRef]
11. Freni, D. *Strongly Transitive Geometric Spaces: Applications to Hypergroups and Semigroups Theory*; Taylor & Francis: Abingdon, UK, 2004.
12. Firouzkouhi, N.; Ameri, R. New Fundamental Relation on Fuzzy Hypersemigroups. *New Math. Nat. Comput.* **2021**, *18*, 209–217. [CrossRef]
13. Davvaz, B.; Firouzkouhi, N. Commutative Rings Derived from Fuzzy Hyperrings. *Honam Math. J.* **2020**, *42*, 219–234.
14. Davvaz, B.; Firouzkouhi, N. Fundamental relation on fuzzy hypermodules. *Soft Comput.* **2019**, *23*, 13025–13033. [CrossRef]
15. Firouzkouhi, N.; Davvaz, B. New fundamental relation and complete part of fuzzy hypermodules. *J. Discreet. Math. Sci. Cryptogr.* **2020**, *25*, 1225–1237. [CrossRef]
16. Firouzkouhi, N.; Davvaz, B. Γ*-Relation on Fuzzy Hyperrings and Fundamental Ring. *New Math. Nat. Comput.* **2021**, *17*, 691–701. [CrossRef]
17. Corsini, P.; Leoreanu-Fotea, V.; Iranmanesh, A. On the Sequence of Hypergroups and Membership Functions Determined by a Hypergraph. *J. Mult.-Valued Log. Soft Comput.* **2008**, *14*, 565–577.
18. Leoreanu-Fotea, V.; Corsini, P. Isomorphisms of hypergroups and of n-hypergroups with applications. *Soft Comput.* **2009**, *13*, 985–994. [CrossRef]
19. Farshi, M.; Davvaz, B.; Mirvakili, S. Hypergraphs and hypergroups based on a special relation. *Commun. Algebra* **2014**, *42*, 3395–3406. [CrossRef]
20. Kalampakas, A.; Spartalis, S. Path hypergroupoids: Commutativity and graph connectivity. *Eur. J. Comb.* **2015**, *44*, 257–264. [CrossRef]
21. Nikkhah, A.; Davvaz, B.; Mirvakili, S. Hypergroups constructed from hypergraphs. *Filomat* **2018**, *32*, 3487–3494. [CrossRef]
22. Shamsi, K.; Ameri, R.; Mirvakili, S. Cayley graph associated to a semihypergroup. *Algebr. Struct. Their Appl.* **2020**, *7*, 29–49.
23. Gross, J.L.; Yellen, J. *Graph Theory and Its Applications*; CRC Press: Boca Raton, FL, USA, 2005.
24. Savinkov, R.; Grebennikov, D.; Puchkova, D.; Chereshnev, V.; Sazonov, I.; Bocharov, G. Graph Theory for Modeling and Analysis of the Human Lymphatic System. *Mathematics* **2020**, *8*, 2236. [CrossRef]
25. Li, Y.; Kuprat, J.; Li, Y.; Liserre, M. Graph-Theory-Based Derivation, Modeling and Control of Power Converter Systems. *IEEE J. Emerg. Sel. Top. Power Electron.* **2022**. [CrossRef]
26. Park, S.; Yuan, Y.; Choe, Y. Application of graph theory to mining the similarity of travel trajectories. *Tour. Manag.* **2021**, *87*, 104391. [CrossRef]
27. Huang, B.; Ge, L.; Chen, G.; Radenkovic, M.; Wang, X.; Duan, J.; Pan, Z. Nonlocal graph theory based transductive learning for hyperspectral image classification. *Pattern Recognit.* **2021**, *116*, 107967. [CrossRef]
28. Amini, A.; Firouzkouhi, N.; Gholami, A.; Gupta, A.R.; Cheng, C.; Davvaz, B. Soft hypergraph for modeling global interactions via social media networks. *Expert Syst. Appl.* **2022**, *203*, 117466. [CrossRef]
29. Mercer, T.R.; Salit, M. Testing at scale during the COVID-19 pandemic. *Nat. Rev. Genet.* **2021**, *22*, 415–426. [CrossRef]
30. Chakraborty, I.; Maity, P. COVID-19 outbreak: Migration, effects on society, global environment, and prevention. *Sci. Total Environ.* **2020**, *728*, 138882. [CrossRef] [PubMed]

31. Massouros, C.G.; Tsitouras, C. Enumeration of hypercompositional structures defined by binary relations. *Ital. J. Pure Appl. Math.* **2011**, *28*, 43–54.
32. Rahman, M.S. *Basic Graph Theory*; Springer International Publishing: Cham, Switzerland, 2017.
33. Imrich, W.; Klavzar, S.; Rall, D.F. *Topics in Graph Theory: Graphs and Their Cartesian Product*; CRC Press: Boca Raton, FL, USA, 2008.
34. Corsini, P. *Prolegomena of Hypergroup Theory*; Aviani Editore: Udine, Italy, 1993.

Article

Hopf Differential Graded Galois Extensions

Bo-Ye Zhang

School of Mathematical Sciences, Zhejiang University, Hangzhou 310027, China; boyezhang@zju.edu.cn

Abstract: We introduce the concept of Hopf dg Galois extensions. For a finite dimensional semisimple Hopf algebra H and an H-module dg algebra R, we show that $\mathcal{D}(R\#H) \cong \mathcal{D}(R^H)$ is equivalent to that R/R^H is a Hopf differential graded Galois extension. We present a weaker version of Hopf differential graded Galois extensions and show the relationships between Hopf differential graded Galois extensions and Hopf Galois extensions.

Keywords: differential graded algebras; derived categories; Galois extensions

MSC: 16T05; 18E30; 18G10

1. Introduction

The Hopf Galois extension was introduced in [1]. It was shown that for a finite dimensional semisimple Hopf algebra H and a left H-module algebra R, the smash product $R\#H$ is Morita equivalent to R^H if and only if R/R^H is an H^*-Galois extension. Now suppose R is a differential graded (dg) algebra and the differential is compatible with the H-module action. The Hopf Galois extension on dg algebra R and the equivalence between dg module categories gr-$(R\#H)$ and gr-R^H follows easily from [1]. However, if we consider the derived categories $\mathcal{D}(R\#H)$ and $\mathcal{D}(R^H)$, then the problem is subtle.

In the present paper, we focus our attention on the relationship between the derived categories $\mathcal{D}(R\#H)$ and $\mathcal{D}(R^H)$. We introduce the concept of Hopf dg Galois extensions and show that $R\#H$ and R^H is derived equivalent to each other if and only if R/R^H is a Hopf dg Galois extension. In some situations, for example, when R is a positive graded algebra, the concept of Hopf dg Galois extensions is precisely equal to the concept of Hopf Galois extensions. Thus, we can consider the Hopf dg Galois extension as a generality of the Hopf Galois extension.

For this purpose, we proceed as follows. We first review the basic facts on derived categories and derived functors. In Section 4, we define the Hopf dg Galois extensions. We show that $R\#H$ and R^H is derived equivalent to each other if and only if R/R^H is a Hopf dg Galois extension in Theorem 2. Finally, we give some conditions for the quotient categories of derived categories $\mathcal{D}(R\#H)$ and $\mathcal{D}(R^H)$ to be equivalent.

2. Preliminaries

Throughout this paper, k is a field of characteristic 0 and all algebras are k-algebras; unadorned $\otimes$ means $\otimes_k$ and Hom means Hom_k. Recall that a differential graded (dg) algebra is a $\mathbb{Z}$-graded algebra $A = \bigoplus_{n \in \mathbb{Z}} A^n$ equipped with a differential d of degree 1 such that $d(ab) = d(a)b + (-1)^{|a|}ad(b)$, where $a, b \in A$ are homogeneous elements and $|a|$ is the degree of a.

Suppose A is an algebra without gradings. We may view A as a dg algebra $\bigoplus_{n \in \mathbb{Z}} A_n$ concentrated in degree zero, where

(1) $A_0 = A$,
(2) $A_n = 0$, for every $n \neq 0$,
(3) the differential $d = 0$.

Unless otherwise stated, all modules in this paper are right modules. Let A and B be dg algebras. A (right) dg A-module M is a (right) A-module M, which has a grading $M = \bigoplus_{n \in \mathbb{Z}} M^n$ and a differential d such that $M^n A^m \in M^{n+m}$ and $d(ma) = d(m)a + (-1)^n m d(a)$, for $m \in M^n$ and $a \in A^m$. We call M a dg (A,B)-bimodule if M, which comes with one grading and one differential, is both a left dg A-module and a right dg B-module.

Let A and B be dg algebras. Let M be a dg (A, B)-module and N be a right dg B-bimodule. Let

$$\mathrm{Hom}_B^\bullet(M, N) = \bigoplus_{n \in \mathbb{Z}} \mathrm{Hom}_B^n(M, N),$$

where $\mathrm{Hom}_B^n(M, N)$ is the set of all graded B-module maps of degree n. Then, $\mathrm{Hom}_B^\bullet(M, N)$ is a right dg A-module with a differential defined by $d(f) = d_N \circ f - (-1)^n f \circ d_M \in \mathrm{Hom}_B^{n+1}(M, N)$, for $f \in \mathrm{Hom}_B^n(M, N)$ and $n \in \mathbb{Z}$. Let T be a right dg A-module. Then, the tensor product $T \otimes_A M$ is a right dg B-module with differential $d(t \otimes m) = d(t) \otimes m + (-1)^n t \otimes d(m)$ for $t \in T^n$ and $m \in M$.

Let A and B be dg algebras. $\mathcal{M}(A)$ will denote the dg module category of dg A-modules. $\mathcal{D}(A)$ will denote the derived category of dg A-modules. For a dg (A, B)-module M, we have two functors:

$$\mathrm{Hom}_B^\bullet(M, -) \colon \mathcal{M}(B) \to \mathcal{M}(A),$$

and

$$- \otimes_A M \colon \mathcal{M}(A) \to \mathcal{M}(B).$$

These two functors compose an adjoint pair $(- \otimes_A M, \mathrm{Hom}_B^\bullet(M, -))$, see ([2], Lemma 19.11).

Let

$$\mathrm{RHom}_B^\bullet(M, -) \colon \mathcal{D}(B) \to \mathcal{D}(A)$$

denote the right derived functor of $\mathrm{Hom}_B^\bullet(M, -)$ and

$$- \otimes_A^L M \colon \mathcal{D}(A) \to \mathcal{D}(B)$$

denote the left derived functor of $- \otimes_A M$. Due to the adjoint above, $(- \otimes_A^L M, R\,\mathrm{Hom}_B^\bullet(M, -))$ is an adjoint pair, see ([3], Section 5.8).

Let H be a finite dimensional semisimple Hopf algebra with counit ε. We say that R is a left H-module algebra, if there is a left H-module action on R such that

(1) $h \cdot a \in R^n$,
(2) $h \cdot (ab) = \Sigma (h_{(1)} \cdot a)(h_{(2)} \cdot b)$,
(3) $h \cdot 1 = \varepsilon(h) \cdot 1$,

for every $a, b \in R$ and $h \in H$.

Let R be a left H-module algebra. For a left H-module M, we write $M^H = \{m \in M \mid h \cdot m = \varepsilon(h)m, \text{ for all } h \in H\}$. Let $\bar{S}$ denote the inverse of the antipode S. It is well known that R^H is a subalgebra of R and R has an $(R^H, R\#H)$-bimodule structure defined by

$$r_1.r.(r_2 \# h) = (\bar{S}h) \cdot (r_1 r r_2),$$

and R has a $(R\#H, R^H)$-bimodule structure defined by

$$(r_1 \# h).r.r_2 = r_1 (h \cdot r) r_2,$$

where the notation "." denotes the multiplication on the module R and the notation "·" denotes the H-module action on the algebra R, see ([4], Sections 1.7 and 4.1).

The Hopf Galois extension is defined in [1]. R/R^H is said to be right H^*-Galois if the map

$$\gamma \colon R \otimes_{R^H} R \to R \otimes H^*, \ r_1 \otimes r_2 \mapsto (r_1 \otimes 1)\rho(r_2)$$

is surjective, where R is considered as a right H^*-comodule and ρ is the comodule structure map. By ([1], Theorem 1.2), R/R^H is right H^*-Galois if and only if the map $R \otimes_{R^H} R \to R \# H$, $r_1 \otimes r_2 \mapsto (r_1 \# t)(r_2 \# 1)$ is surjective.

Let $\mathcal{C}$ be a triangulated category and $\mathcal{B}$ be a full triangulated subcategory of $\mathcal{C}$. We call $\mathcal{B}$ a thick subcategory if the following condition is satisfied:

If $f \colon X \to Y$ is a map in $\mathcal{C}$ which is contained in a distinguished triangle

$$X \to Y \to Z \to X[1]$$

where Z is in $\mathcal{B}$, and if the map f also factors through an object W of $\mathcal{B}$, then X and Y are objects of $\mathcal{B}$.

If $\mathcal{B}$ is a thick subcategory of $\mathcal{C}$, then the quotient category $\frac{\mathcal{C}}{\mathcal{B}}$ is a triangulated category. For the thick subcategory, we have the following proposition.

Proposition 1 ([5], Proposition 1.3). *A full triangulated subcategory $\mathcal{B}$ of a triangulated category $\mathcal{C}$ is thick if and only if every object of $\mathcal{C}$ that is a direct summand of an object of $\mathcal{B}$ is itself an object of $\mathcal{B}$.*

3. The Equivalences of Triangulated Categories

Let B be a dg algebra and e be an idempotent in B^0 such that $d(e) = 0$. Then, $A = eBe$ is a dg algebra, Be is a dg (B, A)-bimodule and eB is a dg (A, B)-bimodule. For the dg (B, A)-bimodule Be, we may find a dg (B, A)-bimodule P and a dg (B, A)-bimodule morphism $p \colon P \to Be$ such that p is a quasi-isomorphism and P is K-projective both as a left dg B-module and as a right dg A-module. Similarly, we may find a dg (A, B)-bimodule Q and a dg (A, B)-bimodule morphism $q \colon Q \to eB$ such that q is a quasi-isomorphism and Q is K-projective both as a left dg A-module and as a right dg B-module. Then, the functor $- \otimes_A^L Be \colon \mathcal{D}(B) \to \mathcal{D}(A)$ is isomorphic to the functor $- \otimes_A P \colon \mathcal{D}(B) \to \mathcal{D}(A)$ and the functor $- \otimes_A^L eB \colon \mathcal{D}(A) \to \mathcal{D}(B)$ is isomorphic to the functor $- \otimes_A Q \colon \mathcal{D}(A) \to \mathcal{D}(B)$.

Since $(- \otimes_A^L eB, \mathrm{RHom}_B^\bullet(eB, -))$ is an adjoint pair between $\mathcal{D}(B)$ and $\mathcal{D}(A)$, we have a bijection

$$\Psi \colon \mathrm{Hom}_{\mathcal{D}(B)}(Be \otimes_A^L eB, B) \to \mathrm{Hom}_{\mathcal{D}(A)}(Be, Be),$$

since $\mathrm{RHom}_B^\bullet(eB, B) \cong Be$ in $\mathcal{D}(A)$. Below, we set

$$\psi = \Psi^{-1}(\mathrm{Id}_{Be}), \tag{1}$$

where Id_{Be} is the identity morphism in $\mathrm{Hom}_{\mathcal{D}(A)}(Be, Be)$.

Similarly, $(- \otimes_A Q, \mathrm{Hom}_B^\bullet(Q, -))$ is an adjoint pair between $\mathcal{D}(B)$ and $\mathcal{D}(A)$. For every $i \in \mathbb{Z}$, there exists an isomorphism of dg (B, B)-bimodules

$$\alpha_i \colon \mathrm{Hom}_B^\bullet(P \otimes_A Q, B[i]) \to \mathrm{Hom}_A^\bullet(P, \mathrm{Hom}_B^\bullet(Q, B[i])),$$

such that for $f \in \mathrm{Hom}_B^n(P \otimes_A Q, B[i])$, $x \in P$, $y \in Q$, we have

$$\alpha_i(f)(x) \colon y \mapsto f(x \otimes y).$$

Note that both Q and eB are K-projective as right dg B-modules. It follows that the quasi-isomorphism $q \colon Q \to eB$, when viewed as a right dg B-module morphism, is indeed a homotopic equivalence. Hence the dg (B, B)-bimodule morphism

$$\mathrm{Hom}_A^\bullet(P, \mathrm{Hom}_B^\bullet(q, B[i])) \colon \mathrm{Hom}_A^\bullet(P, \mathrm{Hom}_B^\bullet(eB, B[i])) \to \mathrm{Hom}_A^\bullet(P, \mathrm{Hom}_B^\bullet(Q, B[i]))$$

is a quasi-isomorphism. Since $\mathrm{Hom}_B^\bullet(eB, B[i]) \cong Be[i]$ as dg (B, A)-bimodules, let β_i denote the quasi-isomorphism from $\mathrm{Hom}_A^\bullet(P, Be[i])$ to $\mathrm{Hom}_A^\bullet(P, \mathrm{Hom}_B^\bullet(Q, B[i]))$. Thus, we have the following isomorphism

$$\Phi_i = (H^0(\beta_i))^{-1} \circ H^0(\alpha_i) \colon \mathrm{Hom}_{\mathcal{D}(B)}(P \otimes_A Q, B[i]) \to \mathrm{Hom}_{\mathcal{D}(A)}(P, Be[i]).$$

Let $\phi := m \circ (p \otimes_A q)$ be the composition

$$P \otimes_A Q \xrightarrow{p \otimes q} Be \otimes_A eB \xrightarrow{m} B, \qquad (2)$$

where m is the multiplication map in B, that is, $m(b_1 \otimes b_2) = b_1 b_2$. Then, ϕ is a dg (B, B)-bimodule morphism. For $b \in B^i$ such that $d(b) = 0$, let l_b denote the map $l_b \colon B \to B[i]$, $a \mapsto ba$ for $a \in B$, and let l'_b denote the map $l'_b \colon Be \to Be[i]$, $ae \mapsto bae$, for $a \in B$. Then, we have the following lemma.

Lemma 1. *Retain the notation above,* $\Phi_i(l_b \circ \phi) = l'_b \circ p$.

Proof. By the definitions, for $x \in P$, $y \in Q$, $\alpha_i(l_b \circ \phi)(x) \colon y \mapsto bp(x)q(y)$ and $\beta_i(l'_b \circ p)(x) \colon y \mapsto bp(x)q(y)$. Thus, $\Phi_i(l_b \circ \phi) = l'_b \circ p$. □

Since $P \cong Be$ in $\mathcal{D}(A)$ and $Be \otimes_A^L eB \cong P \otimes_A Q$ in $\mathcal{D}(B)$, we have the following commutative diagram.

$$\begin{array}{ccc}
\mathrm{Hom}_{\mathcal{D}(B)}(Be \otimes_A^L eB, B) & \xrightarrow{\Psi} & \mathrm{Hom}_{\mathcal{D}(A)}(Be, Be) \\
\cong \downarrow & & \downarrow \cong \\
\mathrm{Hom}_{\mathcal{D}(B)}(P \otimes_A Q, B) & \xrightarrow{\Phi_0} & \mathrm{Hom}_{\mathcal{D}(A)}(P, Be).
\end{array}$$

Hence the morphism ψ may be represented by ϕ as defined in (2). That is, we have $\mathrm{cone}(\phi) \cong \mathrm{cone}(\psi)$ in $\mathcal{D}(B)$. Moreover, we may use Φ_0 to conduct calculations instead of using Ψ.

Let A, B be dg algebras. Let N be a dg (A, B)-bimodule. The bimodule structure implies a natural map $l_A \colon A \to \mathrm{RHom}_B^\bullet(N, N)$, sending $a \in A$ to the left module action on N. In [5], Rickard characterized the Morita equivalence of derived categories. For dg algebras, we have the following lemma.

Lemma 2 ([5], Theorem 6.4). *Let A, B be dg algebras. Let N be a dg (A, B)-bimodule. Then, the functor $- \otimes_A^L N \colon \mathcal{D}(A) \to \mathcal{D}(B)$ gives an equivalence of triangulated categories if and only if*
(1) *N is a compact object of $\mathcal{D}(B)$.*
(2) *N is a weak generator in $\mathcal{D}(B)$.*
(3) *The map $l_A \colon A \to \mathrm{RHom}_B^\bullet(N, N)$ is a quasi-isomorphism.*

Now we can get the following theorem.

Theorem 1. *Let B be a dg algebra and e be an idempotent in B^0 such that $d(e) = 0$. Set $A = eBe$. The following conditions are equivalent.*
(1) *$F = - \otimes_A^L eB \colon \mathcal{D}(A) \to \mathcal{D}(B)$ is an equivalence of triangulated categories.*
(2) *$G = - \otimes_B^L Be \colon \mathcal{D}(B) \to \mathcal{D}(A)$ is an equivalence of triangulated categories.*
(3) *The morphism $\psi \colon Be \otimes_A^L eB \to B$ is an isomorphism in $\mathcal{D}(B)$.*

Proof. (1)⇔(2) F is left adjoint to $G' = \mathrm{RHom}_B^\bullet(eB, -) \colon \mathcal{D}(B) \to \mathcal{D}(A)$. The functors G and G' are naturally isomorphic to each other since eB is a compact K-projective dg module in $\mathcal{D}(B)$ and $\mathrm{Hom}_B^\bullet(eB, B) \cong Be$, see ([6], Section 2.1). Then, (F, G) is an adjoint pair. Therefore F is an equivalence of triangulated categories if and only if G is an equivalence of triangulated categories.

(1)⇒(3) Since F and G are equivalences, the functors $-\otimes_A P\colon \mathcal{D}(B) \to \mathcal{D}(A)$ and $-\otimes_B Q\colon \mathcal{D}(A) \to \mathcal{D}(B)$ are equivalences. For every $n \in \mathbb{Z}$, we have the following morphisms of groups.

$$\mathrm{Hom}_{\mathcal{D}(A)}(P,P[n]) \xrightarrow{-\otimes_A Q} \mathrm{Hom}_{\mathcal{D}(B)}(P\otimes_A Q, P\otimes_A Q[n])$$
$$\xrightarrow{\mathrm{Hom}_{\mathcal{D}(B)}(P\otimes_A Q,\phi[n])} \mathrm{Hom}_{\mathcal{D}(B)}(P\otimes_A Q, B[n])$$
$$\xrightarrow{\Phi_n} \mathrm{Hom}_{\mathcal{D}(A)}(P, Be[n])$$
$$\xrightarrow{\mathrm{Hom}_{\mathcal{D}(A)}(P,p[n])^{-1}} \mathrm{Hom}_{\mathcal{D}(A)}(P,P[n]).$$

By Lemma 1, the composition above is the identity morphism. Since the morphisms $-\otimes_A Q$, Φ_n and $\mathrm{Hom}_{\mathcal{D}(A)}(P,p[n])^{-1}$ are isomorphisms, $\mathrm{Hom}_{\mathcal{D}(B)}(P\otimes_A Q, \phi[n])$ is an isomorphism.

By (1) and (2), (F,G) is an adjoint pair and then $(-\otimes_A P, -\otimes_B Q)$ is an adjoint pair. So, we have $P\otimes_A Q \cong B\otimes_B P\otimes_A Q \cong B$ in $\mathcal{D}(B)$. Thus, $\mathrm{Hom}_{\mathcal{D}(B)}(B, \phi[n])$ is an isomorphism for every $n \in \mathbb{Z}$. Hence, ϕ is an isomorphism in $\mathcal{D}(B)$ and ϕ is a quasi-isomorphism of dg modules. Then, ψ is an isomorphism.

(3)⇒(1) The morphism of dg modules $\phi\colon P\otimes_A Q \to B$ is a quasi-isomorphism since ψ is an isomorphism. Then, $P\otimes_A Q \cong B$ in $\mathcal{D}(B)$. By Lemma 2, the functor $-\otimes_B P\otimes_A Q\colon \mathcal{D}(B) \to \mathcal{D}(B)$ is an equivalence. Since we have isomorphisms $Q\otimes_B P \cong eB\otimes_B Be \cong A$ in $\mathcal{D}(A)$, the functor $-\otimes_A Q\otimes_B P\colon \mathcal{D}(A) \to \mathcal{D}(A)$ is an equivalence. Thus, the functor $-\otimes_A Q\colon \mathcal{D}(A) \to \mathcal{D}(B)$ is an equivalence. Hence, $F = -\otimes_A^L eB\colon \mathcal{D}(A) \to \mathcal{D}(B)$ is an equivalence of triangulated categories. □

4. Hopf DG Galois Extensions

Let H be a finite dimensional semisimple Hopf algebra with integral t such that $\varepsilon(t) = 1$. Suppose that R is a dg algebra with the differential d. We call R a left dg H-module algebra if R is a left graded H-module algebra and the differential of R is compatible with the H-module action, that is,

$$d(h\cdot r) = h\cdot d(r)$$

for $h \in H$ and $r \in R$. Since R is a dg algebra, the smash product $R\#H$ is a dg algebra with the differential $\delta = d\#\mathrm{Id}$ and R^H is a dg subalgebra of R. Let $e = 1_R\#t \in R\#H$. Then, e is an idempotent in R^0 and $\delta(e) = 0$. Thus, $e(R\#H)e$ is a dg algebra with differential δ. By direct calculation, we have the following isomorphisms ([7], Lemma 3.1).

(1) The map $R^H \to e(R\#H)e$, $r \mapsto e(r\#1)e$, is an isomorphism of dg algebras.
(2) The map $R \to (R\#H)e$, $r \mapsto (r\#1)e$, is an isomorphism of dg $(R\#H, R^H)$-bimodules.
(3) The map $R \to e(R\#H)$, $r \mapsto e(r\#1)$, is an isomorphism of dg $(R^H, R\#H)$-bimodules.

Let $B = R\#H$ and $A = eBe \cong R^H$. Let $p\colon P \to R$ be the dg $(R\#H, R^H)$-bimodule quasi-isomorphism such that P is K-projective on both sides. Let $q\colon Q \to R$ be the dg $(R^H, R\#H)$-bimodule quasi-isomorphism such that Q is K-projective on both sides. Recall the dg $(R\#H, R\#H)$-bimodule morphism

$$\phi\colon P\otimes_{R^H} Q \to R\#H, x\otimes y \mapsto p(x)q(y)$$

defined above. Now we can define the Hopf dg Galois extension.

Definition 1. *For a dg left H-module algebra R, R/R^H is called dg H^*-Galois if the morphism $\phi\colon P\otimes_{R^H} Q \to R\#H$ is a quasi-isomorphism.*

Now we have the following theorem for dg H^*-Galois extensions.

Theorem 2. *Let H be a finite dimensional semisimple Hopf algebra with integral t such that $\varepsilon(t) = 1$. Let R be a left dg H-module algebra. The following conditions are equivalent.*
(1) R/R^H *is dg H^*-Galois.*
(2) (a) *The map $l_{R\#H}: R\#H \to \mathrm{RHom}^\bullet_{R^H}(R,R)$ is a quasi-isomorphism,*
    (b) *R is a compact object in $\mathcal{D}(R^H)$.*
(3) *R is a weak generator in $\mathcal{D}(R\#H)$.*

Proof. Let $B = R\#H$, $e = 1_R\#t$, then $A = eBe = R^H$. Thus, the condition (1) is equivalent to Theorem 1 (3). By Lemma 2, the condition (2) is equivalent to Theorem 1 (2) and the condition (3) is equivalent to Theorem 1 (1). Then, by Theorem 1, (1) $\Leftrightarrow$ (2) $\Leftrightarrow$ (3). □

The following results will show the relation between Hopf Galois extensions and Hopf dg Galois extensions.

Lemma 3. *Let H be a finite dimensional semisimple Hopf algebra and R be a dg left H-module algebra. Then, $R_{R\#H}$ is a weak generator in $\mathcal{D}(R\#H)$ if and only if for every dg $R\#H$-module M, $H^n(M^H) = 0$ for every $n \in \mathbb{Z}$ implies $H^n(M) = 0$ for every $n \in \mathbb{Z}$.*

Proof. Given a dg $R\#H$-module (M, d_M), by ([7], Lemma 2.2), for every $n \in \mathbb{Z}$,

$$\mathrm{Hom}_{R\#H}(R, M[n]) \cong \mathrm{Hom}_R(R, M[n])^H \cong (\mathrm{Ker}\, d_M^n)^H.$$

Then, for every $n \in \mathbb{Z}$, we have

$$\mathrm{Hom}_{K(R\#H)}(R, M[n]) \cong (\mathrm{Ker}\, d_M^n / \mathrm{Im}\, d_M^{n-1})^H \cong (H^n(M))^H.$$

Since H is semisimple, $(-)^H \cong \mathrm{Hom}_H(k, -)$ is an exact functor. Therefore, $(H^n(M))^H \cong H^n(M^H)$ for every $n \in \mathbb{Z}$. By [7] Proposition 2.5, R is a K-projective dg $R\#H$-module. Thus, for every $n \in \mathbb{Z}$,

$$\mathrm{Hom}_{\mathcal{D}(R\#H)}(R, M[n]) \cong \mathrm{Hom}_{K(R\#H)}(R, M[n]) \cong (H^n(M))^H \cong H^n(M^H).$$

Hence, $R_{R\#H}$ is a weak generator in $\mathcal{D}(R\#H)$ if and only if for every dg $R\#H$-module M, $\mathrm{Hom}_{\mathcal{D}(R\#H)}(R, M[n]) = 0$ for every $n \in \mathbb{Z}$ implies $M \cong 0$ in $\mathcal{D}(R\#H)$, if and only if for every dg $R\#H$-module M, $H^n(M^H) = 0$ for every $n \in \mathbb{Z}$ implies $H^n(M) = 0$ for every $n \in \mathbb{Z}$. □

Corollary 1. *Let H be a finite dimensional semisimple Hopf algebra with integral t such that $\varepsilon(t) = 1$. Let R be a left H-module dg algebra. If R/R^H is right dg H^*-Galois, then the map $\varphi: R \otimes_{R^H} R \to R\#H, r_1 \otimes r_2 \mapsto (r_1\#t)(r_2\#1)$ is a quasi-isomorphism.*

Proof. Consider the short exact sequence of dg $R\#H$-modules

$$0 \to \mathrm{Ker}\, \varphi \to R \otimes_{R^H} R \to \mathrm{Im}\, \varphi \to 0.$$

Since $(-)^H$ is an exact functor, we have the short exact sequence

$$0 \to (\mathrm{Ker}\, \varphi)^H \to (R \otimes_{R^H} R)^H \to (\mathrm{Im}\, \varphi)^H \to 0.$$

Since $(R \otimes_{R^H} R)^H = t(R \otimes_{R^H} R) = (tR) \otimes_{R^H} R = R^H \otimes_{R^H} R \cong R$, for every $\alpha \in (\mathrm{Ker}\, \varphi)^H$, there exists $r \in R$ such that $\alpha = 1 \otimes r \in (R \otimes_{R^H} R)^H$. Then, $\varphi(\alpha) = \varphi(1 \otimes r) = (1\#t)(r\#1) = 0$. However, by [1] [Lemma 0.5], $(1\#H)(R\#1) \cong H \otimes R$ as vector spaces by

$$\eta: (1\#H)(R\#1) \to H \otimes R,\ (1\#h)(r\#1) \mapsto h_{(2)} \otimes (h_{(1)} \cdot r),$$

and
$$\eta^{-1}: H \otimes R \to (1\#H)(R\#1), \ h \otimes r \mapsto (1\#h_{(2)})((S^{-1}h_{(1)}) \cdot r\#1).$$

Thus, $(1\#t)(r\#1) = 0$ if and only if $r = 0$, which means $(\text{Ker } \varphi)^H = 0$. Therefore, $H^n((\text{Ker } \varphi)^H) = 0$ for every $n \in \mathbb{Z}$. By Lemma 3, $H^n(\text{Ker } \varphi) = 0$ for every $n \in \mathbb{Z}$. Then, $H^n(R \otimes_{R^H} R) \cong H^n(\text{Im } \varphi)$ for every $n \in \mathbb{Z}$.

Consider another short exact sequence of dg $R\#H$-modules
$$0 \to \text{Im } \varphi \to R\#H \to \text{Coker } \varphi \to 0.$$

Since $(-)^H$ is an exact functor, we have the short exact sequence
$$0 \to (\text{Im } \varphi)^H \to (R\#H)^H \to (\text{Coker } \varphi)^H \to 0.$$

By ([1], Lemma 0.5), $(R\#H)^H = (1\#t)(R\#1)$. However,
$$\begin{aligned}
(\text{Im } \phi)^H &= ((R\#t)(R\#1))^H \\
&= (1\#t)(R\#t)(R\#1) \\
&= (R^H\#t)(R\#1) \\
&= (1\#t)(R^H\#1)(R\#1) \\
&= (1\#t)(R\#1).
\end{aligned}$$

Thus, the map $(\text{Im } \varphi)^H \to (R\#H)^H$ is surjective. Then, $(\text{Coker } \varphi)^H = 0$. Therefore, $H^n((\text{Coker } \varphi)^H) = 0$ for every $n \in \mathbb{Z}$. By Lemma 3, $H^n(\text{Coker } \varphi) = 0$ for every $n \in \mathbb{Z}$. So, we have $H^n(\text{Im } \varphi) \cong H^n(R\#H)$ for every $n \in \mathbb{Z}$. Thus, $H^n(R \otimes_{R^H} R) \cong H^n(R\#H)$ for every $n \in \mathbb{Z}$. Hence, φ is a quasi-isomorphism. □

Corollary 2. *Let H be a finite dimensional semisimple Hopf algebra with integral t such that $\varepsilon(t) = 1$. Let $R = \bigoplus_{n \geq 0} R^n$ be a left H-module dg algebra. Then, R/R^H is dg H^*-Galois if and only if R/R^H, forgetting the differentials, is right H^*-Galois.*

Proof. Suppose that R/R^H is dg H^*-Galois. Then, by Corollary 1, the map $\varphi: R \otimes_{R^H} R \to R\#H, r_1 \otimes r_2 \mapsto (r_1\#t)(r_2\#1)$ is a quasi-isomorphism. Since $H^0(R) = \text{Ker } d_R^0$ and $1_R \in \text{Ker } d_R^0$, the map φ is surjective. Thus, R/R^H is right H^*-Galois.

Suppose that R/R^H is right H^*-Galois. Then, by [1] Theorem 1.2, R is a dg finitely generated projective left R^H-module, and for every dg $R\#H$-module M, $M^H \otimes_{R^H} R \cong M$ as dg $R\#H$-modules. Thus, $H^n(M^H) = 0$ for every $n \in \mathbb{Z}$ implies $H^n(M) = 0$ for every $n \in \mathbb{Z}$. By Lemma 3, $R_{R\#H}$ is a weak generator in $\mathcal{D}(R\#H)$. Thus, R/R^H is dg H^*-Galois. □

If R is a dg algebra concentrated in degree 0, then Corollary 2 shows that R/R^H is dg H^*-Galois if and only if R/R^H, forgetting the differentials, is H^*-Galois. Thus, the definition of dg H^*-Galois is an extension of the definition of H^*-Galois.

5. The Equivalences of Quotient Categories

Suppose that B is a dg algebra and e is an idempotent in B^0 such that $d(e) = 0$. Then, eBe is a dg algebra. Let $A = eBe$. Let
$$\mathcal{D}_0(B) = \{M \in \mathcal{D}(B) \mid \text{Hom}_{\mathcal{D}(B)}(M, B[n]) = 0, n \in \mathbb{Z}\}$$

and
$$\mathcal{D}_0(A) = \{N \in \mathcal{D}(A) \mid \text{Hom}_{\mathcal{D}(A)}(N, Be[n]) = 0, n \in \mathbb{Z}\}.$$

By Proposition 1, it is clear that $\mathcal{D}_0(B)$ (resp. $\mathcal{D}_0(A)$) is a thick triangulated subcategory of $\mathcal{D}(B)$ (resp. $\mathcal{D}(A)$). Let $\mathcal{D}_q(B)$ denote the quotient category $\frac{\mathcal{D}(B)}{\mathcal{D}_0(B)}$ and $\mathcal{D}_q(A)$ denote the quotient category $\frac{\mathcal{D}(A)}{\mathcal{D}_0(A)}$. Let π denote the natural quotient functor. Theorem 1 shows that the map $\phi: Be \otimes_A^L eB \to B$ is an isomorphism in $\mathcal{D}(B)$ if and only if $\mathcal{D}(A) \cong \mathcal{D}(B)$. In this section, we will give some equivalent conditions for the quotient categories $\mathcal{D}_q(B)$ and $\mathcal{D}_q(A)$ being equivalent.

Theorem 3. *Let B be a dg algebra and e be an idempotent in B^0 such that $d(e) = 0$. Let $A = eBe$ be a dg algebra. The following conditions are equivalent.*
1. *The map $l_B \colon B \to \mathrm{RHom}_A^\bullet(Be, Be)$ is a quasi-isomorphism.*
2. $\mathrm{Hom}_{\mathcal{D}(B)}(\mathrm{cone}(\psi), B[n]) = 0$ *for all $n \in \mathbb{Z}$.*
3. *The functor $- \otimes_B^L Be \colon \mathcal{D}(B) \to \mathcal{D}(A)$ implies an equivalence of triangulated categories from $\mathcal{D}_q(B)$ to $\mathcal{D}_q(A)$.*

Proof. (1)$\Leftrightarrow$(2) Consider the composition $\Phi_n \circ \mathrm{Hom}_{\mathcal{D}(B)}(\phi, B[n])$,

$$\mathrm{Hom}_{\mathcal{D}(B)}(B, B[n]) \xrightarrow{\mathrm{Hom}_{\mathcal{D}(B)}(\phi, B[n])} \mathrm{Hom}_{\mathcal{D}(A)}(P, P[n]) \xrightarrow{\Phi_n} \mathrm{Hom}_{\mathcal{D}(A)}(P, Be[n]).$$

Since Φ_n is an isomorphism for every n, by Lemma 1, the condition (1) is equivalent to $\mathrm{Hom}_{\mathcal{D}(B)}(\phi, B[n])$ being an isomorphism for every n. Consider the distinguished triangle in $\mathcal{D}(B)$,

$$P \otimes_A Q \xrightarrow{\phi} B \longrightarrow \mathrm{cone}(\phi) \longrightarrow P \otimes_A Q[1].$$

In the following proving process, we write $(-)_n^*$ for the functor $\mathrm{Hom}_{\mathcal{D}(B)}(-, B[n])$ temporarily to simplify the notation. Then, we have the long exact sequence

$$\cdots \to (B)_n^* \xrightarrow{(\phi)_n^*} (P \otimes_A Q)_n^* \to (\mathrm{cone}(\phi))_{n+1}^* \to (B)_{n+1}^* \xrightarrow{(\phi)_{n+1}^*} (P \otimes_A Q)_{n+1}^* \to \cdots.$$

Thus, we have that the functor $(\phi)_n^*$ is an isomorphism for every n if and only if $(\mathrm{cone}(\psi))_n^* \cong (\mathrm{cone}(\phi))_n^* = 0$ for every n.

(2)$\Leftrightarrow$(3) Consider the distinguished triangle in $\mathcal{D}(B)$

$$Be \otimes_A^L eB \xrightarrow{\varphi} B \longrightarrow \mathrm{cone}(\varphi) \longrightarrow Be \otimes_A^L eB[1].$$

Then, we have a distinguished triangle in $\mathcal{D}_q(B)$

$$\pi(Be \otimes_A^L eB) \xrightarrow{\pi(\varphi)} \pi(B) \longrightarrow \pi(\mathrm{cone}(\varphi)) \longrightarrow \pi(Be \otimes_A^L eB[1]).$$

Suppose that $\mathrm{Hom}_{\mathcal{D}(B)}(\mathrm{cone}(\psi), B[n]) = 0$ for all $n \in \mathbb{Z}$, then $\pi(\mathrm{cone}(\varphi)) = 0$ in $\mathcal{D}_q(B)$. Thus, $\pi(Be \otimes_A^L eB) \cong \pi(B)$ in $\mathcal{D}_q(B)$. Since $eB \otimes_B^L Be \cong A$ in $\mathcal{D}(A)$, we have $\pi(eB \otimes_B^L Be) \cong \pi(A)$ in $\mathcal{D}_q(A)$. Therefore, the functor $- \otimes_B^L Be \colon \mathcal{D}(B) \to \mathcal{D}(A)$ implies an equivalence of triangulated categories from $\mathcal{D}_q(B)$ to $\mathcal{D}_q(A)$.

Suppose that the functor $- \otimes_B^L Be \colon \mathcal{D}(B) \to \mathcal{D}(A)$ implies an equivalence of triangulated categories from $\mathcal{D}_q(B)$ to $\mathcal{D}_q(A)$; then, $\pi(Be \otimes_A^L eB) \cong \pi(B)$ in $\mathcal{D}_q(B)$. Thus, $\pi(\mathrm{cone}(\varphi)) = 0$ in $\mathcal{D}_q(B)$, that is, $\mathrm{Hom}_{\mathcal{D}(B)}(\mathrm{cone}(\psi), B[n]) = 0$ for all $n \in \mathbb{Z}$. □

Let H be a finite dimensional semisimple Hopf algebra with integral t such that $\varepsilon(t) = 1$. Let R be a left H-module algebra. Let $B = R\#H$ and $e = 1\#t$ in Theorem 3; then, $A \cong eBe \cong R^H$ as dg algebras. Thus, Theorem 3 shows some equivalent conditions of the quasi-isomorphism $R\#H \to \mathrm{RHom}_{R^H}^\bullet(R, R)$.

Corollary 3. *Let B be a dg algebra and e be an idempotent in B^0 such that $d(e) = 0$. Let $A = eBe$ be a dg algebra. If $\mathrm{Hom}_{\mathcal{D}(B)}(B, B[n]) = 0$, $\mathrm{Hom}_{\mathcal{D}(A)}(Be, Be[n]) = 0$, for $n \leq \alpha$ or $n \geq \beta$, then the following conditions are equivalent.*
1. *The map $l_B \colon B \to \mathrm{RHom}_A^\bullet(Be, Be)$ is a quasi-isomorphism.*

(2) $\text{Hom}_{\mathcal{D}(B)}(cone(\psi), B[n]) = 0$ for $\alpha + 1 \leq n \leq \beta$.

(3) $\text{Hom}_{\mathcal{D}(B)}(cone(\psi), B[n]) = 0$ for all $n \in \mathbb{Z}$.

(4) The functor $- \otimes_B^L Be \colon \mathcal{D}(B) \to \mathcal{D}(A)$ implies an equivalence of triangulated categories from $\mathcal{D}_q(B)$ to $\mathcal{D}_q(A)$.

Proof. By Theorem 3, it is clear that (1) $\Leftrightarrow$ (3) $\Leftrightarrow$ (4) and (3) $\Rightarrow$ (2). It suffices to show (2) $\Rightarrow$ (3).

(2) $\Rightarrow$ (3) By the proof of Theorem 3, if we have $\text{Hom}_{\mathcal{D}(B)}(B, B[n]) = 0$ and $\text{Hom}_{\mathcal{D}(A)}(Be, Be[n]) = 0$, for $n \leq \alpha$ or $n \geq \beta$, then the long exact sequence shows that $\text{Hom}_{\mathcal{D}(B)}(B, B[n]) \cong \text{Hom}_{\mathcal{D}(A)}(P, Be[n])$ for $\alpha + 1 \leq n \leq \beta - 1$. Thus $\text{Hom}_{\mathcal{D}(B)}(B, B[n]) \cong \text{Hom}_{\mathcal{D}(A)}(P, Be[n])$ for all n. □

Remark 1. If $\alpha = -1$ and we let $j(M) = \min\{i \mid \text{Ext}^i_{\mathcal{D}(B)}(M, B) \neq 0\}$ for $M \in \mathcal{D}(B)$, then the condition (2) is equivalent to $j(cone(\psi)) \geq \beta + 1$. Thus, Corollary 3 is a dg version of ([8], Theorem 2.4).

6. Conclusions

The Hopf dg Galois extension shows the relationship between dg algbras R and R^H, which relate to the equivalences of some derived categories. Since the Hopf dg Galois extension is compatible with the usual Hopf Galois extension, we can promote the propositions related to Hopf Galois extension, and relate these to derived categories in a similar way. For an H-comodule algebra and its subalgebras, there exists a kind of Hopf Galois extensions. These may be promoted to dg algebras and derived categories in some way.

Funding: This research received no funding.

Acknowledgments: Many thanks to Jiwei He for giving useful advice on the subject.

Conflicts of Interest: The authors declare no conflict of interest.

References

1. Cohen, M.; Fischman, D.; Montgomery, S. Hopf Galois Extensions, Smash Products, and Morita Equivalence. *J. Algebra* **1990**, *133*, 351–372. [CrossRef]
2. Anderson, F.W.; Fuller, K.R. *Rings and Categories of Modules*; Springer: New York, NY, USA, 1974.
3. Zhang, P. *Triangulated Categories and Derived Categories (Chinese)*; Science Press: Beijing, China, 2015.
4. Montgomery, S. *Hopf Algebras and Their Actions on Rings*; American Mathematical Soc.: Rhode Island, PO, USA, 1992.
5. Rickard, J. Morita Theory for Derived Categories. *J. Lond. Math. Soc.* **1989**, *2*, 436–456. [CrossRef]
6. Jørgensen, P. Recollements for differential graded algebras. *J. Algebra* **1991**, *299*, 589–601. [CrossRef]
7. He, J.-W.; Van Oystaeyen, F.; Zhang, Y.H. Hopf Algebra Actions on Differential Graded Algebras and Applications. *Bull. Belg. Math. Soc.-Simon* **2011**, *18*, 99–111. [CrossRef]
8. Bao, Y.-H.; He, J.-W.; Zhang, J.J. Pertinency of Hopf actions and quotient categories of Cohen-Macaulay algebras. *J. Noncommun. Geometry* **2019**, *13*, 283–319. [CrossRef] [PubMed]

Disclaimer/Publisher's Note: The statements, opinions and data contained in all publications are solely those of the individual author(s) and contributor(s) and not of MDPI and/or the editor(s). MDPI and/or the editor(s) disclaim responsibility for any injury to people or property resulting from any ideas, methods, instructions or products referred to in the content.

Article

Characteristic, C-Characteristic and Positive Cones in Hyperfields

Dawid Edmund Kędzierski [1], Alessandro Linzi [2,*] and Hanna Stojałowska [1]

[1] Institute of Mathematics, University of Szczecin, Wielkopolska 15, 70-451 Szczecin, Poland
[2] Center for Information Technologies and Applied Mathematics, University of Nova Gorica, 5000 Nova Gorica, Slovenia
* Correspondence: alessandro.linzi@ung.si

Abstract: We study the notions of the positive cone, characteristic and C-characteristic in (Krasner) hyperfields. We demonstrate how these interact in order to produce interesting results in the theory of hyperfields. For instance, we provide a criterion for deciding whether certain hyperfields cannot be obtained via Krasner's quotient construction. We prove that any positive integer (larger than 1) can be realized as the characteristic of some infinite hyperfield and an analogous result for the C-characteristic. Finally, we study the (directed) graph associated with the strict partial order induced by a positive cone in a hyperfield in various examples.

Keywords: hyperfield; positive cone; order; characteristic

MSC: 20N20; 06F99; 12J15

1. Introduction

In this paper, we study Krasner hyperfields. These structures are a generalization of the concept of the field, where the addition is allowed to be a multivalued operation, i.e., $x+y$ in general denotes a subset and not only an element. Apart from the applications for which they have been introduced in [1] by Krasner, recently, these structures have arisen naturally in several mathematical contexts. For instance, Viro in [2] used hyperfields in tropical geometry and Lee in [3] studied these structures in connection to the model theory of valued fields. Regarding hyperfields, it is certainly also worth mentioning the work of Connes and Consani in number theory [4,5].

Since hyperfields represent a generalization of the concept of a field, it is natural to ask which classical notions and theorems of the theory of fields can be generalized to the theory of hyperfields. M. Marshall in [6] started the investigation towards a theory of real hyperfields, generalizing the Artin-Schreier theory of real fields (for a general reference on the latter, we refer the reader to [7]). The work of Marshall provided the basis for the investigations made later in [8] and in [9]. In the sections below, real hyperfields will be studied further.

Historically, a subhyperfield L of a hyperfield F is required to be closed under the multivalued addition in the sense that $x+y \subseteq L$ for all $x,y \in L$. However, Jun in (Definition 2.4) [10], felt the need for a less restrictive notion and started to talk about a multivalued operation, which can be "induced" by certain subsets. In Section 3, we take a model theoretical point of view (encoding the multivalued operation $+$ via the ternary relation $z \in x+y$) to justify Jun's feeling and precisely define the notion of the multivalued operation induced by a subset (see also [9]). This leads us to the notion of relational subhyperfields. The interest for relational subhyperfields is motivated by the fact that they correspond to the submodels of hyperfields in a natural first-order language, which we describe in Section 3.

The notion of this characteristic is fundamental in classical field theory and, when it is finite, it can only be a prime number. Moreover, the characteristic of a field is preserved by subfields. In this paper, we will demonstrate that a natural generalization of the notion of characteristic for hyperfields (see Definition 3 below), does not have to behave in the same way: we prove that the characteristic of a hyperfield does not have to be preserved by relational subhyperfields (Example 13), and that any integer greater than 1 can be realized as the characteristic of some hyperfield (Theorem 3).

In addition, we study an alternative notion of the characteristic for hyperfields, known as the C-characteristic, which in the case of fields coincides with the usual characteristic. Moreover, this quantity is not preserved by relational subhyperfields (Example 14), and we prove in Theorem 4 that any positive integer can be realized as the C-characteristic of some real hyperfield. In addition, we demonstrate how useful the interplay between these two notions of characteristic can be by providing a criterion (Theorem 6) for deciding whather certain hyperfields cannot be obtained via Krasner's quotient construction (see [11,12]). We apply this result to the finite real hyperfield of Example 4.

In Section 5, we define a strict partial order relation induced by a positive cone in a real hyperfield and study the associated directed graph in various interesting examples.

2. Preliminaries

In this section, we provide an overview of the definitions and facts that are necessary for the rest of the paper, with several examples.

2.1. Hyperfields

Let H be a non-empty set and $\mathcal{P}(H)$ be its power-set. A *multivalued operation* $+$ on H is a function which associates with every pair $(x,y) \in H \times H$ an element of $\mathcal{P}(H)$, denoted by $x + y$. A *hyperoperation* $+$ on H is a multivalued operation, such that $x + y \neq \emptyset$ for all $x,y \in H$. If $+$ is a multivalued operation on $H \neq \emptyset$, then for $x \in H$ and $A, B \subseteq H$, we set

$$A + B := \bigcup_{a \in A, b \in B} a + b, \tag{1}$$

$A + x := A + \{x\}$ and $x + A := \{x\} + A$. If A or B is empty, then so is $A + B$.

A *hypergroup* can be defined as a non-empty set H with a multivalued operation $+$, which is *associative* (see Definition 1 (CH1) below) and *reproductive* (i.e., $x + H = H + x = H$ for all $x \in H$). This notion was first considered by F. Marty in [13–15]. Let us mention [16] for an extended historical overview and [17,18] for a description of some applications.

If $(H, +)$ is a hypergroup, then it follows that $+$ is a hyperoperation. Indeed, suppose that $x + y = \emptyset$ for some $x, y \in H$. Then,

$$H = x + H = x + (y + H) = (x + y) + H = \emptyset + H = \emptyset,$$

which is excluded (cf. (Theorem 12) in [16]).

The following special class of hypergroups will be of interest for us.

Definition 1. *A* canonical hypergroup *is a tuple* $(H, +, 0)$*, where* $H \neq \emptyset$*,* $+$ *is a multivalued operation on* H *and* 0 *is an element of* H *such that the following axioms hold:*

(CH1) $+$ *is associative, i.e.,* $(x + y) + z = x + (y + z)$ *for all* $x, y, z \in H$,
(CH2) $x + y = y + x$ *for all* $x, y \in H$,
(CH3) *for every* $x \in H$, *there exists a unique* $x' \in H$, *such that* $0 \in x + x'$ *(the element* x' *will be denoted by* $-x$*)*,
(CH4) $z \in x + y$ *implies* $y \in z - x := z + (-x)$ *for all* $x, y, z \in H$.

The axiom (CH4) is known as the *reversibility* axiom.

Remark 1. Some authors (see e.g., (Definition 1.2) in [19]) define canonical hypergroups requiring explicitly that $x + 0 = \{x\}$ for all $x \in H$. However, as already noted in (Section III, (b)) [12], this property follows from (CH3) and (CH4). Indeed, suppose that $y \in x + 0$ for some $x, y \in H$. Then, $0 \in y - x$ by (CH4). Presently, $y = x$ follows from the uniqueness required in (CH3). For this reason, we call 0 the neutral element for $+$.

Remark 2. The multivalued operation of a canonical hypergroup $(H, +, 0)$ is reproductive. To observe this fix, $a \in H$. For $x \in H + a$, there exists $h \in H$, such that $x \in h + a \subseteq H$, demonstrating that $H + a \subseteq H$. For the other inclusion, take $x \in H$, then

$$x \in x + 0 \subseteq x + (a - a) = (x - a) + a,$$

so there exists $h \in x - a \subseteq H$, such that $x \in h + a \subseteq H + a$. It follows, in particular, that $+$ is a hyperoperation.

The following structures have been considered by Krasner in [1,11].

Definition 2. *A hyperfield is a tuple* $(F, +, \cdot, 0, 1)$, *which satisfies the following axioms:*
(HF1) $(F, +, 0)$ *is a canonical hypergroup;*
(HF2) $(F \setminus \{0\}, \cdot, 1)$ *is an abelian group and* $x \cdot 0 = 0 \cdot x = 0$ *for all* $x \in F$;
(HF3) *the operation $\cdot$ is distributive with respect to $+$. That is, for all* $x, y, z \in F$,

$$x \cdot (y + z) = x \cdot y + x \cdot z,$$

where for $x \in F$ *and* $A \subseteq F$, *we have set*

$$xA := \{xa \mid a \in A\}.$$

We denote the multiplicative group of a hyperfield F by $F^\times$.

Remark 3. *One can think about other kinds of hyperfields by modifying the axioms that the additive hypergroup should fulfill. The hyperfields for which the additive hypergroup is a canonical hypergroup (as above) are commonly known as Krasner hyperfields. As we mentioned in the introduction, we will consider only these kinds of structures and call them simply hyperfields, as indicated in the above definition.*

Remark 4. *The double distributivity law, i.e.,*

$$(a + b)(c + d) = a \cdot c + a \cdot d + b \cdot c + b \cdot d,$$

does not hold in general in hyperfields. However, the fact that the inclusion

$$(a + b)(c + d) \subseteq ca + ad + bc + bd$$

holds is not difficult to verify from the definitions and has been known for long time. For instance, it was stated without proof in [12,20]. A proof has been written in (Theorem 4B) of [2].

By induction, it is straightforward to show that

$$(a_1 + \ldots + a_n)(b_1 + \ldots + b_m) \subseteq a_1 b_1 + a_1 b_2 + \ldots + a_1 b_m + a_2 b_1 + \ldots + a_n b_m$$

for any natural numbers n, m.

Examples of hyperfields can be obtained in the following way. Let K be a field and G a subgroup of $K^\times$. For $x \in K^\times$, we denote by $[x]_G$ the coset $xG \in K^\times/G$. Further, let $[0]_G$

denote the singleton containing only $0 \in K$. Then, the *quotient hyperfield* of the K modulo G is the set $K_G := K^\times/G \cup \{[0]_G\}$ with the hyperoperation

$$[x]_G + [y]_G := \{[x+yg]_G \mid g \in G\} \quad (x, y \in K)$$

and the operation

$$[x]_G[y]_G := [xy]_G \quad (x, y \in K).$$

This construction was demonstrated to always yield a hyperfield by Krasner himself in [11].

Not all hyperfields can be obtained in this way, i.e., there are hyperfields that are not quotient hyperfields. This has been demonstrated by Massouros in [12], who then improved his results in [21]. Afterwards, Baker and Jin in [22] have found the following theorem of Bergelson-Shapiro and Turnwald useful to prove that certain hyperfields cannot be obtained via Krasner's quotient construction.

Theorem 1 (Theorem 1.3 in [23]; Theorem 1 in [24]). *If F is an infinite field and G is a subgroup of $F^\times$ of the finite index, then $G - G = F$.*

The next observation gives a necessary condition for a hyperfield to be a field. This fact is an immediate corollary of a result already noted in [20] (p. 369). We wish to state it for later reference and we will take the opportunity to write a quick proof.

Proposition 1 ([20]). *Let F be a hyperfield. If $1 - 1 = \{0\}$, then F is a field.*

Proof. By distributivity (HR3), our assumption implies that $a - a = \{0\}$ for all $a \in F$. Let $a, b \in F^\times$ and take $x, y \in a + b$. Then

$$x - y \subseteq (a+b) - (a+b) = (a-a) + (b-b) = \{0\},$$

so $x = y$ and $a + b$ is always a singleton in F. □

In the literature, one can find different interesting notions of the characteristic for hyperfields. Maybe the oldest is the one introduced by Mittas (a student of Krasner, see [20]). Later, Viro in [2] highlighted two other possible such notions. All these coincide with the usual characteristic of fields when the addition of the hyperfield under consideration is singlevalued (i.e., the hyperfield is in fact a field). However, they can be different in the general case. In this paper, we focus on the latter two notions, highlighted by Viro and which appear also in the work of P. Gładki [8] as well as in [4,5].

Definition 3. *Let F be a hyperfield. We set $1 \times_F 1 := \{1\}$ and for $n \geq 2$*

$$n \times_F 1 := \underbrace{1 + \ldots + 1}_{n \text{ times}}.$$

If there is no risk of confusion, we simply write $n \times 1$ in place of $n \times_F 1$.

(i) *The minimal $n \in \mathbb{N}$ such that $0 \in n \times_F 1$ is called the characteristic of F. We denote this number by char F. If no such number exists, we set char $F = \infty$.*

(ii) *The minimal $n \in \mathbb{N}$ such that $1 \in (n+1) \times_F 1$ is called the C-characteristic of F. We denote this number by C-char F. If no such number exists, we set C-char $F = \infty$.*

Remark 5.

(i) Usually if the characteristic is not finite, then one sets it to be 0. We set it to be ∞ in this case because then some results below can be stated in a clearer way (cf. Proposition 2 and Proposition 4).

(ii) Note that in any field, the C-characteristic is equal to the characteristic.

(iii) For any hyperfield F we have that C-char $F \leq$ char F. Indeed, if $0 \in n \times_F 1$, then

$$(n+1) \times_F 1 = n \times_F 1 + 1 = \bigcup_{a \in n \times_F 1} (a+1) \supseteq 0 + 1 = \{1\},$$

so C-char $F \leq n$.

Let us describe some examples of finite hyperfields that will be of interest for us.

Example 1. *The sign hyperfield $\mathbb{S}$ is the set $\{-1, 0, 1\}$ with the hyperoperation and operation defined by the following tables:*

+	0	1	-1
0	{0}	{1}	{-1}
1	{1}	{1}	{-1,0,1}
-1	{-1}	{-1,0,1}	{-1}

·	0	1	-1
0	0	0	0
1	0	1	-1
-1	0	-1	1

As it was noted in, e.g., [25], (Page 22 (b)) this hyperfield is isomorphic to the quotient hyperfield of an ordered field (e.g., the field $\mathbb{R}$ of real numbers) over the multiplicative subgroup given by its positive cone (in the example of real numbers that is the multiplicative subgroup $\mathbb{R}_{>0}$ of positive real numbers). This hyperfield has the C-characteristic 1 and characteristic ∞.

Example 2. *This example is the hyperfield generated by the algorithm presented in [26] and called HF_{521}. It has five elements $\{0, 1, -1, a, -a\}$ and its multiplicative group is isomorphic to $\mathbb{Z}_4$. The table for the hyperoperation is as follows:*

+	0	1	-1	a	-a
0	{0}	{1}	{-1}	{a}	{-a}
1	{1}	{1,a,-a}	{0,1,-1,a,-a}	{1,-1,a,-a}	{1,-1,a,-a}
-1	{-1}	{0,1,-1,a,-a}	{-1,a,-a}	{1,-1,a,-a}	{1,-1,a,-a}
a	{a}	{1,-1,a,-a}	{1,-1,a,-a}	{1,-1,a}	{0,1,-1,a,-a}
-a	{-a}	{1,-1,a,-a}	{1,-1,a,-a}	{0,1,-1,a,-a}	{1,-1,-a}

As it was noted in [26], this hyperfield is isomorphic to the quotient hyperfield of the finite field with 29 elements $\mathbb{F}_{29}$ over the multiplicative subgroup of $\mathbb{F}_{29}^{\times}$ generated by 7. This hyperfield has C-characteristic 1 and characteristic 4.

Example 3. *This example is the hyperfield generated by the algorithm presented in [26] and called HF_{56}. It has five elements, $\{0, 1, -1, a, -a\}$. The table for the hyperoperation is as follows:*

+	0	1	−1	a	$-a$
0	$\{0\}$	$\{1\}$	$\{-1\}$	$\{a\}$	$\{-a\}$
1	$\{1\}$	$\{-1, a, -a\}$	$\{0, a, -a\}$	$\{1, -1, a, -a\}$	$\{1, -1, a, -a\}$
−1	$\{-1\}$	$\{0, a, -a\}$	$\{1, a, -a\}$	$\{1, -1, a, -a\}$	$\{1, -1, a, -a\}$
a	$\{a\}$	$\{1, -1, a, -a\}$	$\{1, -1, a, -a\}$	$\{1, -1, -a\}$	$\{0, 1, -1\}$
$-a$	$\{-a\}$	$\{1, -1, a, -a\}$	$\{1, -1, a, -a\}$	$\{0, 1, -1\}$	$\{1, -1, a\}$

The multiplicative group is isomorphic to $\mathbb{Z}_2 \times \mathbb{Z}_2$.

Note that this hyperfield is not a quotient hyperfield. Indeed, since its multiplicative group is not cyclic, it cannot be a quotient of a finite field. Moreover, $1 - 1$ does not coincide with the whole hyperfield and thus Theorem 1 ensures that it cannot be a quotient of an infinite field. We observe that this hyperfield has the C-characteristic 2 and characteristic 3.

Example 4. Consider the set $F = \{0, 1, -1, a, -a, a^2, -a^2\}$ and its subset $P := \{1, a, a^2\}$. We define on F the following hyperaddition:

+	0	1	−1	a	$-a$	a^2	$-a^2$
0	$\{0\}$	$\{1\}$	$\{-1\}$	$\{a\}$	$\{-a\}$	$\{a^2\}$	$\{-a^2\}$
1	$\{1\}$	$\{a, a^2\}$	$F \setminus \{1, -1\}$	P	$F \setminus \{0\}$	P	$F \setminus \{0\}$
−1	$\{-1\}$	$F \setminus \{1, -1\}$	$\{-a, -a^2\}$	$F \setminus \{0\}$	$-P$	$F^\times$	$-P$
a	$\{a\}$	P	$F \setminus \{0\}$	$\{1, a^2\}$	$F \setminus \{a, -a\}$	P	$F \setminus \{0\}$
$-a$	$\{-a\}$	$F \setminus \{0\}$	$-P$	$F \setminus \{a, -a\}$	$\{-1, -a^2\}$	$F \setminus \{0\}$	$-P$
a^2	$\{a^2\}$	P	$F \setminus \{0\}$	P	$F \setminus \{0\}$	$\{1, a\}$	$F \setminus \{a^2, -a^2\}$
$-a^2$	$\{-a^2\}$	$F \setminus \{0\}$	$-P$	$F \setminus \{0\}$	$-P$	$F \setminus \{a^2, -a^2\}$	$\{-1, -a\}$

The multiplicative group is isomorphic to $\mathbb{Z}_6$. One can demonstrate that F is a hyperfield with straightforward direct computations. In Section 2.2, below, we will study some properties of the subset P. Moreover, we will prove that this cannot be obtained with Krasner's quotient construction. Note that F has the C-characteristic 2 and its characteristic is ∞.

2.2. Real Hyperfields

The Artin-Schreier theory of ordered fields, which led Artin to his solution of Hilbert's 17th problem (see [7] for details), was generalised to hyperfields in [6]. Let us recall some basic facts and definitions.

Definition 4. Let F be a hyperfield. A subset $P \subseteq F$ is called a positive cone in F if
(P1) $P + P \subseteq P$,;
(P2) $P \cdot P \subseteq P$,;
(P3) $P \cap -P = \emptyset$,;
(P4) $P \cup -P = F^\times$..

A hyperfield F is called real if it admits a positive cone.

Note that $1 \in P$ for every positive cone P in a hyperfield F. Indeed, from the axioms, either 1 or −1 belongs to P, but not both. If $-1 \in P$, then $1 = (-1) \cdot (-1) \in P$ again by the axioms. Hence, -1 cannot be in P, implying our assertion. This implies that the characteristic of a real hyperfield must be ∞, since $P + P \subseteq P$ and $0 \notin P$.

Example 5. *The sign hyperfield $\mathbb{S}$ introduced in Example 1 above is real with the positive cone $\{1\}$. This is clearly the unique possible positive cone of $\mathbb{S}$.*

Example 6. *The hyperfield that we have introduced in Example 4 is real with the positive cone $P := \{1, a, a^2\}$. Again, this can be observed by straightforward computations.*

Example 7. *The hyperfield which we have introduced in Example 2 is not real. Indeed, if P would be a positive cone, then since $1 \in P$ we must have $1 + 1 \subseteq P$. On the other hand, $1 + 1 = \{1, a, -a\}$ and by the axioms only one among a and $-a$ can belong to P.*
A similar reasoning yields that the hyperfield that we have introduced in Example 3 is not real.

Let us now briefly recall some results which have been proved in [9]. As in that paper, we will denote by $\mathcal{X}(F|G)$ the set of all positive cones in F, which contain some subset G of F and by $\mathcal{X}(F)$ the set of all positive cones in F.

Theorem 2 ([9])**.** *Let P be a positive cone of a field K and assume that a multiplicative subgroup G of K is contained in P. Consider the quotient hyperfield $K_G = \{[x]_G \mid x \in K\}$.*

(i) *The set $P_G := \{[x]_G \mid x \in P\}$ is a positive cone of K_G.*
(ii) *If Q is a positive cone in K_G, then $P := \{x \in K \mid [x]_G \in Q\}$ is a positive cone in K.*
(iii) *$|\mathcal{X}(K_G)| = |\mathcal{X}(K|G)|$.*

Example 8. *Consider the quotient hyperfield $F := \mathbb{Q}_{(\mathbb{Q}^\times)^2}$, where $(\mathbb{Q}^\times)^2$ is the set of nonzero squares in $\mathbb{Q}$. The set $\mathbb{Q}^+ := \sum (\mathbb{Q}^\times)^2$ of the sums of nonzero squares in $\mathbb{Q}$ is the unique positive cone of $\mathbb{Q}$. Hence, by assertion (iii) of Theorem 2, F is real and has a unique positive cone $P_{(\mathbb{Q}^\times)^2} = \{[x]_{(\mathbb{Q}^\times)^2} \mid x \in \mathbb{Q}^+\}$.*

3. On Relational Subhyperfields

From the point of view of model theory (for a general reference, see, e.g., [27]), standard operations are usually encoded via binary function symbols. The same is not possible for multivalued operations, since function symbols are classically interpreted in a structure as functions with values in the universe of the structure and not in its power-set. Nevertheless, as it was observed in [3], we can use the ternary relation $z \in x + y$ to encode a multivalued operation $+$. Thus, a hyperfield is naturally a structure on the first-order language having two constant symbols $0, 1$ for the neutral elements, a binary function symbol for the multiplication and a ternary relation symbol to encode the hyperoperation. Considering hyperfields as structures on this language, the general model theoretical notion of the *submodel* leads to the following definition (we provide more details in Remark 7, below).

Definition 5. *Let F be a hyperfield. A subset $L \subseteq F$ is a relational subhyperfield of F if $0 \in L$, $L \setminus \{0\}$ is a (multiplicative) subgroup of $F \setminus \{0\}$ and with the induced multivalued operation, which is defined as*
$$x +_L y := (x +_F y) \cap L \quad (x, y \in L)$$
we have that $(L, 0, 1, \cdot, +_L)$ is a hyperfield.

Remark 6. *Note that a priori, the multivalued operation induced by a subset L of a hyperfield F might not be a hyperoperation, as it may admit empty values, i.e., $(x +_F y) \cap L = \emptyset$ may hold for some $x, y \in L$. If the latter is the case, then $(L, 0, 1, \cdot, +_L)$ is certainly not a hyperfield; in particular, L would not be a relational subhyperfield of F, by definition.*

Remark 7. *Presently, we will motivate the study of the notion introduced in Definition 5 above. A first-order language $\mathcal{L}$ consists of relation, function and constant symbols. A structure on $\mathcal{L}$*

is (informally) a universe (i.e., a non-empty set) where any (well-formed) expression over $\mathcal{L}$ is interpreted (for a formal definition, see, e.g., (Section 1.5) in [27]). A first-order theory $\mathbb{T}$ over $\mathcal{L}$ is a list of axioms, i.e., expressions, which can be true or false when interpreted in a certain structure. A structure in which all the axioms of $\mathbb{T}$ are true is called a **model** of $\mathbb{T}$. For example, the additive group of integers or the cyclic group of order 5 are models of the theory of groups, as is any other group. The field of rational numbers or the field of complex numbers are models of the theory of fields, as is any other field. Complete graphs or star graphs are models of the theory of graphs, as is any other graph.

Given a structure S on $\mathcal{L}$ and a non-empty subset A of S, it is possible to restrict to A the interpretations in S of the symbols of $\mathcal{L}$. In this way, A itself becomes a structure on $\mathcal{L}$, and A is called a substructure of S (cf. (Section 2.3) in [27]). One of the main differences between an n-ary ($n \in \mathbb{N}$) relation symbol, interpreted in S as a relation $R \subseteq S^n$ and an n-ary function symbol, interpreted in S as a function $f : S^n \to S$, is that, when restricted to A, the latter has to satisfy the requirement $f(\bar{a}) \in A$ for all $\bar{a} \in A^n$; because an n-ary function symbol must, by definition, be interpreted on A as a function $f : A^n \to A$. On the other hand, the restricted relation on A is just defined to be $R \cap A^n$, and there are no further requirements to be satisfied.

With the notation introduced above, let us stress that under the assumption that S is a model of $\mathbb{T}$, it does not follow in general that A is a model of $\mathbb{T}$ too. For example, we may restrict the operations of the field of real numbers to the set of integers $\mathbb{Z}$, but we do not obtain a field. If the substructure A happens to be itself a model of $\mathbb{T}$, then it is called a **submodel** of S. For example, the field of rational numbers is a submodel of the field of real numbers. If the axioms of $\mathbb{T}$ are all (equivalent to) universal axioms (i.e., they can be written using only the $\forall$ quantifier), then substructures are automatically submodels (this is a consequence of, e.g., (Theorem 3.3.3) in [27]).

Presently, let $\mathbb{T}_{hf}$ be the theory given by the axioms of hyperfields (see Definitions 2 and 1) written, encoding the symbol $+$ with the ternary relation $z \in x + y$. Then, a hyperfield F is a model of $\mathbb{T}_{hf}$ and L is a relational subhyperfield of F if and only if L is a submodel of F.

In mathematics, it is customary to call subobjects the submodels of a model of the theory of (those) objects. For example, subgroups are submodels of a group; subfields are submodels of a field; subgraphs are submodels of a graph. However, **subhyperfields** are historically defined as subsets L of a hyperfield $(F, 0, 1, \cdot, +)$, such that $0, 1 \in L$, $x^{-1} \in L$ for all $x \in L \setminus \{0\}$, $xy \in L$ and $x + y \subseteq L$ for all $x, y \in L$ (cf. [12,20,21]). This definition can be traced back to the definition of the subhypergroup already present in, e.g., Definition 2 and the subsequent remark in [28].

While it is clear that any subhyperfield L of a hyperfield F is a relational subhyperfield of F with $+_L = +$, there are examples of relational subhyperfields L', which do not satisfy the condition $x + y \subseteq L'$ for all $x, y \in L'$ (see Examples 9 and 10, below). Thus, in this setting and perhaps for historical reasons, the use of the prefix "sub" seems to not match the common practice. Nevertheless, our point of view is based on the choice of encoding hyperoperations with relations; thus, we chose the name **relational** subhyperfield to distinguish our notion from the traditional one.

Example 9. *Consider the hyperfield $F := HF_{521}$ of Example 2 and its subset $L := \{-1, 0, 1\}$. Equip L with the multivalued operation $+_L$, as in Definition 5. Then, L is the sign hyperfield (cf. Example 1); in particular, it is a relational subhyperfield of F. Note that $1 \in L$ but $1 +_F 1 = \{1, a, -a\} \not\subseteq L$.*

Example 10. *Consider the hyperfield $F := HF_{56}$ of Example 3 and its subset $L := \{-1, 0, 1\}$. Equip L with the multivalued operation $+_L$, as in Definition 5. Then, L is the the finite field with 3 elements $\mathbb{F}_3$; in particular, it is a relational subhyperfield of F. Note that $1 \in L$ but $1 +_F 1 = \{-1, a, -a\} \not\subseteq L$.*

One might think that the subset $\{-1, 0, 1\}$ is a relational subhyperfield of any hyperfield. The next examples demonstrate that this is not the case.

Example 11. *Let F be a field (considered as a hyperfield) with $\operatorname{char} F > 3$. Then, $L := \{-1, 0, 1\}$ is not a relational subhyperfield of F, since $-1, 0, 1 \notin 1 + 1$ and so $(1 + 1) \cap L = \emptyset$.*

Example 12. *Consider the hyperfield F from Example 4 and its subset $L := \{-1, 0, 1\}$. Then, $1 +_L 1 = \emptyset$ and thus L is not a relational subhyperfield of F.*

The following easy observation will be useful later.

Lemma 1. *Let F be a hyperfield and L be a relational subhyperfield of F. For all $n \in \mathbb{N}$, we have that*

$$n \times_L 1 \subseteq n \times_F 1.$$

Proof. Let us show this by induction on n. The base step is clear. For the induction step, given $n > 1$, we compute

$$n \times_L 1 = \bigcup_{x \in I_{n-1}(L)} x +_L 1 = \bigcup_{x \in I_{n-1}(L)} (x +_F 1) \cap L$$
$$\subseteq \bigcup_{x \in I_{n-1}(L)} (x +_F 1) \subseteq \bigcup_{x \in I_{n-1}(F)} (x +_F 1) = n \times_F 1,$$

where we have used the induction hypothesis $(n-1) \times_L 1 \subseteq (n-1) \times_F 1$. □

3.1. Characteristic of Relational Subhyperfields

It is not difficult to observe that if K is a field, considered as a hyperfield, then all relational subhyperfields of K are (traditional) subhyperfields of K and they coincide with the subfields of K. In field theory, the characteristic of a subfield coincides with the characteristic of the upper field. Nevertheless, in the multivalued setting, the same might not hold.

Example 13. *Consider the hyperfield $F = HF_{521}$ from Example 2. We have observed that char $F = 4$ and that the sign hyperfield $\mathbb{S}$ is a relational subhyperfield of F (cf. Example 9). Since $\mathbb{S}$ is real, we have that char $\mathbb{S} = \infty$. Thus, the strict inequality char $F <$ char $\mathbb{S}$ holds.*

On the basis of Lemma 1, we can demonstrate that the characteristic of a hyperfield is not greater than the characteristic of any of its relational subhyperfields.

Proposition 2. *Let F be a hyperfield and L be a relational subhyperfield of F. Then, char $F \leq$ char L.*

Proof. Directly from Lemma 1, we can argue that if char $L < \infty$, then also char $F < \infty$ and char $F \leq$ char L. Otherwise, char $L = \infty$ is automatically not smaller than char F. □

As we have observed in Example 13, the strict inequality might occur. In that example, we considered the hyperfield HF_{521}, which has characteristic 4. We now prove that that is the minimal characteristic that a hyperfield can have in order to produce such a situation.

Proposition 3. *Let F be a hyperfield and L be a relational subhyperfield of F. If char $F \in \{2, 3\}$, then char $L =$ char F.*

Proof. As we will observe, this follows from the fact that, since L is a relational subhyperfield of F, we have that $0, -1 \in L$.

Assume first that char $F = 2$. By assumption, we have $0 \in 1 +_F 1$ and since $0 \in L$, we also have $0 \in (1 +_F 1) \cap L = 1 +_L 1$, showing that char $L = 2 =$ char F in this case.

Now assume that char $F = 3$. This means that $0 \in 1 +_F 1 +_F 1$ and hence $-1 \in 1 +_F 1$. Since $-1 \in L$, we obtain that $-1 \in (1 +_F 1) \cap L = 1 +_L 1$ and $0 \in 1 +_L 1 +_L 1$ follows, since $0 \in L$. Hence, char $L \leq 3 =$ char F. By Proposition 2, we obtain the equality. □

3.2. C-Characteristic of Relational Subhyperfields

A result analogous to Proposition 2 for the C-characteristic follows similarly as above from Lemma 1.

Proposition 4. *Let F be a hyperfield and L be a relational subhyperfield of F. Then, C-char F ≤ C-char L.*

Proof. By Lemma 1, we have that if C-char $L < \infty$, then also C-char $F < \infty$ and C-char $F \leq$ C-char L. Otherwise, C-char $L = \infty$ is automatically not smaller than C-char F. □

Also for C-characteristics, the strict inequality might hold, as the following example shows.

Example 14. *Consider the hyperfield $F = HF_{56}$ from Example 3. We have observed that C-char $F = 2$ and that the finite field $\mathbb{F}_3$ of cardinality 3 is a relational subhyperfield of F (cf. Example 10). We have that*
$$\text{C-char } \mathbb{F}_3 = 3 > 2 = \text{C-char } F.$$

Let F be a hyperfield and L be a relational subhyperfield of F. Again, similarly as we observed for the characteristic, since $1 \in L$, we have that if C-char $F = 1$, then C-char $L = 1$ as well. Let us state this result.

Proposition 5. *Let F be a hyperfield with C-char $F = 1$ and L be a relational subhyperfield of F. Then, C-char $L =$ C-char F.*

Thus, in Example 14, the C-characteristic of F has the minimal value which can produce the strict inequality.

4. Realizing Characteristics and C-Characteristics

In this section, we deal with the problem of realizing a given positive integer as the characteristic or the C-characteristic of some hyperfield.

4.1. On the Characteristic

In classical field theory, the characteristic of a field, if finite, can only be a prime number. In contrast, we demonstrate below that any positive integer larger than 1 can be realized as the characteristic of some hyperfield. For a group G and $x_1, ..., x_n \in G$, we will denote by $\langle x_1, ..., x_n \rangle$ the subgroup of G generated by $x_1, ..., x_n$.

Theorem 3. *For every natural number $n \in \mathbb{N}_{>1}$, there exists an infinite quotient hyperfield F, such that char $F = n$.*

Proof. We are going to demonstrate that for every natural number $n \in \mathbb{N}$, the quotient hyperfield $F = \mathbb{Q}_{\langle -n \rangle}$ has characteristic $n+1$. Observe that
$$0 = \underbrace{1 + ... + 1}_{n \text{ times}} - n, \text{ so } [0]_{\langle -n \rangle} \in (n+1) \times [1]_{\langle -n \rangle}.$$

Hence, char $F \leq n+1$. If $n = 1$, then char $F = 2$. Let $n > 1$ and suppose that char $F = k < n+1$, i.e.,
$$[0]_{\langle -n \rangle} \in k \times [1]_{\langle -n \rangle}, \text{ where } 1 < k \leq n.$$
Then, there exist $x_i \in \mathbb{Z}, i \in \{1, ..., k\}$, such that
$$0 = (-n)^{x_1} + ... + (-n)^{x_k}.$$

Let $N = \min\{x_1, ..., x_k\}$ and denote $m_i := x_i - N$. Then
$$0 = (-n)^{m_1} + ... + (-n)^{m_k}, \quad (2)$$
where $m_i \in \mathbb{N} \cup \{0\}$ and $1 < k \leq n$. Observe that
$$(-n)^{2l} \equiv 1 \pmod{n+1} \text{ and } (-n)^{2l+1} \equiv 1 \pmod{n+1}$$
for every $l \in \mathbb{N}$. Thus, the left side of the Equation (2) is congruent to 0 modulo $n + 1$, while the right side is congruent to k modulo $n + 1$. Hence,
$$0 \equiv k \pmod{n+1},$$
which is a contradiction, since $1 < k \leq n$. As a consequence, we obtain that F has characteristic $n + 1$. □

Let us now demonstrate that finite hyperfields of even cardinality must have characteristic 2.

Proposition 6. *Let F be a finite hyperfield of even cardinality. Then, char $F = 2$.*

Proof. Assume that F is a finite hyperfield, such that char $F > 2$. Then, $1 \neq -1$ and thus $a \neq -a$ for all $a \in F^\times$. Therefore, $|F^\times|$ must be even, and hence $|F|$ is odd. □

If we restrict our attention to hyperfields obtained with the quotient construction, then we can make the following observations.

Lemma 2. *Let K be a field and G a subgroup of $K^\times$. Then, char $K_G \leq$ char K.*

If we consider quotient hyperfields K_G, constructed with respect to a finite group G, then by looking at the proper divisors of the cardinality of G, it is possible to bound from above the characteristic of K_G.

Proposition 7. *Let K be a field and G be a finite subgroup of $K^\times$. If $n \in \mathbb{N}_{>1}$ divides the cardinality of G, then char $K_G \leq n$.*

Proof. The group G is a finite subgroup of the multiplicative group of a field; thus, it is cyclic. Therefore, if n divides $|G|$, then there exists an element $g \in G$ of order n. Hence, $\langle g \rangle = \{1, g, ..., g^{n-1}\}$ and $|\langle g \rangle| = n$. Since
$$0 = g^n - 1 = (g-1)(1 + g + g^2 + ... + g^{n-1})$$
and $g \neq 1$, we obtain that
$$1 + g + g^2 + ... + g^{n-1} = 0.$$
We conclude that $0 \in n \times [1]_{K_G}$, so char $F \leq n$. □

Conversely, sometimes from the characteristic of K_G it is possible to deduce information on the divisors of $|G|$.

Lemma 3. *Let K be a field such that $1 \neq -1$ and G are a finite subgroup of $K^\times$. If char $K_G = 2$, then the cardinality of G is even.*

Proof. Since G is a finite subgroup of the multiplicative group of a field, it is cyclic. Let g be a generator of G. If char $K_G = 2$, then $0 \in [1]_G + [1]_G$. Hence $[1]_G = -[1]_G = [-1]_G$, so $-1 \in G$. We conclude that there is a positive integer $k \in \mathbb{N}$, such that $k < |G|$ and $g^k = -1$. Then, $1 = (-1)^2 = g^{2k}$, hence $|G| = 2k$ is even. □

Remark 8. *The above result does not hold if $1 = -1$ in K, since in that case the quotient hyperfield K_G would have characteristic 2 for any multiplicative subgroup G of $K^\times$.*

Combining Proposition 7 and Lemma 3, we derive the following result.

Corollary 1. *Let K be a field such that $1 \neq -1$ and G are a finite subgroup of $K^\times$. Then, the quotient hyperfield K_G has characteristic 2 if and only if $|G|$ is even.*

4.2. On the C-Characteristic

We now demonstrate that any positive integer can be realized as the C-characteristic of some real hyperfield.

Theorem 4. *For every natural number $n \in \mathbb{N}$, there exists an infinite real hyperfield F, such that C-char $F = n$.*

Proof. We are going to demonstrate that for every positive integer $n \in \mathbb{N}$, the quotient hyperfield $F = \mathbb{Q}_{\langle n+1 \rangle}$ is real and has C-char $F = n$. First, observe that

$$n + 1 = \underbrace{1 + \ldots + 1}_{n+1 \text{ times}}, \text{ so } [1]_{\langle n+1 \rangle} \in (n+1) \times [1]_{\langle n+1 \rangle}.$$

Hence, C-char $F \leq n$. If $n = 1$, then C-char $F = 1$. Let $n > 2$. and suppose that C-char $F < n$, i.e.,

$$[1]_{\langle n+1 \rangle} \in k \times [1]_{\langle n+1 \rangle}, \text{ where } 2 \leq k < n.$$

Then, there exist $x_i \in \mathbb{Z}$, $i \in \{1, \ldots, k+1\}$, such that

$$(n+1)^{x_{k+1}} = (n+1)^{x_1} + \ldots + (n+1)^{x_k}.$$

Let $N = \min\{x_1, \ldots, x_{k+1}\}$ and denote $m_i := x_i - N$. Then,

$$(n+1)^{m_{k+1}} = (n+1)^{m_1} + \ldots + (n+1)^{m_k},$$

where $m_i \in \mathbb{N} \cup \{0\}$ and $2 \leq k \leq n$. We obtain that

$$(n+1)^{m_1} + \ldots + (n+1)^{m_k} - (n+1)^{m_{k+1}} = 0. \tag{3}$$

Presently, since for any $m \in \mathbb{N} \cup \{0\}$, we have

$$(n+1)^m \equiv 1 \pmod{n},$$

the left hand side of the Equation (3) is congruent to

$$\underbrace{1 + \ldots + 1}_{(k \text{ times})} - 1 = k - 1$$

modulo n, while the right hand side is congruent to 0 modulo n. Hence,

$$k - 1 \equiv 0 \pmod{n},$$

which is a contradiction, since $2 \leq k \leq n$. Hence, C-char $F = n$ must hold.

Consider now the set of natural numbers

$$S := \{(n+1)^p + 1 \mid p \in \mathbb{N}\}.$$

By definition, for $s, t \in S$ we have that $[s]_{\langle n+1 \rangle} = [t]_{\langle n+1 \rangle}$ if and only if there exists some $g \in \langle n+1 \rangle$ such that $s = gt$. Suppose that $g \neq 1$. Without loss of generality, we can assume that $g = (n+1)^m$ for some $m \in \mathbb{N}$ (if not, we apply the following reasoning to the equality

$t = g^{-1}s$). Since $s \equiv 1 \pmod{n+1}$ and $gt \equiv 0 \pmod{n+1}$, we obtain a contradiction. Hence, $[s]_{\langle n+1 \rangle} = [t]_{\langle n+1 \rangle}$ if and only if $s = t$; thus,

$$\{[(n+1)^p + 1]_{\langle n+1 \rangle} \mid p \in \mathbb{N}\}$$

is an infinite subset of F, implying that F is infinite. Moreover, the set

$$P_{\langle n+1 \rangle} := \{[x]_{\langle n+1 \rangle} \mid x > 0\}$$

is a positive cone in F by Theorem 2. □

Let us now demonstrate that a finite hyperfield F must satisfy C-char $F < \infty$.

Proposition 8. *Let F be a finite hyperfield of cardinality $n > 1$, which is not the field $\mathbb{F}_2$. Then, C-char $F \leq 2n - 3$.*

Proof. Let $F \neq \mathbb{F}_2$ be a finite hyperfield of cardinality $n > 1$. If $1 + 1 = \{0\}$, then by Proposition 1, F is a field of characteristic 2, so $n \geq 4$ and thus C-char $F = 2 \leq 2n - 3$. If $1 \in 1 + 1$, then C-char $F = 1 \leq 2n - 3$. Otherwise, let $a \in 1 + 1$. Since $F^\times$ is an abelian group of cardinality $n - 1$, by Remark 4, we have that

$$1 = a^{n-1} \in (1+1)^{n-1} \subseteq 2(n-1) \times_F 1$$

and hence C-char $F \leq 2n - 3$. □

From Proposition 7 and Remark 5, the following result follows immediately.

Proposition 9. *Let K be a field and G a finite subgroup of $K^\times$. If $n \in \mathbb{N}_{>1}$ divides the cardinality of G, then C-char $K_G \leq n$.*

An ordered field has to be infinite. This is a consequence of the compatibility of the order relation induced by the positive cone, with the addition of the field (see Section 5, below). However, we have observed that there are real hyperfields, which are finite (cf. Example 1 and Example 4). The following result shows that we can construct finite real hyperfields with the C-characteristic 1 of any odd cardinality. Note that a finite real hyperfield has to have an odd number of elements by Proposition 6.

Let p be a prime number. In the proof of the next result, we will use the p-adic valuation v_p on the field of rational numbers $\mathbb{Q}$. Let us briefly recall how is that is defined (for more details on valuations, we refer to [29]). Let $v_p(0) := \infty$, and for $\frac{a}{b} \in \mathbb{Q}^\times$, write

$$\frac{a}{b} := p^v \frac{a'}{b'},$$

where $v \in \mathbb{Z}$ and $a', b' \in \mathbb{Z}$ are not divisible by p. Define $v_p(\frac{a}{b}) := v$. Thus, v_p is a map from $\mathbb{Q}$ to $\mathbb{Z} \cup \{\infty\}$.

Theorem 5. *For every odd number $n \geq 3$, there exists a finite real hyperfield F with C-char $F = 1$, such that $|F| = n$.*

Proof. Consider the field of rational numbers $\mathbb{Q}$, a positive integer $k \in \mathbb{N}$ and the subgroup of $\mathbb{Q}^\times$:

$$G_k := \{x \in \mathbb{Q} \mid v_p(x) \equiv 0 \pmod{k} \text{ and } x > 0\}$$

where v_p is the p-adic valuation on $\mathbb{Q}$, for some prime number p. We are going to show that the quotient hyperfield $\mathbb{Q}_{G_k}$ is a finite, real hyperfield with C-characteristic 1 and cardinality

$2k + 1$. First, observe that $G_k \subseteq \mathbb{Q}^+$; thus, $\mathbb{Q}_{G_k}$ is real by Theorem 2. Moreover, the index $(\mathbb{Q}^+ : G_k) = k$, so $(\mathbb{Q}^\times : G_k) = 2k$ and $|\mathbb{Q}_{G_k}| = 2k + 1$. Observe also that

$$1 = \frac{1}{p^k} + \frac{p^k - 1}{p^k} \in G_k + G_k,$$

hence $[1]_{G_k} \in 2 \times [1]_{G_k}$, which means that C-char $\mathbb{Q}_{G_k} = 1$. □

The following result provides a criterion for deciding whether certain hyperfields cannot be obtained via Krasner's quotient construction.

Theorem 6. *Every finite hyperfield F with char $F = \infty$ and C-char $F > 1$ is not a quotient hyperfield.*

Proof. Consider a finite hyperfield, which is a quotient hyperfield K_G. Observe first that if char $K_G = \infty$, then char $K = \infty$ by Lemma 2. Hence, K must be infinite. On the other hand, since K_G is finite, G has a finite index in $K^\times$. From Theorem 1, we obtain that $G - G = K$, so in particular $[1]_G \in [1]_G - [1]_G$. From the reversibility axiom, we obtain that $[1]_G \in [1]_G + [1]_G$. This shows that C-char $K_G = 1$. □

In particular, the hyperfield that we have introduced in Example 4 cannot be obtained with Krasner's quotient construction.

5. The Strict Partial Order Induced by a Positive Cone

We begin this section recalling the following definition.

Definition 6. *A strict partial order is a set S with an binary relation $<$, which is:*
(O1) *irreflexive ($x \not< x$ for all $x \in S$);*
(O2) *asymmetric ($x < y$ implies $y \not< x$ for all $x, y \in S$);*
(O3) *transitive ($x < y$ and $y < z$ imply $x < z$ for all $x, y, z \in S$).*
A strict partial order $(S, <)$ is called a strict linear order if for all $x, y \in S$ one has $x < y$, $y < x$ or $x = y$.

In the theory of ordered fields, any positive cone P induces a strict linear order. This is defined as follows: $x < y$ if and only if $y - x \in P$. In the hyperfield case, one can define the relation $x < y$ as $y - x \subseteq P$. One then obtains a strict partial order. Indeed, $x \not< x$ because $0 \notin P$ and if $x < y$, then $y < x$ cannot hold since $P \cap -P = \emptyset$. In order to show transitivity, take $x, y, z \in F$, such that $x < y$ and $y < z$. We have to demonstrate that $x < z$. Since $y - x, z - y \subseteq P$ and $P + P \subseteq P$, we obtain that

$$z - x \subseteq y - y + z - x = (y - x) + (z - y) \subseteq P.$$

Nevertheless, this strict partial order does not have to be linear, as the following example shows.

Example 15. *Consider the quotient hyperfield $\mathbb{Q}_{(\mathbb{Q}^\times)^2}$ with its unique positive cone*

$$P := \{[x]_{(\mathbb{Q}^\times)^2} \mid x \in \mathbb{Q}^+\}.$$

Observe that since $1 = 2 \cdot 1^2 - 1 \cdot 1^2$ and $2 = 1 \cdot 2^2 - 2 \cdot 1^2$, we have that

$$[1]_{(\mathbb{Q}^\times)^2} \in [2]_{(\mathbb{Q}^\times)^2} - [1]_{(\mathbb{Q}^\times)^2} \quad \text{and} \quad [2]_{(\mathbb{Q}^\times)^2} \in [1]_{(\mathbb{Q}^\times)^2} - [2]_{(\mathbb{Q}^\times)^2}.$$

Therefore, both $[2]_{(\mathbb{Q}^\times)^2} - [1]_{(\mathbb{Q}^\times)^2}$ and $[1]_{(\mathbb{Q}^\times)^2} - [2]_{(\mathbb{Q}^\times)^2}$ contain elements of P and thus $[1]_{(\mathbb{Q}^\times)^2}$ and $[2]_{(\mathbb{Q}^\times)^2}$ are incomparable with respect to the order relation associated to P, since $P \cap -P = \emptyset$.

In an ordered field K, the order relation $<$ associated to a positive cone P is *compatible with the addition* of K in the sense that $a < b$ implies that $a + c < b + c$ for all $a, b, c \in K$. In the next example, we consider a real hyperfield F, such that $a < b$ and $a + c = b + c$ for some $a, b, c \in F$, where $<$ is the order induced by a positive cone of F.

Example 16. *Consider the sign hyperfield $\mathbb{S}$ and its unique positive cone $P = \{1\}$. Observe that $0 < 1$, but $0 + 1 = \{1\}$ and $1 + 1 = \{1\}$.*

We now note that in the case of the sign hyperfield $\mathbb{S}$, the strict partial order relation induced by its positive cone $P = \{1\}$ is a strict linear order.

At this point, let us consider another example of the real hyperfield.

Example 17 (Example 3.6 in [9]). *Consider the following cartesian product $\{-1, 1\} \times \Gamma$, where $(\Gamma, +, 0, <)$ is an ordered abelian group. Denote $F := \{-1, 1\} \times \Gamma \cup \{0\}$. Then, the tuple $(F, \boxplus, \cdot, 0, (1, 0))$ is a hyperfield, with the hyperaddition $\boxplus$ defined as follows:*

$$x \boxplus 0 = 0 \boxplus x := \{x\} \qquad x \in F$$
$$(1, \gamma_1) \boxplus (1, \gamma_2) := \{(1, \min\{\gamma_1, \gamma_2\})\} \qquad \gamma_1, \gamma_2 \in \Gamma$$
$$(-1, \gamma_1) \boxplus (-1, \gamma_2) := \{(-1, \min\{\gamma_1, \gamma_2\})\} \qquad \gamma_1, \gamma_2 \in \Gamma$$
$$(1, \gamma_1) \boxplus (-1, \gamma_2) := \{(1, \gamma_1)\} \qquad \gamma_1 < \gamma_2 \in \Gamma$$
$$(1, \gamma_1) \boxplus (-1, \gamma_2) := \{(-1, \gamma_2)\} \qquad \gamma_1 > \gamma_2 \in \Gamma$$
$$(1, \gamma) \boxplus (-1, \gamma) := \{(e, \delta) \mid e \in \{-1, 1\}, \delta \geq \gamma\} \cup \{0\} \qquad \gamma \in \Gamma$$

The result of the group multiplication $\cdot$ by 0 is defined to be 0, and for nonzero elements of F, we set:

$$(s_1, \gamma_1) \cdot (s_2, \gamma_2) = (s_1 s_2, \gamma_1 + \gamma_2).$$

Moreover, $\{(1, \gamma) \mid \gamma \in \Gamma\}$ is a positive cone in F.

Remark 9. *The hyperfield F from the previous example is a quotient hyperfield. It is obtained as $R_{E^+(R)}$, where R is a real closed field and $E^+(R)$ is the group of totally positive units with respect to the natural valuation associated with the unique positive cone of R. For more details, we refer the reader to [9].*

One can observe that the positive cone of the real hyperfield that we have introduced in the above example also induces a strict linear order relation.

The property that the sign hyperfield and the real hyperfield of Example 17 have in common is that they are stringent hyperfields.

Definition 7 ([30]). *A Krasner hyperfield F is said to be* stringent *if for all $x, y \in F$, we have that $x + y$ is a singleton unless $y = -x$.*

In fact, we have the following general result.

Proposition 10. *Let F be a stringent real hyperfield with positive cone P. Then, the relation*

$$a < b \quad \Longleftrightarrow \quad b - a \subseteq P \quad (a, b \in F)$$

is a strict linear order relation on F.

Proof. Take two distinct elements $a \neq b$ of F. Then, $b - a = \{c\}$ for some $c \in F^\times$. Therefore, either $c \in P$, in which case $a < b$, or $c \in -P$, in which case $b < a$. Hence, a and b are comparable and $<$ is indeed a linear order. □

To any strict partial order one can easily associate a directed graph. Let $(S, <)$ be a strict partial order. The *(directed) graph associated to* $<$ has S as its set of vertices, and an edge goes from a vertex a to a vertex b precisely when $a < b$. The reader should note that in the following illustrations, we do not draw the edges that can be deduced from the transitivity of $<$. For instance, for $S = \{a, b, c\}$ with $a < b < c$, we draw the following graph

$$a \longrightarrow b \longrightarrow c$$

instead of

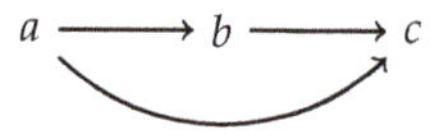

$$a \longrightarrow b \longrightarrow c$$

In the case of stringent real hyperfields, we obtain a linear order by Proposition 10. The directed graph obtained in this case can be found in Figure 1 below.

$$\ldots - x \longrightarrow \cdots \longrightarrow -1 \longrightarrow \cdots \longrightarrow 0 \longrightarrow \cdots \longrightarrow 1 \longrightarrow \cdots \longrightarrow x \ldots$$

Figure 1. The directed graph associated to a strict linear order.

In the case described in Example 15, the directed graph associated to the strict partial order induced by the positive cone is illustrated in Figure 2 below.

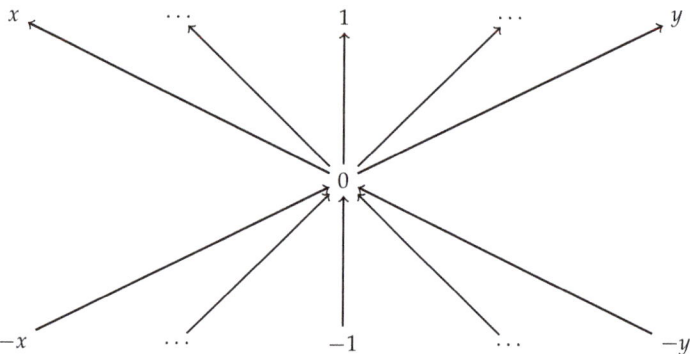

Figure 2. The directed graph associated to the strict partial order induced by the positive cone described in Example 15.

Indeed, in that case, one can demonstrate that if $a, b \in P$ are two distinct elements, then they are not comparable.

In the following example, we consider a more complex situation.

Example 18. *Consider the field of rational functions over the real numbers* $\mathbb{R}(X)$. *This field admits infinitely many positive cones. Below, we define a specific one, but our reasoning would apply to any of them. Every rational function* $h \in \mathbb{R}(X)$ *can be written uniquely in the following form:*

$$h = x^k \frac{f}{g}, \text{ where } f(0), g(0) \neq 0, k \in \mathbb{Z}.$$

We consider the positive cone $P := \{h \in \mathbb{R}(X) \mid \frac{f(0)}{g(0)} > 0\}$. *We set* $U := \{h \in \mathbb{R}(X) \mid k = 0\}$ *and* $\sum(\mathbb{R}(X)^\times)^2 := \{h \in \mathbb{R}(X) \mid h = h_1^2 + \ldots + h_n^2 \text{ for some } h_1, \ldots, h_n \in \mathbb{R}(X)^\times, n \in \mathbb{N}\}$. *Consider the quotient hyperfield* $F := \mathbb{R}(X)_{E^+}$, *where* $E^+ := U \cap \sum(\mathbb{R}(X)^\times)^2$. *In particular,* $E^+ \subseteq P$, *since sums of non-zero squares are contained in any positive cones, and from Theorem 2 we obtain that* F *is real with the positive cone* $P_{E^+} = \{[h]_{E^+} \mid h \in P\}$.

Take two elements $[h_1], [h_2] \in F$, such that

$$h_1 = x^{k_1} \frac{f_1}{g_1}, h_2 = x^{k_2} \frac{f_2}{g_2} \text{ with } \frac{f_1(0)}{g_1(0)} > 0, \frac{f_2(0)}{g_2(0)} > 0.$$

First, assume that $k_1 \neq k_2$. Without a loss of generality, let $k_1 > k_2$. We compute

$$h_2 - hh_1 = x^{k_2} \frac{f_2 g_1 g - x^{k_1-k_2} f_1 g_2 f}{g_1 g_2 g}, \text{ where } h = \frac{f}{g} \in E^+.$$

Then

$$\frac{(f_2 g_1 g - x^{k_1-k_2} f_1 g_2 f)(0)}{(g_1 g_2 g)(0)} = \frac{f_2(0)}{g_2(0)} > 0.$$

Hence $[h_2] - [h_1] \subseteq P_{E^+}$, so $[h_2] > [h_1]$. Now assume that $k := k_1 = k_2$. We have

$$h_2 - hh_1 = x^k \frac{f_2 g_1 g - f_1 g_2 f}{g_1 g_2 g}, \text{ where } h = \frac{f}{g} \in E^+.$$

Take $h = \frac{f}{g}$, such that $g(0) = 1$ and $f(0) < \frac{(f_2 g_1)(0)}{(f_1 g_2)(0)}$. Then

$$\frac{(f_2 g_1 g - f_1 g_2 f)(0)}{(g_1 g_2 g)(0)} > 0.$$

Hence, $([h_2] - [h_1]) \cap P_{E^+} \neq \emptyset$. On the other hand, let $g(0) = 1$ and $f(0) > \frac{(f_2 g_1)(0)}{(f_1 g_2)(0)}$. Then

$$\frac{(f_2 g_1 g - f_1 g_2 f)(0)}{(g_1 g_2 g)(0)} < 0.$$

Hence, $([h_2] - [h_1]) \cap -P_{E^+} \neq \emptyset$, so $([h_1] - [h_2]) \cap P_{E^+} \neq \emptyset$. This means that $[h_1]$ and $[h_2]$ are incomparable with respect to the partial order induced by P_{E^+}.

We illustrate in Figure 3 below the graph associated to the strict partial order induced by P on $\mathbb{R}(X)$.

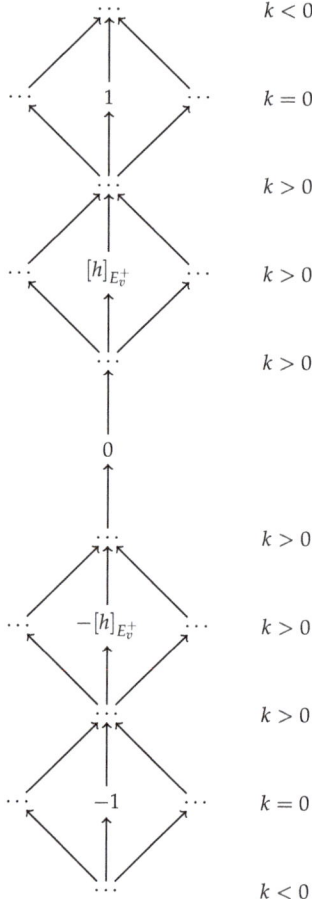

Figure 3. The graph associated to the strict partial order induced by P on $\mathbb{R}(X)$.

The nodes situated above the central node labelled by 0 correspond to elements of P. The nodes below correspond to elements of $-P$. Each level of this graph, which is above the central node labelled by 0, corresponds to an integer k. For instance, the level of the node labelled by 1 consists of all the nodes situated on the left and on the right of the node labelled by 1 and corresponds to the integer $k = 0$. The levels below this level correspond to positive integers $k > 0$ and the levels above to negative integers $k < 0$. Similarly, each level below the central node labelled by 0 correspond to an integer. There are no edges between any two nodes of the same level, as they are incomparable. If two nodes are in different levels, then they are connected in the upwards direction.

6. Further Research

In this paper, we have studied the notion of the positive cone in hyperfields. We have investigated the (directed) graph associated to the strict partial order induced by a positive cone in a hyperfield in some examples. What we have observed suggests a particular structure of this graph (that of a linear order and that of a star, as in Example 15, or a combination of these two, as in Example 18).

Moreover, we have considered the characteristic and the C-characteristic of hyperfields, and we have demonstrated how these interact to produce interesting results in the theory of hyperfields. In particular, we have obtained Theorem 6, which gives a criterion for deciding whether a given finite hyperfield cannot be obtained via Krasner's quotient construction.

We have demonstrated that any positive integer larger than 1 can be realized as the characteristic of an infinite hyperfield (Theorem 3). We ask if it is possible to realize any characteristic with finite hyperfields as well. To try to answer this question, we have initially focused on the finite hyperfields of the form $(\mathbb{F}_p)_G$ and developed an algorithm to compute their characteristic. At this point, we can provide the data in Table 1 below.

Table 1. Finite hyperfields of non-prime characteristic ≤ 12.

char$(\mathbb{F}_p)_G$	p	G	$|G|$
4	29	$\langle 16 \rangle$	7
6	199	$\langle 125 \rangle$	11
8	2179	$\langle 2118 \rangle$	11
9	2707	$\langle 1085 \rangle$	11
10	7151	$\langle 5825 \rangle$	11
12	62,323	$\langle 28,216 \rangle$	13

The characteristic of a hyperfield of the form $(\mathbb{F}_p)_G$ depends on p and on the multiplicative subgroup G of $\mathbb{F}_p^\times$. Since G is a cyclic group, we conclude that char$(\mathbb{F}_p)_G$ depends on the prime number p and on the choice of a divisor d of $p-1$. This means that the number N of possible hyperfields increases very fast with p. For example, in the case of characteristic 12, before finding the hyperfield $(\mathbb{F}_{62323})_{\langle 28216 \rangle}$, the algorithm would a priori have to generate and check the hyperfields corresponding to 6262 prime numbers, which gives a total of $N = 449{,}569$ possibilities. Nevertheless, we can use Proposition 7 to substantially reduce the number of cases to be considered.

For example, if we are looking for a hyperfield of characteristic 12, then we would assume that char$(\mathbb{F}_p)_G = 12$, which by Proposition 7 cannot hold if $|G|$ is divisible by some prime number < 12. Hence, we can restrict our attention to those prime numbers p, such that $p-1$ is divisible at least by one prime number ≥ 13. Moreover, for these primes p, we can restrict our choice of the divisor of $p-1$ to those d, which are not divisible by primes < 12. These restrictions reduce the number of hyperfields to be checked by the algorithm to 9871, which is approximately 2.19% of N.

Remark 10. *We know that* char$(\mathbb{F}_p)_G \neq 14$ *for all prime numbers* $p \leq 160{,}000$.

An analogous problem can be posed for the realization of C-characteristics, as Theorem 5 provides only infinite hyperfields. For the hyperfields of the form $(\mathbb{F}_p)_G$, our algorithm provided the data in Table 2 below.

Table 2. Finite hyperfields of non-prime C-characteristic ≤ 10.

C-char$(\mathbb{F}_p)_G$	p	G	$|G|$
4	151	$\langle 59 \rangle$	5
6	953	$\langle 879 \rangle$	7
8	15137	$\langle 11803 \rangle$	11
9	26951	$\langle 1202 \rangle$	11
10	44221	$\langle 31076 \rangle$	11

Remark 11. *We know that* C-char$(\mathbb{F}_p)_G \neq 12$ *for all prime numbers is* $p \leq 152897$.

Our algorithm does not consider hyperfields of the form $(\mathbb{F}_{p^k})_G$ with $k > 1$.

Another natural question is if it is possible to somehow generalise Example 4 to construct finite real hyperfields of cardinality > 7 with C-characteristic > 1, thus providing further examples of non-quotient hyperfields.

Author Contributions: Conceptualization, D.E.K., A.L. and H.S.; methodology, D.E.K., A.L. and H.S.; investigation, D.E.K., A.L. and H.S.; writing—original draft preparation, D.E.K., A.L. and H.S.; writing—review and editing, D.E.K., A.L. and H.S. All authors have read and agreed to the published version of the manuscript.

Funding: This research received no external funding.

Institutional Review Board Statement: Not applicable.

Informed Consent Statement: Not applicable.

Data Availability Statement: Not applicable.

Acknowledgments: The authors would like to express their gratitude to Franz-Viktor and Katarzyna Kuhlmann and to the anonymous referees for their suggestions and remarks which helped to improve the paper significantly. Many thanks also to Irina Cristea for encouraging us to write this manuscript.

Conflicts of Interest: The authors declare no conflict of interest.

References

1. Krasner, M. Approximation des corps valués complets de caractéristique $p \neq 0$ par ceux de caractéristique 0. In *Colloque D'algèbre Supérieure, tenu à Bruxelles du 19 au 22 Décembre 1956*; Centre Belge de Recherches Mathématiques, Établissements Ceuterick: Louvain, Belgium; Librairie Gauthier-Villars: Paris, France, 1957; pp. 129–206.
2. Viro, O. Hyperfields for Tropical Geometry I. Hyperfields and dequantization. *arXiv* **2010**, arXiv:1006.3034.
3. Lee, J. Hyperfields, truncated DVRs, and valued fields. *J. Number Theory* **2020**, *212*, 40–71. [CrossRef]
4. Connes, A.; Consani, C. From monoids to hyperstructures: in search of an absolute arithmetic. In *Casimir Force, Casimir Operators and the Riemann Hypothesis*; Walter de Gruyter: Berlin, Germany, 2010; pp. 147–198.
5. Connes, A.; Consani, C. The hyperring of adèle classes. *J. Number Theory* **2011**, *131*, 159–194. [CrossRef]
6. Marshall, M. Real reduced multirings and multifields. *J. Pure Appl. Algebra* **2006**, *205*, 452–468. [CrossRef]
7. Lam, T.Y. Orderings, valuations and quadratic forms. In *CBMS Regional Conference Series in Mathematics*; American Mathematical Society: Providence, RI, USA, 1983; Volume 52, pp. vii+143. [CrossRef]
8. Gładki, P.; Marshall, M. Orderings and signatures of higher level on multirings and hyperfields. *J. K-Theory* **2012**, *10*, 489–518. [CrossRef]
9. Kuhlmann, K.; Linzi, A.; Stojałowska, H. Orderings and valuations in hyperfields. *J. Algebra* **2022**, *611*, 399–421. [CrossRef]
10. Jun, J. Algebraic geometry over hyperrings. *Adv. Math.* **2018**, *323*, 142–192. [CrossRef]
11. Krasner, M. A class of hyperrings and hyperfields. *Internat. J. Math. Math. Sci.* **1983**, *6*, 307–311. [CrossRef]
12. Massouros, C.G. Methods of constructing hyperfields. *Internat. J. Math. Math. Sci.* **1985**, *8*, 725–728. [CrossRef]
13. Marty, F. Sur une généralization de la notion de groupe. In Proceedings of the Huitiéme Congrès des Mathématiciens Scand, Stockholm, Sweden, 14–18 August 1934; pp. 45–49.
14. Marty, F. Rôle de la notion de hypergroupe dans l'étude de groupes non abéliens. *C. R. Acad. Sci.* **1935**, *201*, 636–638.
15. Marty, F. Sur les groupes et hypergroupes attaches à une fraction rationnelle. *Ann. Sci. École Norm. Sup.* **1936**, *53*, 83–123. [CrossRef]
16. Massouros, C.; Massouros, G. An Overview of the Foundations of the Hypergroup Theory. *Mathematics* **2021**, *9*, 1014. [CrossRef]
17. Linzi, A.; Cristea, I. Dependence Relations and Grade Fuzzy Set. *Symmetry* **2023**, *15*, 311. [CrossRef]
18. Corsini, P.; Leoreanu, V. *Applications of Hyperstructure Theory*; Advances in Mathematics (Dordrecht); Kluwer Academic Publishers: Dordrecht, The Netherlands, 2003; Volume 5, pp. xii+322. [CrossRef]
19. Tolliver, J. An equivalence between two approaches to limits of local fields. *J. Number Theory* **2016**, *166*, 473–492. [CrossRef]
20. Mittas, J. Sur les hyperanneaux et les hypercorps. *Math. Balkanica* **1973**, *3*, 368–382.
21. Massouros, C.G. On the theory of hyperrings and hyperfields. *Algebra i Logika* **1985**, *24*, 728–742, 749. [CrossRef]
22. Baker, M.; Jin, T. On the structure of hyperfields obtained as quotients of fields. *Proc. Am. Math. Soc.* **2021**, *149*, 63–70. [CrossRef]
23. Bergelson, V.; Shapiro, D.B. Multiplicative subgroups of finite index in a ring. *Proc. Am. Math. Soc.* **1992**, *116*, 885–896. [CrossRef]
24. Turnwald, G. Multiplicative subgroups of finite index in a division ring. *Proc. Am. Math. Soc.* **1994**, *120*, 377–381. [CrossRef]
25. Massouros, G.; Massouros, C. Hypercompositional Algebra, Computer Science and Geometry. *Mathematics* **2020**, *8*, 1338. [CrossRef]
26. Ameri, R.; Eyvazi, M.; Hoskova-Mayerova, S. Advanced results in enumeration of hyperfields. *AIMS Math.* **2020**, *5*, 6552–6579. [CrossRef]

27. Prestel, A.; Delzell, C.N. *Mathematical Logic and Model Theory*; A Brief Introduction, Expanded Translation of the 1986 German; Universitext; Springer: London, UK, 2011; pp. x+193. [CrossRef]
28. Krasner, M. Sur la primitivité des corps $\mathfrak{P}$-adiques. *Mathematika* **1937**, *13*, 72–191.
29. Engler, A.J.; Prestel, A. *Valued Fields*; Springer Monographs in Mathematics; Springer: Berlin, Germany, 2005; pp. x+205.
30. Bowler, N.; Su, T. Classification of doubly distributive skew hyperfields and stringent hypergroups. *J. Algebra* **2021**, *574*, 669–698. [CrossRef]

Disclaimer/Publisher's Note: The statements, opinions and data contained in all publications are solely those of the individual author(s) and contributor(s) and not of MDPI and/or the editor(s). MDPI and/or the editor(s) disclaim responsibility for any injury to people or property resulting from any ideas, methods, instructions or products referred to in the content.

Article

A Novel Method for Generating the M-Tri-Basis of an Ordered Γ-Semigroup

M. Palanikumar [1], Chiranjibe Jana [2,*], Omaima Al-Shanqiti [3] and Madhumangal Pal [2]

[1] Saveetha School of Engineering, Saveetha Institute of Medical and Technical Sciences, Chennai 602105, India
[2] Department of Applied Mathematics with Oceanology and Computer Programming, Vidyasagar University, Midnapore 721102, India
[3] Department of Applied Science, Umm Al-Qura University, Mecca P.O. Box 24341, Saudi Arabia
* Correspondence: jana.chiranjibe7@gmail.com

Abstract: In this paper, we discuss the hypothesis that an ordered Γ-semigroup can be constructed on the M-left(right)-tri-basis. In order to generalize the left(right)-tri-basis using Γ-semigroups and ordered semigroups, we examined M-tri-ideals from a purely algebraic standpoint. We also present the form of the M-tri-ideal generator. We investigated the M-left(right)-tri-ideal using the ordered Γ-semigroup. In order to obtain their properties, we used M-left(right)-tri-basis. It was possible to generate a M-left(right)-tri-basis from elements and their subsets. Throughout this paper, we will present an interesting example of order $\preceq_{mlt}$ ($\preceq_{mrt}$), which is not a partial order of $\mathcal{S}$. Additionally, we introduce the notion of quasi-order. As an example, we demonstrate the relationship between M-left(right)-tri-basis and partial order.

Keywords: left tri-ideal; right tri-ideal; M-left-tri-basis; M-right-tri-basis; quasi-order; partial order

MSC: 06B10; 20M25; 16Y60

Citation: Palanikumar, M.; Jana, C.; Al-Shanqiti, O.; Pal, M. A Novel Method for Generating the M-Tri-Basis of an Ordered Γ-Semigroup. *Mathematics* **2023**, *11*, 893. https://doi.org/10.3390/math11040893

Academic Editors: Irina Cristea and Hashem Bordbar

Received: 4 January 2023
Revised: 6 February 2023
Accepted: 6 February 2023
Published: 9 February 2023

Copyright: © 2023 by the authors. Licensee MDPI, Basel, Switzerland. This article is an open access article distributed under the terms and conditions of the Creative Commons Attribution (CC BY) license (https:// creativecommons.org/licenses/by/ 4.0/).

1. Introduction

Several applications of algebraic structures can be found in mathematics. Generalizing the ideals of algebraic structures and ordered algebraic structures plays an important role, making them available for further study and application. Mathematicians studied bi-ideals, quasi-ideals, and interior ideals during 1950–1980. However, during 1950–2019, it was only mathematicians who studied their applications. In fact, the notion of one-sided ideals of rings and semigroups can be regarded as a generalization of the notion of ideals of rings and semigroups, as is the notion of quasi-ideals of semigroups and rings. In general, semigroups are generalizations of rings and groups. In semigroup theory, certain band decompositions are useful for studying semigroup structure. A new field in mathematics could be opened up by this research, one that aims to use semigroups of bi-ideals of semirings with semilattices that are additively reduced. The many different ideals associated with Γ-semigroups [1] and Γ-semirings [2] have been described by several researchers. Partially ordered relation " $\preceq$ " satisfies the conditions of reflexivity, antisymmetry, and transitivity. There are different classes of semigroups and Γ-semigroups based on bi-ideals that have been described by researchers in [3–6]. Munir [7] introduced new ideals in the form of M-bi-ideals over semigroups in 2018. An ordered semigroup is a generalization of a semigroup with a partially ordered relation constructed on a semigroup, so that the relation fits with the operation. An algebraic structure such as the ordered Γ-semigroup was introduced by Sen et al. in 1993 [8] and has been studied by several authors [9–12].

For an ordered semigroup $\mathcal{S}$ and subsemigroup $\mathcal{A}$ of $\mathcal{S}$, $\mathcal{A}^m = \mathcal{A}.\mathcal{A}\ldots\mathcal{A}(m-times)$, where m is a positive integer. Clearly, for any subsemigroup $\mathcal{A}$ of ordered semigroup $\mathcal{S}$, $\mathcal{A}^n \subseteq \mathcal{A}$ for all positive integers n, which are similarly right case. Hence, $\mathcal{A}^r \subseteq \mathcal{A}^t$ for all positive integers r and t, such that $r \geq t$, but the converse is not true. As a generalization of the bi-ideal of semirings and semigroups, a tri-ideal of semirings and semigroups can be characterized as a generalization of the bi-ideal. In the context of Γ-semigroups, an ordered Γ-semigroup is an extension of the Γ-semigroup. In contrast to the notion of the tri-ideal of semigroups, the notion of the tri-ideal of an ordered semigroup is a general form of the notion of the tri-ideal of semigroups. In semigroup theory, the M-tri-ideal is a generalization of the tri-ideal. Similarly, an ordered M-tri-ideal is a generalization of an ordered tri-ideal. In this paper, we describe the basic properties of the M-tri-basis from an algebraic standpoint. The fact that semigroups can be generalized to Γ-semigroups and Γ-semigroups to ordered Γ-semigroups is a result of these facts. It was work by Jantanan et al. that introduced the concept of bi-basis of ordered Γ-semigroups in 2022. We further describe the relationship between partial order and bi-basis [13]. As recently discussed in Palanikumar et al. [14–16], algebraic structures such as semigroups, semirings, ternary semigroups, and ternary semirings are all ideals and the generators of these structures are ideals. Rao introduced the tri-ideals of semigroups and semirings in [17,18]. Our paper extends a bi-basis of an ordered Γ-semigroup into a M-bi-basis of an ordered Γ-semigroup. We also generalize the tri-ideal of an ordered Γ-semigroup to an M-tri-basis of an ordered Γ-semigroup. The notion of almost bi-ideals and almost quasi-ideals of ordered semigroups is discussed in Sudaporn et al. [19]. The novel concept of M-bi-basis generators of an ordered Γ-semigroup is introduced by Palanikumar et al. [20]. Susmita et al. have discussed some important properties of bi-ideals of an ordered semigroups [21].

This paper discusses several important classical results for M-tri-basis and Γ-semigroups characterized by M-tri-ideals and M-tri-basis. Furthermore, we demonstrate how the elements and subsets of an ordered Γ-semigroup yield the M-tri-ideal and basis. This paper extends the notion of Γ-semigroup information into ordered Γ-semigroup information. The paper is divided into five sections. Section 1 is the introduction. There is a brief description of an ordered Γ-semigroup in Section 2, as well as relevant definitions and results. A numerical example of an M-left-tri-basis generator can be found in Section 3. As part of Section 4, a numerical example is given for the M-right-tri-basis generator concept. Our conclusions are provided in Section 5. In this paper, our purpose is to describe:

1. The generator of the M-tri-ideal for an ordered Γ-semigroup;
2. To interact, the order relation " $\preceq$ " based on the M-tri-basis should not be a partial order.
3. For example, the subset of an M-tri-basis is not an M-tri-basis itself.

2. Basic Concepts

It is assumed throughout this article that $\mathcal{S}$ denotes a Γ-semigroup, unless stated otherwise.

Definition 1 ([1])**.** *Let $\mathcal{S}$ and Γ be any two non-empty sets. Then, $\mathcal{S}$ is called a Γ-semigroup from $\mathcal{S} \cdot \Gamma \cdot \mathcal{S} \to \mathcal{S}$, which maps $(f_1, \pi, f_2) \to f_1 \cdot \pi \cdot f_2$, satisfying the condition $(f_1 \cdot \pi \cdot f_2) \cdot \theta \cdot f_3 = f_1 \cdot \pi \cdot (f_2 \cdot \theta \cdot f_3)$ for all $f_1, f_2, f_3 \in \mathcal{S}$ and $\pi, \theta \in \Gamma$.*

Definition 2 ([8])**.** *The algebraic system $(\mathcal{S}, \Gamma, \preceq)$ is said to be an ordered Γ-semigroup if it satisfies the following conditions:*

1. *$\mathcal{S}$ is a Γ-semigroup,*
2. *" $\preceq$ " is a relation from a partially ordered set (poset) $\mathcal{S}$,*
3. *If $s'' \preceq s'''$, then $s'' \pi s' \preceq s''' \pi s'$ and $s' \pi s'' \preceq s' \pi s'''$ for any $s', s'', s''' \in \mathcal{S}$ and $\pi \in \Gamma$.*

Remark 1 ([8])**.** *Let $\mathcal{G}'$ and $\mathcal{G}''$ be any two subsets of $\mathcal{S}$. Then, the following properties hold:*

1. *$\mathcal{G}' \Gamma \mathcal{G}'' = \{x' \pi x'' | x' \in \mathcal{G}', x'' \in \mathcal{G}'', \pi \in \Gamma\}$,*
2. *$(\mathcal{G}'] = \{s \in \mathcal{S} | s \preceq x' \text{ for some } x' \in \mathcal{G}'\}$,*
3. *$\mathcal{G}' \sqsubseteq (\mathcal{G}']$,*
4. *If $\mathcal{G}' \sqsubseteq \mathcal{G}''$, then $(\mathcal{G}'] \sqsubseteq (\mathcal{G}'']$ and $(\mathcal{G}']\Gamma(\mathcal{G}''] \sqsubseteq (\mathcal{G}' \Gamma \mathcal{G}'']$.*

Definition 3 ([17]). *Let $\mathcal{S}$ be a Γ-semigroup and $\mathcal{G}$ be a subset of $\mathcal{S}$ called left-tri-ideal(right-tri-ideal) (or LTI and RTI, respectively) if it satisfies the following conditions:*
1. *$\mathcal{G}$ is a Γ-subsemigroup,*
2. *$\mathcal{G}\Gamma\mathcal{S}\Gamma\mathcal{G}\Gamma\mathcal{G} \subseteq \mathcal{G}$ ($\mathcal{G}\Gamma\mathcal{G}\Gamma\mathcal{S}\Gamma\mathcal{G} \subseteq \mathcal{G}$).*

Lemma 1 ([18]). *Let $\mathcal{S}$ be a Γ-semiring, $\mathcal{G}$ a subset of $\mathcal{S}$, and $a \in S$. Then, the following statements hold:*
1. *$\langle G \rangle_{lt} = \mathcal{G} \cup \mathcal{G}\Gamma\mathcal{G} \cup \mathcal{G}\Gamma\mathcal{S}\Gamma\mathcal{G}\Gamma\mathcal{G}$ is the smallest Γ-LTI of S containing $\mathcal{G}$,*
2. *$\langle G \rangle_{rt} = \mathcal{G} \cup \mathcal{G}\Gamma\mathcal{G} \cup \mathcal{G}\Gamma\mathcal{G}\Gamma\mathcal{S}\Gamma\mathcal{G}$ is the smallest Γ-RTI of $\mathcal{S}$ containing $\mathcal{G}$,*
3. *$\langle a \rangle_{lt} = a \cup a\Gamma a \cup a\Gamma\mathcal{S}\Gamma a\Gamma a$ is the smallest Γ-LTI of $\mathcal{S}$ containing "a",*
4. *$\langle a \rangle_{rt} = a \cup a\Gamma a \cup a\Gamma a\Gamma\mathcal{S}\Gamma a$ is the smallest Γ-RTI of $\mathcal{S}$ containing "a".*

Definition 4 ([18]). *(i) Let $\mathcal{S}$ be an ordered semigroup. A subsemigroup $\mathcal{G}$ of $\mathcal{S}$ is called an M-left-ideal of $\mathcal{S}$ if $\mathcal{S}^M \mathcal{G} \subseteq \mathcal{G}$ and $(\mathcal{G}] = \mathcal{G}$, where M is a positive integer that is not necessarily one.*
(ii) A subsemigroup $\mathcal{G}$ of $\mathcal{S}$ is called a M-right-ideal of $\mathcal{S}$ if $\mathcal{G}\mathcal{S}^M \subseteq \mathcal{G}$ and $(\mathcal{G}] = \mathcal{G}$, where M is a positive integer that is not necessarily one.

Definition 5 ([18]). *Let $\mathcal{G}$ be a subsemigroup of an ordered semigroup $\mathcal{G}$. Then,*
(i) The M-left-ideal generated by $\mathcal{G}$ is $(\mathcal{G})_{ml} = (\mathcal{G} \cup \mathcal{S}^M \mathcal{G}]$.
(ii) The M-right-ideal generated by $\mathcal{G}$ is $(\mathcal{G})_{mr} = (\mathcal{G} \cup \mathcal{G}\mathcal{S}^M]$.

Definition 6 ([7]). *Let $\mathcal{G}$ be a subset of $\mathcal{S}$, which is called an M-bi-ideal of semigroup S if it satisfies the following conditions:*
1. *$\mathcal{G}$ is a Γ-subsemigroup,*
2. *$\mathcal{G} \cdot \mathcal{S}^M \cdot \mathcal{G} \subseteq \mathcal{G}$, where M is a positive integer.*

Definition 7 ([13]). *Let $\mathcal{G}$ be a subset of $\mathcal{S}$ that is called a bi-basis of $\mathcal{S}$ if it satisfies the following conditions:*
1. *$\mathcal{S} = \langle \mathcal{G} \rangle_b$.*
2. *If $\mathcal{F} \subseteq \mathcal{G}$ such that $\mathcal{S} = \langle \mathcal{F} \rangle_b$, then $\mathcal{F} = \mathcal{G}$.*

Definition 8. *Let $\mathcal{G}$ be the subset of $\mathcal{S}$ that is called an M-bi-basis of $\mathcal{S}$ if satisfies the following conditions:*
1. *$\mathcal{S} = \langle \mathcal{G} \rangle_{mb}$.*
2. *If $\mathcal{F} \subseteq \mathcal{G}$ such that $\mathcal{S} = \langle \mathcal{F} \rangle_{mb}$, then $\mathcal{F} = \mathcal{G}$.*

3. M-LTB Generator

In this paper, we present some results on the M-left-tri-ideal (M-LTI) generator, based on an ordered Γ-semigroup.

Definition 9. *Let $\mathcal{G}$ be the subset of $\mathcal{S}$ called an M-LTI of $\mathcal{S}$ if it satisfies the following conditions:*
1. *$\mathcal{G}$ is a Γ-subsemigroup,*
2. *$\mathcal{G} \cdot \Gamma \cdot (\mathcal{S} \cdot \Gamma \cdot \ldots \cdot \Gamma \cdot \mathcal{S} \ (M - times)) \cdot \Gamma \cdot \mathcal{G} \cdot \Gamma \cdot \mathcal{G} \subseteq \mathcal{G}$, where M is a positive integer,*
3. *If $g \in \mathcal{G}$ and $s \in \mathcal{S}$ such that $s \preceq g$, then $s \in \mathcal{G}$.*

Remark 2. *If $f_1 \in \mathcal{S}$ and $\mathcal{N}$ and M are positive integers, then the following statements hold:*
1. *$\mathcal{N} f_1 = f_1 \cdot \Gamma \cdot f_1 \cdot \Gamma \cdot \ldots \cdot \Gamma \cdot f_1 \ (\mathcal{N} - times)$*
2. *$\mathcal{S} \cdot \Gamma \cdot \mathcal{S} \cdot \Gamma \cdot \ldots \cdot \Gamma \cdot \mathcal{S} \ ((M - times)) \subseteq \mathcal{S} \cdot \Gamma \cdot \ldots \cdot \Gamma \cdot \mathcal{S} \ (M - 1 \ times)$*

Remark 3. 1. *Every M-bi-ideal is an M-LTI.*
2. *Every LTI is an M-LTI.*

Here is an example showing that the converse does not need to be true, as demonstrated by Example 1.

Example 1.

$$\text{Let } \mathcal{S} = \left\{ \begin{pmatrix} 0 & a_1 & a_2 & a_3 & a_4 & a_5 & a_6 & a_7 \\ 0 & 0 & a_8 & a_9 & a_{10} & a_{11} & a_{12} & a_{13} \\ 0 & 0 & 0 & a_{14} & a_{15} & a_{16} & a_{17} & a_{18} \\ 0 & 0 & 0 & 0 & a_{19} & a_{20} & a_{21} & a_{22} \\ 0 & 0 & 0 & 0 & 0 & a_{23} & a_{24} & a_{25} \\ 0 & 0 & 0 & 0 & 0 & 0 & a_{26} & a_{27} \\ 0 & 0 & 0 & 0 & 0 & 0 & 0 & a_{28} \\ 0 & 0 & 0 & 0 & 0 & 0 & 0 & 0 \end{pmatrix} \middle| a_i'^s \in Z^* \right\}$$

and Γ is a unit matrix. Now, we define the partial order relation $\preceq$ on $\mathcal{S}$: for any $A, B \in \mathcal{S}$, $A \preceq_{(S_1, S_2)} B$, if and only if $a_{ij} \preceq b_{ij}$, for all i and j. Then, $\mathcal{S}$ is an ordered Γ-semigroup of matrices over Z^* (non-negative integer) with the partial order relation "$\preceq$".

(i) Clearly,

$$B_1 = \left\{ \begin{pmatrix} 0 & b_1 & 0 & 0 & 0 & 0 & 0 & 0 \\ 0 & 0 & 0 & 0 & 0 & 0 & 0 & 0 \\ 0 & 0 & 0 & b_2 & 0 & 0 & 0 & 0 \\ 0 & 0 & 0 & 0 & 0 & 0 & 0 & 0 \\ 0 & 0 & 0 & 0 & 0 & b_3 & 0 & 0 \\ 0 & 0 & 0 & 0 & 0 & 0 & 0 & 0 \\ 0 & 0 & 0 & 0 & 0 & 0 & 0 & b_4 \\ 0 & 0 & 0 & 0 & 0 & 0 & 0 & 0 \end{pmatrix} \middle| b_i'^s \in Z^* \right\}.$$

Although B_1 is an M-LTI, it is not an M-bi-ideal of $\mathcal{S}$.

(iii) Clearly,

$$B_2 = \left\{ \begin{pmatrix} 0 & b_1 & 0 & 0 & 0 & 0 & 0 & 0 \\ 0 & 0 & 0 & 0 & 0 & 0 & 0 & 0 \\ 0 & 0 & 0 & b_2 & b_3 & b_4 & 0 & 0 \\ 0 & 0 & 0 & 0 & 0 & 0 & 0 & 0 \\ 0 & 0 & 0 & 0 & 0 & b_5 & 0 & 0 \\ 0 & 0 & 0 & 0 & 0 & 0 & 0 & 0 \\ 0 & 0 & 0 & 0 & 0 & 0 & 0 & b_6 \\ 0 & 0 & 0 & 0 & 0 & 0 & 0 & 0 \end{pmatrix} \middle| b_i'^s \in Z^* \right\}.$$

Hence, B_2 is an M-LTI, but not an LTI of $\mathcal{S}$.

Theorem 1. 1. *Let $f_1 \in \mathcal{S}$. The M-LTI generated by an element "f_1" is $\langle f_1 \rangle_{mlt} = \{f_1 \cup \mathcal{N}(f_1 \cdot \Gamma \cdot f_1) \cup f_1 \cdot \Gamma \cdot (\mathcal{S} \cdot \Gamma \cdot \ldots \cdot \Gamma \cdot \mathcal{S} \ (M-times)) \cdot \Gamma \cdot f_1 \cdot \Gamma \cdot f_1\}$ and $\mathcal{N} \succeq M$, where $\mathcal{N}$ and M are positive integers.*

2. *Let $\mathcal{G}$ be a subset of $\mathcal{S}$. The M-LTI generated by set "$\mathcal{G}$" is $\langle \mathcal{G} \rangle_{mlt} = \{\mathcal{G} \cup \mathcal{G} \cdot \Gamma \cdot \mathcal{G} \cup \mathcal{G} \cdot \Gamma \cdot (\mathcal{S} \cdot \Gamma \cdot \ldots \cdot \Gamma \cdot \mathcal{S} \ (M-times)) \cdot \Gamma \cdot \mathcal{G} \cdot \Gamma \cdot \mathcal{G}\}$.*

Definition 10. *Let $\mathcal{G}$ be a subset of $\mathcal{S}$, known as an M-left-tri-basis (LTB) of $\mathcal{S}$ if it meets the criteria listed below:*

1. $\mathcal{S} = \langle \mathcal{G} \rangle_{mlt}.$

2. If $\mathcal{F} \sqsubseteq \mathcal{G}$ such that $\mathcal{S} = \langle \mathcal{F} \rangle_{mlt}$, then $\mathcal{F} = \mathcal{G}$.

Example 2. Let $\mathcal{S} = \{k_1, k_2, k_3, k_4, k_5, k_6\}$ and $\Gamma = \{\pi_1, \pi_2\}$, where π_1 and pi_2 are defined on $\mathcal{S}$ with the following table:

π_1	k_1	k_2	k_3	k_4	k_5	k_6	π_2	k_1	k_2	k_3	k_4	k_5	k_6
k_1	k_1	k_1	k_1	k_1	k_1	k_6	k_1	k_1	k_4	k_1	k_4	k_4	k_6
k_2	k_1	k_1	k_1	k_2	k_3	k_6	k_2	k_1	k_2	k_1	k_4	k_4	k_6
k_3	k_1	k_2	k_3	k_1	k_1	k_6	k_3	k_1	k_4	k_3	k_4	k_5	k_6
k_4	k_1	k_1	k_1	k_4	k_5	k_6	k_4	k_1	k_4	k_1	k_4	k_4	k_6
k_5	k_1	k_4	k_5	k_1	k_1	k_6	k_5	k_1	k_4	k_3	k_4	k_5	k_6
k_6	k_6	k_6	k_6	k_6	k_6	k_6	k_6	k_6	k_6	k_6	k_6	k_6	k_6

$\preceq := \{(k_1, k_1), (k_1, k_6), (k_2, k_2), (k_2, k_6), (k_3, k_3), (k_3, k_6), (k_4, k_4), (k_4, k_6), (k_5, k_5), (k_5, k_6),$
$(k_6, k_6)\}$. Clearly, $(\mathcal{S}, \Gamma, \preceq)$ is an ordered Γ-semigroup.

The covering relation $\preceq := \{(k_1, k_6), (k_2, k_6), (k_3, k_6), (k_4, k_6), (k_5, k_6)\}$ is represented by Figure 1, since $\mathcal{G} = \{k_4, k_5\}$ is a M-LTB of $\mathcal{S}$.

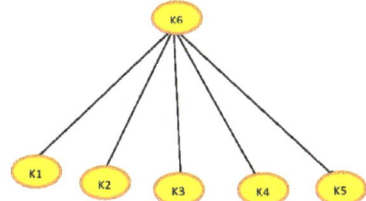

Figure 1. Covering relation.

Theorem 2. Let $\mathcal{G}$ be the M-LTB of $\mathcal{S}$ and $f_1, f_2 \in \mathcal{G}$. If $f_1 \in (\mathcal{N}(f_2 \cdot \Gamma \cdot f_2) \cup f_2 \cdot \Gamma \cdot (\mathcal{S} \cdot \Gamma \cdot \ldots \cdot \Gamma \cdot \mathcal{S}) \cdot \Gamma \cdot f_2 \cdot \Gamma \cdot f_2]$, then $f_1 = f_2$.

Proof. Assume that $f_1 \in (\mathcal{N}(f_2 \cdot \Gamma \cdot f_2) \cup f_2 \cdot \Gamma \cdot (\mathcal{S} \cdot \Gamma \cdot \ldots \cdot \Gamma \cdot \mathcal{S}) \cdot \Gamma \cdot f_2 \cdot \Gamma \cdot f_2]$, and suppose that $f_1 \neq f_2$. Let $\mathcal{F} = \mathcal{G} \setminus \{f_1\}$. Obviously, $\mathcal{F} \subset \mathcal{G}$ since $f_1 \neq f_2, f_2 \in \mathcal{F}$. To show that $\langle \mathcal{F} \rangle_{mlt} = \mathcal{S}$, clearly, $\langle \mathcal{F} \rangle_{mlt} \sqsubseteq \mathcal{S}$. Still, to prove that $\mathcal{S} \sqsubseteq \langle \mathcal{F} \rangle_{mlt}$, let $s \in \mathcal{S}$. By our hypothesis, $\langle \mathcal{G} \rangle_{mlt} = \mathcal{S}$, and hence, $s \in (\mathcal{G} \cup \mathcal{G} \cdot \Gamma \cdot \mathcal{G} \cup \mathcal{G} \cdot \Gamma \cdot (\mathcal{S} \cdot \Gamma \cdot \ldots \cdot \Gamma \cdot \mathcal{S}) \cdot \Gamma \cdot \mathcal{G} \cdot \Gamma \cdot \mathcal{G}]$. We have $s \preceq g$ for some $g \in \mathcal{G} \cup \mathcal{G} \cdot \Gamma \cdot \mathcal{G} \cup \mathcal{G} \cdot \Gamma \cdot (\mathcal{S} \cdot \Gamma \cdot \ldots \cdot \Gamma \cdot \mathcal{S}) \cdot \Gamma \cdot \mathcal{G} \cdot \Gamma \cdot \mathcal{G}$. As a result, the following cases will be discussed.
Case-1: Let $g \in \mathcal{G}$. There are two subcases to examine:
Subcase-1: Let $g \neq f_1$, then $g \in \mathcal{G} \setminus \{f_1\} = \mathcal{F} \sqsubseteq \langle \mathcal{F} \rangle_{mlt}$.
Subcase-2: Let $g = f_1$. We have $g = f_1 \in (\mathcal{N}(f_2 \cdot \Gamma \cdot f_2) \cup f_2 \cdot \Gamma \cdot (\mathcal{S} \cdot \Gamma \cdot \ldots \cdot \Gamma \cdot \mathcal{S}) \cdot \Gamma \cdot f_2 \cdot \Gamma \cdot f_2] \sqsubseteq (\mathcal{F} \cdot \Gamma \cdot \mathcal{F} \cup \mathcal{F} \cdot \Gamma \cdot (\mathcal{S} \cdot \Gamma \cdot \ldots \cdot \Gamma \cdot \mathcal{S}) \cdot \Gamma \cdot \mathcal{F} \cdot \Gamma \cdot \mathcal{F}] \sqsubseteq \langle \mathcal{F} \rangle_{mlt}$.
Case-2: Let $g \in \mathcal{G} \cdot \Gamma \cdot \mathcal{G}$. Then, $g = g_1 \cdot \pi \cdot g_2$, for some $g_1, g_2 \in \mathcal{G}$ and $\pi \in \Gamma$. In addition, there are four subcases to be considered.
Subcase-1: Let $g_1 = f_1$ and $g_2 = f_1$. Now,

$$\begin{aligned}
g &= g_1 \cdot \pi \cdot g_2 \\
&= f_1 \cdot \pi \cdot f_1 \\
&\in (\mathcal{N}(f_2 \cdot \Gamma \cdot f_2) \cup f_2 \cdot \Gamma \cdot (\mathcal{S} \cdot \Gamma \cdot \ldots \cdot \Gamma \cdot \mathcal{S}) \cdot \Gamma \cdot f_2 \cdot \Gamma \cdot f_2] \cdot \Gamma \cdot (\mathcal{N}(f_2 \cdot \Gamma \cdot f_2) \cup \\
&\quad f_2 \cdot \Gamma \cdot (\mathcal{S} \cdot \Gamma \cdot \ldots \cdot \Gamma \cdot \mathcal{S}) \cdot \Gamma \cdot f_2 \cdot \Gamma \cdot f_2] \\
&\sqsubseteq ((\mathcal{N}(f_2 \cdot \Gamma \cdot f_2) \cup f_2 \cdot \Gamma \cdot (\mathcal{S} \cdot \Gamma \cdot \ldots \cdot \Gamma \cdot \mathcal{S}) \cdot \Gamma \cdot f_2 \cdot \Gamma \cdot f_2) \cdot \Gamma \cdot \\
&\quad (\mathcal{N}(f_2 \cdot \Gamma \cdot f_2) \cup f_2 \cdot \Gamma \cdot (\mathcal{S} \cdot \Gamma \cdot \ldots \cdot \Gamma \cdot \mathcal{S}) \cdot \Gamma \cdot f_2 \cdot \Gamma \cdot f_2)] \\
&\sqsubseteq (\mathcal{F} \cdot \Gamma \cdot (\mathcal{S} \cdot \Gamma \cdot \ldots \cdot \Gamma \cdot \mathcal{S}) \cdot \Gamma \cdot \mathcal{F} \cdot \Gamma \cdot \mathcal{F}] \\
&\sqsubseteq \langle \mathcal{F} \rangle_{mlt}.
\end{aligned}$$

Subcase-2: Let $g_1 \neq f_1$ and $g_2 = f_1$. Now,

$$\begin{aligned}
g &= g_1 \cdot \pi \cdot g_2 \\
&\in (\mathcal{G} \setminus \{f_1\}) \cdot \Gamma \cdot (\mathcal{N}(f_2 \cdot \Gamma \cdot f_2) \cup f_2 \cdot \Gamma \cdot (\mathcal{S} \cdot \Gamma \cdot \ldots \cdot \Gamma \cdot \mathcal{S}) \cdot \Gamma \cdot f_2 \cdot \Gamma \cdot f_2] \\
&\sqsubseteq ((\mathcal{G} \setminus \{f_1\}) \cdot \Gamma \cdot (\mathcal{N}(f_2 \cdot \Gamma \cdot f_2) \cup f_2 \cdot \Gamma \cdot (\mathcal{S} \cdot \Gamma \cdot \ldots \cdot \Gamma \cdot \mathcal{S}) \cdot \Gamma \cdot f_2 \cdot \Gamma \cdot f_2)] \\
&\sqsubseteq (\mathcal{F} \cdot \Gamma \cdot (\mathcal{S} \cdot \Gamma \cdot \ldots \cdot \Gamma \cdot \mathcal{S}) \cdot \Gamma \cdot \mathcal{F} \cdot \Gamma \cdot \mathcal{F}] \\
&\sqsubseteq \langle \mathcal{F} \rangle_{mlt}.
\end{aligned}$$

Subcase-3: Let $g_1 = f_1$ and $g_2 \neq f_1$. Now,

$$\begin{aligned}
g &= g_1 \cdot \pi \cdot g_2 \\
&\in (\mathcal{N}(f_2 \cdot \Gamma \cdot f_2) \cup f_2 \cdot \Gamma \cdot (\mathcal{S} \cdot \Gamma \cdot \ldots \cdot \Gamma \cdot \mathcal{S}) \cdot \Gamma \cdot f_2 \cdot \Gamma \cdot f_2] \cdot \Gamma \cdot (\mathcal{G} \setminus \{f_1\}) \\
&\sqsubseteq ((\mathcal{N}(f_2 \cdot \Gamma \cdot f_2) \cup f_2 \cdot \Gamma \cdot (\mathcal{S} \cdot \Gamma \cdot \ldots \cdot \Gamma \cdot \mathcal{S}) \cdot \Gamma \cdot f_2 \cdot \Gamma \cdot f_2) \cdot \Gamma \cdot (\mathcal{G} \setminus \{f_1\})] \\
&\sqsubseteq (\mathcal{F} \cdot \Gamma \cdot (\mathcal{S} \cdot \Gamma \cdot \ldots \cdot \Gamma \cdot \mathcal{S}) \cdot \Gamma \cdot \mathcal{F} \cdot \Gamma \cdot \mathcal{F}] \\
&\sqsubseteq \langle \mathcal{F} \rangle_{mlt}.
\end{aligned}$$

Subcase-4: Let $g_1 \neq f_1, g_2 \neq f_1$, and $\mathcal{F} = \mathcal{G} \setminus \{f_1\}$. Now,

$$\begin{aligned}
g &= g_1 \cdot \pi \cdot g_2 \\
&\in (\mathcal{G} \setminus \{f_1\}) \cdot \Gamma \cdot (\mathcal{G} \setminus \{f_1\}) \\
&\sqsubseteq \langle \mathcal{F} \rangle_{mlt}.
\end{aligned}$$

Case-3: Let $g \in \mathcal{G} \cdot \Gamma \cdot (\mathcal{S} \cdot \Gamma \cdot \ldots \cdot \Gamma \cdot \mathcal{S}) \cdot \Gamma \cdot \mathcal{G} \cdot \Gamma \cdot \mathcal{G}$. Then, $g = g_3 \cdot \pi \cdot (s_1 \cdot \pi_1 \cdot s_2 \cdot \ldots \cdot \pi_n \cdot s_n) \cdot \theta \cdot g_4 \cdot \theta_1 \cdot g_5$ for some $g_3, g_4, g_5 \in \mathcal{G}, s_1, s_2, \ldots, s_n \in \mathcal{S}$ and $\pi, \theta, \theta_1, \pi_1, \pi_2, \ldots, \pi_n \in \Gamma$. We will examine eight subcases.

$g_i \setminus$ Subcase	g_3	g_4	g_5
Subcase $-$ 1	$f_1 = g_3$	$f_1 = g_4$	$f_1 = g_5$
Subcase $-$ 2	$f_1 \neq g_3$	$f_1 = g_4$	$f_1 = g_5$
Subcase $-$ 3	$f_1 = g_3$	$f_1 \neq g_4$	$f_1 = g_5$
Subcase $-$ 4	$f_1 = g_3$	$f_1 = g_4$	$f_1 \neq g_5$
Subcase $-$ 5	$f_1 \neq g_3$	$f_1 \neq g_4$	$f_1 = g_5$
Subcase $-$ 6	$f_1 = g_3$	$f_1 \neq g_4$	$f_1 \neq g_5$
Subcase $-$ 7	$f_1 \neq g_3$	$f_1 = g_4$	$f_1 \neq g_5$
Subcase $-$ 8	$f_1 \neq g_3$	$f_1 \neq g_4$	$f_1 \neq g_5$

Subcase-1: Let $g_3 = f_1, g_4 = f_1$, and $g_5 = f_1$. Now,

$$\begin{aligned}
g &= g_3 \cdot \pi \cdot (s_1 \cdot \pi_1 \cdot s_2 \cdot \ldots \cdot \pi_n \cdot s_n) \cdot \theta \cdot g_4 \cdot \theta_1 \cdot g_5 \\
&= f_1 \cdot \pi \cdot (s_1 \cdot \pi_1 \cdot s_2 \cdot \ldots \cdot \pi_n \cdot s_n) \cdot \theta \cdot f_1 \cdot \theta_1 \cdot f_1 \\
&\in (\mathcal{N}(f_2 \cdot \Gamma \cdot f_2) \cup f_2 \cdot \Gamma \cdot (\mathcal{S} \cdot \Gamma \cdot \ldots \cdot \Gamma \cdot \mathcal{S}) \cdot \Gamma \cdot f_2 \cdot \Gamma \cdot f_2] \cdot \Gamma \cdot (\mathcal{S} \cdot \Gamma \cdot \ldots \cdot \Gamma \cdot \mathcal{S}) \cdot \Gamma \cdot \\
&\quad (\mathcal{N}(f_2 \cdot \Gamma \cdot f_2) \cup f_2 \cdot \Gamma \cdot (\mathcal{S} \cdot \Gamma \cdot \ldots \cdot \Gamma \cdot \mathcal{S}) \cdot \Gamma \cdot f_2 \cdot \Gamma \cdot f_2] \cdot \Gamma \cdot (\mathcal{N}(f_2 \cdot \Gamma \cdot f_2) \cup \\
&\quad f_2 \cdot \Gamma \cdot (\mathcal{S} \cdot \Gamma \cdot \ldots \cdot \Gamma \cdot \mathcal{S}) \cdot \Gamma \cdot f_2 \cdot \Gamma \cdot f_2] \\
&\sqsubseteq \Big(\big\{ (\mathcal{N}(f_2 \cdot \Gamma \cdot f_2) \cup f_2 \cdot \Gamma \cdot (\mathcal{S} \cdot \Gamma \cdot \ldots \cdot \Gamma \cdot \mathcal{S}) \cdot \Gamma \cdot f_2 \cdot \Gamma \cdot f_2) \cdot \Gamma \cdot (\mathcal{S} \cdot \Gamma \cdot \ldots \cdot \Gamma \cdot \mathcal{S}) \cdot \Gamma \cdot \\
&\quad (\mathcal{N}(f_2 \cdot \Gamma \cdot f_2) \cup f_2 \cdot \Gamma \cdot (\mathcal{S} \cdot \Gamma \cdot \ldots \cdot \Gamma \cdot \mathcal{S}) \cdot \Gamma \cdot f_2 \cdot \Gamma \cdot f_2) \big\} \cdot \Gamma \cdot (\mathcal{N}(f_2 \cdot \Gamma \cdot f_2) \cup \\
&\quad f_2 \cdot \Gamma \cdot (\mathcal{S} \cdot \Gamma \cdot \ldots \cdot \Gamma \cdot \mathcal{S}) \cdot \Gamma \cdot f_2 \cdot \Gamma \cdot f_2 \Big] \\
&\sqsubseteq (\mathcal{F} \cdot \Gamma \cdot (\mathcal{S} \cdot \Gamma \cdot \ldots \cdot \Gamma \cdot \mathcal{S}) \cdot \Gamma \cdot \mathcal{F} \cdot \Gamma \cdot \mathcal{F}] \\
&\sqsubseteq \langle \mathcal{F} \rangle_{mlt}.
\end{aligned}$$

Subcase-2: Let $g_3 \neq f_1, g_4 = f_1$, and $g_5 = f_1$. Now,

$$
\begin{aligned}
g &= g_3 \cdot \pi \cdot (s_1 \cdot \pi_1 \cdot s_2 \cdot \ldots \cdot \pi_n \cdot s_n) \cdot \theta \cdot g_4 \cdot \theta_1 \cdot g_5 \\
&\in (\mathcal{G} \setminus \{f_1\}) \cdot \Gamma \cdot (\mathcal{S} \cdot \Gamma \cdot \ldots \cdot \Gamma \cdot \mathcal{S}) \cdot \Gamma \cdot (\mathcal{N}(f_2 \cdot \Gamma \cdot f_2) \cup f_2 \cdot \Gamma \cdot (\mathcal{S} \cdot \Gamma \cdot \ldots \cdot \Gamma \cdot \mathcal{S}) \\
&\quad \cdot \Gamma \cdot f_2 \cdot \Gamma \cdot f_2] \cdot \Gamma \cdot (\mathcal{N}(f_2 \cdot \Gamma \cdot f_2) \cup f_2 \cdot \Gamma \cdot (\mathcal{S} \cdot \Gamma \cdot \ldots \cdot \Gamma \cdot \mathcal{S}) \cdot \Gamma \cdot f_2 \cdot \Gamma \cdot f_2] \\
&\sqsubseteq \Big((\mathcal{G} \setminus \{f_1\}) \cdot \Gamma \cdot (\mathcal{S} \cdot \Gamma \cdot \ldots \cdot \Gamma \cdot \mathcal{S}) \cdot \Gamma \cdot (\mathcal{N}(f_2 \cdot \Gamma \cdot f_2) \cup f_2 \cdot \Gamma \cdot (\mathcal{S} \cdot \Gamma \cdot \ldots \cdot \Gamma \cdot \mathcal{S}) \\
&\quad \cdot \Gamma \cdot f_2 \cdot \Gamma \cdot f_2) \cdot \Gamma \cdot (\mathcal{N}(f_2 \cdot \Gamma \cdot f_2) \cup f_2 \cdot \Gamma \cdot (\mathcal{S} \cdot \Gamma \cdot \ldots \cdot \Gamma \cdot \mathcal{S}) \cdot \Gamma \cdot f_2 \cdot \Gamma \cdot f_2\Big] \\
&\sqsubseteq (\mathcal{F} \cdot \Gamma \cdot (\mathcal{S} \cdot \Gamma \cdot \ldots \cdot \Gamma \cdot \mathcal{S}) \cdot \Gamma \cdot \mathcal{F} \cdot \Gamma \cdot \mathcal{F}] \\
&\sqsubseteq \langle \mathcal{F} \rangle_{mlt}.
\end{aligned}
$$

Subcase-6: Let $g_3 = f_1, g_4 \neq f_1$, and $g_5 \neq f_1$. Now,

$$
\begin{aligned}
g &= g_3 \cdot \pi \cdot (s_1 \cdot \pi_1 \cdot s_2 \cdot \ldots \cdot \pi_n \cdot s_n) \cdot \theta \cdot g_4 \cdot \theta_1 \cdot g_5 \\
&\in (\mathcal{N}(f_2 \cdot \Gamma \cdot f_2) \cup f_2 \cdot \Gamma \cdot (\mathcal{S} \cdot \Gamma \cdot \ldots \cdot \Gamma \cdot \mathcal{S}) \cdot \Gamma \cdot f_2 \cdot \Gamma \cdot f_2] \cdot \Gamma \cdot (\mathcal{S} \cdot \Gamma \cdot \ldots \cdot \Gamma \cdot \mathcal{S}) \\
&\quad \cdot \Gamma \cdot (\mathcal{G} \setminus \{f_1\}) \cdot \Gamma \cdot (\mathcal{G} \setminus \{f_1\}) \\
&\sqsubseteq ((\mathcal{N}(f_2 \cdot \Gamma \cdot f_2) \cup f_2 \cdot \Gamma \cdot (\mathcal{S} \cdot \Gamma \cdot \ldots \cdot \Gamma \cdot \mathcal{S}) \cdot \Gamma \cdot f_2 \cdot \Gamma \cdot f_2) \cdot \Gamma \cdot (\mathcal{S} \cdot \Gamma \cdot \ldots \cdot \Gamma \cdot \mathcal{S}) \\
&\quad \cdot \Gamma \cdot (\mathcal{G} \setminus \{f_1\}) \cdot \Gamma \cdot (\mathcal{G} \setminus \{f_1\})] \\
&\sqsubseteq (\mathcal{F} \cdot \Gamma \cdot (\mathcal{S} \cdot \Gamma \cdot \ldots \cdot \Gamma \cdot \mathcal{S}) \cdot \Gamma \cdot \mathcal{F} \cdot \Gamma \cdot \mathcal{F}] \\
&\sqsubseteq \langle \mathcal{F} \rangle_{mlt}.
\end{aligned}
$$

Subcase-8: Let $g_3 \neq f_1, g_4 \neq f_1, g_5 \neq f_1$, and $\mathcal{F} = \mathcal{G} \setminus \{f_1\}$. Now,

$$
\begin{aligned}
g &= g_3 \cdot \pi \cdot (s_1 \cdot \pi_1 \cdot s_2 \cdot \ldots \cdot \pi_n \cdot s_n) \cdot \theta \cdot g_4 \cdot \theta_1 \cdot g_5 \\
&\in (\mathcal{G} \setminus \{f_1\}) \cdot \Gamma \cdot (\mathcal{S} \cdot \Gamma \cdot \ldots \cdot \Gamma \cdot \mathcal{S}) \cdot \Gamma \cdot (\mathcal{G} \setminus \{f_1\}) \cdot \Gamma \cdot (\mathcal{G} \setminus \{f_1\}) \\
&\sqsubseteq (\mathcal{F} \cdot \Gamma \cdot (\mathcal{S} \cdot \Gamma \cdot \ldots \cdot \Gamma \cdot \mathcal{S}) \cdot \Gamma \cdot \mathcal{F} \cdot \Gamma \cdot \mathcal{F}] \\
&\sqsubseteq \langle \mathcal{F} \rangle_{mlt}.
\end{aligned}
$$

It is similar to prove other subcases. Hence, for all the cases, we have $\mathcal{S} \sqsubseteq \langle \mathcal{F} \rangle_{mlt}$. Thus, $\mathcal{S} = \langle \mathcal{F} \rangle_{mlt}$, which is a contradiction. Hence $f_1 = f_2$. □

Lemma 2. *Let $\mathcal{G}$ be the M-LTB of $\mathcal{S}$ and $f_1, f_2, f_3, f_4 \in \mathcal{G}$. If $f_1 \in (\mathcal{N}(f_3 \cdot \Gamma \cdot f_2) \cup f_3 \cdot \Gamma \cdot (\mathcal{S} \cdot \Gamma \cdot \ldots \cdot \Gamma \cdot \mathcal{S}) \cdot \Gamma \cdot f_2 \cdot \Gamma \cdot f_4]$, then $f_1 = f_2$ or $f_1 = f_3$ or $f_1 = f_4$.*

Proof. Theorem 2 leads to the proof. □

Definition 11. *For any $s_1, s_2 \in \mathcal{S}$, $s_1 \preceq_{mlt} s_2 \iff \langle s_1 \rangle_{mlt} \sqsubseteq \langle s_2 \rangle_{mlt}$ is called a quasi-order on $\mathcal{S}$.*

Remark 4. *The order $\preceq_{mlt}$ is not a partial order of $\mathcal{S}$.*

Example 3. *By Example 2, $\langle k_5 \rangle_{mlt} \sqsubseteq \langle k_6 \rangle_{mlt}$ and $\langle k_6 \rangle_{mlt} \sqsubseteq \langle k_5 \rangle_{mlt}$ but $k_5 \neq k_6$. Hence, the relation $\preceq_{mlt}$ is not a partial order on $\mathcal{S}$.*

If $\mathcal{F}$ is an M-LTB of $\mathcal{S}$, then $\langle \mathcal{F} \rangle_{mlt} = \mathcal{S}$. Let $s \in \mathcal{S}$. Then, $s \in \langle \mathcal{F} \rangle_{mlt}$, and so, $s \in \langle f_1 \rangle_{mlt}$ for some $f_1 \in \mathcal{F}$. This implies $\langle s \rangle_{mlt} \sqsubseteq \langle f_1 \rangle_{mlt}$. Hence, $s \preceq_{mlt} f_1$.

Remark 5. *If $\mathcal{G}$ is a M-LTB of $\mathcal{S}$ then for any $s \in \mathcal{S}$, there exists $f_1 \in \mathcal{G}$ such that $s \preceq_{mlt} f_1$.*

Lemma 3. *Let $\mathcal{G}$ be an M-LTB of $\mathcal{S}$. If $f_1, f_2 \in \mathcal{G}$ such that $f_1 \neq f_2$, then neither $f_1 \preceq_{mlt} f_2$ nor $f_2 \preceq_{mlt} f_1$.*

Proof. Assume that $f_1, f_2 \in \mathcal{G}$, such that $f_1 \neq f_2$. Suppose that $f_1 \preceq_{mlt} f_2$. Let $\mathcal{F} = \mathcal{G} \setminus \{f_1\}$. Then, $f_2 \in \mathcal{F}$. Let $s \in \mathcal{S}$. By Remark 5, there exists $f_3 \in \mathcal{G}$, such that $s \preceq_{mlt} f_3$. We think about two cases to be discussed. If $f_3 \neq f_1$, then $f_3 \in \mathcal{F}$. Thus, $\langle s \rangle_{mlt} \sqsubseteq \langle f_3 \rangle_{mlt} \sqsubseteq \langle \mathcal{F} \rangle_{mlt}$. Hence, $\mathcal{S} = \langle \mathcal{F} \rangle_{mlt}$, which is a contradiction. If $f_3 = f_1$, then $s \preceq_{mlt} f_2$. Hence, $s \in \langle \mathcal{F} \rangle_{mlt}$, since $f_2 \in \mathcal{F}$. Hence, $\mathcal{S} = \langle \mathcal{F} \rangle_{mlt}$, which is a contradiction. A similar argument can be made for other cases. □

Lemma 4. *Let $\mathcal{G}$ be the M-LTB of $\mathcal{S}$ and $f_1, f_2, f_3 \in \mathcal{G}$ and $s \in \mathcal{S}$.*

1. *If $f_1 \in (\{f_2 \cdot \pi \cdot f_3\} \cup \mathcal{N}(\{f_2 \cdot \pi \cdot f_3\} \cdot \Gamma \cdot \{f_2 \cdot \pi \cdot f_3\}) \cup \{f_2 \cdot \pi \cdot f_3\} \cdot \Gamma \cdot (\mathcal{S} \cdot \Gamma \cdot \ldots \cdot \Gamma \cdot \mathcal{S}) \cdot$*
   $\Gamma \cdot \{f_2 \cdot \pi \cdot f_3\} \cdot \Gamma \cdot \{f_2 \cdot \pi \cdot f_3\}]$, then $f_1 = f_2$ or $f_1 = f_3$,
2. *If $f_1 \in (\{f_2 \cdot \pi \cdot (s_1 \cdot \pi_1 \cdot s_2 \cdot \ldots \cdot \pi_n \cdot s_n) \cdot \theta \cdot f_3 \cdot \theta_1 \cdot f_4\} \cup \mathcal{N}(\{f_2 \cdot \pi \cdot (s_1 \cdot \pi_1 \cdot s_2 \cdot \ldots \cdot \pi_n \cdot s_n) \cdot \theta \cdot f_3 \cdot \theta_1 \cdot f_4\} \cdot \Gamma \cdot \{f_2 \cdot \pi \cdot (s_1 \cdot \pi_1 \cdot s_2 \cdot \ldots \cdot \pi_n \cdot s_n) \cdot \theta \cdot f_3 \cdot \theta_1 \cdot f_4\}) \cup \{f_2 \cdot \pi \cdot (s_1 \cdot \pi_1 \cdot s_2 \cdot \ldots \cdot \pi_n \cdot s_n) \cdot \theta \cdot f_3 \cdot \theta_1 \cdot f_4\} \cdot \Gamma \cdot (\mathcal{S} \cdot \Gamma \cdot \ldots \cdot \Gamma \cdot \mathcal{S}) \cdot \Gamma \cdot \{f_2 \cdot \pi \cdot (s_1 \cdot \pi_1 \cdot s_2 \cdot \ldots \cdot \pi_n \cdot s_n) \cdot \theta \cdot f_3 \cdot \theta_1 \cdot f_4\} \cdot \Gamma \cdot \{f_2 \cdot \pi \cdot (s_1 \cdot \pi_1 \cdot s_2 \cdot \ldots \cdot \pi_n \cdot s_n) \cdot \theta \cdot f_3 \cdot \theta_1 \cdot f_4\}]$, then $f_1 = f_2$ or $f_1 = f_3$ or $f_1 = f_4$.*

Proof. (1) Assume that $f_1 \in (\{f_2 \cdot \pi \cdot f_3\} \cup \mathcal{N}(\{f_2 \cdot \pi \cdot f_3\} \cdot \Gamma \cdot \{f_2 \cdot \pi \cdot f_3\}) \cup \{f_2 \cdot \pi \cdot f_3\} \cdot \Gamma \cdot (\mathcal{S} \cdot \Gamma \cdot \ldots \cdot \Gamma \cdot \mathcal{S}) \cdot \Gamma \cdot \{f_2 \cdot \pi \cdot f_3\} \cdot \Gamma \cdot \{f_2 \cdot \pi \cdot f_3\}]$ and suppose that $f_1 \neq f_2$ and $f_1 \neq f_3$. Let $\mathcal{F} = \mathcal{G} \setminus \{f_1\}$. Clearly, $\mathcal{F} \subset \mathcal{G}$, since $f_1 \neq f_2$ and $f_1 \neq f_3$ implies $f_2, f_3 \in \mathcal{F}$. To prove that $\langle \mathcal{G} \rangle_{mlt} \sqsubseteq \langle \mathcal{F} \rangle_{mlt}$, it suffices to determine that $\mathcal{G} \sqsubseteq \langle \mathcal{F} \rangle_{mlt}$. Let $f \in \mathcal{G}$, if $f \neq f_1$ that $f \in \mathcal{F}$, and hence, $f \in \langle \mathcal{F} \rangle_{mlt}$. If $f = f_1$, then

$$\begin{aligned} f = f_1 &\in (\{f_2 \cdot \pi \cdot f_3\} \cup \mathcal{N}(\{f_2 \cdot \pi \cdot f_3\} \cdot \Gamma \cdot \{f_2 \cdot \pi \cdot f_3\}) \cup \{f_2 \cdot \pi \cdot f_3\} \cdot \Gamma \cdot (\mathcal{S} \cdot \Gamma \cdot \ldots \\ & \quad \cdot \Gamma \cdot \mathcal{S}) \cdot \Gamma \cdot \{f_2 \cdot \pi \cdot f_3\} \cdot \Gamma \cdot \{f_2 \cdot \pi \cdot f_3\}] \\ &\sqsubseteq (\mathcal{F} \cdot \Gamma \cdot \mathcal{F} \cup \mathcal{F} \cdot \Gamma \cdot (\mathcal{S} \cdot \Gamma \cdot \ldots \cdot \Gamma \cdot \mathcal{S}) \cdot \Gamma \cdot \mathcal{F} \cdot \Gamma \cdot \mathcal{F} \\ &\sqsubseteq \langle \mathcal{F} \rangle_{mlt}. \end{aligned}$$

Thus, $\mathcal{G} \sqsubseteq \langle \mathcal{F} \rangle_{mlt}$. This implies $\langle \mathcal{G} \rangle_{mlt} \sqsubseteq \langle \mathcal{F} \rangle_{mlt}$, as $\mathcal{G}$ is an M-LTB of $\mathcal{S}$ and $\mathcal{S} = \langle \mathcal{G} \rangle_{mlt} \sqsubseteq \langle \mathcal{F} \rangle_{mlt} \sqsubseteq \mathcal{S}$. Therefore, $\mathcal{S} = \langle \mathcal{F} \rangle_{mlt}$, which is a contradiction. Hence, $f_1 = f_2$ or $f_1 = f_3$.

(2) Assume that $f_1 \in (\{f_2 \cdot \pi \cdot (s_1 \cdot \pi_1 \cdot s_2 \cdot \ldots \cdot \pi_n \cdot s_n) \cdot \theta \cdot f_3 \cdot \theta_1 \cdot f_4\} \cup \mathcal{N}(\{f_2 \cdot \pi \cdot (s_1 \cdot \pi_1 \cdot s_2 \cdot \ldots \cdot \pi_n \cdot s_n) \cdot \theta \cdot f_3 \cdot \theta_1 \cdot f_4\} \cdot \Gamma \cdot \{f_2 \cdot \pi \cdot (s_1 \cdot \pi_1 \cdot s_2 \cdot \ldots \cdot \pi_n \cdot s_n) \cdot \theta \cdot f_3 \cdot \theta_1 \cdot f_4\}) \cup \{f_2 \cdot \pi \cdot (s_1 \cdot \pi_1 \cdot s_2 \cdot \ldots \cdot \pi_n \cdot s_n) \cdot \theta \cdot f_3 \cdot \theta_1 \cdot f_4\} \cdot \Gamma \cdot \mathcal{S} \cdot \Gamma \cdot \{f_2 \cdot \pi \cdot (s_1 \cdot \pi_1 \cdot s_2 \cdot \ldots \cdot \pi_n \cdot s_n) \cdot \theta \cdot f_3 \cdot \theta_1 \cdot f_4\} \cdot \Gamma \cdot \{f_2 \cdot \pi \cdot (s_1 \cdot \pi_1 \cdot s_2 \cdot \ldots \cdot \pi_n \cdot s_n) \cdot \theta \cdot f_3 \cdot \theta_1 \cdot f_4\}]$ and suppose that $f_1 \neq f_2$ and $f_1 \neq f_3$ and $f_1 \neq f_4$. Let $\mathcal{F} = \mathcal{G} \setminus \{f_1\}$. Clearly, $\mathcal{F} \subset \mathcal{G}$, since $f_1 \neq f_2, f_1 \neq f_3$, and $f_1 \neq f_4$ imply that $f_2, f_3, f_4 \in \mathcal{F}$. To prove that $\langle \mathcal{G} \rangle_{mlt} \sqsubseteq \langle \mathcal{F} \rangle_{mlt}$, it remains to prove that $\mathcal{G} \sqsubseteq \langle \mathcal{F} \rangle_{mlt}$. Let $f \in \mathcal{G}$ if $f \neq f_1$ that $f \in \mathcal{F}$, and so, $f \in \langle \mathcal{F} \rangle_{mlt}$. Hence,

$$\begin{aligned} f &= f_1 \\ &\in (\{f_2 \cdot \pi \cdot (s_1 \cdot \pi_1 \cdot s_2 \cdot \ldots \cdot \pi_n \cdot s_n) \cdot \theta \cdot f_3 \cdot \theta_1 \cdot f_4\} \cup \mathcal{N}(\{f_2 \cdot \pi \cdot (s_1 \cdot \pi_1 \cdot s_2 \cdot \ldots \\ & \quad \cdot \pi_n \cdot s_n) \cdot \theta \cdot f_3 \cdot \theta_1 \cdot f_4\} \cdot \Gamma \cdot \{f_2 \cdot \pi \cdot (s_1 \cdot \pi_1 \cdot s_2 \cdot \ldots \cdot \pi_n \cdot s_n) \cdot \theta \cdot f_3 \cdot \theta_1 \cdot f_4\}) \cup \\ & \quad \{f_2 \cdot \pi \cdot (s_1 \cdot \pi_1 \cdot s_2 \cdot \ldots \cdot \pi_n \cdot s_n) \cdot \theta \cdot f_3 \cdot \theta_1 \cdot f_4\} \cdot \Gamma \cdot (\mathcal{S} \cdot \Gamma \cdot \ldots \cdot \Gamma \cdot \mathcal{S}) \cdot \Gamma \cdot \\ & \quad \{f_2 \cdot \pi \cdot (s_1 \cdot \pi_1 \cdot s_2 \cdot \ldots \cdot \pi_n \cdot s_n) \cdot \theta \cdot f_3 \cdot \theta_1 \cdot f_4\} \cdot \Gamma \cdot \{f_2 \cdot \pi \cdot (s_1 \cdot \pi_1 \cdot s_2 \cdot \ldots \\ & \quad \cdot \pi_n \cdot s_n) \cdot \theta \cdot f_3 \cdot \theta_1 \cdot f_4\}] \\ &\sqsubseteq (\mathcal{F} \cdot \Gamma \cdot (\mathcal{S} \cdot \Gamma \cdot \ldots \cdot \Gamma \cdot \mathcal{S}) \cdot \Gamma \cdot \mathcal{F} \cdot \Gamma \cdot \mathcal{F}] \\ &\sqsubseteq \langle \mathcal{F} \rangle_{mlt}. \end{aligned}$$

Thus, $\mathcal{G} \sqsubseteq \langle \mathcal{F} \rangle_{mlt}$. This implies $\langle \mathcal{G} \rangle_{mlt} \sqsubseteq \langle \mathcal{F} \rangle_{mlt}$ as $\mathcal{G}$ is an M-LTB of $\mathcal{S}$ and $\mathcal{S} = \langle \mathcal{G} \rangle_{mlt} \sqsubseteq \langle \mathcal{F} \rangle_{mlt} \sqsubseteq \mathcal{S}$. Therefore, $\mathcal{S} = \langle \mathcal{F} \rangle_{mlt}$, which is a contradiction. Hence, $f_1 = f_2$ or $f_1 = f_3$ or $f_1 = f_4$. □

Lemma 5. *Let $\mathcal{G}$ be an M-LTB of $\mathcal{S}$,*

1. If $f_1 \neq f_2$ and $f_1 \neq f_3$, then $f_1 \npreceq_{mlt} f_2 \cdot \pi \cdot f_3$.
2. If $f_1 \neq f_2, f_1 \neq f_3$ and $f_1 \neq f_4$, then $f_1 \npreceq_{mlt} f_2 \cdot \pi \cdot (s_1 \cdot \pi_1 \cdot s_2 \cdot \ldots \cdot \pi_n \cdot s_n) \cdot \theta \cdot f_3 \cdot \theta_1 \cdot f_4$, for $f_1, f_2, f_3, f_4 \in \mathcal{G}, \pi, \pi_i, \theta, \theta_1 \in \Gamma$ and $s_i \in \mathcal{S}, i = 1, 2, \ldots, n$.

Proof. (1) For any $f_1, f_2, f_3 \in \mathcal{G}$, let $f_1 \neq f_2$ and $f_1 \neq f_3$. Suppose that $f_1 \preceq_{mlt} f_2 \cdot \pi \cdot f_3$ and

$$\begin{aligned}
f_1 &\in \langle f_1 \rangle_{mlt} \\
&\sqsubseteq \{(f_2 \cdot \pi \cdot f_3)\}_{mlt} \\
&= (\{(f_2 \cdot \pi \cdot f_3)\} \cup \mathcal{N}(\{(f_2 \cdot \pi \cdot f_3)\} \cdot \Gamma \cdot \{(f_2 \cdot \pi \cdot f_3)\}) \cup \{(f_2 \cdot \pi \cdot f_3)\} \cdot \Gamma \cdot \\
&\quad (\mathcal{S} \cdot \Gamma \cdot \ldots \cdot \Gamma \cdot \mathcal{S}) \cdot \Gamma \cdot \{(f_2 \cdot \pi \cdot f_3)\} \cdot \Gamma \cdot \{(f_2 \cdot \pi \cdot f_3)\}].
\end{aligned}$$

By Lemma 4 (1), it follows that $f_1 = f_2$ or $f_1 = f_3$, which is a contradiction.
(2) For any $f_1, f_2, f_3, f_4 \in \mathcal{G}$, let $f_1 \neq f_2$, $f_1 \neq f_3$, and $f_1 \neq f_4$. Suppose that $f_1 \preceq_{mlt} f_2 \cdot \pi \cdot (s_1 \cdot \pi_1 \cdot s_2 \cdot \ldots \cdot \pi_n \cdot s_n) \cdot \theta \cdot f_3 \cdot \theta_1 \cdot f_4$, we have

$$\begin{aligned}
f_1 &\in \langle f_1 \rangle_{mlt} \\
&\sqsubseteq \{(f_2 \cdot \Gamma \cdot (s_1 \cdot \pi_1 \cdot s_2 \cdot \ldots \cdot \pi_n \cdot s_n) \cdot \theta \cdot f_3 \cdot \theta_1 \cdot f_4)\}_{mlt} \\
&= (\{(f_2 \cdot \Gamma \cdot (s_1 \cdot \pi_1 \cdot s_2 \cdot \ldots \cdot \pi_n \cdot s_n) \cdot \theta \cdot f_3 \cdot \theta_1 \cdot f_4)\} \cup \mathcal{N}(\{(f_2 \cdot \Gamma \cdot (s_1 \cdot \pi_1 \cdot s_2 \\
&\quad \cdot \ldots \cdot \pi_n \cdot s_n) \cdot \theta \cdot f_3 \cdot \theta_1 \cdot f_4)\} \cdot \Gamma \cdot \{(f_2 \cdot \Gamma \cdot (s_1 \cdot \pi_1 \cdot s_2 \cdot \ldots \cdot \pi_n \cdot s_n) \cdot \theta \cdot f_3 \cdot \theta_1 \cdot f_4)\}) \\
&\quad \cup \{(f_2 \cdot \Gamma \cdot (s_1 \cdot \pi_1 \cdot s_2 \cdot \ldots \cdot \pi_n \cdot s_n) \cdot \theta \cdot f_3 \cdot \theta_1 \cdot f_4)\} \cdot \Gamma \cdot (\mathcal{S} \cdot \Gamma \cdot \ldots \cdot \Gamma \cdot \mathcal{S}) \cdot \Gamma \\
&\quad \cdot \{(f_2 \cdot \Gamma \cdot (s_1 \cdot \pi_1 \cdot s_2 \cdot \ldots \cdot \pi_n \cdot s_n) \cdot \theta \cdot f_3 \cdot \theta_1 \cdot f_4)\}].
\end{aligned}$$

By Lemma 4 (2), it follows that $f_1 = f_2$, $f_1 = f_3$, or $f_1 = f_4$, which contradicts our assumption. □

Theorem 3. *Let $\mathcal{G}$ be the M-LTB of $\mathcal{S}$, if and only if $\mathcal{G}$ satisfies the following*

1. For any $s \in \mathcal{S}$,
   (1.1) there exists $f_2 \in \mathcal{G}$ such that $s \preceq_{mlt} f_2$ (or),
   (1.2) there exists $g_1, g_2 \in \mathcal{G}$ such that $s \preceq_{mlt} g_1 \cdot \pi \cdot g_2$ (or),
   (1.3) there exists $g_3, g_4, g_5 \in \mathcal{G}$ such that $s \preceq_{mlt} g_3 \cdot \pi \cdot (s_1 \cdot \pi_1 \cdot s_2 \cdot \ldots \cdot \pi_n \cdot s_n) \cdot \theta \cdot g_4 \cdot \theta_1 \cdot g_5$;
2. If $f_1 \neq f_2$ and $f_1 \neq f_3$ and $f_1 \neq f_4$, then $f_1 \npreceq_{mlt} f_2 \cdot \pi \cdot f_3$, for any $f_1, f_2, f_3 \in \mathcal{G}$;
3. If $f_1 \neq f_2$ and $f_1 \neq f_3$ and $f_1 \neq f_4$, then $f_1 \npreceq_{mlt} f_2 \cdot \pi \cdot (s_1 \cdot \pi_1 \cdot s_2 \cdot \ldots \cdot \pi_n \cdot s_n) \cdot \theta \cdot \theta_1 \cdot f_4$, for any $f_1, f_2, f_3, f_4 \in \mathcal{G}, s_i \in \mathcal{S}$ and $\pi_i, \pi, \theta, \theta_1 \in \Gamma, i = 1, 2, \ldots, n$.

Proof. Assume that $\mathcal{G}$ is an M-LTB of $\mathcal{S}$, then $\mathcal{S} = \langle \mathcal{G} \rangle_{mlt}$. To prove that (1), let $s \in \mathcal{S}$, $s \in (\mathcal{G} \cup \mathcal{G} \cdot \Gamma \cdot \mathcal{G} \cdot \Gamma \cdot \mathcal{G} \cup \mathcal{G} \cdot \Gamma \cdot (\mathcal{S} \cdot \Gamma \cdot \ldots \cdot \Gamma \cdot \mathcal{S}) \cdot \Gamma \cdot \mathcal{G} \cdot \Gamma \cdot \mathcal{G}]$. As $s \in (\mathcal{G} \cup \mathcal{G} \cdot \Gamma \cdot \mathcal{G} \cdot \Gamma \cdot \mathcal{G} \cup \mathcal{G} \cdot \Gamma \cdot (\mathcal{S} \cdot \Gamma \cdot \ldots \cdot \Gamma \cdot \mathcal{S}) \cdot \Gamma \cdot \mathcal{G} \cdot \Gamma \cdot \mathcal{G}]$, we have $s \preceq g$ for some $g \in \mathcal{G} \cup \mathcal{G} \cdot \Gamma \cdot \mathcal{G} \cdot \Gamma \cdot \mathcal{G} \cup \mathcal{G} \cdot \Gamma \cdot (\mathcal{S} \cdot \Gamma \cdot \ldots \cdot \Gamma \cdot \mathcal{S}) \cdot \Gamma \cdot \mathcal{G} \cdot \Gamma \cdot \mathcal{G}$, we think about the three following cases.
Case-1: Let $g \in \mathcal{G}$. Then, $g = f_2$ for some $f_2 \in \mathcal{G}$. This implies $\langle g \rangle_{mlt} \sqsubseteq \langle f_2 \rangle_{mlt}$. Hence, $g \preceq_{mlt} f_2$. As $s \preceq g$ for some $g \in \langle f_2 \rangle_{mlt}$. To find out $\langle s \rangle_{mlt} \sqsubseteq \langle f_2 \rangle_{mlt}$. Now, $s \cup \mathcal{N}(s \cdot \Gamma \cdot s) \cup \Gamma \cdot (\mathcal{S} \cdot \Gamma \cdot \ldots \cdot \Gamma \cdot \mathcal{S}) \cdot \Gamma \cdot s \cdot \Gamma \cdot s \sqsubseteq \langle f_2 \rangle_{mlt} \cup \mathcal{N}(\langle f_2 \rangle_{mlt} \cdot \Gamma \cdot \langle f_2 \rangle_{mlt}) \cup \langle f_2 \rangle_{mlt} \cdot \Gamma \cdot (\mathcal{S} \cdot \Gamma \cdot \ldots \cdot \Gamma \cdot \mathcal{S}) \cdot \Gamma \cdot \langle f_2 \rangle_{mlt} \sqsubseteq f_2 \cup \mathcal{N}(f_2 \cdot \pi \cdot f_2) \cup f_2 \cdot \pi \cdot (\mathcal{S} \cdot \Gamma \cdot \ldots \cdot \Gamma \cdot \mathcal{S}) \cdot \Gamma \cdot f_2 \cdot \Gamma \cdot f_2]$. We have $(s \cup \mathcal{N}(s \cdot \Gamma \cdot s) \cup \Gamma \cdot (\mathcal{S} \cdot \Gamma \cdot \ldots \cdot \Gamma \cdot \mathcal{S}) \cdot \Gamma \cdot s \cdot \Gamma \cdot s] \sqsubseteq (f_2 \cup \mathcal{N}(f_2 \cdot \pi \cdot f_2) \cup f_2 \cdot \pi \cdot (\mathcal{S} \cdot \Gamma \cdot \ldots \cdot \Gamma \cdot \mathcal{S}) \cdot \Gamma \cdot f_2 \cdot \Gamma \cdot f_2]$. Thus, $\langle s \rangle_{mlt} \sqsubseteq \langle f_2 \rangle_{mlt}$, and hence, $s \preceq_{mlt} f_2$.
Case-2: Let $g \in \mathcal{G} \cdot \Gamma \cdot \mathcal{G}$. Then, $g = g_1 \cdot \pi \cdot g_2$ for some $g_1, g_2 \in \mathcal{G}$ and $\pi \in \Gamma$. This implies $\langle g \rangle_{mlt} \sqsubseteq \langle g_1 \cdot \pi \cdot g_2 \rangle_{mlt}$. Hence, $g \preceq_{mlt} g_1 \cdot \pi \cdot g_2$. As $s \preceq g$ for some $g \in \langle g_1 \cdot \pi \cdot g_2 \rangle_{mlt}$. We have $s \in \langle g_1 \cdot \pi \cdot g_2 \rangle_{mlt}$. We determine that $\langle s \rangle_{mlt} \sqsubseteq \langle g_1 \cdot \pi \cdot g_2 \rangle_{mlt}$. Now, $s \cup \mathcal{N}(s \cdot \Gamma \cdot s) \cup \Gamma \cdot (\mathcal{S} \cdot \Gamma \cdot \ldots \cdot \Gamma \cdot \mathcal{S}) \cdot \Gamma \cdot s \cdot \Gamma \cdot s \sqsubseteq (\{g_1 \cdot \pi \cdot g_2\} \cup \mathcal{N}(\{g_1 \cdot \pi \cdot g_2\} \cdot \Gamma \cdot \{g_1 \cdot \pi \cdot g_2\}) \cup \{g_1 \cdot \pi \cdot g_2\} \cdot \Gamma \cdot (\mathcal{S} \cdot \Gamma \cdot \ldots \cdot \Gamma \cdot \mathcal{S}) \cdot \Gamma \cdot \{g_1 \cdot \pi \cdot g_2\} \cdot \Gamma \cdot \{g_1 \cdot \pi \cdot g_2\}]$. Hence, $(s \cup \mathcal{N}(s \cdot \Gamma \cdot s) \cup \Gamma \cdot (\mathcal{S} \cdot \Gamma \cdot \ldots \cdot \Gamma \cdot \mathcal{S}) \cdot \Gamma \cdot s \cdot \Gamma \cdot s] \sqsubseteq (\{g_1 \cdot \pi \cdot g_2\} \cup \mathcal{N}(\{g_1 \cdot \pi \cdot g_2\} \cdot \Gamma \cdot \{g_1 \cdot \pi \cdot g_2\}) \cup \{g_1 \cdot \pi \cdot g_2\} \cdot \Gamma \cdot (\mathcal{S} \cdot \Gamma \cdot \ldots \cdot \Gamma \cdot \mathcal{S}) \cdot \Gamma \cdot \{g_1 \cdot \pi \cdot g_2\} \cdot \Gamma \cdot \{g_1 \cdot \pi \cdot g_2\}]$. This

implies $\langle s \rangle_{mlt} \sqsubseteq \langle g_1 \cdot \pi \cdot g_2 \rangle_{mlt}$. Hence, $s \preceq_{mlt} g_1 \cdot \pi \cdot g_2$.

Case-3: Let $g \in \mathcal{G} \cdot \Gamma \cdot (\mathcal{S} \cdot \Gamma \cdot \ldots \cdot \Gamma \cdot \mathcal{S}) \cdot \Gamma \cdot \mathcal{G} \cdot \Gamma \cdot \mathcal{G}$. Then, $g = g_3 \cdot \pi \cdot (s_1 \cdot \pi_1 \cdot s_2 \cdot \ldots \cdot \pi_n \cdot s_n) \cdot \theta \cdot g_4 \cdot \theta_1 \cdot g_5$ for some $g_3, g_4 \in \mathcal{G}$. This implies $\langle g \rangle_{mlt} \sqsubseteq \langle g_3 \cdot \pi \cdot (s_1 \cdot \pi_1 \cdot s_2 \cdot \ldots \cdot \pi_n \cdot s_n) \cdot \theta \cdot g_4 \cdot \theta_1 \cdot g_5 \rangle_{mlt}$. Hence, $g \preceq_{mlt} \langle g_3 \cdot \pi \cdot (s_1 \cdot \pi_1 \cdot s_2 \cdot \ldots \cdot \pi_n \cdot s_n) \cdot \theta \cdot g_4 \cdot \theta_1 \cdot g_5 \rangle_{mlt}$. As $s \preceq g$ for some $g \in \langle g_3 \cdot \pi \cdot (s_1 \cdot \pi_1 \cdot s_2 \cdot \ldots \cdot \pi_n \cdot s_n) \cdot \theta \cdot g_4 \cdot \theta_1 \cdot g_5 \rangle_{mlt}$. We have $s \in \langle g_3 \cdot \pi \cdot (s_1 \cdot \pi_1 \cdot s_2 \cdot \ldots \cdot \pi_n \cdot s_n) \cdot \theta \cdot g_4 \cdot \theta_1 \cdot g_5 \rangle_{mlt}$. To prove that $\langle s \rangle_{mlt} \sqsubseteq \langle g_3 \cdot \pi \cdot (s_1 \cdot \pi_1 \cdot s_2 \cdot \ldots \cdot \pi_n \cdot s_n) \cdot \theta \cdot g_4 \cdot \theta_1 \cdot g_5 \rangle_{mlt}$. Now, $s \cup \mathcal{N}(s \cdot \Gamma \cdot s) \cup \Gamma \cdot (\mathcal{S} \cdot \Gamma \cdot \ldots \cdot \Gamma \cdot \mathcal{S}) \cdot \Gamma \cdot s \cdot \Gamma \cdot s \sqsubseteq (\{g_3 \cdot \pi \cdot (s_1 \cdot \pi_1 \cdot s_2 \cdot \ldots \cdot \pi_n \cdot s_n) \cdot \theta \cdot g_4 \cdot \theta_1 \cdot g_5\} \cup \mathcal{N}(\{g_3 \cdot \pi \cdot (s_1 \cdot \pi_1 \cdot s_2 \cdot \ldots \cdot \pi_n \cdot s_n) \cdot \theta \cdot g_4 \cdot \theta_1 \cdot g_5\}) \cup \{g_3 \cdot \pi \cdot (s_1 \cdot \pi_1 \cdot s_2 \cdot \ldots \cdot \pi_n \cdot s_n) \cdot \theta \cdot g_4 \cdot \theta_1 \cdot g_5\} \cdot \Gamma \cdot (\mathcal{S} \cdot \Gamma \cdot \ldots \cdot \Gamma \cdot \mathcal{S}) \cdot \Gamma \cdot \{g_3 \cdot \pi \cdot (s_1 \cdot \pi_1 \cdot s_2 \cdot \ldots \cdot \pi_n \cdot s_n) \cdot \theta \cdot g_4 \cdot \theta_1 \cdot g_5\} \cdot \Gamma \cdot \{g_3 \cdot \pi \cdot (s_1 \cdot \pi_1 \cdot s_2 \cdot \ldots \cdot \pi_n \cdot s_n) \cdot \theta \cdot g_4 \cdot \theta_1 \cdot g_5\}]$. Hence, $(s \cup \mathcal{N}(s \cdot \Gamma \cdot s) \cup \Gamma \cdot (\mathcal{S} \cdot \Gamma \cdot \ldots \cdot \Gamma \cdot \mathcal{S}) \cdot \Gamma \cdot s \cdot \Gamma \cdot s] \sqsubseteq (\{g_3 \cdot \pi \cdot (s_1 \cdot \pi_1 \cdot s_2 \cdot \ldots \cdot \pi_n \cdot s_n) \cdot \theta \cdot g_4 \cdot \theta_1 \cdot g_5\} \cup \mathcal{N}(\{g_3 \cdot \pi \cdot (s_1 \cdot \pi_1 \cdot s_2 \cdot \ldots \cdot \pi_n \cdot s_n) \cdot \theta \cdot g_4 \cdot \theta_1 \cdot g_5\}) \cup \{g_3 \cdot \pi \cdot (s_1 \cdot \pi_1 \cdot s_2 \cdot \ldots \cdot \pi_n \cdot s_n) \cdot \theta \cdot g_4 \cdot \theta_1 \cdot g_5\} \cdot \Gamma \cdot (\mathcal{S} \cdot \Gamma \cdot \ldots \cdot \Gamma \cdot \mathcal{S}) \cdot \Gamma \cdot \{g_3 \cdot \pi \cdot (s_1 \cdot \pi_1 \cdot s_2 \cdot \ldots \cdot \pi_n \cdot s_n) \cdot \theta \cdot g_4 \cdot \theta_1 \cdot g_5\} \cdot \Gamma \cdot \{g_3 \cdot \pi \cdot (s_1 \cdot \pi_1 \cdot s_2 \cdot \ldots \cdot \pi_n \cdot s_n) \cdot \theta \cdot g_4 \cdot \theta_1 \cdot g_5\}]$. This implies $\langle s \rangle_{mlt} \sqsubseteq \langle g_3 \cdot \pi \cdot (s_1 \cdot \pi_1 \cdot s_2 \cdot \ldots \cdot \pi_n \cdot s_n) \cdot \theta \cdot g_4 \cdot \theta_1 \cdot g_5 \rangle_{mlt}$. Hence, $s \preceq_{mlt} g_3 \cdot \pi \cdot (s_1 \cdot \pi_1 \cdot s_2 \cdot \ldots \cdot \pi_n \cdot s_n) \cdot \theta \cdot g_4 \cdot \theta_1 \cdot g_5$. By Lemma 5(1) and Lemma 5(2), we have the proof of (2) and (3), respectively.

Conversely, assume that (1), (2), and (3) hold to prove that $\mathcal{G}$ is an M-LTB of $\mathcal{S}$. To determine that $\mathcal{S} = \langle \mathcal{G} \rangle_{mlt}$, clearly, $\langle \mathcal{G} \rangle_{mlt} \sqsubseteq \mathcal{S}$. By (1), $\mathcal{S} \sqsubseteq \langle \mathcal{G} \rangle_{mlt}$ and $\mathcal{S} = \langle \mathcal{G} \rangle_{mlt}$. It remains to be determined whether $\mathcal{G}$ is a minimal subset of $\mathcal{S}$, $\mathcal{S} = \langle \mathcal{G} \rangle_{mlt}$. Suppose that $\mathcal{S} = \langle \mathcal{F} \rangle_{mlt}$ for some $\mathcal{F} \subset \mathcal{G}$. As $\mathcal{F} \subset \mathcal{G}$, there exists $f_2 \in \mathcal{G} \setminus \mathcal{F}$. As $f_2 \in \mathcal{G} \subseteq \mathcal{S} = \langle \mathcal{F} \rangle_{mlt}$ and $f_2 \notin \mathcal{F}$, it follows that $f_2 \in (\mathcal{F} \cdot \Gamma \cdot \mathcal{F} \cdot \Gamma \cdot \mathcal{F} \cup \mathcal{F} \cdot \Gamma \cdot (\mathcal{S} \cdot \Gamma \cdot \ldots \cdot \Gamma \cdot \mathcal{S}) \cdot \Gamma \cdot \mathcal{F} \cdot \Gamma \cdot \mathcal{F}]$. As $f_2 \in (\mathcal{F} \cdot \Gamma \cdot \mathcal{F} \cdot \Gamma \cdot \mathcal{F} \cup \mathcal{F} \cdot \Gamma \cdot (\mathcal{S} \cdot \Gamma \cdot \ldots \cdot \Gamma \cdot \mathcal{S}) \cdot \Gamma \cdot \mathcal{F} \cdot \Gamma \cdot \mathcal{F}]$, this implies $f_2 \preceq g$ for some $g \in \mathcal{F} \cdot \Gamma \cdot \mathcal{F} \cdot \Gamma \cdot \mathcal{F} \cup \mathcal{F} \cdot \Gamma \cdot (\mathcal{S} \cdot \Gamma \cdot \ldots \cdot \Gamma \cdot \mathcal{S}) \cdot \Gamma \cdot \mathcal{F} \cdot \Gamma \cdot \mathcal{F}$. There are two cases to be observed.

Case-1: Let $g \in \mathcal{F} \cdot \Gamma \cdot \mathcal{F} \cdot \Gamma \cdot \mathcal{F}$. Then, $g = g_1 \cdot \pi \cdot g_2$ for some $g_1, g_2 \in \mathcal{F}$ and $\pi \in \Gamma$. We have $g_1, g_2 \in \mathcal{G}$. As $f_2 \notin \mathcal{F}$, $f_2 \neq g_1$ and $f_2 \neq g_2$. As $g = g_1 \cdot \pi \cdot g_2$, $\langle g \rangle_{mlt} \sqsubseteq \langle g_1 \cdot \pi \cdot g_2 \rangle_{mlt}$. Hence, $g \preceq_{mlt} g_1 \cdot \pi \cdot g_2$. As $f_2 \preceq g$ for some $g \in \langle g_1 \cdot \pi \cdot g_2 \rangle_{mlt}$, we have $f_2 \in \langle g_1 \cdot \pi \cdot g_2 \rangle_{mlt}$ to prove that $\langle f_2 \rangle_{mlt} \sqsubseteq \langle g_1 \cdot \pi \cdot g_2 \rangle_{mlt}$. Now, $f_2 \cup \mathcal{N}(f_2 \cdot \pi \cdot f_2) \cup f_2 \cdot \pi \cdot (\mathcal{S} \cdot \Gamma \cdot \ldots \cdot \Gamma \cdot \mathcal{S}) \cdot \Gamma \cdot f_2 \cdot \Gamma \cdot f_2 \sqsubseteq (\{g_1 \cdot \pi \cdot g_2\} \cup \mathcal{N}(\{g_1 \cdot \pi \cdot g_2\} \cdot \Gamma \cdot \{g_1 \cdot \pi \cdot g_2\}) \cup \{g_1 \cdot \pi \cdot g_2\} \cdot \Gamma \cdot (\mathcal{S} \cdot \Gamma \cdot \ldots \cdot \Gamma \cdot \mathcal{S}) \cdot \Gamma \cdot \{g_1 \cdot \pi \cdot g_2\} \cdot \Gamma \cdot \{g_1 \cdot \pi \cdot g_2\}]$. Hence, $(f_2 \cup \mathcal{N}(f_2 \cdot \pi \cdot f_2) \cup f_2 \cdot \pi \cdot (\mathcal{S} \cdot \Gamma \cdot \ldots \cdot \Gamma \cdot \mathcal{S}) \cdot \Gamma \cdot f_2 \cdot \Gamma \cdot f_2] \sqsubseteq (\{g_1 \cdot \pi \cdot g_2\} \cup \mathcal{N}(\{g_1 \cdot \pi \cdot g_2\} \cdot \Gamma \cdot \{g_1 \cdot \pi \cdot g_2\}) \cup \{g_1 \cdot \pi \cdot g_2\} \cdot \Gamma \cdot (\mathcal{S} \cdot \Gamma \cdot \ldots \cdot \Gamma \cdot \mathcal{S}) \cdot \Gamma \cdot \{g_1 \cdot \pi \cdot g_2\} \cdot \Gamma \cdot \{g_1 \cdot \pi \cdot g_2\}]$. This implies $\langle f_2 \rangle_{mlt} \sqsubseteq \langle g_1 \cdot \pi \cdot g_2 \rangle_{mlt}$. Hence, $f_2 \preceq_{mlt} g_1 \cdot \pi \cdot g_2$. This contradicts (2).

Case-2: Let $g \in \mathcal{F} \cdot \Gamma \cdot (\mathcal{S} \cdot \Gamma \cdot \ldots \cdot \Gamma \cdot \mathcal{S}) \cdot \Gamma \cdot \mathcal{F} \cdot \Gamma \cdot \mathcal{F}$. Then, $g = g_3 \cdot \pi \cdot (s_1 \cdot \pi_1 \cdot s_2 \cdot \ldots \cdot \pi_n \cdot s_n) \cdot \theta \cdot g_4 \cdot \theta_1 \cdot g_5$ for some $g_3, g_4 \in \mathcal{F}, s_i \in \mathcal{S}$ and $\pi_i, \pi, \theta \in \Gamma$, $i = 1, 2, \ldots, n$. We have $g_3, g_4 \in \mathcal{G}$. As $f_2 \notin \mathcal{F}$, so $f_2 \neq g_3$ and $f_2 \neq g_4$. As $g = g_3 \cdot \pi \cdot (s_1 \cdot \pi_1 \cdot s_2 \cdot \ldots \cdot \pi_n \cdot s_n) \cdot \theta \cdot g_4 \cdot \theta_1 \cdot g_5$, $\langle g \rangle_{mlt} \sqsubseteq \langle g_3 \cdot \pi \cdot (s_1 \cdot \pi_1 \cdot s_2 \cdot \ldots \cdot \pi_n \cdot s_n) \cdot \theta \cdot g_4 \cdot \theta_1 \cdot g_5 \rangle_{mlt}$. Hence, $g \preceq_{mlt} g_3 \cdot \pi \cdot (s_1 \cdot \pi_1 \cdot s_2 \cdot \ldots \cdot \pi_n \cdot s_n) \cdot \theta \cdot g_4 \cdot \theta_1 \cdot g_5$. Since $f_2 \preceq g$ for some $g \in \langle g_3 \cdot \pi \cdot (s_1 \cdot \pi_1 \cdot s_2 \cdot \ldots \cdot \pi_n \cdot s_n) \cdot \theta \cdot g_4 \cdot \theta_1 \cdot g_5 \rangle_{mlt}$, we have $f_2 \in \langle g_3 \cdot \pi \cdot (s_1 \cdot \pi_1 \cdot s_2 \cdot \ldots \cdot \pi_n \cdot s_n) \cdot \theta \cdot g_4 \cdot \theta_1 \cdot g_5 \rangle_{mlt}$. We determine that $\langle f_2 \rangle_{mlt} \sqsubseteq \langle g_3 \cdot \pi \cdot (s_1 \cdot \pi_1 \cdot s_2 \cdot \ldots \cdot \pi_n \cdot s_n) \cdot \theta \cdot g_4 \cdot \theta_1 \cdot g_5 \rangle_{mlt}$. Now, $f_2 \cup \mathcal{N}(f_2 \cdot \pi \cdot f_2) \cup f_2 \cdot \pi \cdot (\mathcal{S} \cdot \Gamma \cdot \ldots \cdot \Gamma \cdot \mathcal{S}) \cdot \Gamma \cdot f_2 \cdot \Gamma \cdot f_2 \sqsubseteq (\{g_3 \cdot \pi \cdot (s_1 \cdot \pi_1 \cdot s_2 \cdot \ldots \cdot \pi_n \cdot s_n) \cdot \theta \cdot g_4 \cdot \theta_1 \cdot g_5\} \cup \mathcal{N}(\{g_3 \cdot \pi \cdot (s_1 \cdot \pi_1 \cdot s_2 \cdot \ldots \cdot \pi_n \cdot s_n) \cdot \theta \cdot g_4 \cdot \theta_1 \cdot g_5\} \cdot \Gamma \cdot \{g_3 \cdot \pi \cdot (s_1 \cdot \pi_1 \cdot s_2 \cdot \ldots \cdot \pi_n \cdot s_n) \cdot \theta \cdot g_4 \cdot \theta_1 \cdot g_5\}) \cup \{g_3 \cdot \pi \cdot (s_1 \cdot \pi_1 \cdot s_2 \cdot \ldots \cdot \pi_n \cdot s_n) \cdot \theta \cdot g_4 \cdot \theta_1 \cdot g_5\} \cdot \Gamma \cdot (\mathcal{S} \cdot \Gamma \cdot \ldots \cdot \Gamma \cdot \mathcal{S}) \cdot \Gamma \cdot \{g_3 \cdot \pi \cdot (s_1 \cdot \pi_1 \cdot s_2 \cdot \ldots \cdot \pi_n \cdot s_n) \cdot \theta \cdot g_4 \cdot \theta_1 \cdot g_5\} \cdot \Gamma \cdot \{g_3 \cdot \pi \cdot (s_1 \cdot \pi_1 \cdot s_2 \cdot \ldots \cdot \pi_n \cdot s_n) \cdot \theta \cdot g_4 \cdot \theta_1 \cdot g_5\}]$. Hence, $(f_2 \cup \mathcal{N}(f_2 \cdot \pi \cdot f_2) \cup f_2 \cdot \pi \cdot (\mathcal{S} \cdot \Gamma \cdot \ldots \cdot \Gamma \cdot \mathcal{S}) \cdot \Gamma \cdot f_2 \cdot \Gamma \cdot f_2] \sqsubseteq (\{g_3 \cdot \pi \cdot (s_1 \cdot \pi_1 \cdot s_2 \cdot \ldots \cdot \pi_n \cdot s_n) \cdot \theta \cdot g_4 \cdot \theta_1 \cdot g_5\} \cup \mathcal{N}(\{g_3 \cdot \pi \cdot (s_1 \cdot \pi_1 \cdot s_2 \cdot \ldots \cdot \pi_n \cdot s_n) \cdot \theta \cdot g_4 \cdot \theta_1 \cdot g_5\}) \cup \{g_3 \cdot \pi \cdot (s_1 \cdot \pi_1 \cdot s_2 \cdot \ldots \cdot \pi_n \cdot s_n) \cdot \theta \cdot g_4 \cdot \theta_1 \cdot g_5\} \cdot \Gamma \cdot (\mathcal{S} \cdot \Gamma \cdot \ldots \cdot \Gamma \cdot \mathcal{S}) \cdot \Gamma \cdot \{g_3 \cdot \pi \cdot (s_1 \cdot \pi_1 \cdot s_2 \cdot \ldots \cdot \pi_n \cdot s_n) \cdot \theta \cdot g_4 \cdot \theta_1 \cdot g_5\} \cdot \Gamma \cdot \{g_3 \cdot \pi \cdot (s_1 \cdot \pi_1 \cdot s_2 \cdot \ldots \cdot \pi_n \cdot s_n) \cdot \theta \cdot g_4 \cdot \theta_1 \cdot g_5\}]$. This implies $\langle f_2 \rangle_{mlt} \sqsubseteq \langle g_3 \cdot \pi \cdot (s_1 \cdot \pi_1 \cdot s_2 \cdot \ldots \cdot \pi_n \cdot s_n) \cdot \theta \cdot g_4 \cdot \theta_1 \cdot g_5 \rangle_{mlt}$. Hence, $f_2 \preceq_{mlt} g_3 \cdot \pi \cdot (s_1 \cdot \pi_1 \cdot s_2 \cdot \ldots \cdot \pi_n \cdot s_n) \cdot \theta \cdot g_4 \cdot \theta_1 \cdot g_5$, which is a contradiction to (3). Therefore, $\mathcal{G}$ is an M-LTB of $\mathcal{S}$. □

Theorem 4. *Let $\mathcal{G}$ be an M-LTB of $\mathcal{S}$. Then, $\mathcal{G}$ is an ordered Γ-subsemigroup of $\mathcal{S}$, if and only if $g_1 \cdot \pi \cdot g_2 = g_1$ or $g_1 \cdot \pi \cdot g_2 = g_2$, for any $g_1, g_2 \in \mathcal{G}$ and $\pi \in \Gamma$.*

Proof. If $\mathcal{G}$ is an ordered Γ-subsemigroup of $\mathcal{S}$, then $g_1 \cdot \pi \cdot g_2 \in \mathcal{G}$. As $g_1 \cdot \pi \cdot g_2 \in (\mathcal{N}(g_1 \cdot \Gamma \cdot g_2) \cup g_1 \cdot \Gamma \cdot (\mathcal{S} \cdot \Gamma \cdot \ldots \cdot \Gamma \cdot \mathcal{S}) \cdot \Gamma \cdot g_2 \cdot \Gamma \cdot g_2]$, it follows by Lemma 2 that $g_1 \cdot \pi \cdot g_2 = g_1$ or $g_1 \cdot \pi \cdot g_2 = g_2$. □

4. M-RTB Generator

We present some results on the M-right-tri-ideal (RTI) generator based on an ordered Γ-semigroup.

Definition 12. *Let $\mathcal{S}$ be an ordered Γ-semigroup. $\mathcal{G} \sqsubseteq \mathcal{S}$ is said to be an M-RTI of $\mathcal{S}$ if it meets the criteria listed below:*
1. *$\mathcal{G}$ is a Γ-subsemigroup,*
2. *$\mathcal{G} \cdot \Gamma \cdot \mathcal{G} \cdot \Gamma \cdot (\mathcal{S} \cdot \Gamma \cdot \ldots \cdot \Gamma \cdot \mathcal{S} \ (M-times)) \cdot \Gamma \cdot \mathcal{G} \sqsubseteq \mathcal{G}$,*
3. *If $g \in \mathcal{G}$ and $s \in \mathcal{S}$, such that $s \preceq g$, then $s \in \mathcal{G}$.*

Theorem 5.
1. *For $f_1 \in \mathcal{S}$, the M-RTI generated by "f_1" is $\langle f_1 \rangle_{mrt} = \{f_1 \cup \mathcal{N}(f_1 \cdot \Gamma \cdot f_1) \cup f_1 \cdot \Gamma \cdot f_1 \cdot \Gamma \cdot (\mathcal{S} \cdot \Gamma \cdot \ldots \cdot \Gamma \cdot \mathcal{S} \ (M-times)) \cdot \Gamma \cdot f_1\}$ and $\mathcal{N} \succeq M$, where $\mathcal{N}$ and M are positive integers;*
2. *For $\mathcal{G} \sqsubseteq \mathcal{S}$, the M-RTI generated by "$\mathcal{G}$" is $\langle \mathcal{G} \rangle_{mrt} = \{\mathcal{G} \cup \mathcal{G} \cdot \Gamma \cdot \mathcal{G} \cup \mathcal{G} \cdot \Gamma \cdot \mathcal{G} \cdot \Gamma \cdot (\mathcal{S} \cdot \Gamma \cdot \ldots \cdot \Gamma \cdot \mathcal{S} \ (M-times)) \cdot \Gamma \cdot \mathcal{G}\}$.*

Definition 13. *Let $\mathcal{G}$ be a subset $\mathcal{S}$ called a M-right tri-basis (RTB) of $\mathcal{S}$ if it satisfies the following conditions:*
1. *$\mathcal{S} = \langle \mathcal{G} \rangle_{mrt}$.*
2. *If $\mathcal{F} \sqsubseteq \mathcal{G}$ such that $\mathcal{S} = \langle \mathcal{F} \rangle_{mrt}$, then $\mathcal{F} = \mathcal{G}$.*

Theorem 6. *Let $\mathcal{G}$ be an M-RTB of $\mathcal{S}$ and $f_1, f_2 \in \mathcal{G}$. If $f_1 \in (\mathcal{N}(f_2 \cdot \Gamma \cdot f_2) \cup f_2 \cdot \Gamma \cdot f_2 \cdot \Gamma \cdot (\mathcal{S} \cdot \Gamma \cdot \ldots \cdot \Gamma \cdot \mathcal{S}) \cdot \Gamma \cdot f_2]$, then $f_1 = f_2$.*

Proof. The proof is the same as in Theorem 2. □

Lemma 6. *Let $\mathcal{G}$ be an M-RTB of $\mathcal{S}$ and $f_1, f_2, f_3, f_4 \in \mathcal{G}$. If $f_1 \in (\mathcal{N}(f_3 \cdot \Gamma \cdot f_2) \cup f_2 \cdot \Gamma \cdot f_4 \cdot (\mathcal{S} \cdot \Gamma \cdot \ldots \cdot \Gamma \cdot \mathcal{S}) \cdot \Gamma \cdot f_3]$, then $f_1 = f_2$ or $f_1 = f_3$ or $f_1 = f_4$.*

Proof. Theorem 2 leads to the proof. □

Definition 14. *For any $s_1, s_2 \in \mathcal{S}$, $s_1 \preceq_{mrt} s_2 \iff \langle s_1 \rangle_{mrt} \sqsubseteq \langle s_2 \rangle_{mrt}$ is called a quasi-order on $\mathcal{S}$.*

Remark 6. *The order $\preceq_{mrt}$ is not a partial order of $\mathcal{S}$.*

Example 4. *By Example 2, $\langle k_4 \rangle_{mrt} \sqsubseteq \langle k_6 \rangle_{mrt}$ and $\langle k_6 \rangle_{mrt} \sqsubseteq \langle k_4 \rangle_{mrt}$ but $k_4 \neq k_6$. Hence, the relation $\preceq_{mrt}$ is not a partial order on $\mathcal{S}$.*

If $\mathcal{F}$ is an M-RTB of $\mathcal{S}$, then $\langle \mathcal{F} \rangle_{mrt} = \mathcal{S}$. Let $s \in \mathcal{S}$. Then, $s \in \langle \mathcal{F} \rangle_{mrt}$ and so $s \in \langle f_1 \rangle_{mrt}$ for some $f_1 \in \mathcal{F}$. This implies $\langle s \rangle_{mrt} \sqsubseteq \langle f_1 \rangle_{mrt}$. Hence, $s \preceq_{mrt} f_1$.

Remark 7. *If $\mathcal{G}$ is an M-RTB of $\mathcal{S}$, then for any $s \in \mathcal{S}$, there exists $f_1 \in \mathcal{G}$ such that $s \preceq_{mrt} f_1$.*

Lemma 7. *Let $\mathcal{G}$ be an M-RTB of $\mathcal{S}$. If $f_1, f_2 \in \mathcal{G}$ such that $f_1 \neq f_2$, then neither $f_1 \preceq_{mrt} f_2$ nor $f_2 \preceq_{mrt} f_1$.*

Proof. The proof follows from Lemma 3. □

Lemma 8. *Let $\mathcal{G}$ be the M-RTB of $\mathcal{S}$ and $f_1, f_2, f_3 \in \mathcal{G}$ and $s \in \mathcal{S}$.*
1. *If $f_1 \in (\{f_2 \cdot \pi \cdot f_3\} \cup \mathcal{N}(\{f_2 \cdot \pi \cdot f_3\} \cdot \Gamma \cdot \{f_2 \cdot \pi \cdot f_3\}) \cup \{f_2 \cdot \pi \cdot f_3\} \cdot \Gamma \cdot \{f_2 \cdot \pi \cdot f_3\} \cdot \Gamma \cdot (\mathcal{S} \cdot \Gamma \cdot \ldots \cdot \Gamma \cdot \mathcal{S}) \cdot \Gamma \cdot \{f_2 \cdot \pi \cdot f_3\}]$, then $f_1 = f_2$ or $f_1 = f_3$;*
2. *If $f_1 \in (\{f_2 \cdot \pi \cdot (s_1 \cdot \pi_1 \cdot s_2 \cdot \ldots \cdot \pi_n \cdot s_n) \cdot \theta \cdot f_3 \cdot \theta_1 \cdot f_4\} \cup \mathcal{N}(\{f_2 \cdot \pi \cdot (s_1 \cdot \pi_1 \cdot s_2 \cdot \ldots \cdot \pi_n \cdot s_n) \cdot \theta \cdot f_3 \cdot \theta_1 \cdot f_4\} \cdot \Gamma \cdot \{f_2 \cdot \pi \cdot (s_1 \cdot \pi_1 \cdot s_2 \cdot \ldots \cdot \pi_n \cdot s_n) \cdot \theta \cdot f_3 \cdot \theta_1 \cdot f_4\}) \cup \{f_2 \cdot \pi \cdot (s_1 \cdot \pi_1 \cdot s_2 \cdot \ldots \cdot \pi_n \cdot s_n) \cdot \theta \cdot f_3 \cdot \theta_1 \cdot f_4\} \cdot \{f_2 \cdot \pi \cdot (s_1 \cdot \pi_1 \cdot s_2 \cdot \ldots \cdot \pi_n \cdot s_n) \cdot \theta \cdot f_3 \cdot \theta_1 \cdot f_4\} \cdot \Gamma \cdot (\mathcal{S} \cdot \Gamma \cdot \ldots \cdot \Gamma \cdot \mathcal{S}) \cdot \Gamma \cdot \{f_2 \cdot \pi \cdot (s_1 \cdot \pi_1 \cdot s_2 \cdot \ldots \cdot \pi_n \cdot s_n) \cdot \theta \cdot f_3 \cdot \theta_1 \cdot f_4\}]$, then $f_1 = f_2$ or $f_1 = f_3$ or $f_1 = f_4$.*

Proof. The proof follows from Lemma 4. □

Lemma 9. *Let $\mathcal{G}$ be the M-RTB of $\mathcal{S}$,*
1. *If $f_1 \neq f_2$ and $f_1 \neq f_3$, then $f_1 \npreceq_{mrt} f_2 \cdot \pi \cdot f_3$.*
2. *If $f_1 \neq f_2, f_1 \neq f_3$ and $f_1 \neq f_4$, then $f_1 \npreceq_{mrt} f_3 \cdot \theta \cdot f_4 \cdot \theta_1 \cdot (s_1 \cdot \pi_1 \cdot s_2 \cdot \ldots \cdot \pi_n \cdot s_n) \cdot \pi \cdot f_2$, for $f_1, f_2, f_3, f_4 \in \mathcal{G}, \pi, \pi_i, \theta, \theta_1 \in \Gamma$ and $s_i \in \mathcal{S}, i = 1, 2, \ldots, n$.*

Proof. The proof follows from Lemma 5. □

Theorem 7. *Let $\mathcal{G}$ be the M-RTB of $\mathcal{S}$, if and only if the following conditions are met by $\mathcal{G}$.*
1. *For any $s \in \mathcal{S}$,*
   (1.1) there exists $f_2 \in \mathcal{G}$, such that $s \preceq_{mrt} f_2$ (or),
   (1.2) there exists $g_1, g_2 \in \mathcal{G}$, such that $s \preceq_{mrt} g_1 \cdot \pi \cdot g_2$ (or),
   (1.3) there exists $g_3, g_4, g_5 \in \mathcal{G}$, such that $s \preceq_{mrt} g_4 \cdot \theta \cdot g_5 \cdot \theta_1 \cdot (s_1 \cdot \pi_1 \cdot s_2 \cdot \ldots \cdot \pi_n \cdot s_n) \cdot \pi \cdot g_3$;
2. *If $f_1 \neq f_2, f_1 \neq f_3$, and $f_1 \neq f_4$, then $f_1 \npreceq_{mrt} f_2 \cdot \pi \cdot f_3$, for any $f_1, f_2, f_3 \in \mathcal{G}$,*
3. *If $f_1 \neq f_2, f_1 \neq f_3$, and $f_1 \neq f_4$, then $f_1 \npreceq_{mrt} f_3 \cdot \theta \cdot f_4 \cdot \theta_1 \cdot (s_1 \cdot \pi_1 \cdot s_2 \cdot \ldots \cdot \pi_n \cdot s_n) \cdot \pi \cdot f_2$, for any $f_1, f_2, f_3, f_4 \in \mathcal{G}, s_i \in \mathcal{S}$ and $\pi_i, \pi, \theta, \theta_1 \in \Gamma, i = 1, 2, \ldots, n$.*

Proof. Theorem 3 leads to the proof. □

Theorem 8. *Let $\mathcal{G}$ be an M-RTB of $\mathcal{S}$. Then, $\mathcal{G}$ is an ordered Γ-subsemigroup of $\mathcal{S}$, if and only if $g_1 \cdot \pi \cdot g_2 = g_1$ or $g_1 \cdot \pi \cdot g_2 = g_2$, for any $g_1, g_2 \in \mathcal{G}$ and $\pi \in \Gamma$.*

Proof. The proof is the same as Theorem 4. □

5. Conclusions

Several characterizations of the M-LTB (RTB) of an ordered Γ-semigroup are described in this article. Our discussion has focused on some of their fundamental characteristics and has also examined some of them using the M-tri-ideal generator. We presented the M-LTB (RTB) of an ordered Γ-semigroup, which was constructed from an ordered Γ-semigroup element and subset. At the end of our discussion, we explored the relationship between partial order and the M-LTB (RTB). In the future, we plan to explore a few more types of tri-basis and tri-M-basis. Our study will examine their research on Γ-hyper semigroups using bi-basis and M-bi-basis.

Author Contributions: Conceptualization, M.P. (M. Palanikumar); methodology, M.P. (Madhumangal Pal); writing original draft, M.P. (M. Palanikumar); conceptualization, C.J.; validation, C.J.; conceptualization, O.A.-S.; review and editing, O.A.-S. and writing—review and editing, M.P. (Madhumangal Pal). All authors have read and agreed to the published version of the manuscript.

Funding: This research received no external funding.

Informed Consent Statement: The article does not contain any studies with human participants or animals performed by the author.

Data Availability Statement: Not applicable.

Acknowledgments: The authors declare that the present work is a joint contribution to this paper and was not supported by any financial and material agency.

Conflicts of Interest: The authors declare no conflict of interest.

References

1. Sen, M.K.; Saha, On Γ-semigroup. *Bull. Cal. Math. Soc.* **1996**, *78*, 180–186.
2. Rao, M.M.K. Γ-semirings. *Southeast Asian Bull. Math.* **1995**, *19*, 49–54.
3. Kapp, K.M. On bi-ideals and quasi-ideals in semigroups. *Publ. Math. Debrecen.* **1969**, *16*, 179–185. [CrossRef]
4. Kapp, K.M. Bi-ideals in associative rings and semigroups. *Acta Sci. Math.* **1972**, *33*, 307–314.
5. Kemprasit, Y. Quasi-ideals and bi-ideals in semigroups and rings. In Proceedings of the International Conference on Algebra and its Applications, Bangkok, Thailand, 18–22 March 2002; pp. 30–46.
6. Kwon, Y.I.; Lee, S.K. Some special elements in ordered Γ-semigroups. *Kyungpook Math. J.* **1996**, *35*, 679–685.
7. Munir, M. On M-bi-ideals in semigroups. *Bull. Int. Math. Virtual Inst.* **2018**, *8*, 461–467.
8. Sen, M.K.; Seth, A. On po-Γ-semigroups. *Bull. Cal. Math. Soc.* **1993**, *85*, 445–450.
9. Iampan, A.; Siripitukdet, M. On minimal and maximal ordered left ideals in ordered Γ-semigroups. *Thai J. Math.* **2004**, *2*, 275–282.
10. Iampan, A. Characterizing ordered bi-ideals in ordered Γ-semigroups. *Iran. J. Math. Sci. Inform.* **2009**, *4*, 17–25.
11. Iampan, A. Characterizing ordered Quasi-ideals of ordered Γ-semigroups. *Kragujev. J. Math.* **2011**, *35*, 13–23.
12. Kwon, Y.I.; Lee, S.K. The weakly semi-prime ideals of ordered Γ-semigroups. *Kangweon Kyungki Math. J.* **1997**, *5*, 135–139.
13. Jantanan, W.; Latthi, M.; Puifai, J. On bi-basis of ordered Γ-semigroups. *Naresuan Univ. J. Sci. Technol.* **2022**, *30*, 75–84.
14. Palanikumar, M.; Arulmozhi, K. On new ways of various ideals in ternary semigroups. *Matrix Sci. Math.* **2020**, *4*, 6–9.
15. Palanikumar, M.; Arulmozhi, K. On various tri-ideals in ternary semirings. *Bull. Int. Math. Virtual Inst.* **2021**, *11*, 79–90.
16. Palanikumar, M.; Arulmozhi, K. On various almost ideals of semirings. *Ann. Commun. Math.* **2021**, *4*, 1–17.
17. Rao, M.M.K. Tri-quasi ideals of Γ semigroups. *Bull. Int. Math. Virtual Inst.* **2021**, *11*, 111–120.
18. Rao, M.M.K. Tri-ideals of Γ-semirings. *Analele Univ. Oradea Fasc. Mat. Tom* **2019**, *2*, 51–60.
19. Suebsung, S.; Chinram, R. Yonthanthum, W.; Hila, K.; Iampan, A. On almost bi-ideals and almost quasi-ideals of ordered semigroups and their fuzzifications. *Icic Express Lett.* **2022**, *16*, 127–135.
20. Palanikumar, M.; Iampan, A.; Manavalan, L.J. M-bi-basis generator of ordered gamma-semigroups. *Icic Express Lett. Part B Appl.* **2022**, *13*, 795–802.
21. Mallick, S; Hansda, K. On the semigroup of bi-ideals of an ordered semigroup. *Kragujev. J. Math.* **2023**, *47*, 339–345.

Disclaimer/Publisher's Note: The statements, opinions and data contained in all publications are solely those of the individual author(s) and contributor(s) and not of MDPI and/or the editor(s). MDPI and/or the editor(s) disclaim responsibility for any injury to people or property resulting from any ideas, methods, instructions or products referred to in the content.

Article

Left (Right) Regular Elements of Some Transformation Semigroups

Kitsanachai Sripon [1], Ekkachai Laysirikul [1,*] and Worachead Sommanee [2]

[1] Department of Mathematics, Faculty of Science, Naresaun University, Phitsanulok 65000, Thailand; kitsanachais61@nu.ac.th
[2] Department of Mathematics and Statistics, Faculty of Science and Technology, Chiang Mai Rajabhat University, Chiang Mai 503000, Thailand; worachead_som@cmru.ac.th
* Correspondence: ekkachail@nu.ac.th

Abstract: For a nonempty set X, let $T(X)$ be the total transformation semigroup on X. In this paper, we consider the subsemigroups of $T(X)$ which are defined by $T(X,Y) = \{\alpha \in T(X) : X\alpha \subseteq Y\}$ and $S(X,Y) = \{\alpha \in T(X) : Y\alpha \subseteq Y\}$ where Y is a non-empty subset of X. We characterize the left regular and right regular elements of both $T(X,Y)$ and $S(X,Y)$. Moreover, necessary and sufficient conditions for $T(X,Y)$ and $S(X,Y)$ to be left regular and right regular are given. These results are then applied to determine the numbers of left and right regular elements in $T(X,Y)$ for a finite set X.

Keywords: regular elements; magnifying elements; transformation semigroups

MSC: 20M20

1. Introduction and Preliminaries

Let S be a semigroup. An element x of S is called left regular if $x = yx^2$ for some $y \in S$. A right regular element is defined dually. Denote by $LReg(S)$ and $RReg(S)$ the sets of all left regular elements and right regular elements of S, respectively. Left and right regular elements are important in semigroup theory because they play a key role in the study of regular semigroups, which are semigroups in which every element is both left and right regular. Regular semigroups have many interesting properties and are used in various areas of mathematics, including algebra, topology and theoretical computer science. An element x of S is called left (right) magnifying if there is a proper subset M of S satisfying $xM = S$ ($Mx = S$). In 1963, Ljapin [1] studied the notion of right and left magnifying elements of a semigroup. Several years later, Migliorini introduced the concepts of the minimal subset related to a magnifying element of S in [2,3]. Gutan [4] researched semigroups with strong and nonstrong magnifying elements in 1996. Later, he proved that every semigroup with magnifying elements is factorizable in [5].

Let $T(X)$ be the total transformation semigroup on a nonempty set X. It is well known that $T(X)$ is a regular semigroup. Moreover, every semigroup is isomorphic to a subsemigroup of some total transformation semigroups. The most basic mathematical structures are transformation semigroups. In 1952, Malcev [6] characterized ideals of $T(X)$. Later, Miller and Doss [7] studied its group $\mathcal{H}$-classes and its Green's relations. The generalization of these studies is the focus of this paper.

In 1975, Symons [8] considered a subsemigroup of $T(X)$ defined by

$$T(X,Y) = \{\alpha \in T(X) : X\alpha \subseteq Y\}$$

where Y is a nonempty subset of X. He determined all the automorphisms of $T(X,Y)$. In 2005, Nenthein et al. [9] described regular elements in $T(X,Y)$ and counted the number of all regular elements of $T(X,Y)$ when X is a finite set. They described such a number in terms of $|X|$, $|Y|$ and their related Stirling numbers. A few years later, Sanwong and Sommanee [10] studied regularity and Green's relations for the semigroup $T(X,Y)$. They determined when $T(X,Y)$ becomes a regular semigroup. Moreover, they gave a class of maximal inverse subsemigroups of $T(X,Y)$ in 2008. After that, they proved that the set $F(X,Y) = \{\alpha \in T(X,Y) : X\alpha = Y\alpha\}$ is the largest regular subsemigroup of $T(X,Y)$ and determined its Green's relations. In [11], Sanwong described Green's relations and found all maximal regular subsemigroups of $F(X,Y)$. In 2009, maximal and minimal congruences on $T(X,Y)$ were considered. Sanwong et al. [12] found that $T(X,Y)$ has only one maximal congruence if X is a finite set. They generalized [13] Theorem 3.4 for Y being infinite. Furthermore, characterizations of all minimal congruences on $T(X,Y)$ were given. In the same year, Sun [14] proved that while the semigroup $T(X,Y)$ is not left abundant, it is right abundant. Later in 2016, Lei Sun and Junling Sun [15] investigated the natural partial order on $T(X,Y)$. Moreover, they determined the maximal elements and the minimal elements of $T(X,Y)$.

Consider the semigroup

$$S(X,Y) = \{\alpha \in T(X) : Y\alpha \subseteq Y\}.$$

of transformations that leave Y invariant. In 1966, Magill [16] constructed and discussed the semigroup $S(X,Y)$. In fact, if $Y = X$, then $S(X,Y) = T(X)$. Later in [8], automorphism groups of a semigroup $S(X,Y)$ were given by Symons. In 2005, Nenthein et al. [9] characterized regularity for $S(X,Y)$. In addition, they found the number of regular elements in $S(X,Y)$ for a finite set X. Honyam and Sanwong [17] studied its ideals, group $\mathcal{H}$-classes and Green's relations on $S(X,Y)$. Furthermore, they described when $S(X,Y)$ is isomorphic to $T(A)$ for some set A. A few years later, the left, right regular and intra-regular elements of a semigroup $S(X,Y)$ were discussed by Choomanee, Honyam and Sanwong [18]. Moreover, when X is finite, they calculated the number of left regular elements in $S(X,Y)$. In [19], natural partial orders on the semigroup $S(X,Y)$ were considered by Sun and Wang. Moreover, they investigated left and right compatible elements with respect to this partial oder. Finally, they described the abundance of $S(X,Y)$. In [20], all elements in the semigroup $S(X,Y)$ that are left compatible with the natural partial order were studied. Left and right magnifying elements of $S(X,Y)$ were given by Chiram and Baupradist in [21]. In a recent study, Punkumkerd and Honyam [22] provided a characterization of left and right magnifying elements on the semigroup $\overline{PT}(X,Y)$. $\overline{PT}(X,Y)$ denotes the set of all partial transformations α from a subset of X to X and $(dom\alpha \cap Y)\alpha \subseteq Y$, where $dom\alpha$ is the domain of α. Their results have shown to be more general than the previous findings from [21].

In Section 2, we consider left and right magnifying elements of $T(X,Y)$ and $S(X,Y)$. We prove that each left magnifying element in $T(X,Y)$ is not regular. Furthermore, we show that every left and right magnifying element in $S(X,Y)$ is regular. In Sections 3 and 4, we focus on left and right regularity on $T(X,Y)$ and $S(X,Y)$. We show that every left magnifying element in $T(X,Y)$ is a right regular element. Every right magnifying element is a left regular element. As [10] determined when $T(X,Y)$ becomes a regular semigroup, we also characterize whenever $T(X,Y)$ and $S(X,Y)$ is a left (right) regular semigroup.

Note that throughout this paper, we will write mappings from the right, $x\alpha$ rather than $\alpha(x)$ and compose that the left to the right, $x(\alpha\beta) = (x\alpha)\beta$ rather than $(\alpha\beta)(x) = \alpha(\beta(x))$ where $\alpha, \beta \in T(X)$ and $x \in X$. For each $\alpha \in T(X)$, we denote the set $\{z\alpha^{-1} : z \in X\alpha\}$ by $\pi(\alpha)$ and $\pi_Y(\alpha)$ is the set $\{P \in \pi(\alpha) : P \cap Y \neq \emptyset\}$ for a subset $Y \subseteq X$. Then, it is obvious that $\pi(\alpha)$ is a partition of X.

2. Magnifying Elements

In this section, we focus on characterizations of left magnifying elements and right magnifying elements in $T(X,Y)$ and $S(X,Y)$. The relationships between magnifying elements and regular elements are given.

Theorem 1 ([23]). *Let $\alpha \in T(X, Y)$. Then, α is right magnifying if and only if α is surjective but not injective and is such that $y\alpha^{-1} \cap Y \neq \emptyset$ for all $y \in Y$ and $|y\alpha^{-1} \cap Y| > 1$ for some $y \in Y$.*

Theorem 2 ([23]). *A mapping $\alpha \in T(X)$ is left magnifying if and only if α is injective but not surjective.*

Theorem 3 ([23]). *Let $\alpha \in T(X, Y)$. If $|Y| = |X|$ and $X \neq Y$, then α is left magnifying if and only if α is injective.*

Remark 1. *For each $\alpha \in T(X, Y)$, we note from Theorem 1 that α is right magnifying if and only if $Y\alpha = Y$ and $\alpha|_Y$ is not injective.*

Theorem 4. *Let $\alpha \in T(X, Y)$ and $X \neq Y$. Then, α is left magnifying in a semigroup $T(X, Y)$ if and only if α is injective.*

Proof. Assume that α is left magnifying. Suppose that M is a proper subset of $T(X, Y)$ satisfying $\alpha M = T(X, Y)$. Let $a, b \in X$ be such that $a\alpha = b\alpha$. If $|Y| = 1$, then $T(X, Y)$ contains exactly one element. Thus, $T(X, Y)$ has no proper subset M such that $\alpha M = T(X, Y)$. This is a contradiction. Therefore, $|Y| > 1$. Let $y \in Y \setminus \{a\alpha\}$. Define $\gamma : X \to X$ by

$$x\gamma = \begin{cases} a\alpha & \text{if } x = a, \\ y & \text{otherwise.} \end{cases}$$

It is verifiable that $\gamma \in T(X, Y)$. From $\alpha M = T(X, Y)$, there exists $\beta \in M$ such that $\alpha\beta = \gamma$. Suppose that $a \neq b$. Then, $a\alpha = a\gamma = a\alpha\beta = b\alpha\beta = b\gamma = y$, which is a contradiction. Hence, $a = b$ and so α is injective.

Suppose that α is injective. Then, we choose $y \in Y$ and let M be the set $\{\gamma \in T(X, Y) : (X \setminus Y)\gamma = \{y\}\}$. From $X \neq Y$, we have $M \neq \emptyset$. It follows from our assumption that every $x \in X\alpha$, there is a unique $x' \in X$ satisfying $x'\alpha = x$. Let $\beta \in T(X, Y)$. We define $\gamma : X \to X$ by

$$x\gamma = \begin{cases} x'\beta & \text{if } x \in X\alpha, \\ y & \text{otherwise.} \end{cases}$$

It is verifiable that $\gamma \in M$. Now, let $x \in X$. Thus, $x\alpha\gamma = (x\alpha)'\beta$. From α being injective and $(x\alpha)'\alpha = x\alpha$, we obtain $(x\alpha)' = x$. Therefore, $x\alpha\gamma = (x\alpha)'\beta = x\beta$. Hence, $\beta = \alpha\gamma$ and so $\alpha M = T(X, Y)$. This implies that α is left magnifying. □

The set of natural numbers is represented by the letter $\mathbb{N}$. Additionally, we denote the set of even natural numbers and the set of all odd natural numbers greater than 3 by $2\mathbb{N}$ and $2\mathbb{N} + 1$, respectively.

Example 1. *Let $X = \mathbb{N}$ and $Y = 2\mathbb{N}$. Define $\alpha : X \to Y$ by $x\alpha = 2x$ for all $x \in X$. Then, $X\alpha = Y$. Clearly, α is injective. From Theorem 4, we obtain that α is left magnifying. We will show that α is not a regular element. Suppose that α is a regular element in $T(X, Y)$. Thus, there exists $\beta \in T(X, Y)$ such that $\alpha\beta\alpha = \alpha$. Consider $3\alpha\beta\alpha = 3\alpha = 6$. Thus, $3\alpha\beta \in 6\alpha^{-1} = \{3\}$. Hence, $3\alpha\beta = 3$, which is a contradiction. So α is not regular element.*

From the above example, we will verify that in $T(X, Y)$, each left magnifying element is not a regular element.

Theorem 5. *If $X \neq Y$, then every left magnifying element of $T(X, Y)$ is not regular.*

Proof. Assume that $X \neq Y$. Let α be a left magnifying element. From Theorem 4, we obtain that α is injective. Suppose that α is a regular element in $T(X, Y)$. Then, there exists

$\beta \in T(X,Y)$ such that $\alpha = \alpha\beta\alpha$. Since $X \neq Y$, we choose $x \in X \setminus Y$. Therefore, $x\alpha\beta\alpha = x\alpha$ and then $x\alpha\beta \in (x\alpha)\alpha^{-1}$. Since α is injective, we have $x\alpha\beta = x \notin Y$. This is a contradiction with $X\beta \subseteq Y$. So α is not regular. □

Theorem 6. *Every right magnifying element of a semigroup $T(X,Y)$ is regular.*

Proof. Let α be a right magnifying element of $T(X,Y)$. By Remark 1, we obtain that $Y\alpha = Y$. It follows that $X\alpha \subseteq Y = Y\alpha$. From $Y\alpha \subseteq X\alpha$, we have $X\alpha = Y\alpha$. This means that $\alpha \in F(X,Y)$ and so α is a regular element of $T(X,Y)$. □

The following example shows that there exists an element in some $T(X,Y)$ which is regular but it is not right magnifying.

Example 2. *Let $X = \mathbb{N}$ and $Y = 2\mathbb{N}$. Define $\alpha : X \to Y$ by*

$$x\alpha = \begin{cases} x+2 & \text{if } x \in 2\mathbb{N}, \\ x+3 & \text{otherwise.} \end{cases}$$

Note that $\alpha \in T(X,Y)$. It is easy to see that $Y\alpha = 2\mathbb{N} \setminus \{2\} \neq Y$ and $X\alpha = 2\mathbb{N} \setminus \{2\} = Y\alpha$. Thus, α is regular but it is not right magnifying.

In the rest of this section, we consider magnifying elements in $S(X,Y)$.

Lemma 1 ([21]). *Let $\alpha \in S(X,Y)$. Then, α is a right magnifying element if and only if α is surjective but not injective such that $y\alpha^{-1} \cap Y \neq \emptyset$ for all $y \in Y$.*

Lemma 2 ([21]). *Let $\alpha \in S(X,Y)$. Then, α is a left magnifying element if and only if α is injective but not surjective such that $y\alpha^{-1} \subseteq Y$ for all $y \in Y \cap X\alpha$.*

Lemma 3 ([9]). *Let $\alpha \in S(X,Y)$. Then, α is a regular element if and only if $Y\alpha = X\alpha \cap Y$.*

Theorem 7. *Every left magnifying element of a semigroup $S(X,Y)$ is regular.*

Proof. Suppose that α is left magnifying. We will show that $X\alpha \cap Y = Y\alpha$. Clearly, $Y\alpha \subseteq X\alpha \cap Y$. Let $y \in X\alpha \cap Y$. Then, there exists $y' \in X$ such that $y = y'\alpha$. Thus, $y' \in y\alpha^{-1} \subseteq Y$ by Lemma 2. This implies that $X\alpha \cap Y = Y\alpha$. From Lemma 3, we obtain that α is regular. □

Theorem 8. *Every right magnifying element of a semigroup $S(X,Y)$ is regular.*

Proof. Suppose that α is right magnifying. We will show that $X\alpha \cap Y = Y\alpha$. Clearly, $Y\alpha \subseteq X\alpha \cap Y$. Let $y \in X\alpha \cap Y$. By Lemma 1, we have $y\alpha^{-1} \cap Y \neq \emptyset$. Thus, there exists $y' \in y\alpha^{-1} \cap Y$. Hence, $y = y'\alpha \in Y\alpha$. Therefore, $X\alpha \cap Y = Y\alpha$. From Lemma 3, we obtain α is regular. □

Example 3. *Let α be defined in Example 2. It is clear that $\alpha \in S(X,Y)$ and α is neither injective nor surjective. Since $X\alpha = Y\alpha$, this means that $X\alpha \cap Y = Y\alpha$. Hence, α is regular, while it is neither a left nor right magnifying element in $S(X,Y)$.*

3. Left Regular and Right Regular Elements in $T(X,Y)$

Now, we start with the characterizations of left regular and right regular elements in $T(X,Y)$. Moreover, we determine whenever $T(X,Y)$ becomes a left regular semigroup and a right regular semigroup, respectively.

Theorem 9. *Let $\alpha \in T(X,Y)$. Then, the following statements are equivalent.*

(1) α *is left regular.*

(2) $Y\alpha^2 = X\alpha$.
(3) for any $P \in \pi(\alpha)$, $Y\alpha \cap P \neq \emptyset$.

Proof. (1) $\Rightarrow$ (2). Suppose that α is left regular. Thus, there exists $\beta \in T(X,Y)$ satisfying $\alpha = \beta\alpha^2$. This implies that $Y\alpha^2 \subseteq X\alpha = X\beta\alpha^2 \subseteq Y\alpha^2$. Thus, $Y\alpha^2 = X\alpha$.
(2) $\Rightarrow$ (3). Suppose that $Y\alpha^2 = X\alpha$ and let $P \in \pi(\alpha)$. Thus, there is $y \in X\alpha$ satisfying $y\alpha^{-1} = P$. By assumption, we have $y \in Y\alpha^2$. Thus, there exists $z \in Y$ such that $y = z\alpha^2$; that is, $z\alpha \in y\alpha^{-1} = P$. It follows that $z\alpha \in Y\alpha \cap P$. Therefore, $Y\alpha \cap P \neq \emptyset$.
(3) $\Rightarrow$ (1). Assume that (3) holds. For $x \in X$, there is a unique $P_x \in \pi(\alpha)$ satisfying $x \in P_x$. From assumption, we have $Y\alpha \cap P_x \neq \emptyset$. So there exists $x' \in Y$ such that $x'\alpha \in P_x = (x\alpha)\alpha^{-1}$. Define $\beta : X \to X$ by $x\beta = x'$ for all $x \in X$. It is obvious that $\beta \in T(X,Y)$. Let $x \in X$. This implies that $x\beta\alpha^2 = x'\alpha^2 = (x'\alpha)\alpha = x\alpha$. Hence, $\alpha = \beta\alpha^2$ and so α is left regular. □

If we replace Y with X in Theorem 9, we have the following corollary.

Corollary 1. *Let $\alpha \in T(X)$. Then, α is left regular if and only if for each $P \in \pi(\alpha)$, $X\alpha \cap P \neq \emptyset$.*

Proof. By taking $X = Y$, we obtain $T(X,Y) = T(X,X) = T(X)$ and $X\alpha = Y\alpha$. By Theorem 9(3), we obtain that α is regular if and only if for each $P \in \pi(\alpha)$, $X\alpha \cap P = Y\alpha \cap P \neq \emptyset$. □

Theorem 10. *Let $\alpha \in T(X,Y)$. Then, α is right regular if and only if $\alpha|_{X\alpha}$ is injective.*

Proof. Assume that α is right regular. Then, there exists $\beta \in T(X,Y)$ such that $\alpha = \alpha^2\beta$. We will show that $\alpha|_{X\alpha}$ is injective. Let $x,y \in X\alpha$ be such that $x\alpha = y\alpha$. Thus, there exist $x',y' \in X$ such that $x = x'\alpha$ and $y = y'\alpha$. We obtain that

$$x = x'\alpha = x'\alpha^2\beta = (x'\alpha)\alpha\beta = x\alpha\beta = y\alpha\beta = y'\alpha^2\beta = y'\alpha = y.$$

Therefore, $\alpha|_{X\alpha}$ is injective.
Conversely, suppose $\alpha|_{X\alpha}$ is injective. Let $z \in X\alpha^2$. Then, there exists a unique $z' \in X\alpha$ such that $z'\alpha = z$. We choose $y \in Y$. Define $\beta : X \to X$ by

$$z\beta = \begin{cases} z' & \text{if } z \in X\alpha^2, \\ y & \text{otherwise.} \end{cases}$$

From $X\alpha \subseteq Y$, we obtain that $X\beta \subseteq X\alpha \cup \{y\} \subseteq Y$. Let $x \in X$. Note that $x\alpha^2 \in X\alpha^2$ and $(x\alpha^2)\beta = (x\alpha^2)'$ where $(x\alpha^2)'\alpha = x\alpha^2 = x\alpha\alpha$. Since $\alpha|_{X\alpha}$ is injective, we have $(x\alpha^2)' = x\alpha$. So $x\alpha^2\beta = (x\alpha^2)' = x\alpha$. Hence, $\alpha^2\beta = \alpha$ and so α is right regular. □

Corollary 2. *Let $\alpha \in T(X)$. Then, α is right regular if and only if $\alpha|_{X\alpha}$ is injective.*

From Theorems 4 and 10, we obtain the following corollary immediately.

Corollary 3. *For $X \neq Y$, every left magnifying element of a semigroup $T(X,Y)$ is right regular.*

Corollary 4. *Every right magnifying element of a semigroup $T(X,Y)$ is left regular.*

Proof. Let α be a right magnifying element. By Remark 1, we have $Y\alpha = Y$. It follows that $X\alpha \subseteq Y = Y\alpha = (Y\alpha)\alpha = Y\alpha^2 \subseteq X\alpha$. Hence, $Y\alpha^2 = X\alpha$ and so α is left regular. □

Example 4. *Let $X = \mathbb{N}$ and $Y = 2\mathbb{N}$. Define $\alpha : X \to Y$ by*

$$x\alpha = \begin{cases} x+2 & \text{if } x \in 2\mathbb{N}, \\ x+1 & \text{otherwise.} \end{cases}$$

Then, $\alpha \in T(X,Y)$. Clearly, α is not injective. We see that $X\alpha = 2\mathbb{N}$. This means that $\alpha|_{X\alpha} : X\alpha \to X\alpha$ is an injection. This means that α is right regular but it is not left magnifying.

Example 5. Let $X = \mathbb{N}$ and $Y = 2\mathbb{N}$. Define $\alpha : X \to Y$ by

$$x\alpha = \begin{cases} 4 & \text{if } x \in \{1,2,3,4\}, \\ x - 2 & \text{if } x \in 2\mathbb{N} \setminus \{2,4\}, \\ x + 1 & \text{otherwise}. \end{cases}$$

Clearly, $\alpha \in T(X,Y)$. We see that $X\alpha = 2\mathbb{N} \setminus \{2\} = Y\alpha^2$ and $Y\alpha = 2\mathbb{N} \setminus \{2\} \neq Y$. Hence, α is a left regular element but it is not right magnifying.

Notice that for $|X| \leq 2$, we obtain that $T(X,Y)$ is left and right regular. Now, we consider the other case.

Theorem 11. *Let $X \neq Y$ be such that $|X| \geq 3$. Then, $T(X,Y)$ is a right regular semigroup if and only if $|Y| = 1$.*

Proof. If $|Y| = 1$, then $|T(X,Y)| = 1$ and $T(X,Y)$ is a right regular semigroup. Assume that $T(X,Y)$ is a right regular semigroup and suppose $|Y| \neq 1$. Let $a, b, c \in X$ be distinct elements and $a, b \in Y$. Define $\alpha : X \to X$ by

$$x\alpha = \begin{cases} a & \text{if } x \in \{a,b\}, \\ b & \text{otherwise}. \end{cases}$$

Then, $\alpha \in T(X,Y)$. However, $\alpha|_{X\alpha}$ is not injective. Thus, α is not right regular, which is a contradiction. Hence, $|Y| = 1$. □

Theorem 12. *Let $X \neq Y$ be such that $|X| \geq 3$. Then, $T(X,Y)$ is a left regular semigroup if and only if $|Y| = 1$.*

Proof. If $|Y| = 1$, then $|T(X,Y)| = 1$ and $T(X,Y)$ is a left regular semigroup. Assume that $T(X,Y)$ is a left regular semigroup and $|Y| \neq 1$. Let $a, b, c \in X$ be distinct elements and $a, b \in Y$. Define $\alpha : X \to X$ by

$$x\alpha = \begin{cases} a & \text{if } x \in \{a,b\}, \\ b & \text{otherwise}. \end{cases}$$

Then, $\alpha \in T(X,Y)$. Note that $a, b \notin b\alpha^{-1}$ and $Y\alpha \subseteq \{a,b\}$. So $Y\alpha \cap b\alpha^{-1} = \emptyset$. Therefore, α is not left regular. This is a contradiction. Hence, $|Y| = 1$. □

Corollary 5. *Every left regular element of a semigroup $T(X,Y)$ is regular.*

Proof. We first note that $F(X,Y) = \{\alpha \in T(X,Y) : X\alpha = Y\alpha\}$ is the largest regular subsemigroup of $T(X,Y)$. Let α be a left regular element of $T(X,Y)$. It follows from Theorem 9(2) that $X\alpha = Y\alpha^2 = (Y\alpha)\alpha \subseteq Y\alpha \subseteq X\alpha$. Thus, $X\alpha = Y\alpha$ and so $\alpha \in F(X,Y)$. Therefore, α is a regular element of $T(X,Y)$. □

Example 6. Let $X = \mathbb{N}$ and $Y = 2\mathbb{N}$. Define $\alpha : X \to Y$ by

$$x\alpha = \begin{cases} x + 2 & \text{if } x \in 2\mathbb{N}, \\ x + 3 & \text{otherwise}. \end{cases}$$

Clearly, $\alpha \in T(X,Y)$. Consider $X\alpha = Y\alpha = 2\mathbb{N} \setminus \{2\}$ and $Y\alpha^2 = 2\mathbb{N} \setminus \{2,4\}$. Then, α is regular but it is not left regular.

Example 7. Let $X = \mathbb{N}$ and $Y = 2\mathbb{N} + 1$. Define $\alpha : X \to Y$ by

$$x\alpha = \begin{cases} x+1 & \text{if } x \in 2\mathbb{N}, \\ x+2 & \text{otherwise.} \end{cases}$$

It is verifiable that $\alpha \in T(X,Y)$. We obtain that $\alpha|_{X\alpha}$ is injective and $X\alpha \neq Y\alpha$. Thus, α is right regular but it is not regular.

Finally, we consider the set of all left regular elements $LReg(T(X,Y))$ and the set of all right regular elements $RReg(T(X,Y))$ of $T(X,Y)$. We begin with the following example.

Example 8. Let $X = \{1,2,3,4,5\}$ and $Y = \{1,2,3,4\}$. We consider the mappings $\alpha = \begin{pmatrix} 1 & 2 & 3 & 4 & 5 \\ 3 & 3 & 3 & 4 & 3 \end{pmatrix}$ and $\beta = \begin{pmatrix} 1 & 2 & 3 & 4 & 5 \\ 2 & 1 & 1 & 2 & 1 \end{pmatrix}$. We note that $\alpha, \beta \in T(X,Y)$. Moreover, $\alpha|_{X\alpha}$ and $\beta|_{X\beta}$ are injective; that is $\alpha, \beta \in RReg(T(X,Y))$. Clearly, $\alpha\beta = \begin{pmatrix} 1 & 2 & 3 & 4 & 5 \\ 1 & 1 & 1 & 2 & 1 \end{pmatrix}$. This implies that $\alpha\beta|_{X\alpha\beta}$ is not injective and so $\alpha\beta \notin RReg(T(X,Y))$. In this case, we obtain that $RReg(T(X,Y))$ is not a semigroup.

Theorem 13. Let $|X| \geq 3$. Then, $RReg(T(X,Y))$ is a semigroup if and only if $|Y| \leq 2$.

Proof. Suppose that $|Y| \geq 3$. Let $a, b, c \in Y$ be distict elements. Define $\alpha : X \to Y$ by

$$x\alpha = \begin{cases} a & \text{if } x = b, \\ b & \text{if } x = a, \\ c & \text{otherwise.} \end{cases}$$

We see that $X\alpha = \{a,b,c\}$ and $\alpha|_{X\alpha}$ is injective. Define $\beta : X \to Y$ by

$$x\beta = \begin{cases} a & \text{if } x = a, \\ c & \text{otherwise.} \end{cases}$$

Then, $X\beta = \{a,c\}$ and $\beta|_{X\beta}$ is injective. Since $b\alpha\beta = a$ and $c\alpha\beta = c$, we obtain $a, c \in X\alpha\beta$. From $a\alpha\beta = c = c\alpha\beta$, we conclude that $\alpha\beta|_{X\alpha\beta}$ is not injective. Therefore, $\alpha\beta \notin RReg(T(X,Y))$ and $RReg(T(X,Y))$ is not closed.

Assume that $|Y| \leq 2$. If $|Y| = 1$, then $T(X,Y)$ is a right regular semigroup. Therefore, $RReg(T(X,Y))$ is a semigroup. Suppose that $|Y| = 2$. Let $Y = \{a,b\}$ and $\alpha, \beta \in RReg(T(X,Y))$. We will show that $\alpha\beta|_{X\alpha\beta}$ is injective. Let $x, y \in X\alpha\beta$ be such that $x\alpha\beta = y\alpha\beta$. If α or β is a constant mapping, then $\alpha\beta$ is a constant mapping. Thus, $\alpha\beta|_{X\alpha\beta}$ is injective. Suppose that α and β are not constant mappings. Then, $X\alpha = Y = X\beta$. We observe that $x, y \in X\alpha\beta \subseteq X\beta = X\alpha$. If $x \neq y$, then $x\alpha \neq y\alpha$ since $\alpha|_{X\alpha}$ is injective. Note that $x\alpha, y\alpha \in Y = X\beta$. Then, $(x\alpha)\beta \neq (y\alpha)\beta$ since $\beta|_{X\beta}$ is injective. This is a contradiction. Hence, $x = y$. So $\alpha\beta|_{X\alpha\beta}$ is injective. Therefore, $RReg(T(X,Y))$ is closed. □

Theorem 14. Let $|X| \geq 3$. Then, $LReg(T(X,Y))$ is a semigroup if and only if $|Y| \leq 2$.

Proof. Suppose that $|Y| \geq 3$. Let $a, b, c \in Y$ be distinct elements. Define $\alpha : X \to Y$ by

$$x\alpha = \begin{cases} b & \text{if } x = c, \\ c & \text{otherwise.} \end{cases}$$

And we define $\beta : X \to Y$ by

$$x\beta = \begin{cases} a & \text{if } x = b, \\ b & \text{otherwise.} \end{cases}$$

Then, $Y\alpha^2 = (Y\alpha)\alpha = \{b,c\}\alpha = \{b,c\} = X\alpha$ and $Y\beta^2 = (Y\beta)\beta = \{a,b\}\beta = \{a,b\} = X\beta$. Thus, $\alpha, \beta \in LReg(T(X,Y))$. It is easy to verify that

$$x\alpha\beta = \begin{cases} a & \text{if } x = c, \\ b & \text{otherwise} \end{cases}$$

such that $\pi(\alpha\beta) = \{\{c\}, X \setminus \{c\}\}$ and $Y\alpha\beta = \{a,b\}$. Clearly, $Y\alpha\beta \cap \{c\} = \emptyset$. Hence, $\alpha\beta \notin LReg(T(X,Y))$. So $\alpha\beta$ is not left regular.

Conversely, suppose $|Y| \leq 2$. If $|Y| = 1$, then $T(X,Y)$ is a left regular semigroup. We have $LReg(T(X,Y))$ is a semigroup. Assume that $Y = \{a,b\}$. Let $\alpha, \beta \in LReg(T(X,Y))$. If $|X\alpha| = 1$ or $|X\beta| = 1$, then $\alpha\beta$ is a constant mapping. Suppose that $|X\alpha| = |X\beta| = 2$. Then, $\pi(\alpha) = \{a\alpha^{-1}, b\alpha^{-1}\}$ and so it is a left regular element. From α being a left regular element, we obtain $Y\alpha \cap a\alpha^{-1} \neq \emptyset$ and $Y\alpha \cap b\alpha^{-1} \neq \emptyset$. This implies that $|a\alpha^{-1} \cap Y| = 1 = |b\alpha^{-1} \cap Y|$ and $X\alpha = Y\alpha = \{a,b\}$. Similarly, $|a\beta^{-1} \cap Y| = 1 = |b\beta^{-1} \cap Y|$ and $X\beta = Y\beta = \{a,b\}$. It is easy to verify that $\pi(\alpha\beta) = \pi(\alpha)$ and $Y\alpha\beta = \{a,b\} = Y\alpha$. Hence, $Y\alpha\beta \cap a\alpha^{-1} \neq \emptyset$ and $Y\alpha\beta \cap b\alpha^{-1} \neq \emptyset$. Therefore, $\alpha\beta$ is a left regular element. □

Theorem 15. *If Y is finite, then $RReg(T(X,Y)) = LReg(T(X,Y))$.*

Proof. Assume that Y is finite. Let α be a left regular element of $T(X,Y)$. Then, $Y\alpha^2 = X\alpha$. It follows that $X\alpha = Y\alpha^2 \subseteq X\alpha^2 \subseteq X\alpha$; that is, $X\alpha^2 = X\alpha$. From $(X\alpha)\alpha = X\alpha^2 = X\alpha$, we obtain $\alpha|_{X\alpha} : X\alpha \to X\alpha$ is surjective. Since $X\alpha \subseteq Y$ and Y is a finite set, we obtain $\alpha|_{X\alpha}$ is an injection. So α is right regular.

Assume that α is right regular. Thus, $\alpha|_{X\alpha} : X\alpha \to X\alpha$ is injective and also $\alpha|_{X\alpha} : X\alpha \to X\alpha$ is surjective since $X\alpha$ is finite. This means that $(X\alpha)\alpha = X\alpha$. We see that $X\alpha = (X\alpha)\alpha \subseteq Y\alpha \subseteq X\alpha$. Hence, $X\alpha = Y\alpha$ and so $Y\alpha^2 = X\alpha^2 = X\alpha$. Therefore, α is a left regular element of $T(X,Y)$. □

Next, the cardinality of right regular elements in the semigroup $T(X,Y)$ are investigated when X is finite.

Theorem 16. *Let $|X| = n$ and $|Y| = r$. Then,*

$$|LReg(T(X,Y))| = |RReg(T(X,Y))| = \sum_{k=1}^{r} k!\binom{r}{k}k^{n-k}$$

where $1 \leq k \leq r$.

Proof. By Theorem 15, we have $LReg(T(X,Y)) = RReg(T(X,Y))$. This implies that $|LReg(T(X,Y))| = |RReg(T(X,Y))|$. Let $1 \leq k \leq r$ and $B_k = \{\alpha \in RReg(T(X,Y)) : |X\alpha| = k\}$. From Y being finite, we have $\alpha|_{X\alpha} : X\alpha \to X\alpha$ is bijective for all $\alpha \in B_k$ by Theorem 10. Notice that the number of image sets in Y of cardinality k is equal to $\binom{r}{k}$. Since there are k^{n-k} ways of partitioning the remaining $n - k$ elements into k subsets, we obtain $|B_k| = \binom{r}{k}k! \, k^{n-k}$. Therefore,

$$|RReg(T(X,Y))| = \sum_{k=1}^{r} |B_k| = \sum_{k=1}^{r} k!\binom{r}{k}k^{n-k}.$$

□

4. Left Regular and Right Regular Elements in $S(X,Y)$

Theorem 17 ([18]). *Let $\alpha \in S(X,Y)$. Then, α is left regular if and only if $X\alpha = X\alpha^2$ and $Y\alpha = Y\alpha^2$.*

Theorem 18 ([18]). *Let $\alpha \in S(X,Y)$. Then, α is right regular if and only if $\pi(\alpha) = \pi(\alpha^2)$ and $\sigma(\alpha) = \sigma(\alpha^2)$ where $\sigma(\alpha) = \{y\alpha^{-1} : y \in X\alpha \cap Y\}$.*

Although the left and right regular elements of $S(X,Y)$ were characterized in [18], in this section we obtain the different results; see the following theorems.

Theorem 19. *Let $\alpha \in S(X,Y)$. Then, α is a left regular element if and only if for every $P \in \pi(\alpha)$, $P \cap X\alpha \neq \emptyset$ and for every $P \in \pi_Y(\alpha)$, $P \cap Y\alpha \neq \emptyset$.*

Proof. Assume that α is a left regular element. Thus, $\alpha = \beta\alpha^2$ for some $\beta \in S(X,Y)$. Let $P \in \pi(\alpha)$ and let $x \in P$. Then, $P = (x\alpha)\alpha^{-1}$ and $x\alpha = x\beta\alpha^2 = [(x\beta)\alpha]\alpha$. We see that $(x\beta)\alpha \in (x\alpha)\alpha^{-1} = P$ and $(x\beta)\alpha \in X\alpha$. Therefore, $P \cap X\alpha \neq \emptyset$. Let $P \in \pi_Y(\alpha)$ and let $y \in P \cap Y$. Then, $P = (y\alpha)\alpha^{-1}$, $y\beta \in Y$ and $y\alpha = y\beta\alpha^2 = [(y\beta)\alpha]\alpha$. We note that $(y\beta)\alpha \in Y\alpha$ and $(y\beta)\alpha \in (y\alpha)\alpha^{-1} = P$. Hence, $P \cap Y\alpha \neq \emptyset$.

Conversely, suppose the conditions hold. Let $x \in X$. Since $\pi(\alpha)$ is a partition of X, there is a unique $P_x \in \pi(\alpha)$ satisfying $x \in P_x$. If $P_x \cap Y \neq \emptyset$, then $P_x \in \pi_Y(\alpha)$ and $P_x \cap Y\alpha \neq \emptyset$ by our assumption. So, we choose $x' \in Y$ satisfying $x'\alpha \in P_x$. If $P_x \cap Y = \emptyset$, then since $P_x \cap X\alpha \neq \emptyset$, we choose $x' \in X$ satisfying $x'\alpha \in P_x$. Define $\beta : X \to X$ by $x\beta = x'$ for all $x \in X$. Then, β is well-defined and $x\beta\alpha^2 = x'\alpha^2 = (x'\alpha)\alpha = x\alpha$. Let $y \in Y$. Then, there is a unique $P_y \in \pi(\alpha)$ such that $y \in P_y$. Thus, $P_y \cap Y \neq \emptyset$. Therefore, $y\beta = y' \in Y$; that is, $Y\beta \subseteq Y$. So α is left regular. □

Theorem 20. *Let $\alpha \in S(X,Y)$. Then, α is a right regular element if and only if $\alpha|_{X\alpha}$ is injective and $(X\alpha \setminus Y)\alpha \subseteq X \setminus Y$.*

Proof. Assume that α is right regular. Thus, α is also a right regular element in $T(X)$. From Corollary 2, we have $\alpha|_{X\alpha}$ is injective. Next, we will show that $(X\alpha \setminus Y)\alpha \subseteq X \setminus Y$. Let $z \in X\alpha \setminus Y$. Then, $z = z'\alpha$ for some $z' \in X$. Thus, $z = z'\alpha^2\beta$ for some $\beta \in S(X,Y)$ since α is a right regular element in $S(X,Y)$. If $z'\alpha^2 \in Y$, then $z = (z'\alpha^2)\beta \in Y\beta \subseteq Y$, which is a contradiction. Hence, $z\alpha = z'\alpha^2 \notin Y$ and so $(X\alpha \setminus Y)\alpha \subseteq X \setminus Y$.

Assume that $\alpha|_{X\alpha}$ is injective and $(X\alpha \setminus Y)\alpha \subseteq X \setminus Y$. Let β be defined in the converse part of Theorem 10; we note that $\alpha = \alpha^2\beta$. It is enough to verify that $\beta \in S(X,Y)$. Let $x \in Y$. If $x \notin X\alpha^2$, then by the definition of β, we have $x\beta \in Y$. Assume that $x \in X\alpha^2$. There is a unique $x' \in X\alpha$ satisfying $x'\alpha = x$. If $x' \notin Y$, then $x' \in X\alpha \setminus Y$. By assumption, we have $x'\alpha \in X \setminus Y$ which is a contradiction. This means that $x\beta = x' \in Y$. Hence, $Y\beta \subseteq Y$ and so $\beta \in S(X,Y)$. □

Example 9. *Let $X = \mathbb{N}$ and $Y = 2\mathbb{N}$. Define $\alpha : X \to X$ by*

$$x\alpha = \begin{cases} 2 & \text{if } x \in Y, \\ 4 & \text{if } x = 1, \\ x - 2 & \text{otherwise.} \end{cases}$$

Then, $X\alpha = \{2,4\} \cup \{2n - 1 : n \in \mathbb{N}\}$ and $Y\alpha = \{2\} \subseteq Y$. So $\alpha \in S(X,Y)$. Moreover, we obtain $\pi(\alpha) = \{Y\} \cup \{\{2n - 1\} : n \in \mathbb{N}\}$ and $\pi_Y(\alpha) = \{Y\}$. It is clear that $P \cap X\alpha \neq \emptyset$ for every $P \in \pi(\alpha)$ and $P \cap Y\alpha \neq \emptyset$ for all $P \in \pi_Y(\alpha)$. From Theorem 19, α is left regular. Note that $X\alpha \cap Y = \{2,4\} \neq Y\alpha$. By Theorem 3, α is also not regular.

Example 10. *Let α be defined in Example 2. Then, $Y\alpha \subseteq Y$ and also $X\alpha \cap Y = Y\alpha$. Thus, $\alpha \in S(X,Y)$ and α is regular. Note that $4\alpha^{-1} = \{1,2\}$ and $X\alpha \cap 4\alpha^{-1} = \emptyset$. Hence, α is not a left regular element of $S(X,Y)$.*

Example 11. *Let α be defined in Example 5. Then, $Y\alpha \subseteq Y$ and so $\alpha \in S(X,Y)$. Consider $Y \cap X\alpha = 2\mathbb{N} \setminus \{2\} = Y\alpha$ and $\alpha|_{X\alpha}$ is not injective. Hence, α is regular but not right regular in $S(X,Y)$.*

Example 12. *Recall α from Example 4. Then, $Y\alpha \subseteq Y$ and so $\alpha \in S(X,Y)$. We see that $\alpha|_{X\alpha}$ is injective and $(X\alpha \setminus Y)\alpha = \emptyset \subseteq X \setminus Y$. From Theorem 20, we obtain α is right regular. Consider $X\alpha \cap Y = 2\mathbb{N}$ and $Y\alpha = 2\mathbb{N} \setminus \{2\}$. Hence, $X\alpha \cap Y \neq Y\alpha$. From Theorem 3, we obtain α is not regular.*

Theorem 21. *The following statements are equivalent.*
(1) $S(X,Y) = LReg(S(X,Y))$.
(2) $S(X,Y) = RReg(S(X,Y))$.
(3) $|X| \leq 2$.

Proof. (1) $\Leftrightarrow$ (3). Assume that $S(X,Y) = LReg(S(X,Y))$. We will show that $|X| \leq 2$. Suppose that $|X| \geq 3$. Let $a,b,c \in X$ be distinct elements and $a \in Y$. Define $\alpha : X \to X$ by

$$x\alpha = \begin{cases} b & \text{if } x = c, \\ a & \text{if } x \neq c. \end{cases}$$

Claim that $\alpha \in S(X,Y)$. If $c \notin Y$, then $Y\alpha = \{a\} \subseteq Y$. Moreover, if $b,c \in Y$, then $Y\alpha = \{a,b\} \subseteq Y$. Thus, $\alpha \in S(X,Y)$ when $c \notin Y$ or $b,c \in Y$. Note that $\pi(\alpha) = \{\{c\}, X \setminus \{c\}\}$ and $X\alpha = \{a,b\}$. Clearly, $X\alpha \cap \{c\} = \emptyset$. Hence, α is not left regular. For the case $b \notin Y$ and $c \in Y$, we define $\beta : X \to X$ by

$$x\beta = \begin{cases} c & \text{if } x = b, \\ a & \text{if } x \neq b. \end{cases}$$

Then, $Y\beta = \{a\} \subseteq Y$, $\pi(\beta) = \{\{b\}, X \setminus \{b\}\}$ and $X\beta = \{a,c\}$. It follows that $\beta \in S(X,Y)$ but $\beta \notin LReg(S(X,Y))$. We conclude that $S(X,Y) \neq LReg(S(X,Y))$, which is a contradiction. So $|X| \leq 2$.

Conversely, assume that $|X| \leq 2$. Then, it is easy to verify that $S(X,Y)$ is a left regular semigroup. Hence, $S(X,Y) = LReg(S(X,Y))$.

(2) $\Leftrightarrow$ (3). Assume that $S(X,Y) = RReg(S(X,Y))$. We will show that $|X| \leq 2$. Suppose that $|X| \geq 3$. We consider α, β from condition (1) $\Leftrightarrow$ (3). Then $\alpha, \beta \in S(X,Y)$. Since $X\alpha = \{a,b\}$ and $\alpha|_{X\alpha}$ is not injective, we have α is not right regular. Similarly, $X\beta = \{a,c\}$ and $\beta|_{X\beta}$ is not injective. So β is not right regular, which is a contradiction. Hence, $|X| \leq 2$.

Conversely, suppose that $|X| \leq 2$. Then, it is clear that $S(X,Y)$ is a right regular semigroup. Hence, $S(X,Y) = RReg(S(X,Y))$. □

Theorem 22. *The following statements are equivalent.*
(1) $LReg(S(X,Y))$ is a semigroup.
(2) $RReg(S(X,Y))$ is a semigroup.
(3) $|X| \leq 2$.

Proof. (1) $\Leftrightarrow$ (3). Suppose that $|X| \geq 3$. Let $a,b,c \in X$ be distinct elements and $a \in Y$. It is enough to consider only two cases.

Case 1: $\{b,c\} \subseteq Y$. Recall α, β from Theorem 14, we have $\alpha, \beta \in S(X,Y)$. Note that $\pi(\alpha) = \pi_Y(\alpha) = \{\{c\}, X \setminus \{c\}\}$ and $X\alpha = \{b,c\}$. Clearly, $Y\alpha \cap \{c\} \neq \emptyset$ and $Y\alpha \cap X \setminus \{c\} \neq \emptyset$. Thus, α is left regular; that is, $\alpha \in LReg(S(X,Y))$. Similarly, $\beta \in LReg(S(X,Y))$. Consider $\{c\} \in \pi(\alpha\beta)$ and $X\alpha\beta = \{a,b\}$. Therefore, $X\alpha\beta \cap \{c\} = \emptyset$ and so $\alpha\beta$ is not left regular; that is, $\alpha\beta \notin LReg(S(X,Y))$. Hence, $LReg(S(X,Y))$ is not a semigroup.

Case 2: $\{b,c\} \not\subseteq Y$. Assume that $c \notin Y$. Define $\alpha : X \to X$ by

$$x\alpha = \begin{cases} c & \text{if } x = c, \\ a & \text{otherwise.} \end{cases}$$

Then, $Y\alpha = \{a\} \subseteq Y$. So $\alpha \in S(X,Y)$. Note that $\pi(\alpha) = \{\{c\}, X \setminus \{c\}\}$, $\pi_Y(\alpha) = \{X \setminus \{c\}\}$ and $X\alpha = \{a,c\}$. Clearly, $X\alpha \cap \{c\} \neq \emptyset$, $X\alpha \cap (X \setminus \{c\}) \neq \emptyset$ and $Y\alpha \cap (X \setminus \{c\}) \neq \emptyset$. Therefore, α is a left regular element of $S(X,Y)$. Define $\beta : X \to X$ by

$$x\beta = \begin{cases} b & \text{if } x \in \{b,c\}, \\ a & \text{otherwise.} \end{cases}$$

If $b \in Y$, then $Y\beta = \{a,b\} \subseteq Y$. Thus, $\beta \in S(X,Y)$. If $b \notin Y$, then $Y\beta = \{a\} \subseteq Y$. So $\beta \in S(X,Y)$. We can show that $\beta \in LReg(S(X,Y))$. Then, we note that $\pi(\alpha\beta) = \{\{c\}, X \setminus \{c\}\}$ and $X\alpha\beta = \{a,b\}$. Clearly, $X\alpha\beta \cap \{c\} = \emptyset$. Hence, $\alpha\beta \notin LReg(S(X,Y))$ and so $LReg(S(X,Y))$ is not a semigroup.

Conversely, suppose $|X| \leq 2$. Then, we have $LReg(S(X,Y)) = S(X,Y)$ is a semigroup from Theorem 21(1).

(2) $\Leftrightarrow$ (3). Suppose that $|X| \geq 3$. Let $a,b,c \in X$ be distinct elements and $a \in Y$. Recall α and β from the proof of (1) $\Leftrightarrow$ (3). It is enough to show that α and β are right regular elements of $S(X,Y)$. Clearly, $\alpha|_{X\alpha}$ and $\beta|_{X\beta}$ are injective. Consider $(X\alpha \setminus Y)\alpha = \{c\} \subseteq X \setminus Y$ and $(X\beta \setminus Y)\beta = \begin{cases} \emptyset & \text{if } b \in Y, \\ \{b\} & \text{if } b \notin Y \end{cases} \subseteq X \setminus Y$.

Then, α and β are right regular elements of $S(X,Y)$; that is, $\alpha, \beta \in RReg(S(X,Y))$. Note that $\alpha\beta|_{X\alpha\beta}$ is not injective. We conclude that $\alpha\beta$ is not right regular. Thus, $\alpha\beta \notin RReg(S(X,Y))$ and so $RReg(S(X,Y))$ is not a semigroup.

Conversely, suppose $|X| \leq 2$. Then, we have $RReg(S(X,Y)) = S(X,Y)$ is a semigroup from Theorem 21(2). □

Author Contributions: Conceptualization, E.L.; Methodology, K.S.; Investigation, W.S.; Writing—original draft, K.S. and E.L.; Writing—review & editing, W.S. All authors have read and agreed to the published version of the manuscript.

Funding: This research received no external funding.

Acknowledgments: I would like to thank the referee for his/her valuable suggestions and comments which helped to improve the readability of this paper. Moreover, we thank The 25th Annual Meeting in Mathematics (AMM 2021) for a good comments about our presentation and inspiring to find the number of left and right regular elements of $T(X,Y)$ as well as characterize left and right regular elements of $S(X,Y)$.

Conflicts of Interest: The authors declare no conflict of interest.

References

1. Ljapin, E.S. *Translations of Mathematical Monographs Vol. 3, Semigroups*; American Mathematical Society Providence: Providence, RI, USA, 1963.
2. Migliorini, F. Some research on semigroups with magnifying elements. *Period. Math. Hung.* **1971**, *1*, 279–286. [CrossRef]
3. Migliorini, F. Magnifying elements and minimal subsemigroups in semigroups. *Period. Math. Hung.* **1974**, *5*, 279–288. [CrossRef]
4. Gutan, M. Semigroups with strong and nonstrong magnifying elements. *Semigroup Forum* **1996**, *53*, 384–386. [CrossRef]
5. Gutan, M. Semigroups which contain magnifying elements are factorizable. *Commun. Algebra* **1997**, *25*, 3953–3963. [CrossRef]
6. Malcev, A.I. Symmetric groupoids. *Mat. Sb. N. S.* **1952**, *31*, 136–151.
7. Doss, C.G. Certain Equivalence Relations in Transformation Semigroups. Master's Thesis, University of Tennessee, Knoxville, TN, USA, 1955.
8. Symons, J.S.V. Some result concerning a transformation semigroup. *J. Aust. Math. Soc.* **1975**, *19*, 135–141. [CrossRef]
9. Nenthein, S.; Youngkhong, P.; Kemprasit, Y. Regular elements of some transformation semigroups. *Pure Math. Appl.* **2005**, *16*, 307–314.
10. Sanwong, J.; Sommanee, W. Regularity and Green's relation on a semigroup of transformations with restricted range. *Int. J. Math. Math. Sci.* **2008**, *2008*, 794013. [CrossRef]
11. Sanwong, J. The regular part of a semigroup of transformations with restricted range. *Semigroup Forum* **2011**, *83*, 134–146. [CrossRef]
12. Sanwong, J.; Singha, B.; Sullivan, R.P. Maximal and Minimal Congruences on Some Semigroups. *Acta Math. Sin. Engl. Ser.* **2009**, *25*, 455–466. [CrossRef]
13. Sanwong, J.; Sullivan, R.P. Maximal congruences on some semigroups. *Algebra Colloq.* **2007**, *14*, 255–263. [CrossRef]
14. Sun, L. A note on abundance of certain semigroups of transformations with restricted range. *Semigroup Forum* **2013**, *87*, 681–684. [CrossRef]
15. Sun, L.; Sun, J. A natual partial order on certain on certain semigroups of transformations with restricted range. *Semigroup Forum* **2016**, *92*, 135–141. [CrossRef]
16. Magill, K.D., Jr. Subsemigroups of $S(X)$. *Math. Japon.* **1966**, *11*, 109–115.
17. Honyam, P.; Sanwong, J. Semigroups of transformations with invariant set. *J. Korean Math. Soc.* **2011**, *48*, 289–300. [CrossRef]

18. Choomanee, W.; Honyam, P.; Sanwong, J. Regularity in semigroups of transformations with invariant sets. *Int. J. Pure Appl. Math.* **2013**, *87*, 151–164. [CrossRef]
19. Sun, L.; Wang, L. Natural partial order in semigroups of transformations with invariant set. *Bull. Aust. Math. Soc.* **2013**, *87*, 94–107. [CrossRef]
20. Sun, L.; Sun, J. A note on naturally ordered semigroups of transformations with invariant set. *Bull. Aust. Math. Soc.* **2015**, *91*, 264–267. [CrossRef]
21. Chinram, R.; Baupradist, S. Magnifying elements in semigroups of transformations with invariant set. *Asian-Eur. J. Math.* **2019**, *12*, 1950056. [CrossRef]
22. Punkumkerd, C.; Honyam, P. Magnifying elements of some semigroups of partial transformations. *Quasigroups Relat. Syst.* **2021**, *29*, 123–132.
23. Chinram, R.; Baupradist, S. Magnifying elements in a semigroup of transformations with restricted range. *Mo. J. Math. Sci.* **2018**, *30*, 54–58. [CrossRef]

Disclaimer/Publisher's Note: The statements, opinions and data contained in all publications are solely those of the individual author(s) and contributor(s) and not of MDPI and/or the editor(s). MDPI and/or the editor(s) disclaim responsibility for any injury to people or property resulting from any ideas, methods, instructions or products referred to in the content.

MDPI
St. Alban-Anlage 66
4052 Basel
Switzerland
www.mdpi.com

Mathematics Editorial Office
E-mail: mathematics@mdpi.com
www.mdpi.com/journal/mathematics

Disclaimer/Publisher's Note: The statements, opinions and data contained in all publications are solely those of the individual author(s) and contributor(s) and not of MDPI and/or the editor(s). MDPI and/or the editor(s) disclaim responsibility for any injury to people or property resulting from any ideas, methods, instructions or products referred to in the content.

www.ingramcontent.com/pod-product-compliance
Lightning Source LLC
LaVergne TN
LVHW070138100526
838202LV00015B/1844

9 7 8 3 0 3 6 5 8 4 4 0 9